(D'ARCY'S DEVIZES)
Boyde £2·00

Fundamentals of
ELECTROMAGNETIC
FIELD THEORY

ENGINEERING SCIENCE MONOGRAPHS
GENERAL EDITORS
Professor M. W. HUMPHREY DAVIES and Professor H. TROPPER
both of Queen Mary College, University of London

Matrix Theory for Electrical Engineering Students
A. Mary Tropper, M.Sc., Ph.D.

Introduction to Semiconductor Devices
M. J. Morant, B.Sc.(Eng.), Ph.D.

Engineering Systems Analysis
A. G. J. MacFarlane, B.Sc., Ph.D., D.Sc., C.Eng., F.I.E.E.

Science of Materials
Professor T. J. Lewis, D.Sc.(Eng.), F.Inst.P., M.I.E.E. and
P. E. Secker, B.Sc.(Eng.), Ph.D.

Electromechanical Energy Conversion
A. J. Ellison, D.Sc.(Eng.), C.Eng., F.I.Mech.E., F.I.E.E., Sen.Mem.I.E.E.E.

Generalized Electric Machines
A. J. Ellison, D.Sc.(Eng.), C.Eng., F.I.Mech.E., F.I.E.E., Sen.Mem.I.E.E.E.

Principles of Algol 60 Programming
J. S. Collins, Ph.D. and M. Almond, Ph.D., A.Inst.P.

Computational Methods and Algol
G. Bull, M.Sc., Grad.F.I.M.A.

Reactor Physics (2 Volumes)
R. Schulten and W. Güth
Translated by W. K. Mansfield, Ph.D.

Electric Power Transmission and Distribution
P. J. Freeman, C.Eng., M.I.E.E.

Introduction to Logic Circuit Theory
I. Aleksander, B.Sc.(Eng.), Ph.D., M.I.E.E.

Transistors in Linear Circuits
R. W. A. Scarr, B.Sc.(Eng.), Ph.D., M.I.E.E.

Physical Applications of Vectors and Tensors
H. Teichmann
Translated by C. W. Kilmister

Introduction to Fortran Programming
Heather M. Liddell, Ph.D., A.Inst.P. and Anthony J. Powell, M.A.

Worked Examples in Physical Electronics
P. Lynch, B.Sc., C.Eng., M.I.E.E., L.Inst.P. and
A. Nicolaides, B.Sc.(Eng.), A.W.P.

Fundamentals of Electromagnetic Field Theory
A. A. Zaky, B.Sc., M.S., Ph.D., M.Inst.P., C.Eng., M.I.E.E. and
R. Hawley, B.Sc., Ph.D., F.Inst.P., C.Eng., F.I.E.E.

Fundamentals of
ELECTROMAGNETIC FIELD THEORY

A. A. ZAKY,
B.Sc., M.S., Ph.D., M.Inst.P., C.Eng., M.I.E.E.
Professor of Electrical Engineering
Alexandria University, Egypt

and

R. HAWLEY,
B.Sc., Ph.D., F.Inst.P., C.Eng., F.I.E.E.
Director and Chief Electrical Engineer
C. A. Parsons & Co. Ltd, Newcastle upon Tyne

HARRAP   LONDON

First published in Great Britain 1974
by GEORGE G. HARRAP & CO. LTD
182–184 High Holborn, London WC1V 7AX

ISBN 0 245 52023 6

Phototypeset by Oliver Burridge Filmsetting Ltd, Crawley
and printed by J. W. Arrowsmith Ltd, Bristol
Made in Great Britain

PREFACE

Introductory and advanced courses in electromagnetic field theory are among the key subjects in the syllabus of most engineering and science colleges. In this book we have tried to present the physical concepts and mathematical formulations of stationary and quasi-stationary electric and magnetic fields in a manner suitable for a first course in electromagnetic field theory. The preparation of such a course imposes stringent requirements on the teacher both in the selection and weighting of topics, and in the level of presentation. That no author can expect to satisfy all potential readers is perhaps the reason for the proliferation of texts in this field. In the present book the level and the choice of the contents have been primarily determined by (1) the experience that it is generally in the fundamental concepts that beginners find great difficulties rather than in the mathematical level and (2) the fact that in most introductory courses the treatment of static electric and magnetic fields is usually inadequate and serves either as a stepping stone to the Maxwell equations or to give the latter their physical meaning; for this reason students generally find it difficult to pursue these topics in such advanced classical works as Jeans's *The Mathematical Theory of Electricity and Magnetism,* Stratton's *Electromagnetic Theory,* Smythe's *Static and Dynamic Electricity* and Durand's encyclopaedic works *Electrostatique* (3 volumes) and *Magnétostatique.* In the present book considerable emphasis has been placed on the basic laws, concepts and definitions associated with stationary and quasi-stationary electric and magnetic fields and the greater part of the book is devoted to an extended fundamental treatment of static electric and magnetic fields both in free space and in matter.

There is no doubt that Maxwell's equations together with the Lorentz force are the quintessence of electromagnetic theory and as such form the basis of all advanced treatments on the subject. However, the use of these equations as basic postulates in a first course in field theory renders it difficult for the beginning student to fully comprehend their subtle physical meanings. We have found it more logical therefore to follow the historical sequence of events and present the formal mathematical expression of Maxwell's equations as a reformulation of the three basic experimental laws viz. Coulomb's law, Biot-Savart's law and Faraday's law of electromagnetic induction, together with Maxwell's displacement current hypothesis.

E and **B** are considered as the force fields which have as their sources both free and bound charges or current loops. The vectors **D** and **H**

are considered as auxiliary vectors which are mathematically convenient to use but which have no physical significance; if desired they may be regarded as alternative descriptions of free charges and free currents.

It is assumed that the reader is familiar with differential and integral calculus. A prior knowledge of vector analysis is useful but not necessary since most of the vector concepts and vector theorems are developed as and when they are needed; this may not appeal to all readers but it does help the student to relate vector theory to something specific.

The first chapter covers the algebra of vectors. Chapter 2 starts with Coulomb's law and develops the electrostatics of point charges and charge distributions for free space. Chapter 3 deals with the electrostatics of charged conductors in free space; the concept of potential and capacitance coefficients is introduced and the physical meaning of these coefficients is discussed at length. Chapter 4 discusses the behaviour of dielectrics in an electric field from both the macroscopic and the microscopic points of view; this gives the reader an insight into the relationship between macroscopic phenomena and atomic behaviour. The concept of a generalized complex permittivity is also developed to familiarize the student with the fact that the permittivity is a function of frequency and how and why electromagnetic energy is absorbed in a material.

Chapter 5 contains a discussion of the energy and forces in the electrostatic field and Chapter 6 discusses the steady current, Ohm's law and the equation of continuity; the concept of displacement current is introduced and the student's attention is drawn to the fact that a good conductor at one frequency can become a good dielectric at other frequencies.

Chapter 7 discusses the magnetic field of steady currents in free space, starting with the force between current elements and the Biot-Savart law. The concepts of magnetic vector and scalar potentials and of magnetic dipoles are introduced. The coefficients of self and mutual inductance are defined for both filamentary and non-filamentary current distributions and the derivation of these coefficients for some familiar geometric configurations is given.

Chapter 8 gives a comprehensive account of the magnetostatic field in the presence of material bodies. As in Chapter 4 on dielectrics, both the macroscopic and microscopic points of view are presented. The distinction between $\mathbf{B}$ and $\mathbf{H}$ vectors is discussed at length since there still exists a strong controversy on the nature of these two fields. The chapter ends with a brief survey of permanent magnets. Chapter 9 discusses the variational and motional laws of electromagnetic induction, the quasi-stationary state, and concludes with a short discussion of all four of Maxwell's equations. Chapter 10 discusses the energy and forces in the magnetic field (stationary and quasi-stationary fields only) and gives an elementary discussion of hysteresis and eddy current losses.

In Chapter 11 some of the methods used for solving field problems are discussed. The method of images, which is unconvincingly treated in most texts, is here given special attention by closely examining the basic principles underlying the use of this method. The method of separation of variables and complex functions are covered briefly and mention is made of other existing methods which have been omitted on grounds of space. Sufficient references, however, have been given of some excellent works which deal almost exclusively with such methods and to which the interested student can now turn with confidence armed with a good understanding of the basic principles of field theory.

A collection of graded problems will be found at the end of each chapter.

In the past an apology or a justification was usually necessary when discussing the subject of the units used in a text. It has not been found necessary to do this here because the SI units used throughout the text have received international approval. By completely omitting any reference to other existing systems of units we believe ourselves to be rendering the reader a great service.

The authors wish to express their thanks and appreciation to Professor H. Tropper for his encouragement and his valuable suggestions and criticisms; to C. A. Parsons & Co. Ltd for facilities to produce this book and in particular to Miss L. M. Walker for her careful and patient typing and to Mr J. Henderson and Miss O. Horden for all their work on the figures. One of the authors (R.H.) wishes to thank C. A. Parsons & Co. Ltd for permission to publish.

A.A.Z.
R.H.

CONTENTS

Chapter 10   Energy and Forces in the Magnetic Field

Chapter 11   Methods of Solving Field Problems with Specified Boundary Conditions

CHAPTER ONE

VECTORS

1.1 DEFINITION AND REPRESENTATION OF SCALAR AND VECTOR QUANTITIES

Any physical quantity which can be represented algebraically by a single number is known as a *scalar**. Examples of scalar quantities are electric charge, potential, electric current, mass, density, temperature, pressure and speed. A scalar quantity is completely specified by assigning a suitable unit to the number, and it obeys all the laws of ordinary algebraic analysis.

A physical quantity which is specified by both magnitude and direction is known as a *vector*. Examples of vector quantities are force, electric field intensity, current density, velocity and acceleration. A vector quantity is completely specified by its direction and by assigning a suitable unit to the number representing its magnitude.

Any vector quantity can be represented graphically by a straight line of finite length drawn in a definite direction; the length of the line specifies the magnitude of the vector; the sense of the vector is indicated by an arrowhead placed either at the tip of the line or at some convenient point along it (Fig. 1.1). Two vectors are said to be equal if

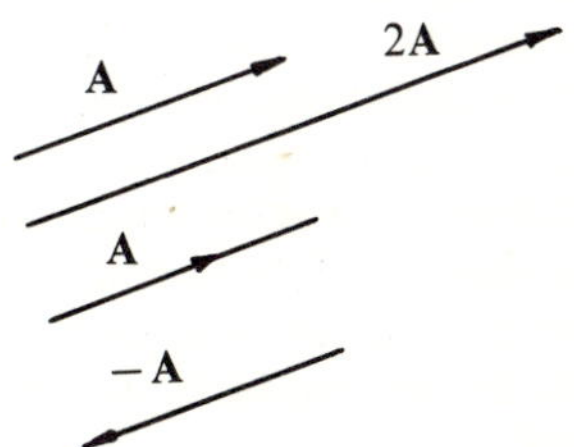

Fig. 1.1 Graphical representation of a vector

they have the same magnitude, the same direction and the same sense; thus they need not be collinear but only parallel. If two vectors have the same magnitude and the same direction but have opposite senses, either vector is said to be the negative of the other.

Arithmetically a vector quantity can be specified by three coordinate

---

* The complex representation of a sinusoidal quantity (e.g. voltage or current) produces a *complex scalar* or *phasor*.

numbers because if we choose the initial point (point at which vector starts) as the origin of a fixed coordinate system, then the three numbers which represent the coordinates of the vector extremity uniquely specify both the length and the direction of the vector.

In the text a vector quantity will be represented by means of a letter in **bold-faced type**, for example **A**. The magnitude of a vector will be represented by the italic form of the bold-faced letter which denotes the vector.

A *unit vector* is defined as a vector having a magnitude of unity but no physical dimensions. If **A** is any vector, then

$$\mathbf{a}_A = \frac{\mathbf{A}}{A} \qquad (1-1)$$

is a unit vector having the same direction as **A**. Throughout this text the symbol **a** will be used to denote unit vectors; subscripts added to **a** will be used to denote a specified direction—e.g., $\mathbf{a}_x$ denotes a unit vector in the positive x-direction. Any vector **A** can be represented by the product of its magnitude and its unit vector,

$$\mathbf{A} = A\mathbf{a}_A$$

1.2 SCALAR AND VECTOR FIELDS

In general the position of a point in space, with respect to a given fixed coordinate system, is uniquely determined by the three coordinates of the point. For example the position of a point P referred to a cartesian coordinate system is given by P (x, y, z). If to each point (x, y, z) of a region R in space there is assigned a number which represents a scalar physical quantity, and if the variation of this number from point to point in space is described by a mathematical function $\phi (x, y, z)$ of the space coordinates, then the function ϕ is said to be a *scalar point function* and ϕ represents a *scalar field* in R. For example the spatial distribution of temperature inside a non-uniformly heated body constitutes a scalar field.

If to each point (x, y, z) of a region R in space there is assigned a vector quantity and if the variation of the magnitude and direction of this vector from point to point in space is described by a mathematical function **A** (x, y, z) of the space coordinates, then the function **A** (x, y, z) is said to be a *vector point function* and it represents a *vector field* in R. For example the velocity vectors at all points of a fluid moving inside a pipe of non-uniform cross-section constitute a vector field.

If the physical quantities described by scalar or vector fields vary with time at each point, the configuration of these fields will be time-dependent. Such fields are called *dynamic* fields and are in general functions of three space variables and one time variable—e.g., $\phi(x, y, z, t)$, $\mathbf{A}(r, \theta, z, t)$. Scalar or vector fields which are independent of time are called *static* or *stationary* fields.

1.3 LAWS OF VECTOR ALGEBRA

The sum of two vectors $\mathbf{A}$ and $\mathbf{B}$ is obtained geometrically by placing the initial point of $\mathbf{B}$ at the terminal point of $\mathbf{A}$ and then drawing a vector from the initial point of $\mathbf{A}$ to the terminal point of $\mathbf{B}$ (Fig. 1.2). This process is written as a vector equation,

$$\mathbf{A} + \mathbf{B} = \mathbf{C}$$

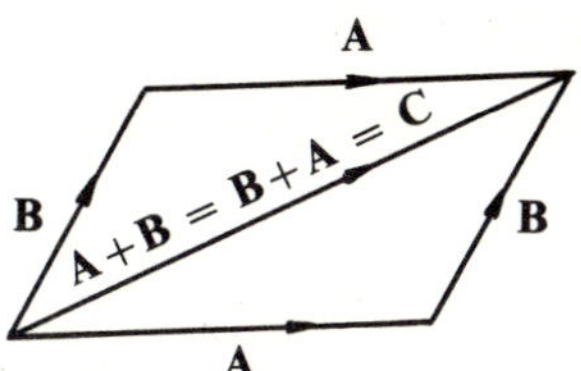

Fig. 1.2 Addition of two vectors

It is evident from Fig. 1.2 that we can obtain the same vector $\mathbf{C}$ by laying off $\mathbf{A}$ from the end point of $\mathbf{B}$ and joining the initial point of $\mathbf{B}$ to the end point of $\mathbf{A}$; that is,

$$\mathbf{B} + \mathbf{A} = \mathbf{C}$$

so that $\quad\quad\quad \mathbf{A} + \mathbf{B} = \mathbf{B} + \mathbf{A}$ (Commutative Law)

The rule for the addition of two vectors is also known as the parallelogram law since the vector $\mathbf{C}$ is the diagonal of the parallelogram of which the two vectors $\mathbf{A}$ and $\mathbf{B}$ are the sides.

The extension of the above rule to the addition of more than two vectors is evident. In this case the vector joining the initial point of the first to the terminal point of the last is the vector sum of them all. Thus in Fig. 1.3

$$\mathbf{R} = \mathbf{A} + \mathbf{B} + \mathbf{C} + \mathbf{D}$$

The vectors $\mathbf{A}$, $\mathbf{B}$, $\mathbf{C}$, $\mathbf{D}$ are known as the *vector components* of $\mathbf{R}$.

The vector summation process is associative, that is, the sum of any number of vectors is independent of the order in which they are taken;

$$\mathbf{A} + (\mathbf{B} + \mathbf{C}) = (\mathbf{A} + \mathbf{B}) + \mathbf{C}$$ (Associative Law)

Since the negative of a vector is defined as a vector of the same

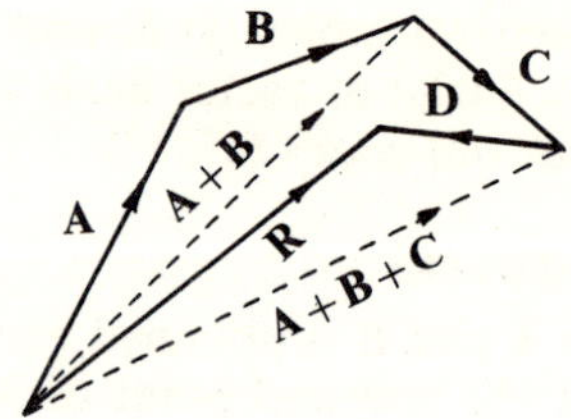

Fig. 1.3 Addition of several vectors

magnitude but in the opposite sense, it follows that subtraction of vectors, like addition, can be carried out geometrically by the parallelogram law (Fig. 1.4); thus

$$\mathbf{A} - \mathbf{B} = \mathbf{A} + (-\mathbf{B})$$

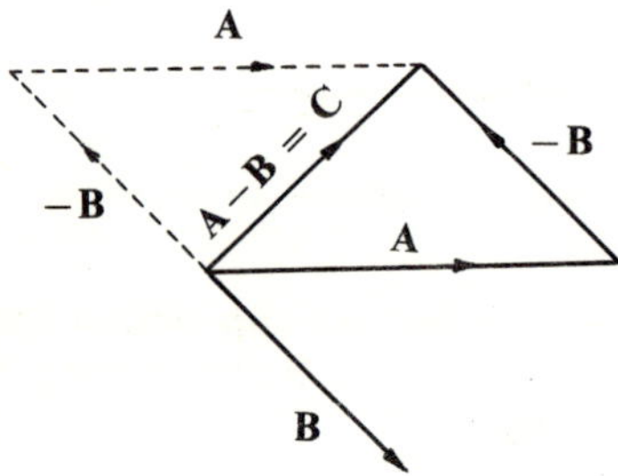

Fig. 1.4 Subtraction of two vectors

The following laws of the algebra of ordinary real numbers are applicable to vectors

$$\mathbf{A} + \mathbf{B} = \mathbf{B} + \mathbf{A} \tag{1-2}$$

$$\mathbf{A} + (\mathbf{B} + \mathbf{C}) = (\mathbf{A} + \mathbf{B}) + \mathbf{C} \tag{1-3}$$

$$m(n\mathbf{A}) = mn\mathbf{A} \tag{1-4}$$

$$(m + n)\mathbf{A} = m\mathbf{A} + n\mathbf{A} \tag{1-5}$$

$$m(\mathbf{A} + \mathbf{B}) = m\mathbf{A} + m\mathbf{B} \tag{1-6}$$

1.4 SCALAR PRODUCT OF TWO VECTORS

The *scalar* or *dot* product of two vectors $\mathbf{A}$ and $\mathbf{B}$ (written $\mathbf{A} \cdot \mathbf{B}$) is by definition a scalar equal to the product of the magnitudes of $\mathbf{A}$ and $\mathbf{B}$ and the cosine of the angle θ between them,

$$\mathbf{A} \cdot \mathbf{B} = AB \cos \theta \tag{1-7}$$

It follows that the dot product of two vectors is equal to the magnitude of either of them multiplied by the projection of the other upon it (Fig. 1.5). If θ lies between $\pi/2$ and $3\pi/2$ then $\cos\theta$ and $\mathbf{A}\cdot\mathbf{B}$ will be negative numbers. If $\mathbf{A}=\mathbf{B}$ then,

$$\mathbf{A}\cdot\mathbf{A}=A^2 \tag{1-8}$$

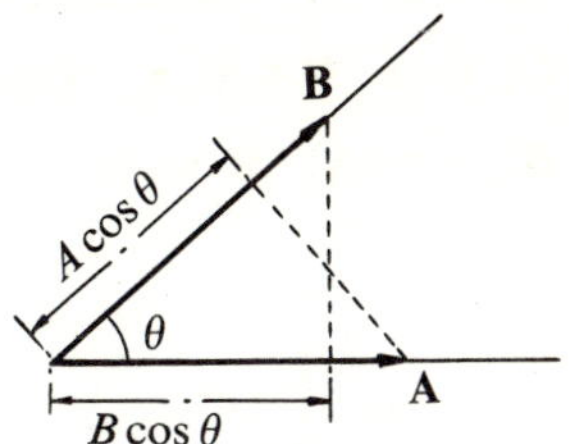

Fig. 1.5 Scalar product of two vectors

If $\mathbf{A}$ and $\mathbf{B}$ are perpendicular, then

$$\mathbf{A}\cdot\mathbf{B}=0 \quad (\mathbf{A}\perp\mathbf{B}) \tag{1-9}$$

If $\mathbf{A}$, $\mathbf{B}$ and $\mathbf{C}$ are three vectors such that $\mathbf{A}-\mathbf{B}=\mathbf{C}$, then

$$(\mathbf{A}-\mathbf{B})\cdot(\mathbf{A}-\mathbf{B})=\mathbf{C}\cdot\mathbf{C}$$

or

$$A^2+B^2-2\mathbf{A}\cdot\mathbf{B}=C^2$$

which is the trigonometric law of cosines.

If $\mathbf{a}$ represents a unit vector in any direction, then

$$\mathbf{A}\cdot\mathbf{a}=A\cos\theta_{Aa}$$

is the component of $\mathbf{A}$ in the direction of the unit vector $\mathbf{a}$. The components of a vector can thus be found by taking its scalar product with unit vectors in the direction of these components. For example the three rectangular components of a vector $\mathbf{A}$ are $A_x=\mathbf{A}\cdot\mathbf{a}_x$, $A_y=\mathbf{A}\cdot\mathbf{a}_y$, $A_z=\mathbf{A}\cdot\mathbf{a}_z$.

The components, referred to a rectangular coordinate system, of an arbitrary unit vector $\mathbf{a}$ are

$$\mathbf{a}\cdot\mathbf{a}_x=\cos\theta_x,\ \mathbf{a}\cdot\mathbf{a}_y=\cos\theta_y,\ \mathbf{a}\cdot\mathbf{a}_z=\cos\theta_z$$

where θ_x, θ_y, θ_z are the angles between $\mathbf{a}$ and the x, y, z axes respectively. The cosines of these angles are called the *direction cosines* of $\mathbf{a}$. Since the magnitude of $\mathbf{a}$ is unity we have that

$$\cos^2\theta_x+\cos^2\theta_y+\cos^2\theta_z=1 \tag{1-10}$$

It is evident that the direction cosines of any vector $\mathbf{A}$ are given by,

$$\cos\theta_x = \frac{A_x}{A}, \cos\theta_y = \frac{A_y}{A}, \cos\theta_z = \frac{A_z}{A} \qquad (1\text{--}11)$$

A little consideration will show that the following laws are valid,

$$\mathbf{A}\cdot\mathbf{B} = \mathbf{B}\cdot\mathbf{A} \quad \text{(commutative)} \qquad (1\text{--}12)$$

$$\mathbf{A}\cdot(\mathbf{B}+\mathbf{C}) = \mathbf{A}\cdot\mathbf{B}+\mathbf{A}\cdot\mathbf{C} \quad \text{(distributive)} \qquad (1\text{--}13)$$

$$m(\mathbf{A}\cdot\mathbf{B}) = (m\mathbf{A})\cdot\mathbf{B} = \mathbf{A}\cdot(m\mathbf{B}) \qquad (1\text{--}14)$$

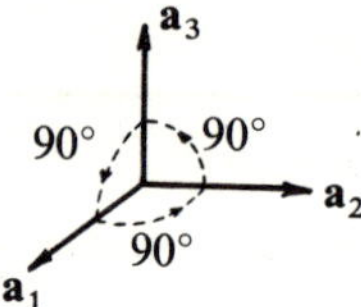

Fig. 1.6 Right-handed system of orthogonal unit vectors

For three unit vectors $\mathbf{a}_1$, $\mathbf{a}_2$, $\mathbf{a}_3$ forming a right-handed orthogonal system (Fig. 1.6)

$$\mathbf{a}_1\cdot\mathbf{a}_1 = \mathbf{a}_2\cdot\mathbf{a}_2 = \mathbf{a}_3\cdot\mathbf{a}_3 = 1; \mathbf{a}_1\cdot\mathbf{a}_2 = \mathbf{a}_2\cdot\mathbf{a}_3 = \mathbf{a}_3\cdot\mathbf{a}_1 = 0$$

If $\mathbf{A} = A_1\mathbf{a}_1+A_2\mathbf{a}_2+A_3\mathbf{a}_3$ and $B = B_1\mathbf{a}_1+B_2\mathbf{a}_2+B_3\mathbf{a}_3$ then

$$\mathbf{A}\cdot\mathbf{B} = A_1B_1+A_2B_2+A_3B_3 \qquad (1\text{--}15)$$

1.5 VECTOR PRODUCT OF TWO VECTORS

The *vector* or *cross* product of two vectors $\mathbf{A}$ and $\mathbf{B}$ (written $\mathbf{A}\times\mathbf{B}$) is by definition a vector whose magnitude is equal to the product of the

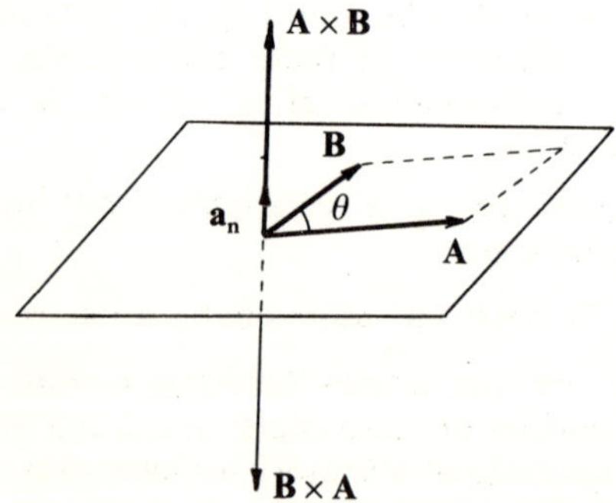

Fig. 1.7 Cross product of two vectors

magnitudes of $\mathbf{A}$ and $\mathbf{B}$ and the sine of the angle between them. The direction of this new vector is perpendicular to the plane containing $\mathbf{A}$ and $\mathbf{B}$ and points in the direction of advance of a right-handed screw as $\mathbf{A}$ is turned into $\mathbf{B}$ through the smaller of the angles between them (Fig. 1.7). Thus,

$$\mathbf{A} \times \mathbf{B} = AB \sin \theta \mathbf{a_n} \tag{1-16}$$

where $\mathbf{a_n}$ is a unit vector normal to the plane of $\mathbf{A}$ and $\mathbf{B}$. The magnitude of the product represents the area of a parallelogram with sides $\mathbf{A}$ and $\mathbf{B}$; we may thus think of $\mathbf{A} \times \mathbf{B}$ as the *vector area* of the parallelogram.

Since the direction of $\mathbf{A} \times \mathbf{B}$ is determined by the right-hand rule, it is clear that interchanging $\mathbf{A}$ and $\mathbf{B}$ reverses the direction or sign of their product, that is,

$$\mathbf{A} \times \mathbf{B} = -\mathbf{B} \times \mathbf{A} \tag{1-17}$$

If the vectors $\mathbf{A}$ and $\mathbf{B}$ are parallel, then

$$\mathbf{A} \times \mathbf{B} = 0 \quad (A \| B) \tag{1-18}$$

The following laws (distributive) are valid,

$$\mathbf{A} \times (\mathbf{B} + \mathbf{C}) = \mathbf{A} \times \mathbf{B} + \mathbf{A} \times \mathbf{C} \tag{1-19}$$

$$m(\mathbf{A} \times \mathbf{B}) = (m\mathbf{A}) \times \mathbf{B} = \mathbf{A} \times (m\mathbf{B}) \tag{1-20}$$

For three unit vectors forming a right-handed orthogonal system

$$\mathbf{a_1} \times \mathbf{a_1} = \mathbf{a_2} \times \mathbf{a_2} = \mathbf{a_3} \times \mathbf{a_3} = 0; \; \mathbf{a_1} \times \mathbf{a_2} = \mathbf{a_3}, \mathbf{a_2} \times \mathbf{a_3} = \mathbf{a_1},$$

$$\mathbf{a_3} \times \mathbf{a_1} = \mathbf{a_2}$$

If $\mathbf{A} = A_1\mathbf{a_1} + A_2\mathbf{a_2} + A_3\mathbf{a_3}$ and $\mathbf{B} = B_1\mathbf{a_1} + B_2\mathbf{a_2} + B_3\mathbf{a_3}$

$$\mathbf{A} \times \mathbf{B} = \begin{bmatrix} \mathbf{a_1} & \mathbf{a_2} & \mathbf{a_3} \\ A_1 & A_2 & A_3 \\ B_1 & B_2 & B_3 \end{bmatrix} \tag{1-21}$$

1.6 TRIPLE PRODUCTS

The product $\mathbf{A} \cdot (\mathbf{B} \times \mathbf{C})$ is known as the *scalar triple product* and it represents the volume of the parallelepiped of which $\mathbf{B} \times \mathbf{C}$ is the base area and $\mathbf{A}$ is the slant edge (Fig. 1.8). Since the volume of a parallelepiped is independent of the face chosen as its base it follows that*

$$\mathbf{A} \cdot \mathbf{B} \times \mathbf{C} = \mathbf{C} \cdot \mathbf{A} \times \mathbf{B} = \mathbf{B} \cdot \mathbf{C} \times \mathbf{A} \tag{1-22}$$

---

*The use of parentheses is superfluous since the alternative possible interpretation $(\mathbf{A} \cdot \mathbf{B}) \times \mathbf{C}$ is meaningless since $\mathbf{A} \cdot \mathbf{B}$ is a scalar and the vector product is defined only for two vectors.

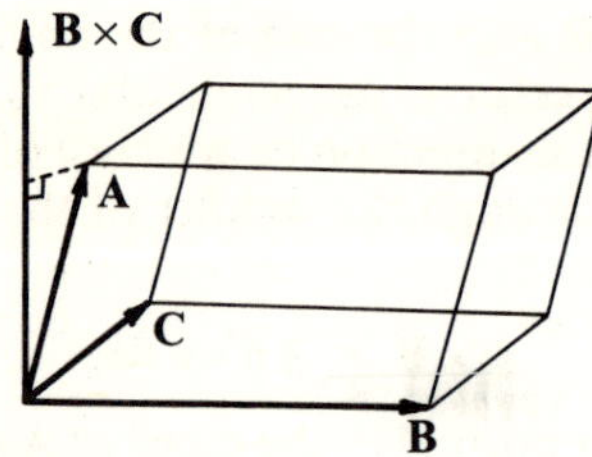

Fig. 1.8 Graphical representation of scalar triple product

Thus any cyclic permutation of vectors in a scalar triple product leaves this product unchanged. Since the scalar product of two vectors is commutative it follows from the first and second terms in equation 1–22 that

$$\mathbf{A} \cdot \mathbf{B} \times \mathbf{C} = \mathbf{A} \times \mathbf{B} \cdot \mathbf{C} \tag{1-23}$$

so that in any scalar triple product the dot and the cross may be interchanged without altering the value of the product. In terms of the orthogonal components of **A**, **B** and **C** we may write

$$\mathbf{A} \cdot \mathbf{B} \times \mathbf{C} = \begin{bmatrix} A_1 & A_2 & A_3 \\ B_1 & B_2 & B_3 \\ C_1 & C_2 & C_3 \end{bmatrix} \tag{1-23}$$

The product $\mathbf{A} \times (\mathbf{B} \times \mathbf{C})$ is called the *vector triple product* and it is a vector perpendicular to **A** and to $\mathbf{B} \times \mathbf{C}$. Since $\mathbf{B} \times \mathbf{C}$ is itself perpendicular to the plane of **B** and **C**, it follows that $\mathbf{A} \times (\mathbf{B} \times \mathbf{C})$ must lie in the plane of **B** and **C** (Fig. 1.9). The position of the parentheses in

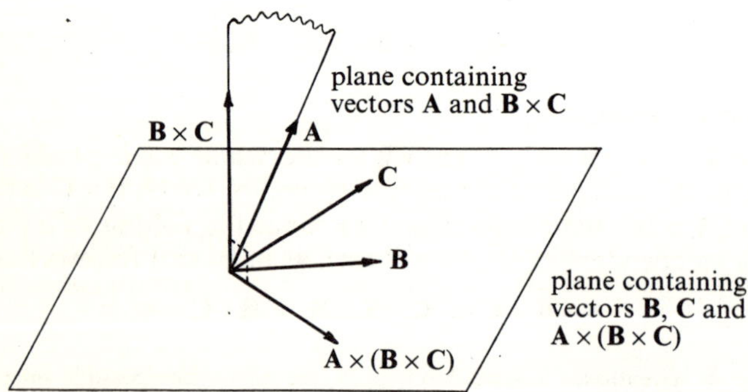

Fig. 1.9 Graphical representation of vector triple product

a vector triple product is critical because $(\mathbf{A} \times \mathbf{B}) \times \mathbf{C}$ is completely different from $\mathbf{A} \times (\mathbf{B} \times \mathbf{C})$.

The following vector identities will be found useful

$$\mathbf{A} \times (\mathbf{B} \times \mathbf{C}) = (\mathbf{A} \cdot \mathbf{C})\mathbf{B} - (\mathbf{A} \cdot \mathbf{B})\mathbf{C} \qquad (1\text{–}24)$$

$$(\mathbf{A} \times \mathbf{B}) \times \mathbf{C} = (\mathbf{A} \cdot \mathbf{C})\mathbf{B} - (\mathbf{B} \cdot \mathbf{C})\mathbf{A} \qquad (1\text{–}25)$$

$$\mathbf{A} \times \mathbf{B} \cdot \mathbf{C} \times \mathbf{D} = (\mathbf{A} \cdot \mathbf{C})(\mathbf{B} \cdot \mathbf{D}) - (\mathbf{A} \cdot \mathbf{D})(\mathbf{B} \cdot \mathbf{C}) \qquad (1\text{–}26)$$

$$(\mathbf{A} \times \mathbf{B})^2 = A^2 B^2 - (\mathbf{A} \cdot \mathbf{B})^2 \qquad (1\text{–}27)$$

$$(\mathbf{A} \times \mathbf{B}) \times (\mathbf{C} \times \mathbf{D}) = [\mathbf{A} \cdot \mathbf{B} \times \mathbf{D}]\mathbf{C} - [\mathbf{A} \cdot \mathbf{B} \times \mathbf{C}]\mathbf{D} \qquad (1\text{–}28)$$

1.7 COORDINATE COMPONENTS OF A VECTOR

One great advantage of vector methods is that we can obtain general results without reference to fixed axes of reference. However, for the purpose of calculation or of carrying out mathematical operations— e.g., differentiation or integration—it is necessary to refer all vector quantities under consideration to a convenient frame of reference, the coordinate system.

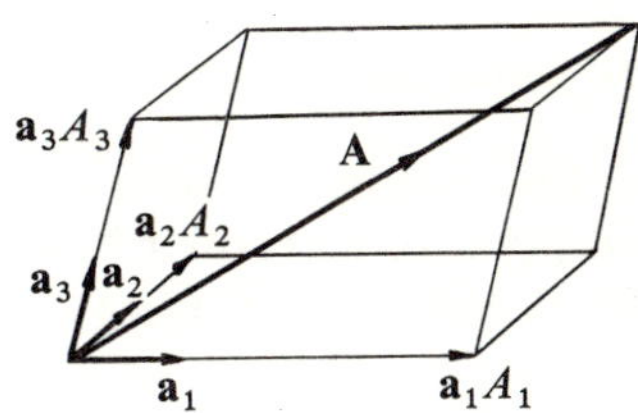

Fig. 1.10 Representation of a vector by three non-coplanar components

Since three non-coplanar vector components are necessary and sufficient to locate a vector in three-dimensional space, then any vector $\mathbf{A}$ can be expressed as

$$\mathbf{A} = A_1 \mathbf{a}_1 + A_2 \mathbf{a}_2 + A_3 \mathbf{a}_3$$

where $\mathbf{a}_1$, $\mathbf{a}_2$ and $\mathbf{a}_3$ are three arbitrary non-coplanar unit vectors and A_1, A_2 and A_3 are the scalar components of the vector $\mathbf{A}$ in the direction of the unit vectors (Fig. 1.10). The directions of the unit vectors $\mathbf{a}_1$, $\mathbf{a}_2$ and $\mathbf{a}_3$ at the initial point of vector $\mathbf{A}$ determine the coordinate system. If these vectors are mutually orthogonal we have an orthogonal coordinate system. The most commonly used orthogonal coordinate systems are the rectangular or cartesian system, the circular cylindrical system and the spherical system.

1.7.1 RECTANGULAR COORDINATES

Consider a right-handed rectangular coordinate system* consisting of three mutually perpendicular straight lines (axes) Ox, Oy and Oz (Fig. 1.11). Let $\mathbf{a}_x$, $\mathbf{a}_y$ and $\mathbf{a}_z$ be unit vectors having the directions of the positive x, y and z axes respectively. Any point P is specified by

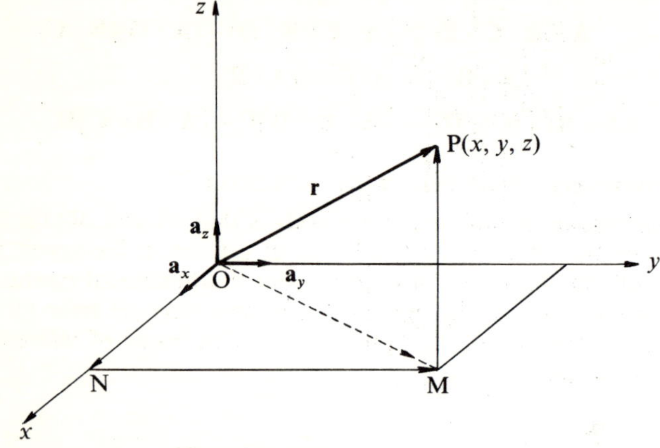

Fig. 1.11 Rectangular components of a vector

the intersection of the three planes $x =$ constant, $y =$ constant, $z =$ constant. Let the initial point of a vector $\mathbf{r}$ be chosen at the origin O and its terminal point at the point P (x, y, z). It is evident from Fig. 1.10 that

$$\overrightarrow{OP} = \overrightarrow{OM} + \overrightarrow{MP} = \overrightarrow{ON} + \overrightarrow{NM} + \overrightarrow{MP}$$

which may be written as

$$\mathbf{r} = x\mathbf{a}_x + y\mathbf{a}_y + z\mathbf{a}_z$$

$x\mathbf{a}_x$, $y\mathbf{a}_y$ and $z\mathbf{a}_z$ are the three component vectors of $\mathbf{r}$ along the three orthogonal axes.

More generally, if $\mathbf{A}$ is any vector with an arbitrary initial point, we may write (Fig. 1.12),

$$\mathbf{A} = (x_2 - x_1)\mathbf{a}_x + (y_2 - y_1)\mathbf{a}_y + (z_2 - z_1)\mathbf{a}_z$$

$$= A_x\mathbf{a}_x + A_y\mathbf{a}_y + A_z\mathbf{a}_z \tag{1-29}$$

* Throughout the book the rectangular coordinate system used is right-handed, that is if a right-handed screw is rotated through 90° from Ox to Oy, it will advance in the positive z-direction.

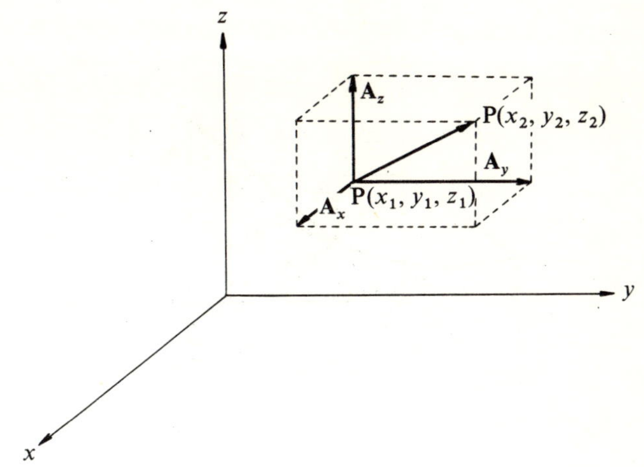

Fig. 1.12 Rectangular components of a vector with an arbitrary initial point

where A_x, A_y and A_z are the scalar components of **A**. The magnitude of **A** is given by

$$A = (A_x^2 + A_y^2 + A_z^2)^{1/2} \qquad (1\text{--}30)$$

If **B** is a second vector given by $\mathbf{B} = B_x\mathbf{a}_x + B_y\mathbf{a}_y + B_z\mathbf{a}_z$, it is evident that,

$$\mathbf{A} + \mathbf{B} = (A_x + B_x)\mathbf{a}_x + (A_y + B_y)\mathbf{a}_y + (A_z + B_z)\mathbf{a}_z$$

$$\mathbf{A} \cdot \mathbf{B} = A_x B_x + A_y B_y + A_z B_z$$

$$\mathbf{A} \times \mathbf{B} = \begin{bmatrix} \mathbf{a}_x & \mathbf{a}_y & \mathbf{a}_z \\ A_x & A_y & A_z \\ B_x & B_y & B_z \end{bmatrix}$$

1.7.2 CIRCULAR CYLINDRICAL COORDINATES

In this system (Fig. 1.13) the point P (r, ϕ, z) is specified by the intersection of the following three orthogonal coordinate surfaces: (i) a cylindrical surface r = constant, (ii) a plane ϕ = constant, which contains the cylinder axis, (iii) a plane z = constant, which is normal to the cylinder axis. The three unit vectors $\mathbf{a}_r$, $\mathbf{a}_\phi$ and $\mathbf{a}_z$ are tangents to the coordinate lines at P, they point in the directions of increasing r, ϕ and z and they form a right-handed orthogonal system. It should be noted that the directions of $\mathbf{a}_r$ and $\mathbf{a}_\phi$ change from point to point in

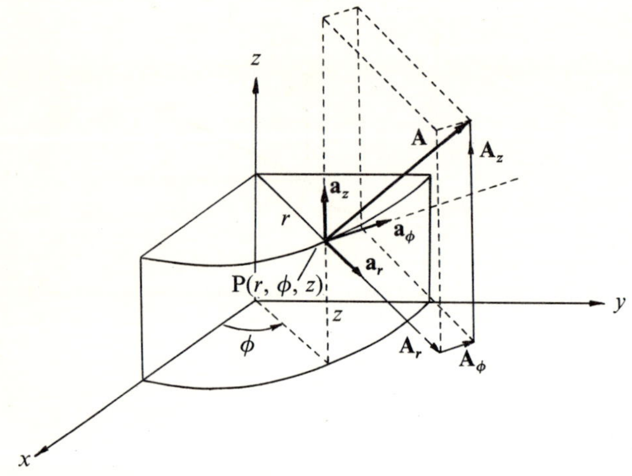

Fig. 1.13 Cylindrical components of a vector

space. The vector $\mathbf{A}$ in cylindrical coordinates is given by

$$\mathbf{A} = A_r\mathbf{a}_r + A_\phi\mathbf{a}_\phi + A_z\mathbf{a}_z \tag{1-31}$$

$$A = (A_r^2 + A_\phi^2 + A_z^2)^{1/2} \tag{1-32}$$

The cylindrical coordinates r, ϕ, z are related to the rectangular coordinates x, y, z by the following relationships,

$$x = r\cos\phi, \qquad y \doteq r\sin\phi, \qquad z = z \tag{1-33}$$

$$r^2 = x^2 + y^2, \qquad \tan\phi = y/x, \qquad z = z \tag{1-34}$$

The unit vectors $\mathbf{a}_r$, $\mathbf{a}_\phi$, $\mathbf{a}_z$ are related to the unit vectors $\mathbf{a}_x$, $\mathbf{a}_y$, $\mathbf{a}_z$ by the following relationships,

$$\left.\begin{aligned}
\mathbf{a}_x &= \cos\phi\,\mathbf{a}_r - \sin\phi\,\mathbf{a}_\phi \\
\mathbf{a}_y &= \sin\phi\,\mathbf{a}_r + \cos\phi\,\mathbf{a}_\phi \\
\mathbf{a}_z &= \mathbf{a}_z
\end{aligned}\right\} \tag{1-35}$$

$$\left.\begin{aligned}
\mathbf{a}_r &= \cos\phi\,\mathbf{a}_x + \sin\phi\,\mathbf{a}_y \\
\mathbf{a}_\phi &= -\sin\phi\,\mathbf{a}_x + \cos\phi\,\mathbf{a}_y \\
\mathbf{a}_z &= \mathbf{a}_z
\end{aligned}\right\} \tag{1-36}$$

Equations 1–33 to 1–36 may be used to convert the representation of a vector in cartesian coordinates to its representation in cylindrical coordinates and vice versa. For example if

$$\mathbf{A} = A_r\mathbf{a}_r + A_\phi\mathbf{a}_\phi + A_z\mathbf{a}_z$$

then

$$\mathbf{A} = (A_r \cos \phi - A_\phi \sin \phi)\mathbf{a}_x + (A_r \sin \phi + A_\phi \cos \phi)\mathbf{a}_y + A_z\mathbf{a}_z$$

and

$$A_x = \frac{A_r x}{(x^2+y^2)^{1/2}} - \frac{A_\phi y}{(x^2+y^2)^{1/2}}$$

$$A_y = \frac{A_r y}{(x^2+y^2)^{1/2}} + \frac{A_\phi x}{(x^2+y^2)^{1/2}}$$

$$A_z = A_z$$

1.7.3 SPHERICAL COORDINATES

In this system (Fig. 1.14) the point P (r, θ, ϕ) is specified by the intersection of the following three orthogonal coordinate surfaces: (i) a sphere, r = constant, (ii) a cone, θ = constant, with apex at the centre of the sphere, (iii) a plane, ϕ = constant. The three unit vectors $\mathbf{a}_r$, $\mathbf{a}_\theta$ and $\mathbf{a}_\phi$ are tangents to the coordinate lines at P; they point in the directions of increasing r, θ and ϕ and form a right-handed orthogonal system. All three unit vectors have different directions at different points. The vector $\mathbf{A}$ in spherical coordinates is given by

$$\mathbf{A} = A_r\mathbf{a}_r + A_\theta\mathbf{a}_\theta + A_\phi\mathbf{a}_\phi \tag{1-37}$$

$$A = (A_r^2 + A_\theta^2 + A_\phi^2)^{1/2} \tag{1-38}$$

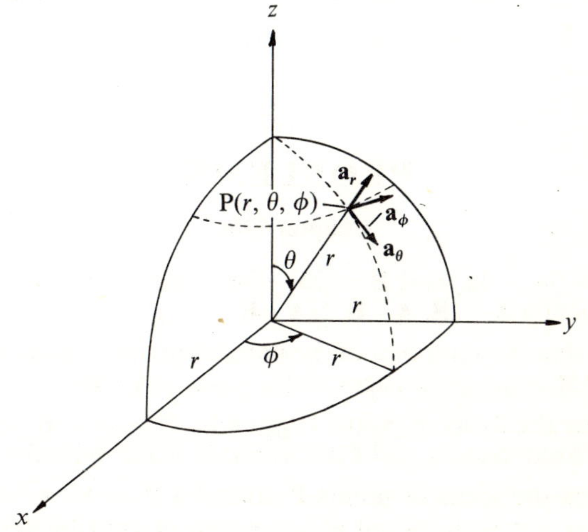

Fig. 1.14 Spherical components of a vector

The spherical coordinates r, θ, ϕ are related to the rectangular coordinates x, y, z by the relationships

$$x = r\sin\theta\cos\phi, \qquad y = r\sin\theta\sin\phi, \qquad z = r\cos\theta \qquad (1\text{-}39)$$

$$r^2 = x^2+y^2+z^2, \cos\theta = \frac{z}{(x^2+y^2+z^2)^{1/2}}, \tan\phi = \frac{y}{x} \qquad (1\text{-}40)$$

The unit vectors $\mathbf{a}_r$, $\mathbf{a}_\theta$, $\mathbf{a}_\phi$ are related to the unit vectors $\mathbf{a}_x$, $\mathbf{a}_y$, $\mathbf{a}_z$ by,

$$\left.\begin{aligned}
\mathbf{a}_x &= \sin\theta\cos\phi\,\mathbf{a}_r+\cos\theta\cos\phi\,\mathbf{a}_\theta-\sin\phi\,\mathbf{a}_\phi \\
\mathbf{a}_y &= \sin\theta\sin\phi\,\mathbf{a}_r+\cos\theta\sin\phi\,\mathbf{a}_\theta+\cos\phi\,\mathbf{a}_\phi \\
\mathbf{a}_z &= \cos\theta\,\mathbf{a}_r-\sin\theta\,\mathbf{a}_\theta
\end{aligned}\right\} \qquad (1\text{-}41)$$

$$\left.\begin{aligned}
\mathbf{a}_r &= \sin\theta\cos\phi\,\mathbf{a}_x+\sin\theta\sin\phi\,\mathbf{a}_y+\cos\theta\,\mathbf{a}_z \\
\mathbf{a}_\theta &= \cos\theta\cos\phi\,\mathbf{a}_x-\cos\theta\sin\phi\,\mathbf{a}_y-\sin\theta\,\mathbf{a}_z \\
\mathbf{a}_\phi &= -\sin\phi\,\mathbf{a}_x+\cos\phi\,\mathbf{a}_y
\end{aligned}\right\} \qquad (1\text{-}42)$$

Equations 1-39 to 1-42 can be used to convert the representation of a vector in rectangular coordinates to its representation in spherical coordinates and vice versa. For example if

$$\mathbf{A} = A_x\mathbf{a}_x+A_y\mathbf{a}_y+A_z\mathbf{a}_z$$

then

$$\begin{aligned}
\mathbf{A} = &(A_x\sin\theta\cos\phi+A_y\sin\theta\sin\phi+A_z\cos\theta)\mathbf{a}_r \\
&+(A_x\cos\theta\cos\phi+A_y\cos\theta\sin\phi-A_z\sin\theta)\mathbf{a}_\theta \\
&+(-A_x\sin\phi+A_y\cos\phi)\mathbf{a}_\phi
\end{aligned}$$

PROBLEMS

Chapter One

1.1   If $\mathbf{A} = 3\mathbf{a}_x+4\mathbf{a}_y$ and $\mathbf{B} = 2\mathbf{a}_x-5\mathbf{a}_y$, find both algebraically and graphically $\mathbf{A}+2\mathbf{B}$, $\mathbf{A}-\mathbf{B}$, $\frac{1}{2}(\mathbf{A}+\mathbf{B})$.

1.2   Show that the midpoint of the parallelogram formed with vectors $\mathbf{A}$ and $\mathbf{B}$ as adjacent edges, is the point $\frac{1}{2}(\mathbf{A}+\mathbf{B})$.

1.3   What is the locus of points $\mathbf{P}$ given by $\mathbf{P} = \mathbf{A}+l\mathbf{B}$, where $\mathbf{A}$ and $\mathbf{B}$ are fixed vectors and l is a variable scalar quantity?

1.4   What is the locus of points $\mathbf{P}$ given by $\mathbf{P} = \mathbf{A}+\mathbf{r}$, where $\mathbf{A}$ is a fixed vector with its tail at the origin of coordinates and $\mathbf{r}$ is a vector of constant magnitude but variable orientation?

1.5   Given that $\mathbf{A} = 4\mathbf{a}_x - 3\mathbf{a}_z$ and $\mathbf{B} = 2\mathbf{a}_x - 2\mathbf{a}_y + \mathbf{a}_z$, find,
        (a) the unit vectors in the directions of $\mathbf{A}$ and $\mathbf{B}$ respectively,
        (b) the projection of $\mathbf{A}$ on $\mathbf{B}$,
        (c) the angle between $\mathbf{A}$ and $\mathbf{B}$.

1.6   Given that $\mathbf{A} = 2\mathbf{a}_x + 3\mathbf{a}_y + \mathbf{a}_z$, $\mathbf{B} = \mathbf{a}_x + 2\mathbf{a}_y + 3\mathbf{a}_z$, find (a) $\mathbf{A} \cdot \mathbf{B}$, (b) $\mathbf{A} \times \mathbf{B}$, (c) the direction cosines of $\mathbf{A}$, $\mathbf{B}$ and $\mathbf{A} \times \mathbf{B}$.

1.7   Given that $\mathbf{A} = A_x\mathbf{a}_x + 6\mathbf{a}_y + 4\mathbf{a}_z$, $\mathbf{B} = 2\mathbf{a}_x - 2\mathbf{a}_y + 6\mathbf{a}_z$, find A_x if $\mathbf{A}$ and $\mathbf{B}$ are to be perpendicular.

1.8   Find the component of the vector $\mathbf{A} = 4\mathbf{a}_x - 2\mathbf{a}_y + 4\mathbf{a}_z$ in the direction of the vector $\mathbf{B} = 2\mathbf{a}_x - 3\mathbf{a}_y - 2\mathbf{a}_z$.

1.9   The magnitude of a vector $\mathbf{A}$ is 5 and that of $\mathbf{B}$ is 2. Find $(\mathbf{A} \times \mathbf{B})^2 + (\mathbf{A} \cdot \mathbf{B})$.

1.10 Given that
        $$\mathbf{A} = \mathbf{a}_x + \mathbf{a}_y + \mathbf{a}_z$$
        $$\mathbf{B} = 2\mathbf{a}_x - \mathbf{a}_y + 3\mathbf{a}_z$$
        $$\mathbf{C} = 3\mathbf{a}_x + 2\mathbf{a}_y + \mathbf{a}_z$$
        $$\mathbf{D} = 4\mathbf{a}_x - 2\mathbf{a}_y$$
        find the angle between the intersection of the plane of $\mathbf{A}$ and $\mathbf{B}$ and the plane of $\mathbf{C}$ and $\mathbf{D}$.

1.11 For the vectors given in problem 1.10 find $\mathbf{A} \times \mathbf{B} \cdot \mathbf{C}$, $\mathbf{A} \cdot \mathbf{B} \times \mathbf{C}$, $(\mathbf{A} \times \mathbf{B}) \times \mathbf{C}$, $\mathbf{A} \times (\mathbf{B} \times \mathbf{C})$.

1.12 Given the vectors $\mathbf{A} = -2\mathbf{a}_x + \mathbf{a}_y - 2\mathbf{a}_z$, $\mathbf{B} = \mathbf{a}_x - 3\mathbf{a}_y + \mathbf{a}_z$, find (a) the cylindrical components of $\mathbf{A}$ and $\mathbf{B}$, (b) the spherical components of $\mathbf{A}$ and $\mathbf{B}$. Show that the products $\mathbf{A} \cdot \mathbf{B}$ and $\mathbf{A} \times \mathbf{B}$ are independent of the choice of the coordinate system.

1.13 Represent the vector $\mathbf{A} = 2\mathbf{a}_x + 2xy\mathbf{a}_y + y\mathbf{a}_z$ in cylindrical and in spherical coordinates.

1.14 Represent the vector $\mathbf{A} = 4\mathbf{a}_r + 3\mathbf{a}_\theta - 2\mathbf{a}_\phi$ in rectangular coordinates.

CHAPTER TWO

ELECTROSTATICS OF FREE SPACE

2.1 COULOMB'S LAW OF FORCE

The beginnings of electrostatics can be traced back to the ancient Greeks who discovered, almost 600 years B.C., that if amber were rubbed with a piece of wool or fur it acquired the peculiar property of attracting light bodies such as bits of hair or dust particles. For centuries this phenomenon was looked upon as one of the many wonders ascribed to the occult forces of nature. Towards the end of the sixteenth century Gilbert showed that this property, acquired by the frictional process of rubbing, was not restricted to amber but could be acquired by a large number of materials such as glass, wool, silk, sulphur, resins and even metals. He christened this property 'electricity', a name derived from the Greek word for amber, $\eta\lambda\varepsilon\kappa\tau\rho o\nu$. During the seventeenth and eighteenth centuries the phenomenon of frictional electricity, or tribo-electricity as it is sometimes called, was systematically and extensively studied by a large number of investigators the most prominent amongst whom were Gilbert, DuFay and Franklin. Here it is only necessary to summarize the most important conclusions drawn from their experiments.* These are:

(a) There exist two kinds of electricity, positive and negative;

(b) Bodies charged with opposite kinds of electricity attract each other whilst bodies charged with the same kind of electricity repel each other;

(c) Electricity is not created during the rubbing process but is merely transferred from one body to the other; one of the bodies gains as much as the other loses.

Conclusion (c) is the principle of conservation of charge.

In 1785 the French physicist Charles Augustin Coulomb carried out his series of decisive and historical experiments to determine the law which governs the forces of attraction and repulsion between charged bodies. As a result he was able to enunciate his inverse square law which remains one of the corner-stones of both the macroscopic and the microscopic theories of electricity. The complete statement of Coulomb's law is as follows: The force F between two bodies carrying charges Q_1 and Q_2 and separated by a distance r is directly proportional to the

---

* For a detailed historical account the reader should consult *A History of the Theories of the Aether and Electricity*, Vol. I by E. T. Whittaker (Nelson, 1951).

product of the charges and inversely proportional to the square of the distance between them. The direction of the force is along the line joining the charged bodies; the force is repulsive if Q_1 and Q_2 are of the same sign and attractive if they are of opposite sign. Mathematically the magnitude of the force F is given by

$$F = K \frac{Q_1 Q_2}{r^2} \qquad (2\text{--}1)$$

where K is a constant of proportionality. This constant plays an important role in determining the possible dimensions of the charge Q. If K is so chosen that it has a value of unity and no dimensions, then the dimensional equation for the charge would be

$$[Q] = [(ML^3 T^{-2})^{1/2}] = [M^{1/2} L^{3/2} T^{-1}]$$

and the unit of charge corresponding to $F = 1$ and $r = 1$ would depend on the units chosen for force and distance. The above dimensional equation for charge has no physical significance since charge is revealed as a quantity derived from the basic mechanical units of mass, length and time. It is more logical to consider charge as a fourth fundamental unit $[Q]$, in which case the coefficient K will have the dimensions

$$[K] = [ML^3 T^{-2} Q^{-2}]$$

In the m.k.s.c. system of units the basic unit of charge is the coulomb. In 1950, however, the International Electrotechnical Commission adopted the *ampere* as the fourth fundamental unit giving the m.k.s.a. (or Giorgi) System. The Conférence Générale des Poids et Mesures (CGPM) defines the ampere as follows:

The ampere is the constant current which, if maintained in two straight parallel conductors of infinite length, of negligible cross-section and placed at a distance of 1 metre apart in vacuum, will produce a force between them equal to 2×10^{-7} newton per metre length.

The coulomb is then a derived unit and is defined as the quantity of electricity transported in one second by a current of one ampere. In 1954 the CGPM adopted a rationalized and coherent system of units based on the four m.k.s.a. units (metre, kilogram, second, ampere), the unit of thermodynamic temperature (kelvin) and the unit of luminous intensity (candela). This system has been formally given the title 'Système International d'Unités' for which the abbreviation is SI in all languages (see Appendix III).

In SI units the dimensions of K are

$$[K] = [ML^3 I^{-2}]$$

For convenience the constant K is defined through another constant ϵ_0

such that

$$K = 1/4\pi\epsilon_0$$

where ϵ_0 is an 'electric constant'* whose numerical value is $1/(36\pi \times 10^9)$ $(= 8 \cdot 854 \times 10^{-12})$ and whose dimensions are the reciprocal of those of K,

$$[\epsilon_0] = [M^{-1}L^{-3}I^2]$$

The SI units of ϵ_0 are coulomb^2 per newton metre^2 ($C^2\,N^{-1}\,m^{-2}$); these, as will be shown later, are equivalent to farad/metre. In the SI system of units Coulomb's law now has the form

$$F = \frac{1}{4\pi\epsilon_0} \frac{Q_1 Q_2}{r^2} \tag{2-2}$$

where the force F is in newtons, the charge separation r is in metres and the electric charges Q_1 and Q_2 are in coulombs.

The validity of Coulomb's law, as given by equation 2–1 or equation 2–2 is subject to the following conditions:

(a) the distance r separating the two charged bodies must be large compared to the linear dimensions of these bodies. In other words the charges Q_1 and Q_2 have to be charged particles† or point charges.

(b) the electric charges must be at rest.

On the atomic scale we know from the Rutherford experiments on the scattering of α-particles that Coulomb's law is valid down to distances of the order of 10^{-14} metre from the centres of atomic nuclei. Failure of the law at smaller distances is due partly to the fact that the charged particles may no longer be considered as point charges and partly to the fact that at such distances the strong but short-range nuclear attractive forces predominate. The nature of these nuclear forces is still imperfectly understood at the present stage of the development of physics.

The smallest known quantity of charge is that carried by an electron; in SI units its value is

$$e = 1 \cdot 602 \times 10^{-19} \text{ coulomb}$$

The electric force between two electrons 1 metre apart is therefore of the order of 10^{-28} newton. It is instructive to compare this force with the force of gravitational attraction between the two electrons. The gravita-

---

* ϵ_0 is also very often called the 'permittivity of free space'. This name, however, is not consistent with the concept of permittivity which, as will be shown in Chapter 4, is a characteristic electric property of a medium which contains atoms or molecules.
† By definition, a particle is a body whose position in space is completely determined by a single position vector.

tional attraction between any two particles of masses m_1 and m_2 separated by a distance r is given by Newton's law

$$F_G = G\,\frac{m_1 m_2}{r^2}$$

where $G = 66{\cdot}7 \times 10^{-12}\ \mathrm{m^3\,kg^{-1}\,s^{-2}}$ is the gravitational constant. The rest mass of the electron is

$$m_0 = 9{\cdot}108 \times 10^{-31}\ \mathrm{kg}$$

so that the attractive force between two electrons 1 metre apart is of the order of 10^{-71} newton, which is negligible compared to the electric force.

Since force is a vector quantity it is more convenient to express equation 2–2 in vector form. Bearing in mind that each of the two charges experiences a force of the same magnitude but opposite in direction we have,

$$\mathbf{F}_1 = \frac{Q_1 Q_2}{4\pi\epsilon_0 r^2}\,\mathbf{a}_{r21}$$

$$\mathbf{F}_2 = \frac{Q_1 Q_2}{4\pi\epsilon_0 r^2}\,\mathbf{a}_{r12}$$

with
$$\mathbf{F}_1 = -\mathbf{F}_2$$

where $\mathbf{F}_1$ and $\mathbf{F}_2$ are the forces acting on the charges Q_1 and Q_2 respectively and $\mathbf{a}_{r21}$ and $\mathbf{a}_{r12}$ are unit vectors along the line segment r (see Fig. 2.1). The sign of the charges Q_1 and Q_2 will determine whether the forces are repulsive or attractive.

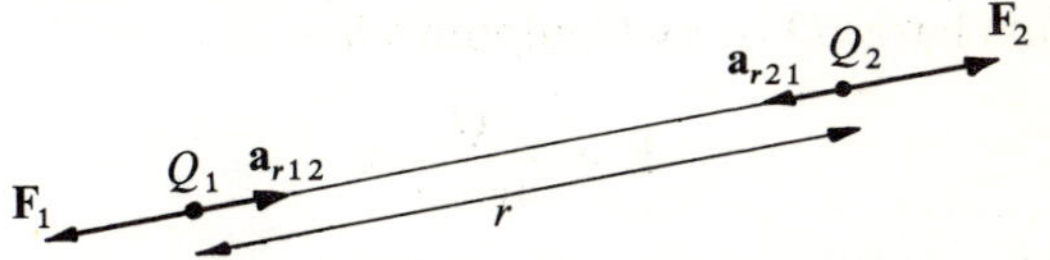

Fig. 2.1 Force between two point charges

For a system of N point charges the total force acting on a given charge is the vector sum of the individual forces acting on it (principle of superposition); the force on the ith charge is given by

$$\mathbf{F}_i = \frac{Q_i}{4\pi\epsilon_0} \sum_{j\neq i}^{N} \frac{Q_j}{r_{ji}^2}\,\mathbf{a}_{r_{ji}} \tag{2-3}$$

2.2 THE ELECTRIC FIELD AND FIELD INTENSITY

If we have a point charge Q fixed at any point in space then Coulomb's

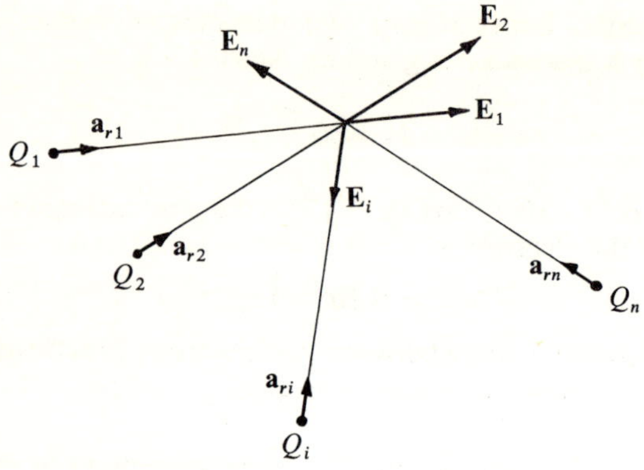

Fig. 2.2 Electrical field intensity at a point due to a system of point charges

law tells us that at any and every other point in space a second point charge will experience an electric force defined by equation 2–2. This means that the space around the fixed point charge has acquired the property of exerting a force on any other charge brought into that space. This property is described by saying that an electrostatic field of force exists everywhere in the space surrounding a fixed or static charge. The only possible way of detecting the presence of such a field in any region of space is to determine whether or not a force is exerted on a 'test' charge brought into that region.

The strength of an electric field on the electric field intensity at any point is defined as the force which would act on a unit *positive* test charge placed at that point. Thus the strength of the field at a distance r from a point charge Q is, by Coulomb's law

$$\mathbf{E} = \frac{Q}{4\pi\epsilon_0 r^2}\, \mathbf{a}_r \qquad (2\text{--}4)$$

It follows that the force on a charge Q placed in an electric field of strength $\mathbf{E}$ is given by

$$\mathbf{F} = Q\mathbf{E} \text{ newton} \qquad (2\text{--}5)$$

Clearly then, $\mathbf{E}$ is a *force per unit charge*; its units are newton/coulomb which are equivalent to the practical units of volt/metre (see Section 2.6).

Although the stipulation that the unit test charge be positive is completely arbitrary, it now forms part of the universally accepted definition of $\mathbf{E}$. An important consequence of this stipulation is that the vector $\mathbf{E}$ will always point away from a positive charge and towards a negative charge. In equation 2–4 therefore, the unit vector $\mathbf{a}_r$ will always be directed from the point charge Q to the point at which $\mathbf{E}$ is measured; the actual direction of $\mathbf{E}$ will be determined by the sign of Q.

For a system of N point charges the resultant field at any point is the vector sum of the individual component fields at that point (Fig. 2.2),

$$\mathbf{E} = \Sigma\, \mathbf{E}_i = \frac{1}{4\pi\epsilon_0} \Sigma \frac{Q_i}{r_i^2}\, \mathbf{a}_{ri} \tag{2-6}$$

Thus if point charges $Q_1, \ldots Q_N$ are located at the points $(x_1, y_1, z_1), \ldots$ (x_N, y_N, z_N) of a cartesian system of coordinates, the field at any point (x, y, z) is given by

$$\mathbf{E} = \frac{1}{4\pi\epsilon_0} \sum_{i=1}^{N} \frac{Q_i\big[(x-x_i)\mathbf{a}_x + (y-y_i)\mathbf{a}_y + (z-z_i)\mathbf{a}_z\big]}{\big[(x-x_i)^2 + (y-y_i)^2 + (z-z_i)^2\big]^{3/2}}$$

2.3 LINES OF FORCE

Suppose we place a unit test charge at some arbitrary point A in an electric field and then proceed to move it an incremental distance ΔL in the direction of the field at A; we then repeat this displacement always moving in the direction of the force acting at the point immediately preceding the new one to which we move. It is evident that in the limit when $\Delta L \to 0$ the test charge will trace a continuous curve. The direction of the tangent to that curve at any point gives the direction of the electric field intensity at that point. Such a curve is called a line of force. If we repeat the above process, starting every time from a different point, we obtain a family of curves which constitute a 'picture' of the electric field.

Since the test charge is positive the lines of force will always point away from a positive charge and towards a negative charge. This means that lines of force will always start at positive charges and terminate at negative charges. Positive charges are therefore considered as *sources* and negative charges as *sinks* of electric lines of force. In those cases where we have only positive or only negative charges, the corresponding sinks and sources are assumed to be equal amounts of negative or of positive charges located at infinity. For an isolated point charge $(+Q)$ for example, the lines of force would terminate on a charge $(-Q)$ at infinity. Figure 2.3 shows the lines of force for a positive and negative point charge and for two positive point charges.

Since there can be only one resultant direction of the electric intensity at any point in space it follows that no two lines of force can intersect. It is possible, however, for lines of force to intersect at points where the resultant intensity is zero; these points are known as singular or equilibrium points.

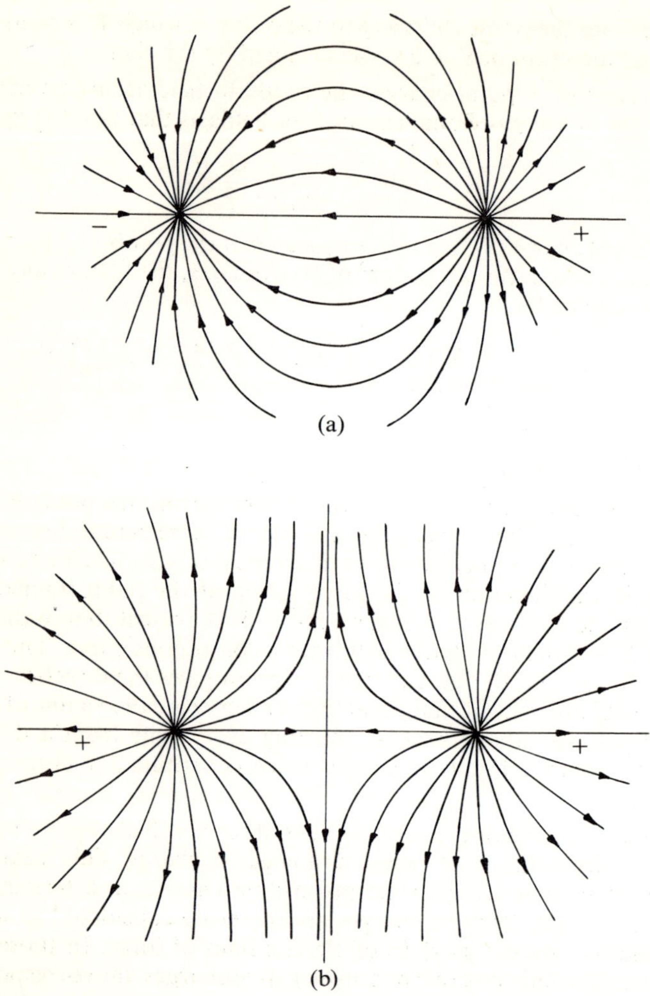

(a)

(b)

Fig. 2.3 Electric lines of force for (a) two equal and opposite point charges, (b) two equal positive point charges

Let the electric field intensity at any point (x, y, z) in an electrostatic field be given by

$$\mathbf{E} = E_x\mathbf{a}_x + E_y\mathbf{a}_y + E_z\mathbf{a}_z$$

The direction cosines of this vector are E_x/E, E_y/E, E_z/E. Since these must be equal to the direction cosines of the tangent to the line of force at the point (x, y, z) (Fig. 2.4), it follows that the analytic expression for the lines of force is given by

$$\frac{dx}{E_x} = \frac{dy}{E_y} = \frac{dz}{E_z} \qquad (2\text{--}7)$$

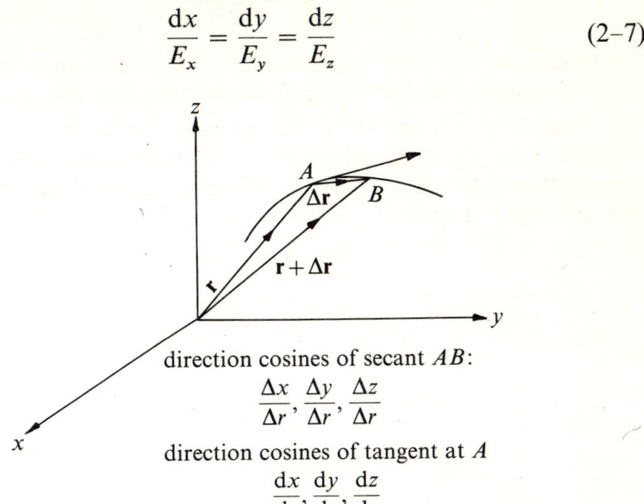

Fig. 2.4 Direction cosines of the tangent to a line of force

2.4 CHARGE DISTRIBUTIONS

(a) Volume distribution. Although all electric charges are essentially multiples of the elementary electronic charge, the number of elementary charges involved on the macroscopic scale is so large that the discreteness of charge can be neglected. Thus a charge of 1 microcoulomb, for example, consists of $6 \cdot 25 \times 10^{12}$ elementary charges; a difference of $\pm 10^{10}$ in the number of elementary charges would change the total charge by only $\pm 0 \cdot 1 \%$.

Because of the large number of charges involved, we can assume that charge is continuously distributed throughout a charged volume. The charge ΔQ within an element of volume Δv may be considered as continuously and uniformly distributed throughout the volume element with an average volume charge density $\rho_{\mathrm{av}} = \Delta Q/\Delta v$. Clearly, the smaller the volume element Δv the truer the value of ρ_{av}. The volume charge density at a point is defined by the limit

$$\rho = \lim_{\Delta v \to 0} \frac{\Delta Q}{\Delta v} = \frac{dQ}{dv} \; \mathrm{C\,m^{-3}} \qquad (2\text{--}8)$$

In general the density ρ at a point will be a scalar function of the coordinates of that point. The total charge Q within a volume v is given by the volume integral

$$Q = \int_v \rho \, dv \qquad (2\text{–}9)$$

If the volume v is small compared with all other macroscopically significant dimensions, the charge Q may be represented as a point charge.

It is sometimes convenient to regard a point charge Q whose position vector is $\mathbf{r}'$ as a volume charge density. This may be done by defining a special 'function', known as the Dirac delta function, and which has the following properties,

$$\delta(\mathbf{r} - \mathbf{r}') = 0 \quad \text{for} \quad \mathbf{r} \neq \mathbf{r}'$$
$$= \infty \quad \text{for} \quad \mathbf{r} = \mathbf{r}' \qquad (2\text{–}10)$$

$$\int \delta(\mathbf{r} - \mathbf{r}') \, dv = 1 \quad \mathbf{r}' \text{ in } v$$
$$= 0 \quad \mathbf{r}' \text{ not in } v \qquad (2\text{–}11)$$

$$\int_v f(\mathbf{r}) \delta(\mathbf{r} - \mathbf{r}') \, dv = f(\mathbf{r}') \quad \mathbf{r}' \text{ in } v$$
$$= 0 \qquad \mathbf{r}' \text{ not in } v \qquad (2\text{–}12)$$

Equation 2–12 is a direct consequence of equations 2–10 and 2–11; by equation 2–10 the integrand in equation 2–12 is zero for all values of $\mathbf{r}$ except $\mathbf{r} = \mathbf{r}'$; for this value $f(\mathbf{r}) = f(\mathbf{r}')$ and may be taken out of the integral sign and application of equation 2–11 gives equation 2–12.

The delta function is not a function in the usual sense but merely a notation having the properties defined by the above equations.

The point charge Q at $\mathbf{r}'$ may now be written as a volume charge density

$$\rho(\mathbf{r}) = Q\delta(\mathbf{r} - \mathbf{r}') \qquad (2\text{–}13)$$

The total charge inside any volume v which includes the point $\mathbf{r}'$ is, by direct application of equations 2–10 and 2–11

$$\int_v \rho(\mathbf{r}) \, dv = \int_v Q\delta(\mathbf{r} - \mathbf{r}') \, dv = Q \qquad (2\text{–}14)$$

(b) Surface distribution. The definition of a charge distribution in terms of a charge per unit area or a surface charge density is a special case of the volume charge distribution. If we have a sufficiently large number of charges of density ρ distributed in such a way that they are contained in a volume made up of a large surface area S of macroscopic dimensions and a thickness δL of atomic dimensions, then the charge in an element of volume $\Delta v = \Delta S \delta L$ is,

$$\Delta Q = \rho \Delta v = \rho \Delta S \, \delta L$$

Since δL is extremely small compared with ΔS it is convenient to replace the product $\rho\delta L$ by a surface charge density σ defined by the limit

$$\sigma = \lim_{\Delta S \to 0} \frac{\Delta Q}{\Delta S} = \frac{dQ}{dS} \text{ C m}^{-2} \qquad (2\text{-}15)$$

The density σ at any point will be a scalar function of the coordinates of that point. The total charge Q on a surface S is given by the surface integral

$$Q = \int_S \sigma \, dS \qquad (2\text{-}16)$$

(c) Linear distribution. A second special case of volume charge distribution is encountered when we have a large number of charges of density ρ distributed throughout a region in such a way that they are contained in a volume made up of a large macroscopic length L and a surface area δS of atomic dimensions. The charge in an element of volume $\Delta v = \Delta L \delta S$ is,

$$\Delta Q = \rho \, \Delta v = \rho \, \Delta L \, \delta S$$

Since δS is extremely small compared with ΔL it is convenient to replace the product $\rho \, \delta S$ by a linear charge density λ defined by the limit

$$\lambda = \lim_{\Delta L \to 0} \frac{\Delta Q}{\Delta L} = \frac{dQ}{dL} \text{ C m}^{-1} \qquad (2\text{-}17)$$

Like ρ and σ, λ is a scalar point function. The total charge Q on a line segment of length L is given by the line integral

$$Q = \int_L \lambda \, dL \qquad (2\text{-}18)$$

2.5 CENTRE OF CHARGE

Suppose that we have a system of N point charges and that the position of the ith charge Q_i is given by the position vector $\mathbf{r}_i$ drawn from some fixed origin to the charge Q_i. The position vector $\mathbf{r}_G$ of the centre of charge of the system is then defined by the vector equation

$$\mathbf{r}_G \sum^N Q_i = \mathbf{r}_G Q_T = \sum^N Q_i \mathbf{r}_i \qquad (2\text{-}19)$$

In cartesian coordinates the coordinates of the centre of charge would be given by

$$x_G Q_T = \sum^N Q_i x_i$$

$$y_G Q_T = \sum^N Q_i y_i$$

$$z_G Q_T = \sum^N Q_i z_i$$

In the case of a continuous volume distribution of charge the centre of charge is defined by the equation

$$r_G \int_v \rho(\mathbf{r})\, dv = r_G Q_T = \int_v \rho(\mathbf{r})\, \mathbf{r}\, dv \qquad (2\text{–}20)$$

At points which are at a large distance from all the charges of a system, the lines of force are asymptotic to straight lines passing through the centre of charge of the system whilst the equipotential surfaces approach spheres whose centres coincide with the centre of charge.

If the total charge Q_T is zero then a centre of charge does not exist. In such a case it is only possible to define a centre of positive and a centre of negative charge. The field and potential at large distances then approach that of two equal and opposite point charges located at these respective centres of charge. This case will be discussed in detail in Section 2.10.

2.6 ELECTRIC SCALAR POTENTIAL

The potential difference between two arbitrary points A and B in an electric field of strength E is defined as the *external* work required to move a unit positive test charge from one point to the other. In the SI system the units of potential difference are newton metre coulomb^{-1} $\equiv$ joule coulomb^{-1}; one joule coulomb^{-1} is equivalent to the practical unit of one volt.

A positive amount of work means that energy to move the charge is supplied by an external source, whilst a negative amount of work means that the required energy is supplied by the electric field itself. We denote the potential difference by V_{AB} if the test charge is moved from point B

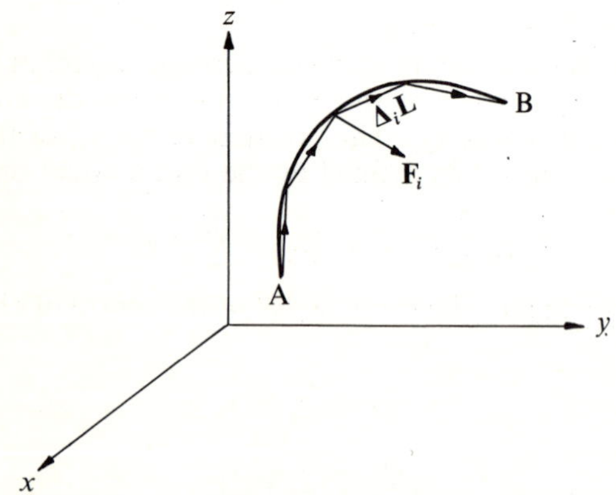

Fig. 2.5 To calculate work done when a variable force moves a particle from point B to point A along a given path

to point A and by V_{BA} if it is moved from point A to point B. The first subscript refers to the final point and the second to the initial point. For example V_{AB} equals 3 volts represents the fact that 3 joules of energy are necessary to move the unit positive charge from point B to point A so that the potential energy at point A is 3 volts higher than that at point B.

If a force **F** acts on a particle and displaces it a distance $\varDelta$**L**, the work done by **F** is defined by the scalar product

$$\Delta W = \mathbf{F} \cdot \varDelta\mathbf{L}$$

To calculate the work done when a variable force **F** moves a particle from point B to point A (Fig. 2.5) along a given path C, the total displacement may be considered as the sum of many incremental displacements along line segments $\varDelta_1\mathbf{L}, \varDelta_2\mathbf{L}, \dots \varDelta_n\mathbf{L}$. The total work done will be given by the sum

$$\sum_1^n \mathbf{F}_i \cdot \varDelta_i\mathbf{L}$$

where $\mathbf{F}_i$ is the force acting over the ith displacement. For an infinite number of segments the work will be given by the integral

$$W = \int_B^A \mathbf{F} \cdot d\mathbf{L}$$

Now if the field of force is the electric intensity E, then the potential difference between the points A and B is given by

$$V_{AB} = -\int_B^A \mathbf{E} \cdot d\mathbf{L} \tag{2-21}$$

The minus sign appears before the integral because the potential difference is defined as the external work done so that $\mathbf{F} = -\mathbf{E}$. The vector element of length d**L** is always positive—i.e., points in the direction of increasing coordinates—since the actual direction of motion is determined by the limits of integration.

We shall now calculate the potential difference between any two points A and B in the field of an isolated point charge Q. It will be assumed that the unit charge is moved from one point to the other along any path C of arbitrary shape (Fig. 2.6). Because of the spherical symmetry of the point charge it is simpler to use spherical coordinates. Using equations 2–21 and 2–4 we have

$$V_{AB} = -\int_{r_B}^{r_A} (E\mathbf{a}_r) \cdot (dr\,\mathbf{a}_r + r\,d\theta\,\mathbf{a}_\theta + r\sin\theta\,d\phi\,\mathbf{a}_\phi)$$

$$= -\int_{r_B}^{r_A} E_r\,dr = -\frac{Q}{4\pi\epsilon_0}\int_{r_B}^{r_A} \frac{dr}{r^2}$$

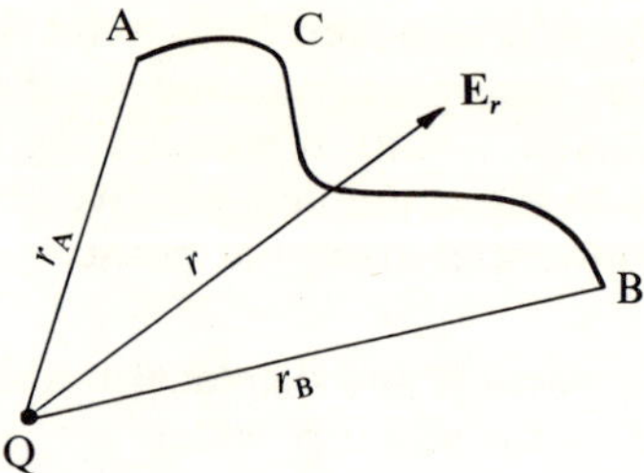

Fig. 2.6 To calculate the potential difference between points A and B in the field of a point charge Q

so that

$$V_{AB} = \frac{Q}{4\pi\epsilon_0}\left(\frac{1}{r_A} - \frac{1}{r_B}\right) \qquad (2\text{--}22)$$

and it follows that

$$V_{BA} = -\int_{r_A}^{r_B} E_r\, dr = \frac{Q}{4\pi\epsilon_0}\left(\frac{1}{r_B} - \frac{1}{r_A}\right) = -V_{AB}$$

Equation 2–22 expresses the important and fundamental result that the potential difference between two points is a scalar quantity which depends only on the distance of each point from the charge and not on the path chosen to carry the unit charge from one point to the other. A direct consequence of this result is that the work required to carry a unit charge around any closed path in a static field is zero. This may be expressed mathematically by the equation

$$\oint_C \mathbf{E} \cdot d\mathbf{L} = 0 \qquad (2\text{--}23)$$

which states that the line integral of the electrostatic field intensity around a closed path* is zero. Fields of force which satisfy equation 2–23 are called conservative fields.

In equation 2–22, if the point B is removed to infinity then the potential difference between the point A and infinity becomes,

$$V_A = \frac{Q}{4\pi\epsilon_0 r_A}$$

Similarly the potential difference between the point B and infinity is,

$$V_B = \frac{Q}{4\pi\epsilon_0 r_B}$$

* The integral is sometimes referred to as the circulation of the electric field.

The potential difference V_{AB} may therefore be written as

$$V_{AB} = V_A - V_B$$

V_A and V_B are known as the 'absolute' potentials of the points A and B with respect to a common point located at infinity and whose absolute potential is zero. The choice of infinity as a zero-potential reference is merely a matter of convenience, in many cases it is more convenient to take the potential of the earth as zero reference. If no particular zero reference is chosen, then it is possible to express the absolute potential of any point in the field of a point charge as

$$V = \frac{Q}{4\pi\epsilon_0 r} + C \tag{2-24}$$

where the constant C may be chosen so that $V = 0$ for any specific value of r. In the present text C will be chosen as zero unless otherwise specified and the term potential will be used to signify absolute potential. The potential at any point distant r from a point charge Q is

$$V = \frac{Q}{4\pi\epsilon_0 r} \tag{2-25}$$

If instead of a single point charge we have a set of N point charges $Q_1, Q_2 \ldots Q_N$, then the total potential at any point P is given simply by the scalar sum of the separate absolute potentials at P,

$$V_P = \frac{1}{4\pi\epsilon_0} \sum \frac{Q_i}{r_{QP}} \tag{2-26}$$

In the case of continuous charge distributions the potential is given by

$$V_P = \frac{1}{4\pi\epsilon_0} \int_v \frac{\rho \, dv}{r_{QP}} \tag{2-27}$$

$$V_P = \frac{1}{4\pi\epsilon_0} \int_S \frac{\sigma \, dS}{r_{QP}} \tag{2-28}$$

$$V_P = \frac{1}{4\pi\epsilon_0} \int_L \frac{\lambda \, dL}{r_{QP}} \tag{2-29}$$

for volume, surface and linear charge distributions respectively. In the most general case where all types of charge distributions are present, the potential at any point would be given by the sum of equations 2–26 to 2–29. It should be noted that V_P is a function of the coordinates of the point P alone, ρ, σ or λ are scalar point functions of the coordinates of the point Q alone and all integrations are to be carried out with respect to the coordinates of the point Q. In the case of a volume

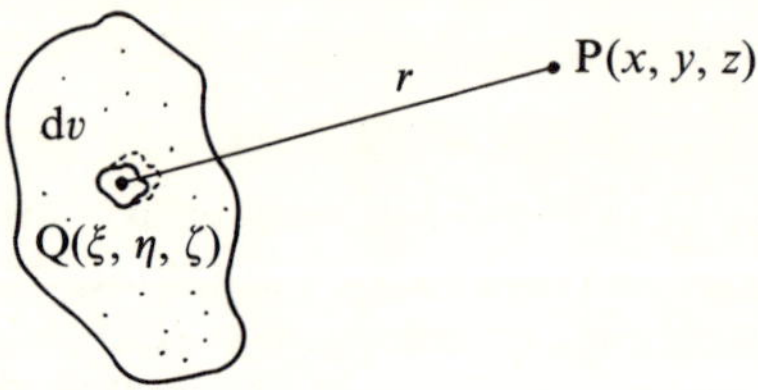

Fig. 2.7 Volume distribution of charge

distribution of charge for example, (Fig. 2.7), we have, using cartesian coordinates,

$$V_P(x, y, z) = \frac{1}{4\pi\epsilon_0} \iiint \frac{\rho(\xi, \eta, \zeta)\, d\xi\, d\eta\, d\zeta}{[(x-\xi)^2 + (y-\eta)^2 + (z-\zeta)^2]^{1/2}}$$

2.7 POTENTIAL GRADIENT

It has been shown in the foregoing section that for a given charge distribution the potential at any point in space will be a function of the coordinates of that point; in rectangular coordinates, for instance, the potential function can be written as $V(x, y, z)$. The equation $V(x, y, z) =$ constant then defines an 'equipotential' surface in space; the potential at all points which lie on an equipotential surface has the same constant value. In the case of a single point charge the equipotentials are, from equation 2–25, concentric spheres $r =$ constant having the charge as centre.

Consider two equipotential surfaces the potentials of which are V and $V+dV$ (Fig. 2.8). Let A be any point on the surface of potential V; if we move from A to any point on the surface of potential $V+dV$, the potential increases by dV. Now the shortest distance between the two surfaces is along the vector $\mathbf{AC} = dn\, \mathbf{a}_n$, which is normal to the surface V at the point A; in moving from A to C therefore, the potential charges at the maximum rate of dV/dn. If B is any point in the vicinity of C such that $\mathbf{AB} = dL$ and $dL \cos\theta = dn$, then

$$dV/dL = (dV/dn)\cos\theta$$

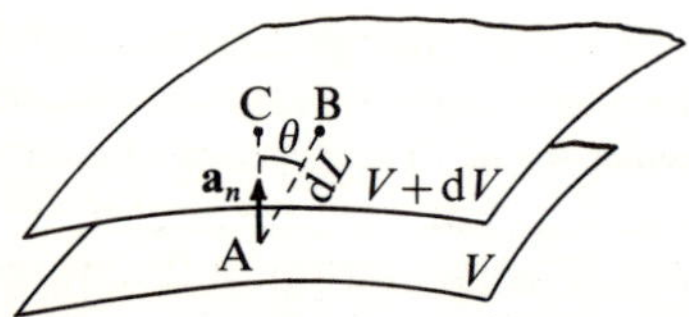

Fig. 2.8 Two equipotential surfaces of potentials V and $V+dV$

or

$$dV = \frac{dV}{dn}\,dL\cos\theta = \frac{dV}{dn}\,\mathbf{a}_n\cdot\mathbf{dL} \qquad (2\text{–}30)$$

In cartesian coordinates V is a function of x, y and z and its differential is

$$dV = \frac{\partial V}{\partial x}\,dx + \frac{\partial V}{\partial y}\,dy + \frac{\partial V}{\partial z}\,dz$$

$$= \left(\frac{\partial V}{\partial x}\,\mathbf{a}_x + \frac{\partial V}{\partial y}\,\mathbf{a}_y + \frac{\partial V}{\partial z}\,\mathbf{a}_z\right)\cdot(dx\,\mathbf{a}_x + dy\,\mathbf{a}_y + dz\,\mathbf{a}_z)$$

or

$$dV = \nabla V\cdot\mathbf{dL} \qquad (2\text{–}31)$$

Where the symbol ∇ (del) represents a vector differential operator which in cartesian coordinates is given by

$$\nabla = \frac{\partial}{\partial x}\,\mathbf{a}_x + \frac{\partial}{\partial y}\,\mathbf{a}_y + \frac{\partial}{\partial z}\,\mathbf{a}_z \qquad (2\text{–}32)$$

In cylindrical and spherical coordinates the corresponding expressions are,

$$\nabla = \frac{\partial}{\partial r}\,\mathbf{a}_r + \frac{1}{r}\frac{\partial}{\partial \phi}\,\mathbf{a}_\phi + \frac{\partial}{\partial z}\,\mathbf{a}_z \qquad \text{(cylindrical)} \qquad (2\text{–}33)$$

$$\nabla = \frac{\partial}{\partial r}\,\mathbf{a}_r + \frac{1}{r}\frac{\partial}{\partial \theta}\,\mathbf{a}_\theta + \frac{1}{r\sin\theta}\frac{\partial}{\partial \phi}\,\mathbf{a}_\phi \qquad \text{(spherical)} \qquad (2\text{–}34)$$

From equations 2–30 and 2–31 we see that

$$\frac{dV}{dn}\,\mathbf{a}_n = \nabla V = \operatorname{grad} V \qquad (2\text{–}35)$$

The vector quantity ∇V is known as the gradient of the potential, usually abbreviated grad V. Thus the gradient of a scalar field V is a vector field; the vector at any point in this field has a magnitude equal to the maximum rate of increase (dV/dn) of V at that point and a direction perpendicular to the surface $V = $ constant through that point; this direction is that of the maximum rate of increase of V.

From the definition of potential given in Section 2.5, the potential difference between two points separated by an infinitesimal distance dL is

$$dV = -\mathbf{E}\cdot\mathbf{dL}$$

Comparing this expression with equation 2–31 it follows that

$$\mathbf{E} = -\nabla V \qquad (2\text{--}36)$$

In cartesian coordinates the components of $\mathbf{E}$ are

$$E_x = -\frac{\partial V}{\partial x}, \quad E_y = -\frac{\partial V}{\partial y}, \quad E_z = -\frac{\partial V}{\partial z}$$

In other coordinate systems the components of $\mathbf{E}$ may be obtained from the del operator as given by equations 2–33 and 2–34.

Equation 2–36 embodies two important results: (i) the magnitude of the electric field intensity at any point is equal to the maximum rate of change of the potential at that point and (ii) the direction of the field intensity is everywhere normal to the equipotential surfaces. The negative sign appears because in moving in the direction of the gradient from one equipotential surface to another which is at a higher potential we need to do external work; this means that we must move against the electric field. Accordingly, $\mathbf{E}$ must point in the direction of maximum potential fall. The following example is a simple illustration of the use of the gradient.

It is required to find the electric intensity at some point P$(x, 0, 0)$ due to a point charge Q placed at the point Q$(x', 0, 0)$; $x > x'$. We have

$$V_P = \frac{Q}{4\pi\epsilon_0 r} = \frac{Q}{4\pi\epsilon_0(x - x')}$$

and

$$E_P = -\nabla_P V_P = \frac{Q}{4\pi\epsilon_0(x - x')^2}$$

Now if the charge Q is placed at the point P$(x, 0, 0)$ we have

$$V_Q = \frac{Q}{4\pi\epsilon_0(x - x')}$$

and

$$E_Q = -\nabla_Q V_Q = -\frac{Q}{4\pi\epsilon_0(x - x')^2}$$

It follows that

$$\frac{\partial V_P}{\partial x} = -\frac{\partial V_Q}{\partial x'}$$

The reader may readily show that in general, for any function $f(r)$ of r where $r = [(x - x')^2 + (y - y')^2 + (z - z')^2]^{1/2}$,

$$\nabla_P f(r) = -\nabla_Q f(r) \qquad (2\text{--}37)$$

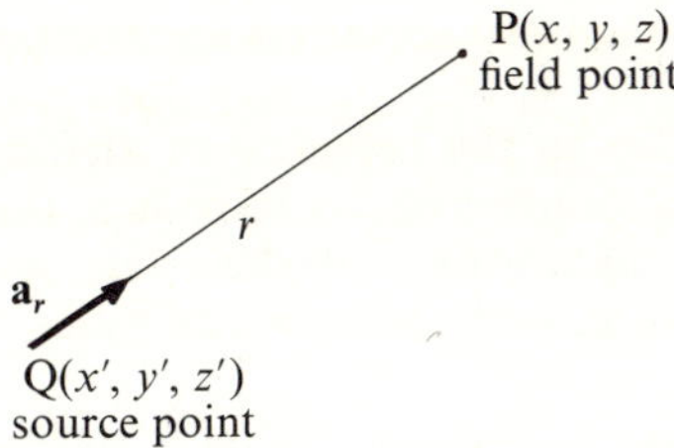

Fig. 2.9 Unit vector $\mathbf{a}_r$ pointing from source point to field point

The subscripts Q and P indicate that the differentiations are to be carried out with respect to the coordinates of the source and of the field point respectively. In particular if $f(r) = 1/r$, then (Fig. 2.9),

$$\nabla_Q\left(\frac{1}{r}\right) = \frac{\mathbf{a}_r}{r^2}, \quad \nabla_P\left(\frac{1}{r}\right) = -\frac{\mathbf{a}_r}{r^2} \tag{2-38}$$

Consider the case in which we want to find the three rectangular components of the electric field intensity at the origin due to a point charge Q at any point (x', y', z'). We have

$$V_0 = \frac{Q}{4\pi\epsilon_0(x'^2 + y'^2 + z'^2)^{1/2}}$$

It is obvious that we cannot obtain $\mathbf{E}_0$ directly from V_0 since the coordinates of the field point do not appear in the potential expression. However, we can obtain $\mathbf{E}_0$ with the aid of equation 2–37,

$$\mathbf{E}_0 = -\nabla_P V_0 = \nabla_Q V_0 = -\frac{Q(x'\,\mathbf{a}_x + y'\,\mathbf{a}_y + z'\,\mathbf{a}_z)}{4\pi\epsilon_0(x'^2 + y'^2 + z'^2)^{3/2}}$$

Points or lines in the electric field where the intensity is zero are called *singular* or *equilibrium* points or lines. At such points or lines the electric equipotential surface which passes through the point or line crosses itself at least twice, depending upon the character of the charge distribution.

2.8 ELECTRIC FLUX DENSITY AND GAUSS'S LAW

So far we have defined two fundamental field quantities: the vector electric field intensity $\mathbf{E}$ and the scalar electric potential V. A third fundamental quantity of importance is the electric flux density or electric displacement vector $\mathbf{D}$. For free space it is defined as

$$\mathbf{D} = \mathbf{D}_0 = \epsilon_0 \mathbf{E} \tag{2-39}$$

$\mathbf{D}$ has the dimensions of charge per unit area—i.e., coulombs per metre square in SI units.

As far as free space is concerned the introduction of the new vector

D is of little significance and serves no apparent purpose; however, as we shall see in Chapter 4, this quantity does play a useful role when studying electric fields in the presence of dielectric bodies. We have found it particularly convenient to introduce the reader to it at this stage, because the equations and theorems given below are valid whether we are dealing with charges and fields in free space or in a dielectric.

The flux of any vector **A** through a surface S is defined as the integral of the normal component of the vector over that surface, that is

$$\text{flux of } \mathbf{A} \text{ through } S = \int_S \mathbf{A} \cdot \mathbf{dS}$$

The flux of the vector **D** through any surface is called the electric flux and is given by

$$\Psi = \int_S \mathbf{D} \cdot \mathbf{dS} \tag{2-40}$$

The SI unit of flux is the coulomb. The word flux signifies flow; in fact the whole concept of electric flux is based on an analogy with fluid flow in hydraulics. This is because the early investigations of static electrification associated this phenomenon with the presence (in those bodies capable of being electrified) of some imponderable 'electric fluid'. Today electric flux is regarded solely as a convenient mathematical quantity; unlike the hydraulic flux, electric flux has no basic physical significance.

Gauss's law states that the total flux of the vector $\mathbf{D}_0$ through any closed surface is equal to the total charge enclosed by that surface. Thus for a closed surface S enclosing N point charges

$$\Psi = \oint_S \mathbf{D}_0 \cdot \mathbf{dS} = \oint_S \epsilon_0 \mathbf{E} \cdot \mathbf{dS} = \sum_1^N Q_i \tag{2-41}$$

To prove Gauss's law consider first a single point charge Q_1 placed at some given point P inside any closed surface S (Fig. 2.10). The electric flux density at any point on the surface is, by definition,

$$\mathbf{D}_0 = \epsilon_0 \mathbf{E} = \frac{Q_1}{4\pi r^2} \mathbf{a}_r$$

and the flux through an element of surface of area dS is

$$d\Psi_1 = \mathbf{D}_0 \cdot \mathbf{dS} = \frac{Q_1 \, dS}{4\pi r^2} \mathbf{a}_r \cdot \mathbf{a}_n$$

$$= \frac{Q_1 \, dS \cos\theta}{4\pi r^2} = \frac{Q_1}{4\pi} d\Omega$$

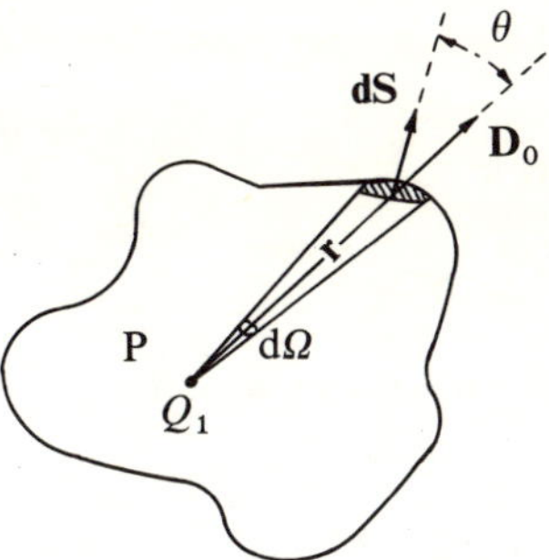

Fig. 2.10 To calculate the flux through a closed surface enclosing a point charge (Gauss's law)

where

$$d\Omega = \frac{dS \cos \theta}{r^2}$$

is the solid angle subtended by the element of area dS at the point P (see Appendix II). The total flux through the closed surface is

$$\Psi_1 = \frac{Q_1}{4\pi} \oint_S d\Omega = \frac{Q_1}{4\pi} \Omega = Q_1 \qquad (2\text{--}42)$$

where Ω is the solid angle subtended by the whole surface S at the point P. Since S is closed $\Omega = 4\pi$ and $\Psi_1 = Q_1$. We have thus shown that for the single point charge the result is independent of the position of that charge inside the surface and also independent of the size and shape of the closed surface S. It follows that the total flux passing through the surface S due to any number of enclosed charges is simply the algebraic sum of the flux produced by each individual charge. For the most general case in which the closed surface encloses a continuous charge distribution, equation 2–41 becomes

$$\Psi = \int_v \rho \, dv \qquad (2\text{--}43)$$

where v is the volume bounded by the surface S.

If through every peripheral point of an element of area dS of an equipotential surface we draw the lines of force, we obtain a tubular region in space which is known as a 'tube of force' or 'tube of electric flux'. Consider now a charge-free volume of space consisting of an elementary tube of force bounded by two equipotential surface elements dS_1 and dS_2 (Fig. 2.10). Applying Gauss's law to this volume we have

$$\oint_S \epsilon_0 \mathbf{E} \cdot \mathbf{dS} = \epsilon_0 \mathbf{E}_1 \cdot \mathbf{dS}_1 + \epsilon_0 \mathbf{E}_2 \cdot \mathbf{dS}_2 = 0$$

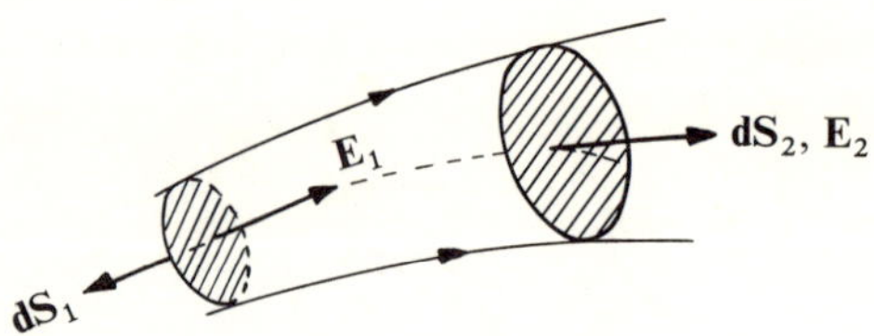

Fig. 2.11 Elementary tube of force

There is no contribution to the integral from the cylindrical surface of the tube since lines of force (and hence of flux density) cannot cross. It is clear from Fig. 2.11 that $d\mathbf{S}_1$ and $\mathbf{E}_1$ point in opposite directions whilst $d\mathbf{S}_2$ and $\mathbf{E}_2$ point in the same direction; hence

$$\epsilon_0 E_1\,dS_1 = \epsilon_0 E_2\,dS_2 = \text{constant} \tag{2-44}$$

Equation 2–44 states the fact that the flux of the vector $\mathbf{D}_0$ is conserved in charge-free space. Since the product $D_0\,dS$ is a constant, it follows that the flux density and hence also the electric field intensity increases as dS decreases and vice versa. Thus the greater the number of lines crossing a unit area—i.e., the closer the lines—the stronger the field. We make use of this result in the pictorial representation or mapping of fields by means of lines of force. By drawing lines of force such that their spacing is inversely proportional to the field strength, a field plot can provide at a glance valuable information on the topography of the field.

2.9 FIELD AND POTENTIAL OF SOME SIMPLE CHARGE DISTRIBUTIONS

We may summarize the foregoing sections by stating that, given a certain distribution of charge, we have the choice of three methods for obtaining the electric field and the electric potential functions:

(a) Use Coulomb's law to find the electric field intensity $\mathbf{E}$ and then the integral equation 2–21 to obtain the potential V.

(b) Use equations 2–26 to 2–29 to find the potential V and then the differential equation 2–36 to obtain the electric field $\mathbf{E}$.

(c) Use Gauss's law to find the electric flux density $\mathbf{D}_0$ (and hence $\mathbf{E}$) and then the integral equation 2–21 to obtain the potential V.

Method (c) is the simplest of the three but its application is limited to those cases in which the charge distribution shows sufficient symmetry to enable the direction of $\mathbf{D}_0$ (or $\mathbf{E}$) and its functional dependence on the position coordinates to be intuitively known. In general it will be found that method (b) is much more convenient to use than method (a). This is because the potential V, being a scalar, is completely defined by a single function whereas three functions are necessary to define the vector field $\mathbf{E}$; moreover, it is mathematically simpler to obtain $\mathbf{E}$ from V by differentiation than to obtain V from $\mathbf{E}$ by integration.

In this section we shall illustrate the use of the above three methods in the derivation of the potential and field functions of a number of simple but important charge distributions.

2.9.1 COLLINEAR POINT CHARGES

Consider a set of collinear point charges $Q_1, Q_2, \ldots Q_N$ placed at points $z_1, z_2, \ldots z_N$ on the Oz-axis. To find the equation of the lines of force we first note the cylindrical symmetry of the charge configuration

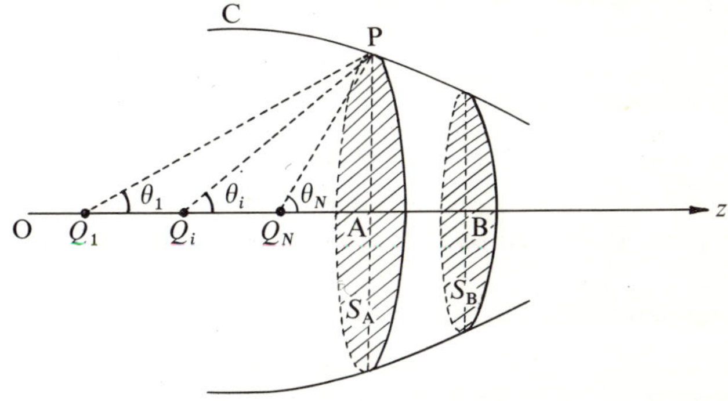

Fig. 2.12 To find the lines of force of a set of collinear point charges

about the Oz-axis; if a curve such as C (Fig. 2.12) represents the line of force passing through an arbitrary point P, then no line of force can cross through the surface of revolution formed by rotating the line C about Oz. If we apply Gauss's law to a charge-free volume bounded by the planes $z = z_A$, $z = z_B$ and the surface of revolution, then the total flux entering surface S_A must be equal to the total flux leaving surface S_B. From equation 2–42 we have

$$\Psi_i = \frac{Q_i}{4\pi}\,\Omega_i = \tfrac{1}{2}Q_i(1-\cos\theta)$$

and
$$\sum^N \Psi_i = \sum^N \tfrac{1}{2}Q_i(1-\cos\theta_i) = \text{constant}$$

The equation of the lines of force is therefore

$$\sum^N Q_i \cos\theta_i = \text{constant} \tag{2–45}$$

The reader should verify that the same equation could be arrived at by using the fact that the resultant field intensity perpendicular to the line of force must be zero.

For the simple case of two point charges, equation 2–45 becomes

$$Q_1 \cos \theta_1 + Q_2 \cos \theta_2 = K \qquad (2\text{–}46)$$

The constant K can be determined for any arbitrary point P by

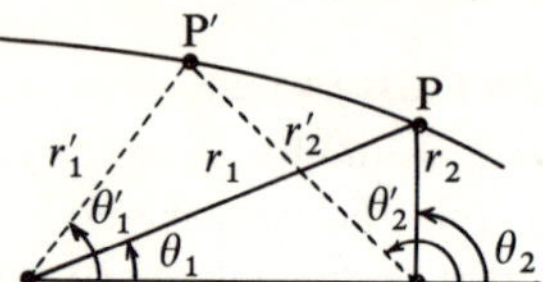

Fig. 2.13 To find the lines of force of two point charges

measuring the angles θ_1 and θ_2 (Fig. 2.13). If P′ is to be a point on the field line through P then

$$Q_1 \cos \theta_1' + Q_2 \cos \theta_2' = K$$

and for any value of θ_1' the corresponding value of θ_2' can be calculated and the point P′ located by the intersection of the two radius vectors r_1' and r_2'.

The equipotential lines are given by

$$V_P = \frac{1}{4\pi\epsilon_0} \left(\frac{Q_1}{r_1} + \frac{Q_2}{r_2} \right) \qquad (2\text{–}47)$$

By assigning a specific value to V_P we obtain a relationship between r_1 and r_2 by means of which the corresponding equipotential line can be drawn.

At large distances from either charge, $r_1 \simeq r_2 \simeq r$, $\theta_1 \simeq \theta_2 \simeq \theta$; the lines of force are the straight lines

$$(Q_1 + Q_2) \cos \theta = \text{constant}$$

and the equipotentials are the spheres

$$V = \frac{1}{4\pi\epsilon_0} \frac{Q_1 + Q_2}{r}$$

The field and equipotential lines for $Q_1 = Q_2$ and $Q_1 = -Q_2$ are shown in Fig. 2.3.

2.9.2 SPHERICAL CHARGED SHELL

Suppose that a spherical shell of radius a carries a uniform surface charge distribution of density σ C m^{-2}. We shall use the three methods referred to above to find the electric field intensity and potential at any point P at a distance z from the centre of the sphere (Fig. 2.14).

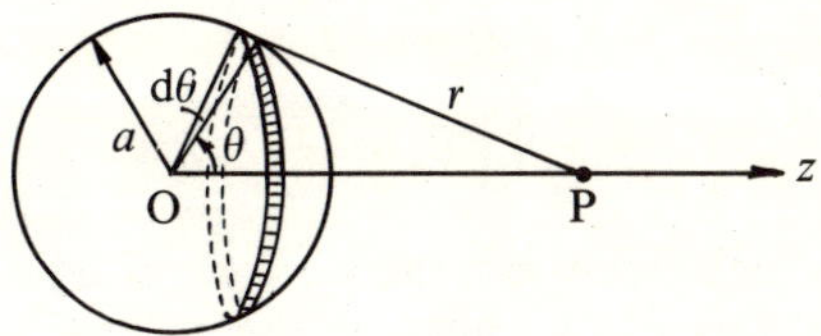

Fig. 2.14 Uniformly charged spherical shell

Method (a). Let P lie outside the sphere so that $z > a$. Consider the field produced at P by an elementary ring of radius $a \sin \theta$ and width $a\,d\theta$; it is clear that because of symmetry there can only be a field component in the direction of the axis Oz. The field due to the charge on this elementary ring of area dS is

$$dE_2 = \frac{\sigma\,dS\cos\alpha}{4\pi\epsilon_0 r^2}$$

where

$$dS = 2\pi a^2 \sin\theta\,d\theta$$

$$\cos\alpha = (z - a\cos\theta)/r$$

$$r = (z^2 + a^2 - 2az\cos\theta)^{1/2}$$

Introducing the total charge $Q = 4\pi a^2\sigma$ we have

$$dE_z = -\frac{Q}{4\pi\epsilon_0}\frac{1}{2}\frac{(z - a\cos\theta)\,d\cos\theta}{(z^2 + a^2 - 2az\cos\theta)^{3/2}}$$

and the total field at P is

$$E_z = -\int_0^\pi dE_z$$

The result of this integration gives

$$E_z = \frac{Q}{4\pi\epsilon_0 z^2} \qquad (z > a) \tag{2-48}$$

The reader should now verify that if the point P lies inside the sphere, the field intensity at P is zero*; that is

$$E_z = 0 \qquad (z < a) \tag{2-49}$$

For $z > a$ the potential at the point P is given by

$$V_P = -\int_\infty^z E_z\,dz = \frac{Q}{4\pi\epsilon_0 z} \qquad (z > a) \tag{2-50}$$

* *Hint:* the direction of dE_z due to an elementary charged ring is either along the positive or the negative direction of Oz depending upon which side of the ring the point P lies.

For $z < a$ the potential at P is given by

$$V_P = -\int_\infty^a E_z\,dz - \int_a^z E_z\,dz$$

the second term, which represents the work required to move a unit test charge from the surface of the sphere to the internal point P, is zero by virtue of equation 2–49; hence

$$V_P = \frac{Q}{4\pi\epsilon_0 a} \qquad (z < a) \tag{2–51}$$

The above results show that the field and the potential at any point outside a uniformly charged spherical shell are identical to the field and potential that would be produced by the same total charge concentrated as a point charge at the centre.

Method (b). The potential at point P is given by equation 2–28

$$V_P = \int_S \frac{\sigma\,dS}{4\pi\epsilon_0 r}$$

where

$$dS = 2\pi a^2 \sin\theta\,d\theta$$

$$r^2 = z^2 + a^2 - 2az\cos\theta$$

and

$$2r\,dr = 2az\sin\theta\,d\theta$$

Letting $Q = 4\pi a^2\sigma$ we have

$$V_P = \frac{Q}{4\pi\epsilon_0}\frac{1}{2az}\int dr$$

If P lies outside the sphere the limits of integration are from $(z-a)$ to $(z+a)$; if P lies inside the sphere the limits are from $(a-z)$ to $(a+z)$; hence

$$V_P = \frac{Q}{4\pi\epsilon_0 z} \qquad (z > a)$$

and

$$V_P = \frac{Q}{4\pi\epsilon_0 a} \qquad (z < a)$$

The corresponding values of the field intensity are given by

$\mathbf{E} = -\operatorname{grad} V$ so that

$$E_z = \frac{Q}{4\pi\epsilon_0 z^2} \qquad (z > a)$$

$$E = 0 \qquad (z < a)$$

Method (c). Because of the spherical symmetry of the charge distribution we expect the electric field to be everywhere radial. We therefore choose as our Gaussian surfaces spheres concentric with the charged shell. Over the Gaussian sphere of radius $z > a$ the field has the same strength and the sphere encloses a total charge $Q = 4\pi a^2 \sigma$; hence

$$\oint \epsilon_0 \mathbf{E} \cdot \mathbf{dS} = \sum Q_i = Q$$

$$\epsilon_0 E 4\pi z^2 = Q$$

or

$$E = \frac{Q}{4\pi\epsilon_0 z^2} \qquad (z > a)$$

The Gaussian sphere of radius $z < a$ encloses zero charge and hence

$$E = 0 \qquad (z < a)$$

The corresponding values of the potential are then obtained as described in method (a). It is evident from the above that the simplest method is (c) and the most laborious one is (a). However, application of method (c) requires a foreknowledge of the field characteristics and since such knowledge is possible only for a very limited number of charge distribution geometries, method (b) will in general be the most convenient one to use.

2.9.3 UNIFORMLY CHARGED DISC

Consider a disc of radius a and of negligible thickness uniformly charged with a surface density σ C m^{-2}. It is required to find the potential and electric field intensity at any point P on the axis of the disc (Fig. 2.15). The potential is given by equation 2–28

$$V_P = \frac{1}{4\pi\epsilon_0} \int \frac{\sigma\, dS}{d}$$

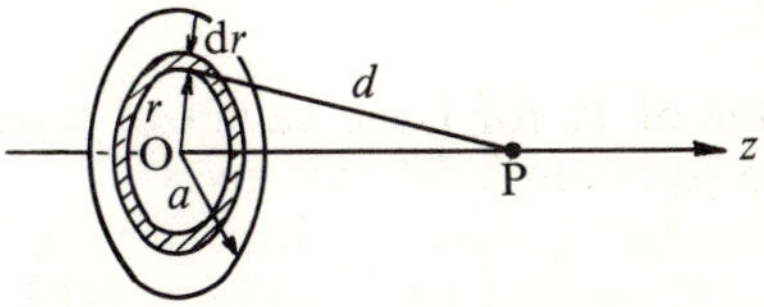

Fig. 2.15 Uniformly charged disc

with

$$dS = 2\pi r \, dr$$

and

$$d = (z^2 + r^2)^{1/2}$$

we have

$$V_P = \frac{\sigma}{2\epsilon_0} \int_0^a r(z^2 + r^2)^{-1/2} \, dr$$

$$= \frac{\sigma}{2\epsilon_0} \left[(z^2 + a^2)^{1/2} + (z^2)^{1/2} \right]$$

$$= \frac{\sigma}{2\epsilon_0} \left[(z^2 + a^2)^{1/2} - |z| \right] \tag{2-52}$$

where $|z| = \pm z$ according to whether $z > 0$ or $z < 0$. The potential at the centre of the disc, whether we approach it from the positive ($z = 0^+$) or the negative ($z = 0^-$) side, is

$$V_0 = \frac{\sigma a}{2\epsilon_0} \tag{2-53}$$

and is therefore finite and continuous as we cross the disc from one side to the other (Fig. 2.16).

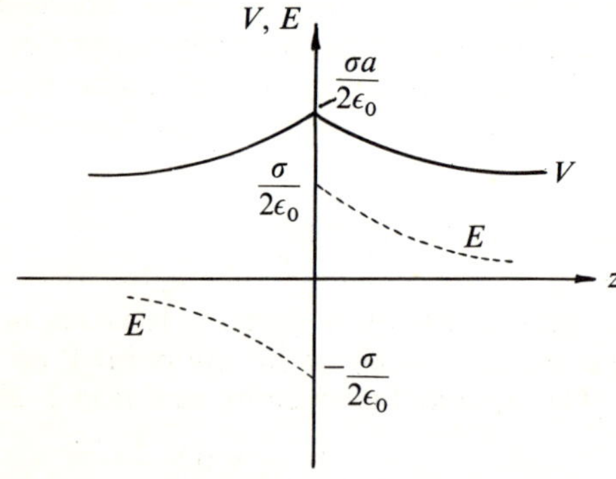

Fig. 2.16 Variation of potential and electric field intensity along the central axis of a uniformly charged disc

To obtain the value of V_P for large values of z we use the binomial theorem to expand equation 2-52, this gives

$$V_P = \frac{\sigma z}{2\epsilon_0} \left(1 + \frac{1}{2} \frac{a^2}{z^2} + \cdots - 1 \right)$$

and when $z \gg a$ we have

$$V_{\mathrm{P}} = \frac{\pi a^2 \sigma}{4\pi\epsilon_0 z} = \frac{Q}{4\pi\epsilon_0 z}$$

so that the charged disc acts as a point charge $Q = \pi a^2 \sigma$ placed at the disc centre.

The electric field intensity at P is given by

$$E_{\mathrm{P}} = -\nabla_{\mathrm{P}} V = -\frac{\partial V}{\partial z} \mathbf{a}_z$$

and

$$E_z = -\frac{\sigma}{2\epsilon_0}\left[\frac{z}{\sqrt{(z^2+a^2)}} - \frac{z}{\sqrt{(z^2)}}\right] \tag{2-54}$$

It is apparent from equation 2–54 that E_z is positive for $z > 0$ and negative for $z < 0$. When $z \to 0^+$, $z/\sqrt{(z^2)} \to +1$ and $E_z \to \sigma/2\epsilon_0$; when $z \to 0^-$, $z/\sqrt{(z^2)} \to -1$ and $E_z \to -\sigma/2\epsilon_0$. Thus in crossing the charged disc the electric field intensity suffers a discontinuity by changing abruptly by $[E_z(0^+) - E_z(0^-)] = \sigma/\epsilon_0$.

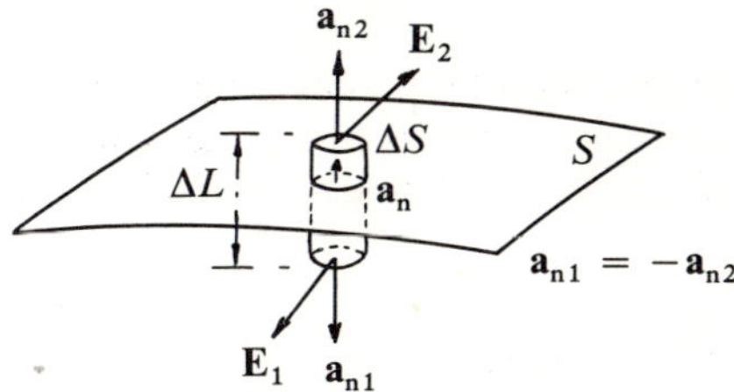

Fig. 2.17 Gaussian pillbox to demonstrate the discontinuity of the normal component of the field intensity across a charged surface

The above results may be extended to any surface S carrying a continuous surface charge density. On crossing such a surface the potential changes continuously but the *normal* component of the electric intensity changes abruptly by σ/ϵ_0. This can be readily shown by applying Gauss's law to the volume enclosed by the pillbox shown in Fig. 2.17.

$$\lim_{\Delta L \to 0}\left[\int_{\Delta S} \epsilon_0 \mathbf{E}_1 \cdot \mathbf{a}_{n1}\, \mathrm{d}S + \int_{\Delta S} \epsilon_0 \mathbf{E}_2 \cdot \mathbf{a}_{n2}\, \mathrm{d}S\right] = \int_{\Delta S} \sigma\, \mathrm{d}S$$

but since

$$\mathbf{a}_{n1} = -\mathbf{a}_{n2}; \; \mathbf{a}_{n2} = \mathbf{a}_n$$

then

$$\int_{\Delta S} (\mathbf{E}_2 - \mathbf{E}_1) \cdot \mathbf{a}_n \, \mathrm{d}S = \int_{\Delta S} (\sigma/\epsilon_0) \, \mathrm{d}S$$

and

$$(\mathbf{E}_2 - \mathbf{E}_1) \cdot \mathbf{a}_n = \sigma/\epsilon_0 \qquad (2\text{-}55)$$

2.9.4 INFINITE PLANE OF CHARGE

If the radius a of the uniformly charged disc is allowed to increase to infinity the absolute potential at any point would according to equation 2–52 become infinite. On the other hand the electric field intensity assumes the value

$$\mathbf{E} = \frac{\sigma}{2\epsilon_0} \mathbf{a}_n \qquad (2\text{-}56)$$

so that the field intensity is independent of the distance from the plane and is constant in magnitude and direction; the latter is always the direction of the outward normal $\mathbf{a}_n$ so that the field always points away from the plane. Starting from equation 2–56 we may express the potential at any point with respect to an arbitrary zero-potential reference by an equation analogous to equation 2–24—viz.,

$$V = -\frac{\sigma}{2\epsilon_0} |z| + C \qquad (2\text{-}57)$$

If we have two infinite parallel planes separated by a distance a and carrying equal but opposite uniform charge densities σ the total field is obtained by adding the contribution of each sheet (Fig. 2.18); thus

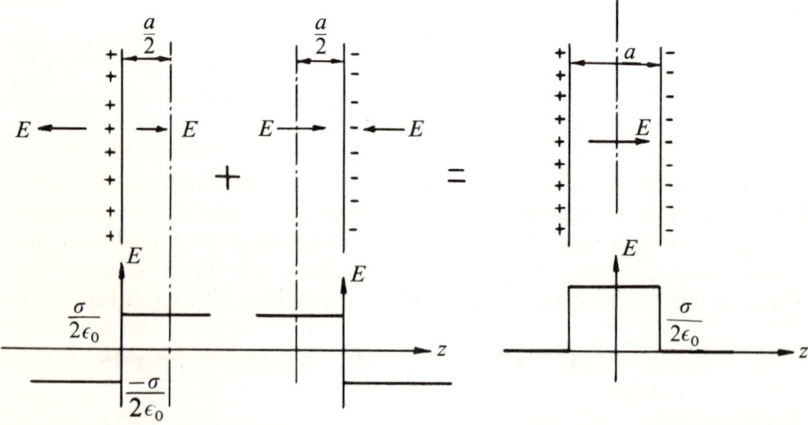

Fig. 2.18 Illustrating the resultant field intensity between two infinite parallel planes with charge densities $\pm\sigma$

although the assumption of infinite planes is hypothetical the results remain valid in practice for finite uniformly charged planes at points whose distance from the plane is small compared to their distance from the edges.

$$\mathbf{E} = \mathbf{E}_+ + \mathbf{E}_- = 0 \qquad (a/2 < z < -a/2)$$

$$= \frac{\sigma}{\epsilon_0}\,\mathbf{a}_z \qquad (-a/2 < z < a/2) \qquad (2\text{--}58)$$

The corresponding potentials may be written as

$$V_+ = -\frac{\sigma}{2\epsilon_0}(|z + a/2|) + C$$

$$V_- = \frac{\sigma}{2\epsilon_0}(|a/2 - z|) + C'$$

and

$$V = V_+ + V_- = -\sigma z/\epsilon_0 \qquad (-a/2 \le z \le a/2) \qquad (2\text{--}59)$$

The constants of integration being chosen such that $V_+ = 0$ at $z = -a/2$ and $V_- = 0$ at $z = a/2$ (see Fig. 2.19).

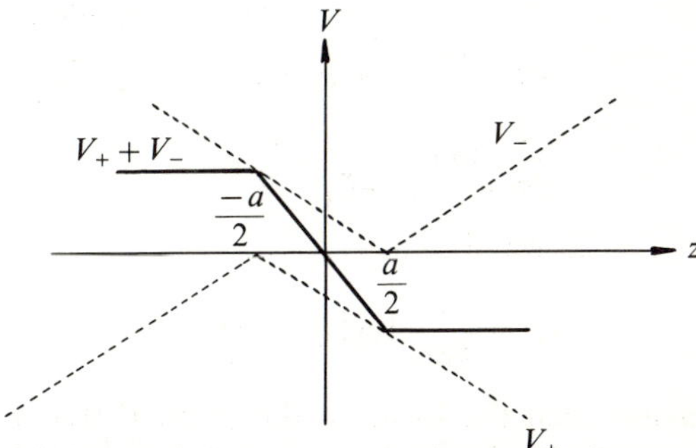

Fig. 2.19 Variation of potential for the case of two infinite parallel planes with charge densities $\pm\sigma$

2.9.5 UNIFORMLY CHARGED LINE SEGMENT

Suppose that a finite line of negligible diameter carries a uniform linear charge density $\lambda\,\mathrm{C\,m^{-1}}$. The segment has cylindrical symmetry around the z-axis (Fig. 2.20). The potential at any point P (z, r) is given by

$$V_\mathrm{P} = \frac{\lambda}{4\pi\epsilon_0}\int_{-l}^{l}\frac{\mathrm{d}z'}{\zeta} = \frac{\lambda}{4\pi\epsilon_0}\int_{-l}^{l}[r^2 + (z - z')^2]^{-1/2}\,\mathrm{d}z'$$

$$= \frac{\lambda}{4\pi\epsilon_0}\log\frac{(z+l) + \sqrt{[r^2 + (z+l)^2]}}{(z-l) + \sqrt{[r^2 + (z-l)^2]}} \qquad (2\text{--}60)$$

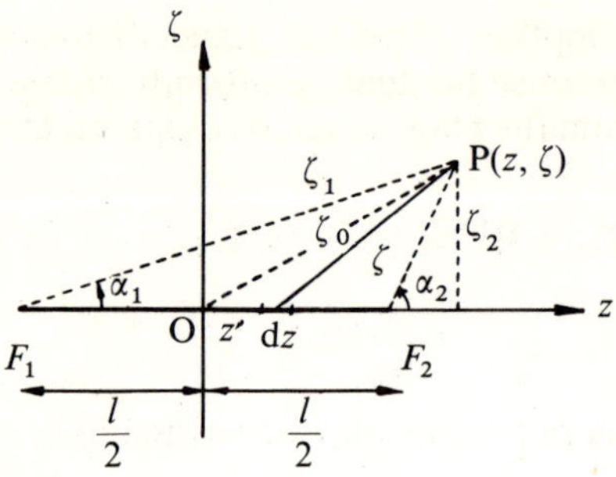

Fig. 2.20 Finite line segment of uniform linear charge density λ

Now from Fig. 2.20

$$\zeta_1 = \sqrt{[r^2 + (z+l)^2]}; \quad \zeta_2 = \sqrt{[r^2 + (z-l)^2]}$$

so that

$$\zeta_1^2 - \zeta_2^2 = 4zl$$

$$z = (\zeta_1^2 - \zeta_2^2)/4l$$

substituting the above values in equation 2–60 we obtain

$$V_P = \frac{\lambda}{4\pi\epsilon_0} \log\frac{(\zeta_1 + \zeta_2 + 2l)(\zeta_1 - \zeta_2 + 2l)}{(\zeta_1 + \zeta_2 - 2l)(\zeta_1 - \zeta_2 + 2l)}$$

$$= \frac{\lambda}{4\pi\epsilon_0} \log\frac{\zeta_1 + \zeta_2 + 2l}{\zeta_1 + \zeta_2 - 2l} \tag{2-61a}$$

If we let $\zeta_1 + \zeta_2 = 2a$ we obtain

$$V_P = \frac{\lambda}{4\pi\epsilon_0} \log[(a+l)/(a-l)] \tag{2-61b}$$

If a is a constant then the locus of the point P traces equipotential surfaces the potentials of which are determined by the values of V_P. It is evident that the equipotential surface will be a family of confocal prolate spheroids (ellipses rotated about their major axis) whose foci are the terminal points F_1 and F_2 of the line segment. At large distances from the segment $\zeta_1 + \zeta_2 \simeq 2\zeta_0 \gg 2l$ and*

$$\log[(a+l)/(a-l)] \simeq 2l/a = 2l/\zeta_0$$

so that

$$V_P = \frac{2\lambda l}{4\pi\epsilon_0\zeta_0} = \frac{Q}{4\pi\epsilon_0\zeta_0}$$

* $\log[1+x/1-x] = 2[x + x^3/3 + x^5/5 + \cdots], (x^2 < 1)$

$$\simeq 2x$$

which is the potential of an equivalent point charge $Q = 2\lambda l$ placed at the midpoint of the line segment.

The field lines are given by an orthogonal family of confocal hyperbolas. The two components E_z and E_r can be readily determined from equation 2–60,

$$E_z = -\frac{\partial V}{\partial z} = \frac{\lambda}{4\pi\epsilon_0}\frac{1}{r}(\sin\alpha_2 - \sin\alpha_1) \qquad (2\text{–}62)$$

$$E_r = -\frac{\partial V}{\partial r} = \frac{\lambda}{4\pi\epsilon_0}\frac{1}{r}(\cos\alpha_1 - \cos\alpha_2) \qquad (2\text{–}63)$$

2.9.6 INFINITE LINE CHARGE

In order to find the potential at any point P due to an infinitely long charged line we first examine equation 2–61a for the case in which $l \gg z$ and $l \gg r$. Bearing in mind that ζ_1 and ζ_2 are always positive the reader can readily verify that,

$$\zeta_1 + \zeta_2 + 2l \simeq 4l \quad \text{and} \quad \zeta_1 + \zeta_2 - 2l \simeq lr^2/l^2 - z^2$$

Equation 2–61 thus becomes

$$V_P \simeq \frac{\lambda}{4\pi\epsilon_0}\log\frac{4(l^2 - z^2)}{r^2}$$

$$\simeq \frac{\lambda}{4\pi\epsilon_0}\log\left[2\log\left(\frac{1}{r}\right) + \log 4(l^2 + z^2)\right]$$

It is clear that if $l \to \infty$ the potential also tends to infinity so that no significance can be attached to the absolute potential of any point in the field of an infinitely long line charge.

The electric field intensity is obtained by substituting $\alpha_1 = 0$, $\alpha_2 = \pi$ in equations 2–62 and 2–63 this gives

$$E = 0$$

$$E = \lambda/2\pi\epsilon_0 r \qquad (2\text{–}64)$$

so that the field lines are radial.

The potential difference between any two points at distances r_1 and r_2 from the line is given by

$$V_{21} = -\int_{r_1}^{r_2} \mathbf{E}\cdot\mathbf{dr} = -\frac{\lambda}{2\pi\epsilon_0}\log\left(\frac{r_2}{r_1}\right)$$

this may be rewritten as

$$V_2 - V_1 = -\frac{\lambda}{2\pi\epsilon_0}(\log r_2 - \log r_1)$$

The reader should now compare this equation with equation 2–22. If we choose $r_1 = 1$ we obtain

$$V_1 = \frac{\lambda}{2\pi\epsilon_0}\log\left(\frac{1}{r_2}\right) \tag{2-65}$$

This is known as the *logarithmic potential* of a line charge. The essential difference between this potential and the Coulomb potential of a point charge is that at infinity the latter becomes zero whilst the former becomes infinite.

The potential at any point at a distance r from an infinite line charge may be expressed as

$$V = \frac{\lambda}{2\pi\epsilon_0}\log\left(\frac{r_0}{r}\right) + V_0 \tag{2-66}$$

where V_0 is an arbitrary chosen reference potential at a distance r_0 from the line charge.

2.9.7 TWO PARALLEL AND INFINITELY LONG LINE CHARGES CARRYING OPPOSITE CHARGE

The two infinite lines are assumed to be parallel to the z-axis. Let $-\lambda\,\mathrm{C\,m^{-1}}$ be the uniform linear charge density of the line at the point $x = -a$, $y = 0$ and $\lambda\,\mathrm{C\,m^{-1}}$ be that of the line at the point $x = a$, $y = 0$ (Fig. 2.21). The potential at a point P due to the positive line charge is

$$V_+ = \frac{\lambda}{2\pi\epsilon_0}\log\left(\frac{r_{01}}{r_1}\right) + V_{01}$$

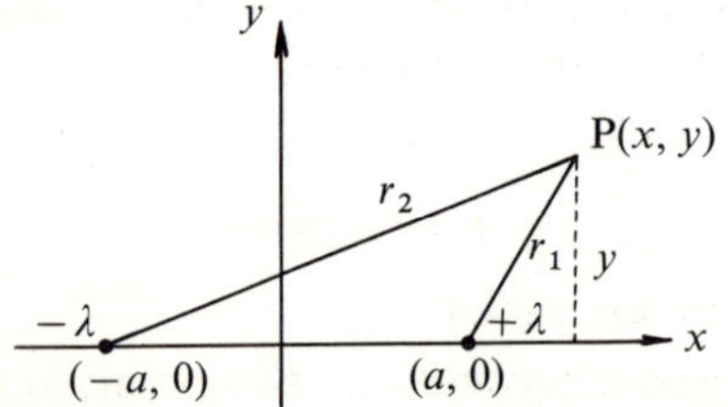

Fig. 2.21 Two parallel and infinitely long line charges of uniform density $\pm\lambda$

and that due to the negative line charge is

$$V_- = \frac{\lambda}{2\pi\epsilon_0}\log\left(\frac{r_2}{r_{02}}\right) + V_{02}$$

The total potential at the point P is therefore,

$$V_P = V_+ + V_- = \frac{\lambda}{2\pi\epsilon_0} \log\left(\frac{r_2}{r_1}\right) + C \tag{2-67}$$

where C is an arbitrary constant. It is evident that the plane midway ($r_1 = r_2$) between the two line charges is an equipotential surface to which we may assign any convenient potential C; here we choose $C = 0$. The equipotential surfaces are the loci of the points for which the ratio r_2/r_1 is a constant. For any fixed value of V_P

$$r_2/r_1 = \exp(2\pi\epsilon_0 V_P/\lambda) = K \tag{2-68}$$

with

$$K = K_+ \ (V_P > 0; K_+ > 1)$$
$$K = K_- \ (V_P < 0; K_- < 1)$$

and

$$K_- = 1/K_+$$

From Fig. 2.21 we have

$$r_1^2 = (a-x)^2 + y^2$$
$$r_2^2 = (a+x)^2 + y^2$$

Substituting in equation 2–68 and rearranging terms we obtain

$$[x - a(K^2+1)/(K^2-1)]^2 + y^2 = [a2K/(K^2-1)]^2$$

This is the equation of a circle of radius

$$R_K = 2aK/(K^2-1) \tag{2-69}$$

and centred at

$$x_K, y_K = a(K^2+1)/(K^2-1), 0 \tag{2-70}$$

The equipotential surfaces are therefore a set of cylinders having their axes parallel to and collinear with the line charges $\pm\lambda$. For each value of $K = K_+$ corresponding to a positive potential V_P there is a value $K = K_- = 1/K_+$ corresponding to a negative potential $-V_P$. The radii of the corresponding equipotential cylinders will be unchanged but the location of the centres will change from x_K for V_P to $-x_K$ for $-V_P$.

The field lines, which are orthogonal to the equipotential surfaces, are the arcs of the circles passing through the line charges and having their centres along the y-axis. Figure 2.22 shows the equipotentials (full circles) and the field lines (broken circles).

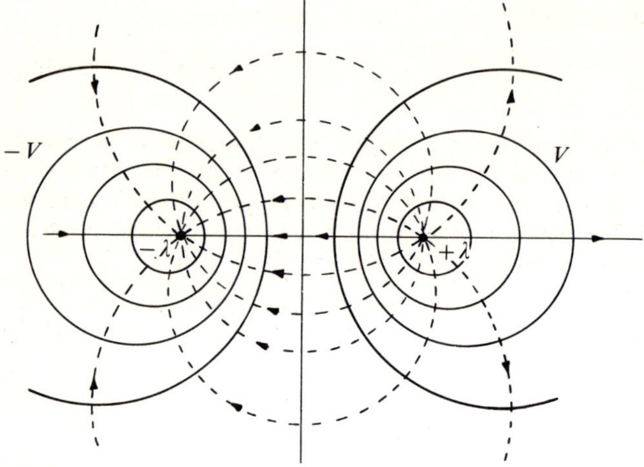

Fig. 2.22 Equipotentials (full circles) and field lines (broken circles) for line charges of Fig. 2.21

From equations 2–69 and 2–70 we have that

$$x_K^2 - R_K^2 = a^2$$

$$x_K^2 = R_K^2 + a^2$$

Thus if P is the point of intersection of an equipotential circle and the circle drawn with OO′ as diameter (Fig. 2.23), then the lines PE and PM are at right angles to one another. From the geometry of the figure it is readily seen that the triangles EPO′ and EOP are similar; thus

$$\frac{\text{EO}'}{\text{EP}} = \frac{\text{EP}}{\text{EO}}$$

$$\frac{\delta}{R_K} = \frac{R_K}{d}$$

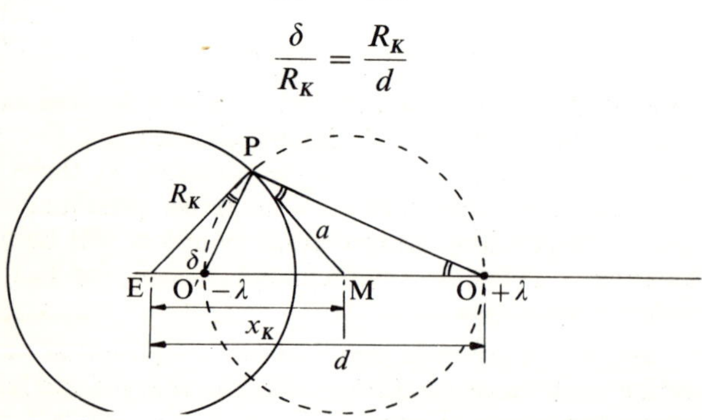

Fig. 2.23 Location of inverse points with respect to a circle

so that

$$\delta = R_K^2/d \qquad (2\text{-}71)$$

The points O and O′ are called inverse points with respect to the circle. We thus see that the positions of the two line charges are inverse points with respect to the family of equipotential circles.

2.9.8 UNIFORMLY CHARGED SPHERE

Consider a sphere of radius R throughout which charge is distributed with a uniform volume density ρ C m^{-3}. It is required to find the potential and field intensity at points both inside and outside the sphere.

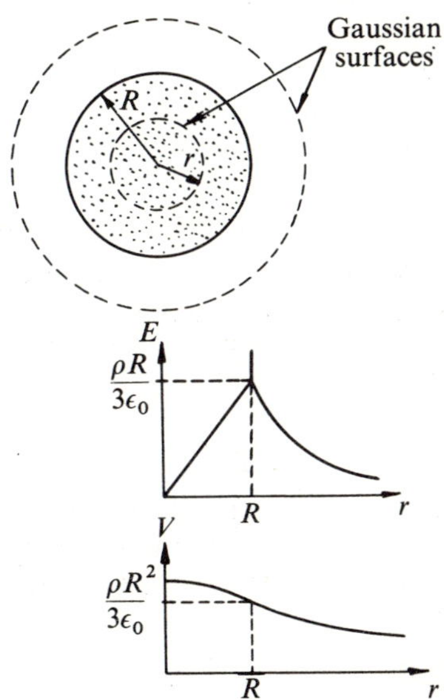

Fig. 2.24 Variations of the field and potential as a function of the radial distance from the centre of a uniformly charged spherical volume

Because of the spherical symmetry of the distribution it is apparent that the field everywhere will be radial. Applying Gauss's theorem over the surface of a concentric sphere of radius r (Fig. 2.24) we obtain for $r < R$,

$$\epsilon_0 E_{ri} 4\pi r^2 = \tfrac{4}{3}\pi r^3 \rho$$

so that

$$\mathbf{E}_{ri} = (\rho r/3\epsilon_0)\mathbf{a}_r \qquad (2\text{-}72)$$

For $r > R$ we obtain

$$\epsilon_0 E_{ro} 4\pi r^2 = \tfrac{4}{3}\pi R^3 \rho$$

$$E_{ro} = (\rho R^3/3\epsilon_0 r^2)\mathbf{a_r} \tag{2-73}$$

The corresponding potentials are given by

$$V_i = -(\rho/3\epsilon_0)\left(\int_\infty^R R^3 r^{-2}\,dr + \int_R^r r\,dr\right)$$

$$V_i = \rho(3R^2 - r^2)/6\epsilon_0 \tag{2-74}$$

and

$$V_o = -(\rho/3\epsilon_0)\int_\infty^r R^3 r^{-2}\,dr = \rho R^3/3\epsilon_0 r \tag{2-75}$$

Both the field and the potential at points outside the spherical charge distribution are the same as though the total charge $Q = \tfrac{4}{3}\pi R^3 \rho$ were concentrated at the centre of the sphere.

2.10 ELECTRIC MOMENT AND ELECTRIC DIPOLE MOMENT

The electric moment of a point charge Q about any fixed point O is defined as the product $Q\mathbf{r}$ where $\mathbf{r}$ is the radius vector from the point O to the charge Q. The electric moment of a system of N point charges relative to a fixed origin O is defined as

$$\mathbf{p} = \sum_{}^{N} Q_i \mathbf{r}_i \tag{2-76}$$

If the net charge of the system is zero, it can be readily shown that the electric moment of the system is independent of the choice of the origin O. Consider two arbitrary points O and O′ (Fig. 2.25); the electric moment about the point O′ is

$$\mathbf{p}' = \sum Q_i \mathbf{r}_i'$$

$$= \sum Q_i(\mathbf{r}_i + \mathbf{r}_0) = \sum Q_i \mathbf{r}_i + \sum Q_i \mathbf{r}_0$$

$$= \sum Q_i \mathbf{r}_i + \mathbf{r}_0 \sum Q_i$$

$$= \mathbf{p} + \Delta\mathbf{p}$$

if $\sum Q_i = 0$, then $\Delta\mathbf{p} = 0$ and $\mathbf{p}' = \mathbf{p}$. In this case the total positive charge of the system must be equal to the total negative charge. The centres of positive and negative charge of the system are by definition from equation 2–19

$$\mathbf{r}_p \sum Q_i^+ = \sum Q_i^+ \mathbf{r}_i^+ = \mathbf{r}_p Q$$

$$\mathbf{r}_n \sum Q_i^- = \sum Q_i^- \mathbf{r}_i^- = -\mathbf{r}_n Q$$

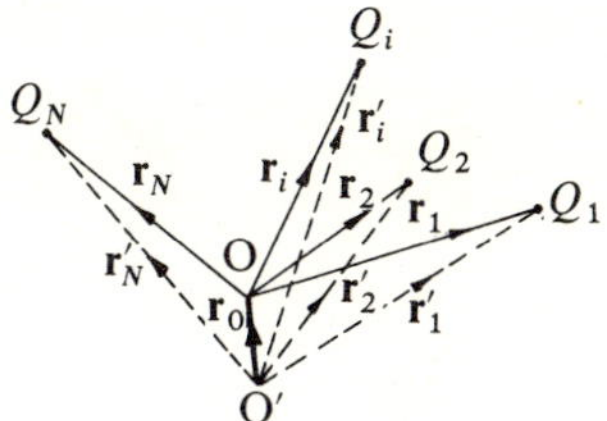

Fig. 2.25 Electric moment of a system of charges with respect to origin O or O′

so that

$$\mathbf{p} = Q(\mathbf{r_p} - \mathbf{r_n}) = Q\mathbf{l} \tag{2-77}$$

where $\mathbf{l}$ is the vectorial distance between the centres of positive and negative charge and points from the negative charge to the positive charge (Fig. 2.26). Equation 2–77 expresses the important result that the electric moment of a system of charges whose net charge is zero depends only on the distance between the negative and positive charge centres (or poles) of the system. The product $Q\mathbf{l}$ is known as the dipole moment of the system.

The above analysis can be readily generalized for the case of a charge distribution. Thus the electric moment of a static volume charge distribution of density $\rho(r)$, with respect to a point O is

$$\mathbf{p} = \int_v \mathbf{r}\rho(r)\,dv \tag{2-78}$$

If the total charge is zero—i.e.,

$$Q = \int_v \rho(r)\,dv = 0$$

then equation 2–78 defines the dipole moment of the charge distribution.

2.11 POTENTIAL AND FIELD OF A DIPOLE

(a) Approximate expressions. Consider a point charge $+Q$ at a point $z = l/2$ on the z-axis and an equal negative charge $-Q$ at the point

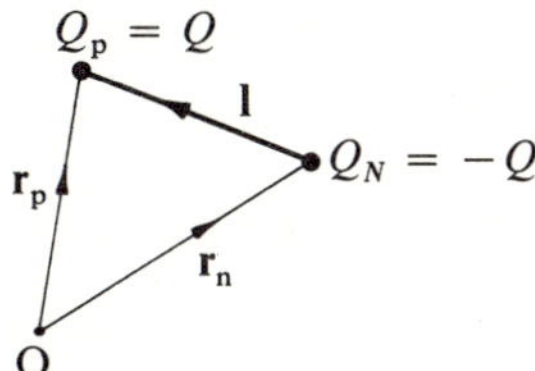

Fig. 2.26 Electric dipole

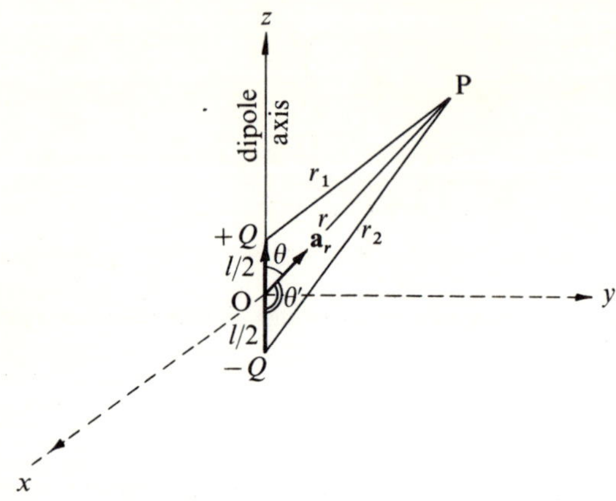

Fig. 2.27 To find potential and field of a dipole

$z = -l/2$. Referring to Fig. 2.27, the potential at a point P is

$$V_P = \frac{Q}{4\pi\epsilon_0}\left(\frac{1}{r_1} - \frac{1}{r_2}\right) \qquad (2\text{--}79)$$

If the distance r of the point P from the origin is large compared with the separation l of the charges we have

$$r_1 = r - \tfrac{1}{2}l\cos\theta;\ r_2 = r + \tfrac{1}{2}l\cos\theta$$

and

$$V_P = \frac{Q}{4\pi\epsilon_0}\left(\frac{l\cos\theta}{r^2 - [\tfrac{1}{2}l\cos\theta]^2}\right) = \frac{Ql\cos\theta}{4\pi\epsilon_0 r^2}\left(1 - \frac{l^2\cos^2\theta}{2r^2}\right)^{-1}$$

$$= \frac{Ql\cos\theta}{4\pi\epsilon_0 r^2} \qquad (r \gg l)$$

$$= \frac{p\cos\theta}{4\pi\epsilon_0 r^2} \qquad (2\text{--}80)$$

where p is the dipole moment of the charges. It is sometimes convenient to express the potential V_P as follows. If $\mathbf{a}_r$ is the unit vector from the origin to the point P we have

$$V_P = \frac{\mathbf{p}\cdot\mathbf{a}_r}{4\pi\epsilon_0 r^2} = -\frac{1}{4\pi\epsilon_0}\mathbf{p}\cdot\nabla\left(\frac{1}{r}\right) = -\nabla\cdot\left(\frac{\mathbf{p}}{4\pi\epsilon_0 r}\right) \qquad (2\text{--}81)$$

when $z > 0$ $(\tfrac{3}{2}\pi < \theta < \tfrac{1}{2}\pi)$, $V_P > 0$ and when $z < 0$ $(\tfrac{3}{2}\pi > \theta > \tfrac{1}{2}\pi)$, $V_P < 0$.

The electric field intensity at the point P is given by the negative gradient of the potential. Thus

$$\mathbf{E_P} = \frac{1}{4\pi\epsilon_0}\nabla\left[\mathbf{p}\cdot\nabla\left(\frac{1}{r}\right)\right]$$

(2–82)

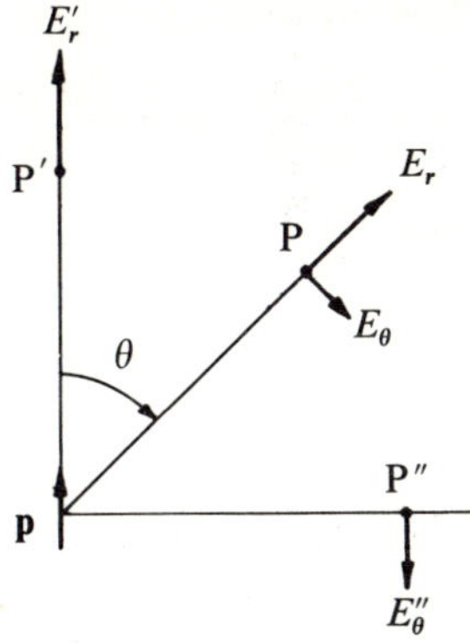

Fig. 2.28 Radial and angular component of electric field of a dipole

Because of the cylindrical symmetry of the dipole about the z-axis (dipole axis) the field has only a radial component and a component in the θ-direction (Fig. 2.28). Thus

$$E_r = -\frac{\partial V}{\partial r} = \frac{2p\cos\theta}{4\pi\epsilon_0 r^3}$$

$$E_\theta = -\frac{1}{r}\frac{\partial V}{\partial \theta} = \frac{p\sin\theta}{4\pi\epsilon_0 r^3}$$

and the resultant field is

$$\mathbf{E} = \frac{p}{4\pi\epsilon_0 r^3}\left[2\cos\theta\,\mathbf{a_r} + \sin\theta\,\mathbf{a_\theta}\right]$$

(2–83)

An alternate expression for the field is

$$\mathbf{E} = \frac{1}{4\pi\epsilon_0}\left[\frac{3(\mathbf{p}\cdot\mathbf{r})\mathbf{r}}{r^5} - \frac{\mathbf{p}}{r^3}\right]$$

(2–84)

When $\theta = 0$ we have

$$E_r' = p/2\pi\epsilon_0 r^3; \; E_\theta' = 0$$

and when $\theta = \pi/2$

$$E_r'' = 0; \; E_\theta'' = p/4\pi\epsilon_0 r^3$$

From equation 2–80 the equation of the equipotentials is given by

$$\frac{\cos\theta}{r^2} = \text{constant} \tag{2–85}$$

The equation of the lines of force are given by (see Section 2–3)

$$E_\theta/r \; d\theta = E_r/dr$$

or

$$\frac{dr}{r} = \frac{2\cos\theta}{\sin\theta}\, d\theta = \frac{2d(\sin\theta)}{\sin\theta}$$

and

$$\frac{\sin^2\theta}{r} = \text{constant} \tag{2–86}$$

The field lines and equipotentials for a point dipole are shown in Fig. 2.29.

(b) Exact expressions. We shall now reconsider equation 2–79 which is the exact expression for the potential of the two point charges constituting the dipole. From Fig. 2.27 we have

$$r_1^{-1} = \frac{1}{r}\left[1 + \left(\frac{l}{2r}\right)^2 - 2\left(\frac{l}{2r}\right)\cos\theta\right]^{-1/2} \tag{2–87}$$

since $r \gg l$

$$\left|(l/2r)^2 - 2(l/2r)\cos\theta\right| < 1$$

and we may use the binomial theorem to expand the expression within the square brackets in equation 2–87; after a rearrangement of terms

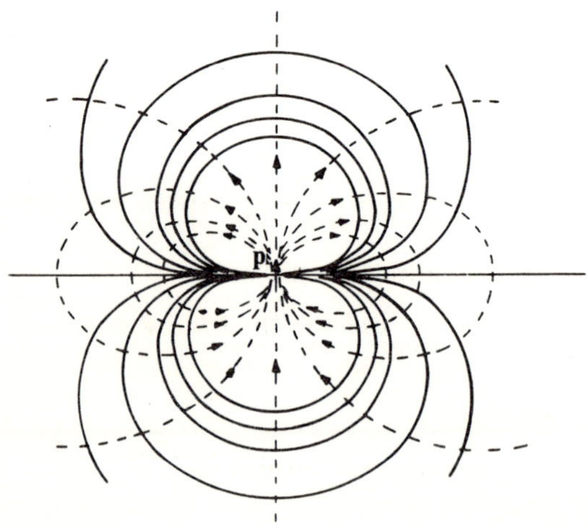

Fig. 2.29 Field lines and equipotentials of a point dipole

we obtain

$$\frac{1}{r_1} = \frac{1}{r}\left[1+\left(\frac{l}{2r}\right)\cos\theta + \left(\frac{l}{2r}\right)^2 \tfrac{1}{2}(3\cos^2\theta - 1) + \cdots\right]$$

$$= \frac{1}{r}\sum_{n=0}^{\infty} P_n(\cos\theta)\left(\frac{l}{2r}\right)^n \qquad (2\text{--}88)$$

where the coefficients $P_n(\cos\theta)$ of $(l/2r)$ are polynomials in $\cos\theta$ known as the Legendre polynomials. The first four are given by

$$P_0(\cos\theta) = 1$$

$$P_1(\cos\theta) = \cos\theta$$

$$P_2(\cos\theta) = \tfrac{1}{2}(3\cos^2\theta - 1)$$

$$P_3(\cos\theta) = \tfrac{1}{2}(5\cos^3\theta - 3\cos\theta)$$

In general the nth polynomial $P_n(\cos\theta)$ may be obtained from the Rodrigues formula

$$P_n(\cos\theta) = \frac{1}{2^n n!}\frac{d^n}{d(\cos\theta)^n}(\cos^2\theta - 1)^n \qquad (2\text{--}89)$$

Referring again to Fig. 2.25 we have,

$$r_2 = [r^2 + (l/2)^2 - 2r(l/2)\cos\theta']^{1/2}$$

Following the same steps as for r_1 the reciprocal distance $1/r_2$ may be expressed as

$$\frac{1}{r_2} = \frac{1}{r}\sum_{n=0}^{\infty} P_n(\cos\theta')\left(\frac{l}{2r}\right)^n \qquad (2\text{--}90)$$

where $\cos\theta' = \cos(\pi - \theta)$. Substituting equations 2–88 and 2–90 into equation 2–79 we obtain

$$V_P = \frac{Q}{4\pi\epsilon_0}\frac{1}{r}\left\{\sum_{n=0}^{\infty} P_n(\cos\theta)\left(\frac{l}{2r}\right)^n - \sum_{n=0}^{\infty} P_n(\cos\theta')\left(\frac{l}{2r}\right)^n\right\} \qquad (2\text{--}91)$$

Using equation 2–91 the potential may be expressed to any desired degree of accuracy by choosing the appropriate number of terms. If $r \gg l$ it is easily shown that equation 2–91 reduces, as a first approximation, to equation 2–79. Now if we let the magnitude $|Q|$ of the two charges increase as their distance apart decreases, in such a manner that their product $p = Ql$ remains constant, then, in the limit when $l \to 0$, we obtain a so-called *point* or *ideal* dipole of moment p whose potential and field intensity are now *exactly* given by equations 2–79 and 2–83. Ideal dipoles do not exist in nature. However, dipoles are often encountered in many neutral atoms and molecules whose electrons are not symmetrically distributed around the positively charged nuclei so that the centres of positive and negative charge do not coincide. Such asymmetry may be either an inherent property of molecules or it may be induced by an external field. The displacement of the charge centres is of the order of atomic dimensions and at distances which are large compared with this displacement the potential and field are accurately described by the formulae of the ideal dipole. As we shall see in later chapters, atomic and molecular dipoles play an important role in determining the behaviour of dielectric materials in static and alternating electric fields.

2.12 MULTIPOLE EXPANSION OF THE POTENTIAL OF AN ARBITRARY CHARGE DISTRIBUTION

Consider an arbitrary volume charge distribution $\rho(r_1)$ contained within a finite volume of space v. Suppose that it is required to find the potential at some point P outside this distribution such that the distance between P and any point within the volume v is large compared with the dimensions of v. In other words we may assume that the charge distribution lies entirely within a sphere of finite radius a such that $r \gg a$ (Fig. 2.30). The potential at P is given by

$$V_{\mathrm{P}} = \frac{1}{4\pi\epsilon_0} \int_v \frac{\rho(r_1)\,\mathrm{d}v}{r_2} = \frac{1}{4\pi\epsilon_0} \int \frac{\rho(r_1)\,\mathrm{d}v}{|\mathbf{r}-\mathbf{r}_1|} \qquad (2\text{–}92)$$

Now it is possible to write

$$\begin{aligned}
|\mathbf{r}-\mathbf{r}_1| &= \left[(\mathbf{r}-\mathbf{r}_1)\cdot(\mathbf{r}-\mathbf{r}_1)\right]^{1/2} \\
&= \left[r^2 - 2\mathbf{r}\cdot\mathbf{r}_1 + r_1^2\right]^{1/2} \\
&= r\left[1 - 2r^{-2}(\mathbf{r}\cdot\mathbf{r}_1) + r^{-2}r_1^2\right]
\end{aligned}$$

Using the binomial theorem to expand $|\mathbf{r}-\mathbf{r}_1|^{-1}$ and rearranging terms

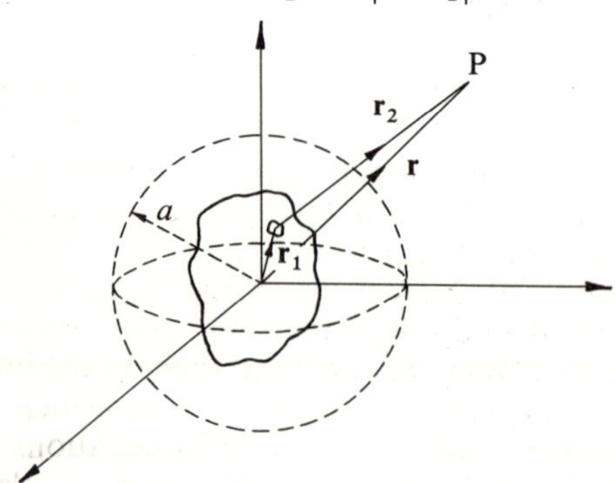

Fig. 2.30 Arbitrary volume charge distribution with a finite volume

we obtain

$$|\mathbf{r}-\mathbf{r}_1|^{-1} = \frac{1}{r}\left\{1+\frac{\mathbf{r}\cdot\mathbf{r}_1}{r^2}+\frac{1}{2r^4}[3(\mathbf{r}\cdot\mathbf{r}_1)^2-r^2r_1^2]+\cdots\right\}$$

Substituting this expression in equation 2–92 we have

$$V_P = \frac{1}{4\pi\epsilon_0}\left\{\frac{1}{r}\int\rho(r_1)\,dv+\frac{\mathbf{r}}{r^3}\cdot\int\mathbf{r}_1\rho(r_1)\,dv+\right.$$

$$\left.\frac{1}{2r^5}\int[3(\mathbf{r}\cdot\mathbf{r}_1)^2-r^2r_1^2]\rho(r_1)\,dv+\cdots\right\} \quad (2\text{–}93)$$

$$= V_0+V_1+V_2+\cdots.$$

The integral $\int\rho(r)\,dv$ in the first term of equation 2–93 represents the total charge of the distribution so that V_0 is the potential produced at P by an equivalent point charge $Q = \int\rho(r)\,dv$ placed at the origin. At sufficiently large distances from the distribution V_0 is the dominant term, it is called the monopole term. The second term need only be considered if $Q = 0$; if we compare the integral in the second term with equation 2–78 we see that this is the dipole moment ($Q = 0$) of the charge distribution about the origin. Thus if $Q = 0$

$$V_P = V_1 = \frac{1}{4\pi\epsilon_0}\frac{\mathbf{r}}{r^3}\cdot\int\mathbf{r}_1\rho(r_1)\,dv = \frac{\mathbf{p}\cdot\mathbf{r}}{4\pi\epsilon_0 r^3}$$

is the potential at the point P produced by a point dipole placed at the origin. The second term is therefore referred to as the dipole term. The third term, which becomes the dominant one only if the first two are zero, is called the quadrupole term; higher order terms are referred to as multipoles. For a detailed discussion of multipoles the reader should consult more advanced texts.*

2.13 DIPOLE IN AN EXTERNAL FIELD

2.13.1 POTENTIAL ENERGY

The potential energy U of a dipole in an external electric field of strength E is given by the algebraic sum of the potential energies of its two point charges. If V^+ and V^- are the potentials at the locations of the positive and negative charges respectively and if the charges are separated by a small distance dl, we have

$$U = (Q)V^+ +(-Q)V^- = Q(V^+ -V^-)$$

$$= Q\,dV = -Q\mathbf{E}\cdot\mathbf{dl}$$

$$= -\mathbf{p}\cdot\mathbf{E} = -pE\cos\theta \quad (2\text{–}94)$$

* e.g., J. A. Stratton, *Electromagnetic Theory*, McGraw-Hill, New York, 1941.

where θ is the angle between the external field and the dipole axis. The energy is minimum when the dipole is parallel to the field ($\theta = 0$) and maximum when the dipole is antiparallel to the field ($\theta = \pi$).

2.13.2 FORCES

In any conservative field the translation force acting on a particle is given by the negative gradient of its potential energy. For a dipole placed in an external electric field the translational force acting on it is

$$\mathbf{F} = -(\nabla U)_{\theta = \text{const}} = \nabla(\mathbf{p} \cdot \mathbf{E})_{\theta = \text{const}} \qquad (2\text{-}95)$$

If the field $\mathbf{E}$ is uniform the net translational force is zero. It is clear, however, that the external field will exert a torque on the dipole given by

$$T = -\partial U/\partial \theta = -pE \sin \theta \qquad (2\text{-}96)$$

This may be expressed in vector form as

$$\mathbf{T} = \mathbf{p} \times \mathbf{E} \qquad (2\text{-}97)$$

The torque will always be in such a direction as to tend to align the dipole in the direction of the field, that is it will always tend to decrease θ.

2.14 ELECTRIC DOUBLE LAYER

Consider two continuous sheets of charge having an equal and opposite surface charge density σ and separated by a small distance dl. Because dl is small it is more convenient to represent this charge configuration by a surface S carrying a continuous distribution of dipoles. Such a distribution is called an electric double layer; the strength τ of the double layer is defined as the dipole moment per unit area; thus

$$\tau = \lim_{\substack{dl \to 0 \\ \sigma \to \infty}} (\sigma \, dl) \qquad (2\text{-}98)$$

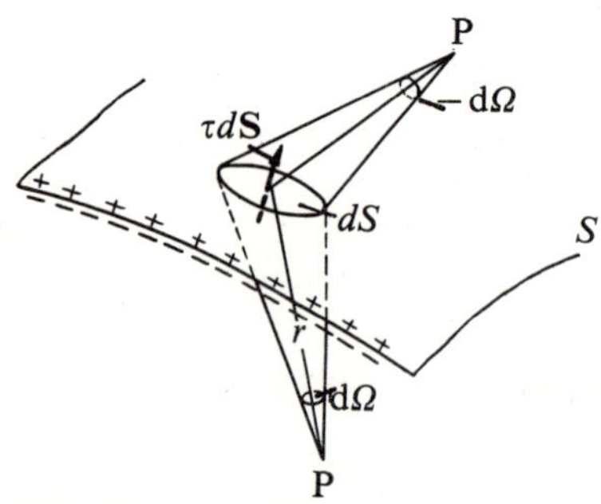

Fig. 2.31 Solid angles subtended by an element of area at two points on either side of a double layer

such that the product $\sigma \, dl$ remains constant. τ is assumed positive if the dipole direction at any point coincides with the direction of the outward normal to the surface at that point. Referring to Fig. 2.31, the potential at some point P not on the surface produced by a double layer of area dS (and hence by an equivalent dipole of moment $\tau \, dS$) is given by

$$dV_P = -\frac{\tau \, \mathbf{dS}}{4\pi\epsilon_0} \cdot \nabla\left(\frac{1}{r}\right)$$

As was shown in Section 2.11, dV_P will be negative if the point P lies on the same side as the negative charge and positive if P lies on the same side as the positive charge. Thus

$$V_P = -\frac{1}{4\pi\epsilon_0} \int_S \tau \, \mathbf{dS} \cdot \nabla\left(\frac{1}{r}\right) = -\frac{1}{4\pi\epsilon_0} \int_\Omega \tau \, d\Omega \qquad (2\text{–}99)$$

where $d\Omega$ is the solid angle subtended by dS at P; it is positive if dS is external with respect to the point P and negative if dS is internal with respect to P (Fig. 2.31). If τ is constant over the whole surface

$$V_P = -\frac{\tau\,\Omega}{4\pi\epsilon_0} \qquad (2\text{–}100)$$

If the surface S is closed then Ω is zero if P lies outside the surface and 4π if it lies inside (see Appendix II); thus

$$\left.\begin{array}{l} V_P = 0 \text{ outside closed surface} \\ V_P = -\tau/\epsilon_0 \text{ inside closed surface} \end{array}\right\} \qquad (2\text{–}101)$$

We shall now consider the particular case of a uniform distribution of dipoles of strength τ over a circular disc of radius a and find the potential and field intensity at any point on the disc axis (see Fig. 2.32a).

The potential at a point P at a distance $\pm z$ from the disc centre is, by equation 2–100,

$$V_P = \pm\frac{\tau\,|\Omega|}{4\pi\epsilon_0} = \pm\frac{\tau}{4\pi\epsilon_0} 2\pi(1-\cos\theta)$$

$$= \pm\frac{\tau}{2\epsilon_0}\left[1 - \frac{|z|}{(z^2+a^2)^{1/2}}\right] \qquad (2\text{–}102)$$

At points very near the disc centre $\theta \simeq \pi/2$, and

$$V_P = \pm\frac{\tau}{2\epsilon_0}$$

The sign of V_P depends upon whether the point P lies on the positive side $(z = 0^+)$ or the negative side $(z = 0^-)$ of the disc centre. In crossing

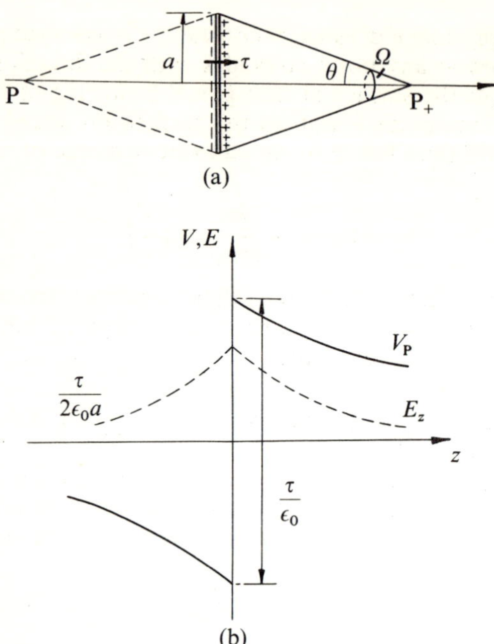

(a)

(b)

Fig. 2.32 Variations of the potential and electric field intensity along the axis of a circular disc carrying a uniform distribution of dipoles of strength τ

from the positive to the negative side of the disc the potential changes by

$$V_P^+ - V_P^- = \tau/\epsilon_0 \tag{2-103}$$

This is the same as the change in potential suffered in moving from a point outside a closed double layer surface to a point inside equation 2–101.

The field intensity at the point P is given by

$$E_z = -\frac{\partial V}{\partial z} = \frac{\tau}{2\epsilon_0} \frac{a^2}{(a^2 + z^2)^{3/2}} \tag{2-104}$$

The direction of E_z is along the positive direction of the z-axis irrespective of whether the point P lies on the positive or the negative side of the disc. It follows that E_z is continuous at $z = 0$.

A little consideration will show that the above results may be extended to any surface carrying a continuous double layer of strength τ. On crossing such a surface the potential changes abruptly by τ/ϵ_0 but the normal component of the electric field intensity changes continuously (Fig. 2.32b).

In nature electric double layers are formed when atoms of some foreign element are adsorbed on a metal surface. This is because the electronic charge distribution between an adsorbed atom and a metal atom is such that a net dipole moment can be associated with each adsorbed atom. The dipole layer may either have a positive outer layer or a negative outer layer depending upon the electron affinity of the adsorbed atom. When a double layer is formed on the surface of a metal, the work function (the minimum amount of work required to remove an electron from the metal) of the metal is altered. If the dipole layer has the positive side outward the work function is lowered whilst if the negative side is outward the work function is increased. In general it has been found that the presence of electric double layers on solid surfaces greatly influences the electrical and electrochemical properties of these surfaces.

2.15 **POINT FORM OF GAUSS'S LAW—DIVERGENCE AND DIVERGENCE THEOREM**

Suppose that a vector field $\mathbf{D}_0$ exists in some region of space and suppose that a charge of volume density ρ is distributed in this region. It will be assumed that $\mathbf{D}_0$ is continuous and that it has continuous spatial derivatives. We now apply Gauss's law to an infinitesimal volume Δv by calculating the total flux leaving the surface enclosed by this volume and equating it to the total charge $\Delta Q = \int \rho \, dv$ within the volume. For simplicity we shall work with rectangular coordinates.

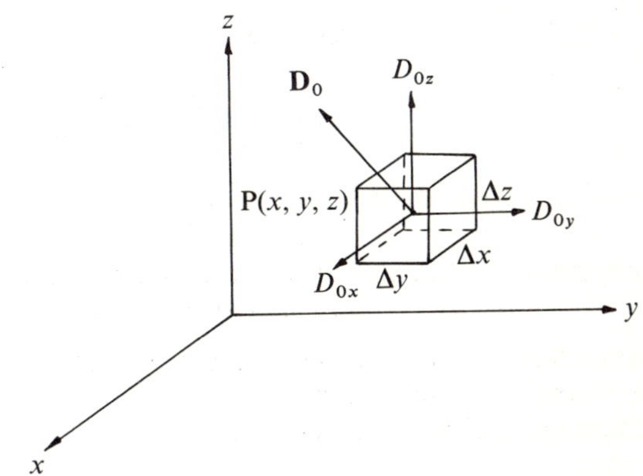

Fig. 2.33 Application of Gauss's law to an element of volume

Let the element of volume $\Delta v = \Delta x \Delta y \Delta z$ be centred around any point P (x, y, z) (Fig. 2.33). At this point the electric flux density may be expressed by its rectangular components as

$$\mathbf{D}_0 = D_{ox}\mathbf{a}_x + D_{oy}\mathbf{a}_y + D_{oz}\mathbf{a}_z$$

The integral $\int \mathbf{D_0} \cdot \mathbf{dS}$ over the closed surface of the volume element is the sum of the element flux contributions over the six faces of the parallelepiped. Consider first the flux leaving the front and back faces which are parallel to the zy-plane and at distances $x+\Delta x/2$ and $x-\Delta x/2$ respectively from the point P. Using Taylor's expansion, the x-components of the flux density at the points $\mathbf{P}^+$ and $\mathbf{P}^-$ are given by

$$D_{ox}(\mathbf{P}^+) = D_{ox} + \frac{\Delta x}{2} \frac{\partial}{\partial x}(D_{ox}) + \left(\frac{\Delta x}{2}\right)^2 \frac{1}{2} \frac{\partial^2}{\partial x^2}(D_{ox})$$

$$+ \cdots \text{(terms with higher powers of } \Delta x)$$

$$D_{ox}(\mathbf{P}^-) = D_{ox} - \frac{\Delta x}{2} \frac{\partial}{\partial x}(D_{ox}) + \left(\frac{\Delta x}{2}\right)^2 \frac{1}{2} \frac{\partial^2}{\partial x^2}(D_{ox})$$

$$+ \cdots \text{(terms with higher powers of } \Delta x)$$

The total outward flux through the front face is

$$\left[D_{ox} + \frac{\Delta x}{2} \frac{\partial}{\partial x}(D_{ox}) + \cdots\right] \mathbf{a}_x \cdot \Delta y \Delta z \, \mathbf{a}_x$$

$$= \left[D_{ox} + \frac{\Delta x}{2} \frac{\partial}{\partial x}(D_{ox}) + \cdots\right] \Delta y \Delta z$$

and the total outward flux through the back face is

$$\left[D_{ox} - \frac{\Delta x}{2} \frac{\partial}{\partial x}(D_{ox}) + \cdots\right] \mathbf{a}_x \cdot \Delta y \Delta z(-\mathbf{a}_x)$$

$$= -\left[D_{ox} - \frac{\Delta x}{2} \frac{\partial}{\partial x}(D_{ox}) + \cdots\right] \Delta y \Delta z$$

The net outward flux from these two faces is

$$\frac{\partial D_{ox}}{\partial x} \Delta x \Delta y \Delta z + \cdots \text{(high order terms containing the factor}$$
$$\Delta x^n \Delta y \Delta z \text{ where } n = 2, 3, \ldots)$$

The net outward flux passing through the other two pairs of parallel faces may be found in a manner analogous to the above. The flux through the faces parallel to the zx-plane is

$$\frac{\partial D_{oy}}{\partial y} \Delta x \Delta y \Delta z + \cdots \text{(high order terms containing the factor}$$
$$\Delta x \Delta y^n \Delta z \text{ where } n = 2, 3, \ldots)$$

and that through the faces parallel to the xy-plane is

$$\frac{\partial D_{oz}}{\partial z} \Delta x \Delta y \Delta z + \cdots \text{(high order terms containing the factor}$$
$$\Delta x \Delta y \Delta z^n \text{ where } n = 2, 3, \ldots)$$

The total flux passing through the surface of the incremental volume is the sum of the above three contributions. Gauss's law for the volume element is

$$\left(\frac{\partial D_{ox}}{\partial x}+\frac{\partial D_{oy}}{\partial x}+\frac{\partial D_{oz}}{\partial z}\right)\Delta x\Delta y\Delta z+\left\{\begin{matrix}\text{higher order}\\ \text{terms}\end{matrix}\right\}=\Delta Q$$

If we divide both sides of this equation by $\Delta v = \Delta x\Delta y\Delta z$ and then let Δv shrink to zero around the point P—i.e., we let $\Delta x \to 0$, $\Delta y \to 0$ and $\Delta z \to 0$—the higher order terms become zero and we have

$$\frac{\partial D_{ox}}{\partial y}+\frac{\partial D_{oy}}{\partial y}+\frac{\partial D_{oz}}{\partial z}=\lim_{\Delta v\to 0}\frac{\Delta Q}{\Delta v}=\rho \qquad (2\text{--}105)$$

Equation 2–105 is the point form of Gauss's law. Recalling the definition of the vector operator ∇ (equation 2–32) we may rewrite equation 2–105 as

$$\left.\begin{matrix}\nabla \cdot \mathbf{D}_0 = \rho \\ \nabla \cdot \mathbf{E} = \rho/\epsilon_0\end{matrix}\right\} \qquad (2\text{--}106)$$

or

The dot product between the operator ∇ and the vector $\mathbf{D}_0$ (or $\mathbf{E}$) is referred to as the *divergence* of $\mathbf{D}_0$ and is usually designated as div $\mathbf{D}_0$. We conclude from the foregoing that the divergence of the electric flux density at a point P gives the density of charge source (positive divergence) or the charge sink (negative divergence) of the flux density at that point.

Although in the above analysis we have confined ourselves to cartesian coordinates, a more general definition for the divergence at a point P of any continuous vector point function $\mathbf{F}$ which has continuous first derivatives is given by

$$\nabla \cdot \mathbf{F} = \lim_{v\to 0}\frac{\oint_S \mathbf{F}\cdot \mathbf{dS}}{v} \qquad (2\text{--}107)$$

where S is a simple closed surface surrounding P and v is the volume enclosed by S.

By definition the divergence of a vector function is *invariant*—i.e., it is independent of the particular coordinate system chosen, and equation 2–107 may therefore be used to derive the explicit form of the divergence of a vector in any convenient coordinate system by expressing the volume and surface elements in terms of the coordinates of the chosen system. The expressions in cylindrical and in spherical coordinates are given below for convenience.

Cylindrical $$\nabla \cdot \mathbf{F} = \frac{1}{r}\frac{\partial}{\partial r}(r\,F_r)+\frac{1}{r}\frac{\partial F_\phi}{\partial \phi}+\frac{\partial F_z}{\partial z} \qquad (2\text{--}108)$$

Spherical $$\mathbf{\nabla} \cdot \mathbf{F} = \frac{1}{r^2} \frac{\partial}{\partial r}(r^2 F_r) + \frac{1}{r \sin \theta} \frac{\partial}{\partial \theta}(F_\theta \sin \theta)$$

$$+ \frac{1}{r \sin \theta} \frac{\partial F_\phi}{\partial \phi} \quad (2\text{–}109)$$

2.15.1 SURFACE DIVERGENCE

The definition of the divergence as given by equation (2–107) is valid only as long as the vector point function $\mathbf{F}$ is continuous in the region under consideration. When $\mathbf{F}$ is discontinuous—e.g., when it has

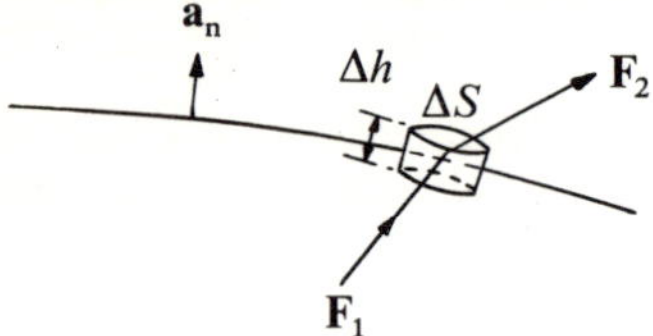

Fig. 2.34 To illustrate surface divergence

different values on two sides of a boundary surface between two regions—the limit in equation 2–107 has no meaning. Consider the volume element shown in Fig. 2.34. The limit of the function

$$\lim_{\substack{\Delta S \to 0 \\ \Delta h \to 0}} \frac{(\mathbf{F}_2 - \mathbf{F}_1) \cdot \mathbf{a}_n \Delta S}{\Delta S \Delta h}$$

will always be infinite since as $\Delta h \to 0$, $\mathbf{F}_2 - \mathbf{F}_1$ remains finite and represents the difference between the values of $\mathbf{F}$ on the two sides of the surface. If, however, we define a *surface divergence* as

$$^S\mathbf{\nabla} \cdot \mathbf{F} = \lim_{\Delta S \to 0} \frac{\oint \mathbf{F} \cdot \mathbf{dS}}{\Delta S} \quad (2\text{–}110)$$

then it is evident that

$$^S\mathbf{\nabla} \cdot \mathbf{F} = (\mathbf{F}_2 - \mathbf{F}_1) \cdot \mathbf{a}_n \quad (2\text{–}111)$$

2.16 DIVERGENCE (GAUSS–OSTROGRADSKY) THEOREM

From Gauss's law we have

$$\Psi = \oint_S \mathbf{D}_0 \cdot \mathbf{dS} = \int_v \rho \, dv$$

but since $\mathbf{\nabla} \cdot \mathbf{D}_0 = \rho$, we have

$$\oint_S \mathbf{D}_0 \cdot \mathbf{dS} = \int_v \mathbf{\nabla} \cdot \mathbf{D}_0 \, dv \quad (2\text{–}112)$$

This is known as the divergence theorem or the Gauss–Ostrogradsky theorem. Although we have derived it for the electric flux density $\mathbf{D}_0$, the theorem is valid for any continuous vector function $\mathbf{F}$. Thus

$$\oint_S \mathbf{F} \cdot \mathbf{dS} = \int_v \nabla \cdot \mathbf{F}\, dv \qquad (2\text{-}113)$$

To show this, let S be any smooth closed surface enclosing a volume v. We can subdivide this volume into N elemental volumes Δv_k with surfaces ΔS_k. For any point inside the kth element we may write,

$$\nabla \cdot \mathbf{F} = \frac{\oint_{S_k} \mathbf{F} \cdot \mathbf{dS}}{\Delta v_k} + \delta_k$$

where δ_k tends to zero as $\Delta v_k \to 0$. We may write the above expression in the form,

$$\oint_{S_k} \mathbf{F} \cdot \mathbf{dS} = (\nabla \cdot \mathbf{F})\,\Delta v_k - \delta_k\,\Delta v_k$$

If we sum over all N volume elements we obtain

$$\sum_1^N \oint_{S_k} \mathbf{F} \cdot \mathbf{dS} = \sum_1^N (\nabla \cdot \mathbf{F})\,\Delta v_k - \sum_1^N \delta_k\,\Delta v_k \qquad (2\text{-}114)$$

Now the sum of the surface integrals over $S_1, S_2, \ldots S_N$ is the integral over the surface S which bounds v. This is because at the common surface of two adjacent volume elements the outward normals are always oppositely directed, so that all contributions to the surface integral from elements internal to v cancel in pairs and the only contribution will be that from the integral over those elements of area which do not constitute a common surface; these are precisely the ones which together make up the entire closed surface S. In the limiting case when $N \to \infty$ and $\Delta v_k \to 0$ equation 2–114 becomes equation 2–113.

The divergence theorem states that the surface integral of the normal component of a vector function of position $\mathbf{F}$ over a closed surface S,

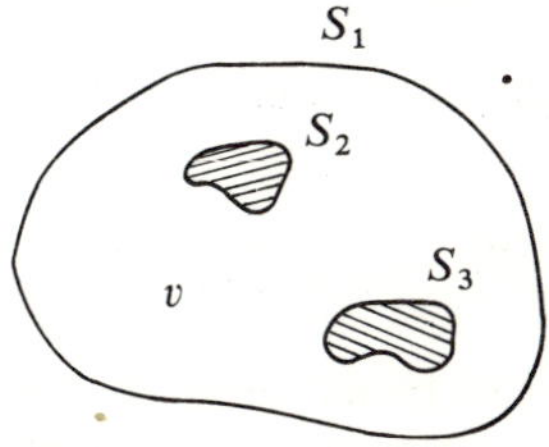

Fig. 2.35 Finite volume bounded by more than one surface

is equal to the integral of the divergence of $\mathbf{F}$ over the volume v enclosed by the surface S. In some cases the surface S enclosing a volume v consists of more than one surface (Fig. 2.35); we must then be careful to include all surfaces bounding the volume v,

$$\oint_S \mathbf{F} \cdot \mathbf{dS} = \oint_{S_1+S_2+S_3} \mathbf{F} \cdot \mathbf{dS} = \int_v \mathbf{\nabla} \cdot \mathbf{F}\, dv \qquad (2\text{--}115)$$

2.17 GREEN'S THEOREM

This theorem is an important extension of the divergence theorem. It states that if ϕ and ψ are any two scalar point functions which within any volume v are finite, continuous and twice differentiable, then

$$\oint_S \psi\nabla\phi \cdot \mathbf{dS} = \int_v (\psi\nabla^2\phi + \nabla\psi \cdot \nabla\phi)\, dv \qquad (2\text{--}116)$$

and

$$\oint_S (\psi\nabla\phi - \phi\nabla\psi) \cdot \mathbf{dS} = \int_v (\psi\nabla^2\phi - \phi\nabla^2\psi)\, dv \qquad (2\text{--}117)$$

where S is the surface bounding the volume v.

Let a vector field $\mathbf{F}$ be given by

$$\mathbf{F} = \psi\nabla\phi$$

then by equation 2–113 we have,

$$\oint_S \mathbf{F} \cdot \mathbf{dS} = \int_v \nabla \cdot \mathbf{F}\, dv = \int_v \nabla \cdot (\psi\nabla\phi)\, dv$$

but

$$\nabla \cdot (\psi\nabla\phi) = \psi\nabla \cdot \nabla\phi + \nabla\psi \cdot \nabla\phi$$
$$= \psi\nabla^2\phi + \nabla\psi \cdot \nabla\phi \qquad (2\text{--}118)$$

Hence

$$\oint_S \psi\nabla\phi \cdot \mathbf{dS} = \int_v (\psi\nabla^2\phi + \nabla\psi \cdot \nabla\phi)\, dv$$

which is known as Green's first identity. If we interchange ψ and ϕ we obtain

$$\oint_S \phi\nabla\psi \cdot \mathbf{dS} = \int_v (\phi\nabla^2\psi + \nabla\psi \cdot \nabla\phi)\, dv \qquad (2\text{--}119)$$

Subtracting equation 2–119 from equation 2–116 we obtain the symmetrical form given by equation 2–117 which is known as Green's second identity.

2.18 POISSON AND LAPLACE EQUATIONS

By combining the point form of Gauss's law

$$\nabla \cdot \mathbf{E} = \rho/\epsilon_0$$

with the point relationship between electric field intensity and potential

$$\mathbf{E} = -\operatorname{grad} V = -\nabla V$$

we obtain

$$\nabla^2 V = -\rho/\epsilon_0 \qquad (2\text{–}120)$$

which is *Poisson's equation*. In a region of free space where there is no volume charge $\rho = 0$ and

$$\nabla^2 V = 0 \qquad (2\text{–}121)$$

which is *Laplace's equation*.

In equations 2–120 and 2–121 the operator ∇^2 ($=\nabla \cdot \nabla$) is known as the *Laplacian*. The Laplacian of V in the three main coordinate system is

cartesian
$$\frac{\partial^2 V}{\partial x^2} + \frac{\partial^2 V}{\partial y^2} + \frac{\partial^2 V}{\partial z^2} = 0 \qquad (2\text{–}122)$$

cylindrical
$$\frac{1}{r}\frac{\partial}{\partial r}\left(r\frac{\partial V}{\partial r}\right) + \frac{1}{r^2}\left(\frac{\partial^2 V}{\partial \phi^2}\right) + \frac{\partial^2 V}{\partial z^2} = 0 \qquad (2\text{–}123)$$

spherical
$$\frac{1}{r^2}\frac{\partial}{\partial r}\left(r^2\frac{\partial V}{\partial r}\right) + \frac{1}{r^2 \sin\theta}\frac{\partial}{\partial \theta}\left(\sin\theta\frac{\partial V}{\partial \theta}\right)$$
$$+ \frac{1}{r^2 \sin^2\theta}\frac{\partial^2 V}{\partial \phi^2} \qquad (2\text{–}124)$$

Some methods of solving these equations will be discussed in Chapter 11. It is important to bear in mind that the function

$$\phi = 1/r$$

where

$$r = \left[(x-x')^2 + (y-y')^2 + (z-z')^2\right]^{1/2}$$

satisfies Laplace's equation $\nabla^2(1/r)$ everywhere in space except at the point (x', y', z') where $1/r$ becomes infinite. The logarithmic potential function

$$\phi = \log 1/r$$

where

$$r = \left[(x-x')^2 + (y-y')^2\right]$$

satisfies Laplace's equation in the plane, that is

$$\left(\frac{\partial^2}{\partial x^2} + \frac{\partial^2}{\partial y^2}\right) \log\left(\frac{1}{r}\right) = 0$$

except at the point (x', y').

2.18.1 PHYSICAL SIGNIFICANCE OF $\nabla^2 V = 0$

Let the value of the scalar field V at a certain point O be V_0. With O as centre construct a sphere of radius r. The average value $\langle V \rangle$ of V over the surface of this sphere is,

$$\langle V \rangle = \frac{1}{4\pi r^2} \int_S V \, dS$$

$$= \frac{1}{4\pi} \iint V \sin\theta \, d\theta \, d\phi$$

The rate of change of $\langle V \rangle$ with r is given by,

$$\frac{\partial}{\partial r} \langle V \rangle = \frac{1}{4\pi} \iint \frac{\partial V}{\partial r} \sin\theta \, d\theta \, d\phi$$

$$= \frac{1}{4\pi r^2} \int_S \frac{\partial V}{\partial r} \, dS$$

$$= \frac{1}{4\pi r^2} \int_S \nabla V \cdot \mathbf{dS}$$

$$= \frac{1}{4\pi r^2} \int_v \nabla^2 V \, dv = 0$$

This means that if $\nabla^2 V = 0$ then $\langle V \rangle$ is independent of the radius of the sphere. If we choose $r = 0$

$$\langle V \rangle = V_0$$

We thus see that if a potential function V satisfies Laplace's equation in a given region of space, this means that in this region the function V varies from point to point in such a way that its average value taken over the surface of any sphere is equal to the value of V at the centre of the sphere.

PROBLEMS

Chapter Two

2.1 A point charge of 5 μC is placed at the point $(0, 0, 1)$. A second charge of -1 μC is located at the point $(1, 2, 3)$. Find the magnitude and direction of the force acting on each charge.

2.2 A point charge of 1 μC and one of -10 μC are 0·05 m apart. Find the point at which no force would be exerted on a third point charge.

2.3 Two point charges Q are placed at two diametrically opposite points of a square of side l. Find the values of the charges which must be placed at the other two corners such that the resultant force on the charges Q is zero.

2.4 Two particles each of mass m and charge Q are suspended at the ends of two strings of negligible mass and length l; the other string ends are attached to a common fixed point. Show that at equilibrium the inclination θ of each string to the vertical is given by

$$16\pi\epsilon_0 m g\, l^2 \sin^3\theta = Q^2 \cos\theta$$

2.5 Three point charges of 1 μC, 3 μC and -5 μC are placed at the corners A, B and C respectively of an equilateral triangle of side 0·1 m. If L, M and N are the midpoints of sides AB, BC and CA respectively, find the potentials of L, M and N and the potential difference between the points L and M, L and N, and M and N.

2.6 If the potential difference between two points at distances of 0·1 m and 0·2 m from a point charge Q is 10 volts, find Q.

2.7 Find the electric field associated with each of the following potential fields,

$$V(x, y, z) = 3x^2 y - y^3 z^2$$

$$V(r, \phi, z) = r(1 - b/r^2)\cos\phi$$

$$V(r, \theta, \phi) = ar\cos\theta + b^3 r^{-1}\cos\theta$$

2.8 If r is the distance between a fixed point $Q(x_0, y_0, z_0)$ and a variable point $P(x, y, z)$, show that ∇r is a unit vector in the direction of the vector $\mathbf{r}$.

2.9 Given that $V(x, y) = x^2 - yz$, find ∇V at the point $P(3, 4, 1)$ and the unit vector normal to the surface $V = 5$, at P.

2.10 For the field corresponding to the potential given in problem 2.9, calculate the integral along the following paths,
(a) straight lines from $(0, 0, 0)$ to $(1, 0, 0)$ then to $(1, 1, 0)$ and then to $(0, 0, 0)$.

(b) straight lines from (0, 0, 0) to (0, 0, 1) then to (0, 1, 1) and then to (0, 0, 0).

2.11 A rectangle of sides a and b lies in the xy-plane with one of its corners at the origin. Show that the solid angle subtended at the point P(0, 0, Z) by the rectangle is given by

$$\Omega = \tan^{-1}(ab/zd)$$

where d is the distance from P to the corner of the rectangle diametrically opposite to the origin.

2.12 Show that if θ is the plane angle between two intersecting planes, then the solid angle between the planes is 2θ.

2.13 Two point charges Q and $-Q$ are located at points A and B respectively at a distance $2a$ apart. A line of force leaving either charge at right angles to AB meets the plane bisecting AB at right angles in a point P. Calculate the angle PAB and find the distance of P from AB.

2.14 Two point charges Q and $-Q$ are placed at A and B respectively. A line of force which leaves A at an angle θ with AB meets the line which bisects AB at right angles, in P. Show that θ is given by

$$\sin\tfrac{1}{2}\theta = \sqrt{2}\sin\tfrac{1}{2}\text{PAB}$$

2.15 A, B, C and D are four collinear points such that $AB/BC = AD/CD = 1/m$, $(m < 1)$. A point charge Q is placed at A and a point charge $-mQ$ is placed at C. Show that the sphere drawn on BD as diameter is an equipotential surface and find its potential.

2.16 A point charge $-Q/\sqrt{2}$ is placed at a point M midway between two charges Q at points A and B. If $AB = \sqrt{2}a$ show that the spheres of radius a and with A and B as centres form an equipotential surface. Find the potential of this surface.

2.17 A charge Q_1 is placed at a distance d from a charge $-Q_2 (Q_1 > Q_2)$. Show that the equipotential surface $V = 0$ is a sphere whose radius is $a = Q_1 Q_2 d (Q_1^2 - Q_2^2)^{-1}$ and whose centre is on the line joining the two charges at a distance aQ_2/Q_1 from Q_2.

2.18 Consider a system of n infinitely long uniform and parallel line charges of densities $\lambda_1, \lambda_2, \ldots \lambda_n$. Let $A_1, A_2, \ldots A_n$ be the traces of these lines in a plane perpendicular to them, and let P be any point in this plane. If C is a line of force which passes through P, show that the equation of this line is given by

$$\sum_{i=1}^{n} \lambda_i \theta_i = \text{constant}$$

where θ_i is the angle between A_iP and any fixed line in the plane.

2.19 Draw the lines of force for the following point charge configurations,
(a) $2Q$ and $-Q$ at a distance a apart,
(b) $4Q$ and $-Q$ at a distance a apart,
(c) Charges Q on the diagonally opposite corners of a square and charges $-Q$ on the other two corners.

2.20 According to wave mechanics the charge distribution corresponding to the electron in the ground state of a hydrogen atom is an exponential function of the type $\rho(r) = A\exp(-2r/r_1)$ where A and r_1 are constants and r represents the distance from the nucleus. If the total charge must be equal to $-e$, show that $A = -e/\pi r_1^3$.

2.21 A charge is distributed with a uniform density $\rho = \rho_0\exp(\alpha r^3)$ within a spherical shell $b \geq r \geq a$. Find the electric field intensity at any point inside the shell.

2.22 A positive charge is distributed with a uniform density ρ throughout the volume of a sphere of radius a. Show that if an electron of charge $-e$ and mass m is placed in the sphere it will execute simple harmonic oscillations through the centre of the sphere with a period $2\pi(3\epsilon_0 m/\rho e)^{1/2}$.

2.23 A straight wire of length L is uniformly charged over one half of its length with a charge density λ and over the other half with a density $-\lambda$. Find the electric field intensity at points along the axis of the wire and along the perpendicular bisector of the wire.

2.24 A thin circular ring of radius a carries a uniform charge of density λ. Show that the potential at any point on the axis of the ring and at a distance z from its centre is

$$V = \lambda a/2\epsilon_0(a^2 + z^2)^{1/2}$$

2.25 A thin loop in the form of a square of side a carries a uniform charge of density λ. Show that the potential at the centre of the loop is given by

$$V = 2\lambda/\pi\epsilon_0 \log(1 + \sqrt{2})$$

2.26 Two equal and opposite line charges of density $1\,\mu C\,m^{-1}$ are $8\,m$ apart. Draw the equipotential surfaces for $V = 10, 15$ and $20\,kV$. Plot the variation of the electric field intensity along the line joining the charge centres.

2.27 Show that the electric field of a dipole of moment $\mathbf{p}$ may be expressed as

$$\mathbf{E} = \frac{1}{4\pi\epsilon_0}\left[\frac{3(\mathbf{p}\cdot\mathbf{r})\mathbf{r}}{r^5} - \frac{\mathbf{p}}{r^3}\right]$$

where r is the distance from the dipole to any point P. Hence show that the mutual energy of two dipoles $\mathbf{p}_1$ and $\mathbf{p}_2$ is given by

$$\frac{1}{4\pi\epsilon_0}\left[\frac{\mathbf{p}_1 \cdot \mathbf{p}_2}{r^3} - \frac{3(\mathbf{p}_1 \cdot \mathbf{r})(\mathbf{p}_2 \cdot \mathbf{r})}{r^5}\right]$$

2.28 Two identical dipoles 1 and 2 are placed so that the axis of dipole 1 passes through the centre of dipole 2 and makes an angle θ with the axis of 2. Derive an expression for the torque acting on the dipoles and show that the torque tending to rotate one dipole is twice that tending to rotate the other.

2.29 The magnitude and positions of four point charges are as follows: Q at $(0, 0)$, $3Q$ at $(0, a)$, $-2Q$ at (a, a) and $-Q$ at $(a, 0)$. Find the electric moment of the charge configuration about the origin.

2.30 Find the dipole moment of a spherical shell carrying a surface charge density $\sigma = \sigma_0 \cos \theta$.

2.31 A quadrupole consists of front equal point charges Q of alternate sign placed at the corners of a rectangle of sides a and b. Show that at distances r from the rectangle centre which are large compared with a and b the potential is given by

$$V = Qab \sin \theta \cos \theta / 2\pi\epsilon_0 r^3$$

2.32 Find the charge density required to produce the following fields:
(a) $x^2 z \mathbf{a}_x + 2y^3 z^2 \mathbf{a}_y - xy^2 z \mathbf{a}_z$
(b) $(3x^2 yz - 3y)\mathbf{a}_x + (x^3 z - 3x)\mathbf{a}_y + (x^3 y + 2z)\mathbf{a}_z$.

2.33 Given the vector $\mathbf{A} = (x^2 - y^2)\mathbf{a}_x + 2xy\mathbf{a}_y + (y^2 - xy)\mathbf{a}_z$, verify the divergence theorem by evaluating $\int \nabla \mathbf{A}\, dv$ throughout a cube of side a and centre at the origin and $\oint \mathbf{A} \cdot d\mathbf{S}$ over the surface of the cube.

2.34 Given the vector $\mathbf{A} = r\cos^2 \phi\, \mathbf{a}_r + r\sin \phi\, \mathbf{a}_\phi$, find its flux over the surface of a cylinder of unit height and cross-section given by $x^2 + y^2 = 16$. Verify the divergence theorem for this case.

2.35 Find the Laplacian of the fields given in Problem 2.7.

CHAPTER THREE

CONDUCTORS IN THE ELECTROSTATIC FIELD

3.1 FIELD INSIDE A CONDUCTOR—MACROSCOPIC DEFINITION OF A CONDUCTOR

In this chapter we shall be primarily concerned with the general properties of solid conductors (metals) under equilibrium conditions in an electrostatic field, and with the charges and potentials of conductors and systems of conductors. Although we shall adopt throughout the chapter a macroscopic point of view, it is pertinent to recall here some essential microscopic details of a metallic conductor.

In a solid conductor each atom is closely surrounded by many similar atoms. The overlap between the outer electron shells of these atoms is great enough for the loosely held electrons in these shells, the valence or free electrons, to be so completely shared that they may no longer be uniquely associated with a particular atom. Such electrons (about one per atom $\sim 10^{29}$ per m^3) may be visualized as free to move throughout the whole volume of the solid thus forming a negatively charged electron gas in which are embedded the (quasi) fixed positive metal ions forming the lattice. In the absence of an external field the free electrons, because of thermal agitation, move in all possible directions so that on average there are as many electrons crossing a fictitious surface element within the conductor in one direction as there are crossing it in the opposite direction.

If an isolated conductor is placed in an external electrostatic field of strength $\mathbf{E}$, every electron experiences a force $\mathbf{F} = -e\mathbf{E}$ and the free electrons move against the field. This migration of electrons produces surface concentrations of charge, and continues until at equilibrium the field $\mathbf{E}_i$ due to these surface charges is equal and opposite to the applied field at every point inside the conductor. At equilibrium, therefore, the electric field inside a conductor is everywhere zero. The surface charge distribution is said to consist of *induced charges*. These charges may be considered as forming a shield on the surface which prevents the penetration of any field lines inside the conductor; thus all lines either terminate at or start from a conductor surface (Fig. 3.1). Inside a conductor we have,

$$\mathbf{E} = 0$$

and

$$\nabla \cdot \mathbf{E} = 0$$

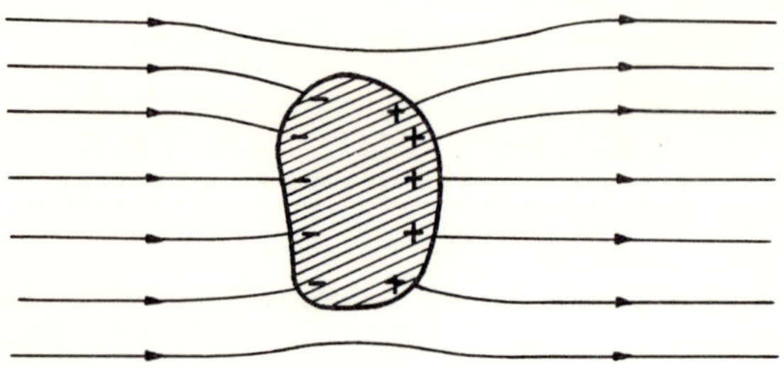

Fig. 3.1 Induced charges on a conductor placed in an external field

also since $\epsilon_0 \nabla \cdot \mathbf{E} = \rho$ (equation 2–106), it follows that at equilibrium the charge density everywhere inside a conductor must be zero. We also have,

$$\mathbf{E} = -\nabla V = 0$$

$$V = \text{constant}$$

so that the whole interior volume of a conductor must be at the same potential.

Because the volume charge density is everywhere zero inside a conductor, any charge on the conductor must reside entirely on its outer surface. This charge constitutes the 'surface charge'; it actually occupies a layer only a few atoms thick so that it may be conveniently represented by a surface charge density σ (see Section 2.4b). Since the potential varies continuously when we cross any surface carrying a continuous surface charge density (Section 2.9), it follows that the surface of a conductor is an equipotential surface whose potential is the same as that of the bulk of the conductor.

It would appear from the foregoing that a suitable definition for a conductor from a macroscopic point of view would be that it is a body which when placed in an external field must, at equilibrium, satisfy the condition that the field inside be everywhere zero. This is a necessary condition but it is by no means sufficient for defining a conductor. This is because the term 'at equilibrium' implies that a certain amount of time must elapse from the instant in which the conductor finds itself in the external field until the free charges distribute themselves along the surface in such a manner as to reduce the field inside the conductor to zero. For a body to be classified as a conductor it is essential that this time interval be infinitesimally small compared with any experimentally significant time interval. For metals in general this time interval is so short (of the order of 10^{-17} to 10^{-19} s) that for all practical purposes we may assume that a state of equilibrium is attained instantaneously. In semiconductors and insulators the time necessary to attain equilibrium may range from several seconds to several months. The question of the rapidity with which a steady state is reached in different media will be discussed in greater detail in Chapter 6.

3.2 FIELD AT THE SURFACE OF A CONDUCTOR

Since the surface of a conductor is an equipotential, the lines of force must be everywhere normal to that surface. If equation 2–55 is applied to the boundary surface of a conductor carrying a charge density σ we have

$$(\mathbf{E}_2 - \mathbf{E}_1) \cdot \mathbf{a}_n = \frac{\sigma}{\epsilon_0}$$

Since the internal field is zero, $\mathbf{E}_1 = 0$ and

$$\mathbf{E}_2 \cdot \mathbf{a}_n = E_2 = E = \frac{\sigma}{\epsilon_0}$$

or

$$\mathbf{E} = \frac{\sigma}{\epsilon_0} \mathbf{a}_n$$

it follows that,

$$\mathbf{D}_0 = \epsilon_0 \mathbf{E} = \sigma \mathbf{a}_n \tag{3–1}$$

and

$$D_0 = \sigma \tag{3–2}$$

Thus the flux density at any point on the surface of a conductor in free space is normal to the surface at that point and is equal to the surface charge density at that point. This charge density is often referred to as the free charge density.

Because the electric field is zero everywhere inside a conductor, it follows that the electric field is also zero inside a charge-free cavity of

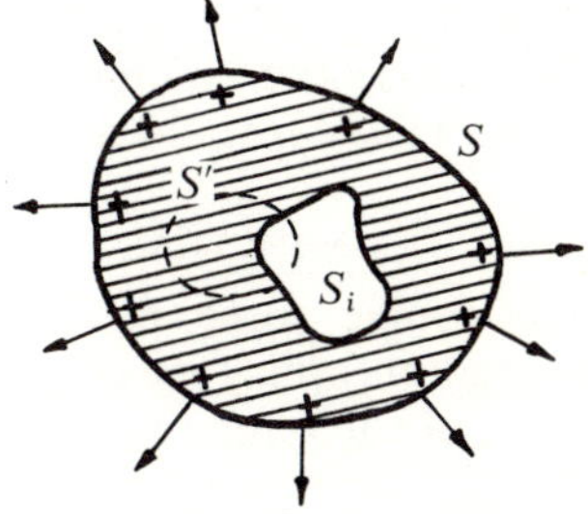

Fig. 3.2 Charged conductor with charge-free hollow cavity

a hollow closed conductor. Since there are no charges inside the cavity the potential in this region can have no maximum or minimum and since the potential is constant over the whole internal surface S_i (Fig. 3.2), it must also be constant at every point of the cavity; hence the field must be zero inside the cavity.

We now apply Gauss's law to the surface S' of Fig. 3.2. Since the field both in the conductor and in the cavity is zero, the total charge enclosed by S' must be zero and therefore no charge can exist on the internal surface S_i of the conductor. Thus if the conductor carries any charges, then these must reside entirely on the *external* surface.

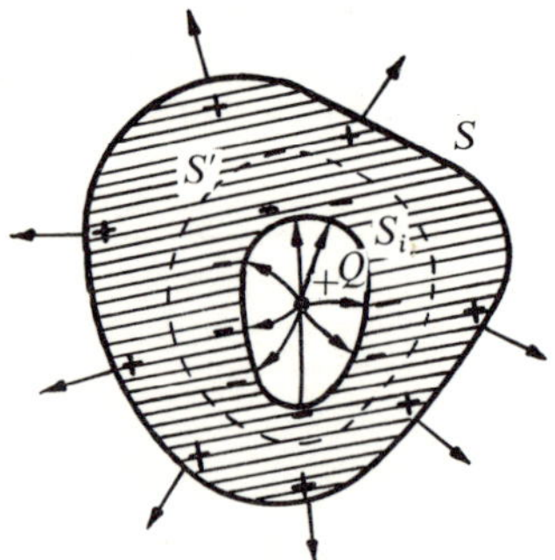

Fig. 3.3 Uncharged hollow conductor with a charge inside cavity

Suppose that a charge $+Q$ is placed inside the cavity of an uncharged hollow conductor. If we apply Gauss's law to a closed surface S' (Fig. 3.3) lying entirely within the conductor, then since $E = 0$ over the whole surface, it follows that the total charge enclosed by S' must be zero. Thus the internal surface S_i of the conductor must carry a total induced charge $-Q$ distributed in such a way that all lines of force leaving the charge Q must end on S_i. Since the conductor is uncharged, a charge $+Q$ must be distributed over its outer surface S.*

The surface charges induced on S and S_i are precisely those required to create at every point inside the conductor a field which exactly opposes the field produced by the free charge Q.

3.3 CONCENTRIC SPHERICAL CONDUCTORS

In this section we shall solve some problems involving charged spherical concentric conductors; this type of problem provides a good exercise in the application of the foregoing results. It should be noted that since any charges must necessarily lie either on the internal or

* Faraday's ice pail experiment.

external surfaces of the spheres, a system of such conductors may be treated as a set of concentric spherical charged shells (Section 2.9) whose radii are those of the inner and outer surfaces of the spheres.

3.3.1 TWO CONCENTRIC SPHERICAL CONDUCTORS

Let the radius of the inner conductor be a_1 and the inner and outer radii of the outer conductor be a_2 and a_2' respectively. Suppose that the conductors are at potentials V_1 and V_2 and that the corresponding charges are Q_1 and Q_2. If we assume that $V_1 > V_2$ and let $K = 4\pi\epsilon_0$, then (see Fig. 3.4a)

$$V_{12} = -\int_{a_2}^{a_1} \frac{Q_1 \, dr}{4\pi\epsilon_0 r^2}$$

or

$$V_{12} = V_1 - V_2 = \frac{Q_1}{K}\left[\frac{1}{a_1} - \frac{1}{a_2}\right] = \frac{Q_1}{K}\left[\frac{a_2 - a_1}{a_1 a_2}\right] \qquad (3\text{--}3)$$

Since the field is zero inside the outer conductor then its inner surface must carry an induced charge $-Q_1$; its outer surface must therefore carry a charge $Q_2 + Q_1$. Hence the potential V_2 is given by

$$V_2 = \frac{Q_2 + Q_1}{Ka_2'} = \frac{Q_1}{Ka_2'} + \frac{Q_2}{Ka_2'} \qquad (3\text{--}4)$$

so that

$$\begin{aligned} V_1 &= \frac{Q_2 + Q_1}{Ka_2'} + \frac{Q_1}{K}\left[\frac{1}{a_1} - \frac{1}{a_2}\right] \\ &= \frac{Q_1}{K}\left[\frac{1}{a_1} - \frac{1}{a_2} + \frac{1}{a_2'}\right] + \frac{Q_2}{K}\frac{1}{a_2'} \end{aligned} \qquad (3\text{--}5)$$

We see from equation 3–3 that if $V_2 > V_1$, then the charge Q_1 on the inner conductor must be negative (Fig. 3.4b). In particular if $V_1 = 0$ (inner sphere earthed) we have from equation 3–5

$$Q_1 = -\frac{Q_2 a_1 a_2}{a_1 a_2 + a_2 a_2' - a_1 a_2'} \qquad (3\text{--}6)$$

If the outer conductor is earthed then $V_2 = 0$, and $Q_1 = -Q_2$ (Fig. 3.4c). If the outer conductor is uncharged, $Q_2 = 0$ and,

$$V_2 = \frac{Q_1}{Ka_2'}; \; V_1 = \frac{Q_1}{K}\left[\frac{1}{a_1} - \frac{1}{a_2} + \frac{1}{a_2'}\right] \qquad (3\text{--}7)$$

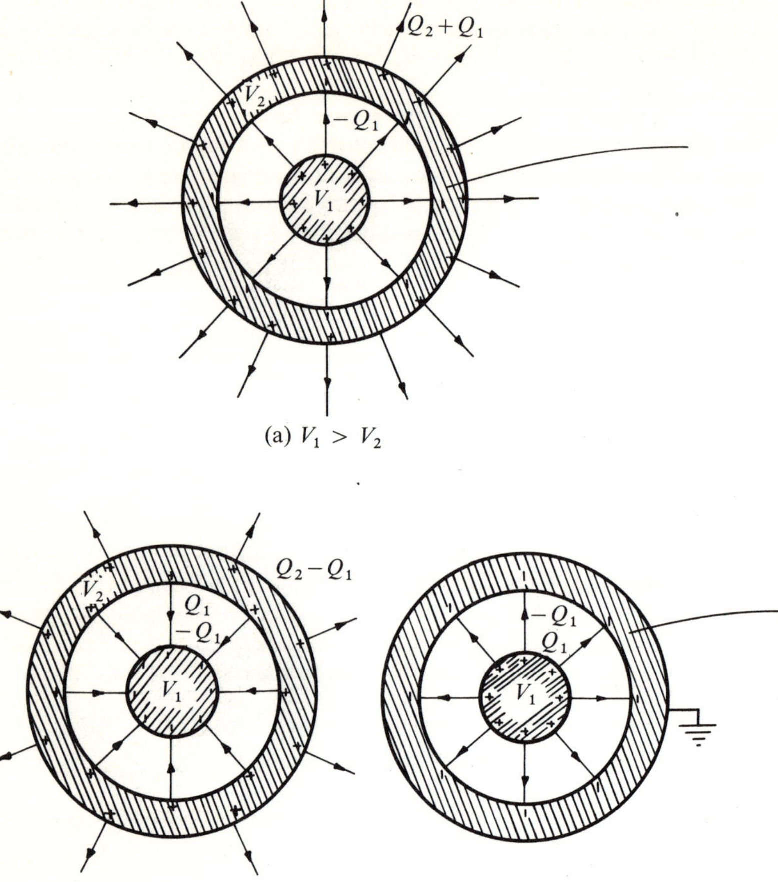

Fig. 3.4 Charge distributions on concentric spherical conductors maintained at potentials V_1 and V_2

If the inner conductor is uncharged, $Q_1 = 0$ and

$$V_1 = V_2 = \frac{Q_2}{Ka_2'} \tag{3-8}$$

3.3.2 N CONCENTRIC SPHERICAL CONDUCTORS

Suppose that we have a system of N concentric spheres whose charges are $Q_1, Q_2 \ldots Q_N$; let the inner and outer radii of the ith conductor be a_i and a_i' respectively. If $-q_i$ is the charge induced on the inner surface of the ith sphere the charge on its outer surface will be $Q_i + q_i$ (Fig. 3.5). Since the field inside the ith conductor must be zero, then by

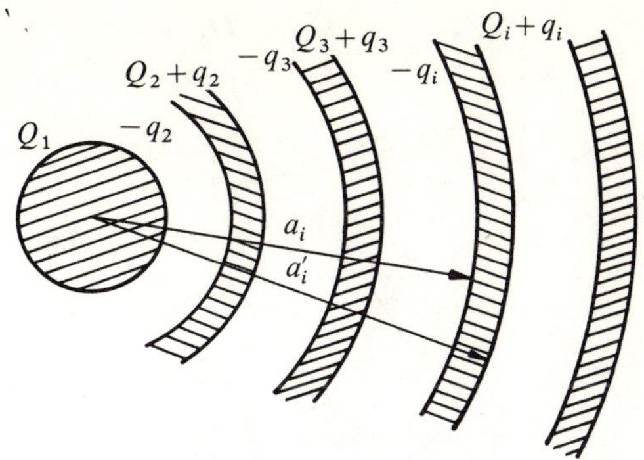

Fig. 3.5 Charged concentric spherical conductors

applying Gauss's law to the surface of a concentric sphere of radius r where $a_i < r < a'_i$ we have

$$q_i = \sum_{j=1}^{i-1} Q_j \tag{3-9}$$

The potential at some point P at a distance r from the origin such that $a'_{i-1} < r < a_i$ consists of two parts: (a) the potential produced at P by all conductors to its left and (b) the potential produced at P by all conductors to its right. The first part is simply the potential at P produced by all the charges on the $i-1$ conductors concentrated at the origin; the second part is the sum of the potentials inside $N-i+1$ shells each of charge $Q_j + q_j$ and radius a'_j plus the sum of the potentials inside $N-i+1$ shells each of charge $-q_j$ and radius a_j. Thus

$$V_P = \frac{1}{K}\left[\frac{1}{r}\sum_{j=1}^{i-1} Q_j + \sum_{j=i}^{N} \frac{Q_j+q_j}{a'_j} - \sum_{j=i}^{N} \frac{q_j}{a_j}\right] \tag{3-10}$$

where $K = 4\pi\epsilon_0$.

If the point P lies on the inner surface of the ith conductor equation 3-10 becomes

$$V_i = \frac{1}{K}\left[\frac{1}{a_i}\sum_{1}^{i-1} Q_j + \sum_{i}^{N} \frac{Q_j+q_j}{a'_j} - \frac{q_i}{a_i} - \sum_{i+1}^{N} \frac{q_j}{a_j}\right]$$

and substituting from equation 3-9 we obtain

$$V_i = \frac{1}{K}\left[\sum_{i}^{N} \frac{Q_j+q_j}{a'_j} - \sum_{i+1}^{N} \frac{q_j}{a_j}\right] \tag{3-11}$$

If the point **P** lies on the outer surface of the ith conductor then its potential may be written as

$$V_i' = \frac{1}{K}\left[\frac{1}{a_i'}\sum_1^{i-1}Q_j - \frac{q_i}{a_i'} + \sum_i^N \frac{Q_j+q_j}{a_j'} - \sum_{i-1}^N \frac{q_j}{a_j}\right]$$

$$= \frac{1}{K}\left[\sum_i^N \frac{Q_j+q_j}{a_j'} - \sum_{i+1}^N \frac{q_j}{a_j}\right] = V_i$$

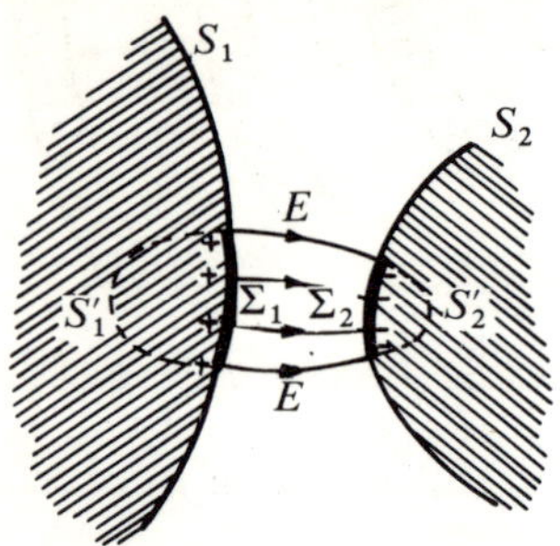

Fig. 3.6 Corresponding surface elements

3.4 CORRESPONDING SURFACE ELEMENTS

Consider two neighbouring conducting surfaces S_1 and S_2 (Fig. 3.6). Suppose that surface S_1 carries a positive charge and that surface S_2 carries a negative charge; some of the lines of force leaving surface S_1 will terminate on surface S_2. Consider the closed surface formed by the tube of force bounded by the two field lines E and by the surfaces S_1 and S_2 which lie entirely within conductor 1 and 2 respectively. If we apply Gauss's law to this surface it is readily seen that, since the field inside the conductors is zero and no lines of force can cross the tube of force, the total flux crossing it must be zero. It follows that the total charge inside this surface is also zero; hence the surface elements Σ_1 and Σ_2 must carry equal and opposite charges. Such surface elements are referred to as *corresponding elements*.

In the case of an isolated charged conductor, all lines of force extend to infinity where it is assumed that they start from or terminate at neutralizing charges distributed over the surface of a conducting sphere of infinite radius. Any element of the conductor surface will then have its corresponding element on the surface of this infinite sphere. For a system of conductors the sum of whose charges is not zero, some of the lines of force will go to infinity and the corresponding elements of the entire surface of any one of the conductors will lie in part over the surfaces of the other conductors and in part on the surface of the sphere at infinity (Fig. 3.7). The infinite sphere will always carry a charge equal and opposite to the resultant charge of the system.

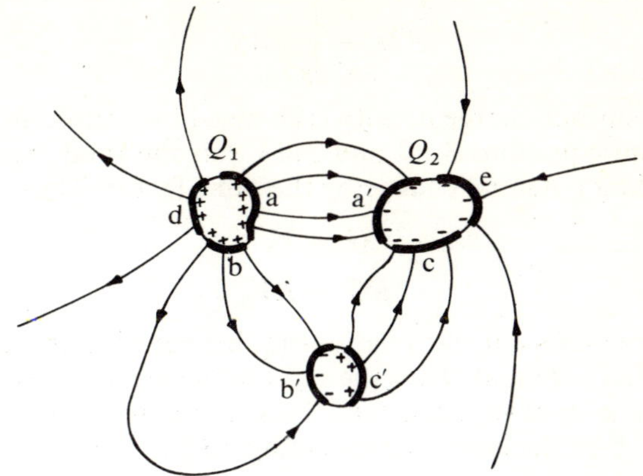

Fig. 3.7 Corresponding surface elements for three conductors. Elements corresponding to d and e lie on the surface of a sphere at infinity

The concept of corresponding elements will be particularly useful in giving a physical significance to the capacitance coefficients of a system of conductors (Section 3.6.6).

3.5 CAPACITANCE OF A CONDUCTOR

Consider an isolated conductor carrying a surface charge density σ. The potential over the surface is constant and finite and may be expressed by

$$V_S = \frac{1}{4\pi\epsilon_0} \int_S \frac{\sigma\,\mathrm{d}S_i}{r_i} + A\sigma \qquad (3\text{--}12)$$

where A is a constant. The first term on the right is the potential produced at any point P on the surface by the charges over the whole surface less a small disc having P as centre (Fig. 3.8). This is finite since

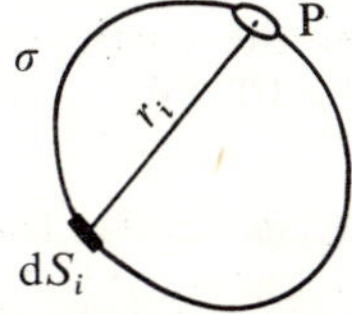

Fig. 3.8 Conductor carrying a surface charge density σ. The potential at point P is given by Equation 3–12

r_i remains finite. The second term is the potential at P produced by the charged disc; as shown in Section 2.9 this is finite and proportional to σ. The total charge on the conductor surface is given by:

$$Q = \int_S \sigma \, \mathrm{d}S \tag{3–13}$$

Now if the surface charge density is changed by a constant factor K, it is clear from equations 3–12 and 3–13 that the total charge and the corresponding potential change by the same factor. Thus,

$$Q \to V_S$$

$$KQ \to KV_S$$

so that there exists a linear relationship between the charge on a conductor and its potential. Thus the ratio of the magnitude of the charge to the potential is always constant. This constant ratio is defined as the *capacitance C* of the isolated conducting body.

$$C = \frac{Q}{V} \text{ farad} \tag{3–14}$$

and it depends only on the size and shape of the conductor. In SI units the unit of capacitance is the farad $[\text{F} \equiv \text{CV}^{-1}]$. Since this is a very large unit, in practice the submultiples microfarad ($\mu\text{F} = 10^{-6}\text{F}$) and picofarad ($\text{pF} = 10^{-12}\text{F}$) are commonly used. In equation 3–14, the potential V is the absolute potential of the conductor and C is therefore its capacitance with respect to infinity or its absolute capacitance.*

As a simple example consider an isolated charged conducting sphere of radius a; since the charge can only be a surface charge, its potential is by equation 2–51,

$$V = \frac{Q}{4\pi\epsilon_0 a}$$

The capacitance of the sphere is therefore

$$C = 4\pi\epsilon_0 a \tag{3–15}$$

In general the capacitance of isolated conductors is very small. As an example, the capacitance of the earth, considered as a conducting sphere, is approximately $6 \cdot 8 \times 10^{-4}\text{F}$.

---

* 'When the capacity of a conductor is spoken of without specifying the form and position of any other conductor in the same system, it is to be interpreted as the capacity of the conductor when no other conductor or electrified body is within a finite distance of the conductor referred to'—J. C. Maxwell, *A Treatise on Electricity and Magnetism*. Volume 1, Dover Publications, New York, 1954.

3.6 RELATIONSHIP BETWEEN POTENTIAL AND CHARGE FOR A SYSTEM OF *N* CONDUCTORS

3.6.1 UNIQUENESS OF THE ELECTROSTATIC PROBLEM

Suppose that we have an arbitrary number N of conductors each insulated from the others. We shall assume that either the potential or the charge of each of the conductors is known and that the whole system is in a state of electrostatic equilibrium in a charge-free space. The problem of determining the potential (and hence the field strength) at each point in space is essentially that of solving Laplace's equation $\nabla^2 V = 0$, subject to certain boundary conditions. But, having obtained a solution, how do we know that other solutions, corresponding to other possible states of equilibrium, do not also exist? In fact we shall show that if a solution of the problem exists, then it is the unique solution. This means that there can be only one value for the potential (and hence field) at each and every point in space and that there is only one possible way in which the charge on each conductor can distribute itself.

Let us assume that the potentials $V_1, V_2 \ldots V_N$ of the N conductors are given. Suppose that there exists two potential functions V' and V'' each of which defines the actual potential at every point in the space surrounding the conductors. It follows that both functions must individually satisfy Laplace's equation so that,

$$\nabla^2 V' = 0 \text{ and } \nabla^2 V'' = 0 \tag{3-16}$$

both functions must also satisfy the boundary conditions which are

$$V' = V'' = V_1 \text{ on conductor 1}$$

$$V' = V'' = V_2 \text{ on conductor 2} \tag{3-17}$$

$$V' = V'' = V_N \text{ on conductor } N$$

from equation 3–16 we have that

$$\nabla^2(V' - V'') = 0 \tag{3-18}$$

and from equation 3–17 that, on the boundaries,

$$V' - V'' = 0 \tag{3-19}$$

Let S' be the surface of a large sphere enclosing all the conductors (Fig. 3.9). Making use of Green's first identity (equation 2–116) and setting

$$\psi = \phi = V' - V''$$

we have

$$\oint_S \psi \nabla \psi \cdot \mathbf{dS} = \int_v [\psi \nabla^2 \psi + (\nabla \psi)^2] \, dv \tag{3-20}$$

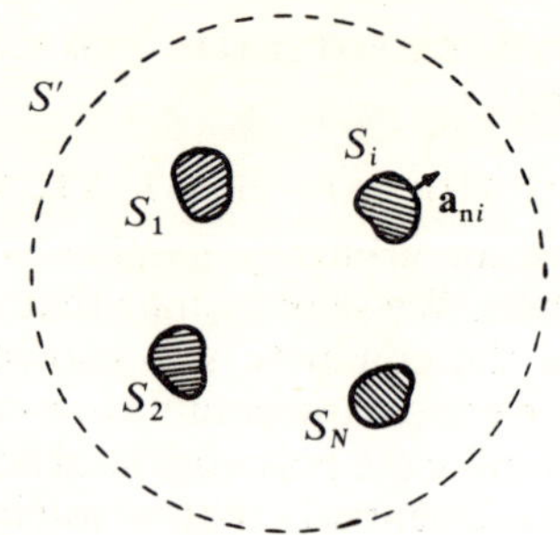

Fig. 3.9 Spherical surface S' enclosing a system of N conductors

where the surface integral extends over the boundary surfaces (ΣS_i) of all the conductors plus the surface (S') of the enclosing sphere, that is

$$\oint_S \psi \nabla \psi \cdot d\mathbf{S} = \oint_{\Sigma S_i} \psi \nabla \psi \cdot d\mathbf{S} + \oint_{S'} \psi \nabla \psi \cdot d\mathbf{S}$$

Because of equation 3–19 the integral over the conductor surfaces must be zero. Now let the sphere S' recede to infinity, then because the charged conductors are confined to a finite region the potential function $V' - V''$ must approach zero as r^{-1}; hence grad ψ will be of the order of r^{-2} and since dS is of the order of r^2 it follows that the integral over S' goes to zero as $r \to \infty$. Equation 3–20 reduces to

$$\int_v (\nabla \psi)^2 \, dv = 0$$

Since the integrand is always positive it follows that

$$(\nabla \psi)^2 = [\nabla(V' - V'')]^2 = 0$$

and

$$\nabla(V' - V'') = 0$$

everywhere so that $V' - V''$ cannot be a function of any coordinates and

$$V' - V'' = \text{constant}$$

Since V' must be equal to V'' on all the conductor boundaries the constant must be zero so that $V' = V''$ everywhere, giving rise to identical solutions. Hence both the potential and the field distribution are unique. It follows that the surface charge densities on the conductor surfaces are uniquely distributed. It is left as an exercise to the reader to show that if the charges $Q_1, Q_2 \ldots Q_N$ on the N conductors are given then again the potential and the charge distribution are uniquely determined.

It should be pointed out that the uniqueness theorem also applies to Poisson's equation $\nabla^2 V = -\rho/\epsilon_0$. With V' and V'' as solutions we still have $\nabla^2(V' - V'') = 0$ and the boundary conditions still require that $V' - V'' = 0$.

3.6.2 GREEN'S RECIPROCATION THEOREM AND THE PRINCIPLE OF SUPERPOSITION

Green's reciprocation theorem states that if charges $Q_1, Q_2 \ldots Q_N$ on conductors $1, 2 \ldots N$ raise them to potentials $V_1, V_2 \ldots V_N$, and charges $Q_1', Q_2' \ldots Q_N'$ raise them to potentials $V_1', V_2' \ldots V_N'$ then,

$$\sum_{i=1}^{N} V_i Q_i' = \sum_{i=1}^{N} V_i' Q_i \qquad (3\text{--}21)$$

Let S' be a spherical surface enclosing all the conductors (Fig. 3.9) and let the potentials at any point within the volume v (bounded by S' and the surfaces ΣS_i of the conductors) be V and V' corresponding to the unprimed and primed equilibrium states respectively. In Green's second identity

$$\oint_{S'+\Sigma S_i} (\psi \nabla \phi - \phi \nabla \psi) \cdot \mathbf{dS} = \int_v (\psi \nabla^2 \phi - \phi \nabla^2 \psi)\, dv \qquad (3\text{--}22)$$

let $\psi = V$ and $\phi = V'$. Since both V and V' are potential functions and satisfy Laplace's equation, the right-hand side of the above equation is zero. Also, by arguments similar to those given in the previous section, the integral over the surface S' of the sphere tends to zero as its radius tends to infinity (to include all space). Hence equation 3–22 may be written as

$$\sum_{i=1}^{N} \oint_{S_i} V_i \nabla V_i' \cdot \mathbf{dS}_i = \sum_{i=1}^{N} \oint_{S_i} V_i' \nabla V_i \cdot \mathbf{dS}_i \qquad (3\text{--}23)$$

where the values V_i, V_i', grad V_i and grad V_i' are all to be evaluated at the surface S_i of the ith conductor. Now

$$\nabla V_i' = -\mathbf{E}_i' = -\frac{\sigma_i'}{\epsilon_0} \mathbf{a}_{ni}$$

and

$$\nabla V_i = -\mathbf{E}_i = -\frac{\sigma_i}{\epsilon_0} \mathbf{a}_{ni}$$

Also since V_i and V_i' are constant over the conductor surfaces, they may be brought out of the integration sign. Equation 3–23 becomes

$$\sum_{1}^{N} V_i \oint_{S_i} \sigma_i'\, dS_i = \sum_{1}^{N} V_i' \oint_{S_i} \sigma_i\, dS_i$$

or

$$\sum_{1}^{N} V_i Q_i' = \sum_{1}^{N} V_i' Q_i \qquad (3\text{-}21)$$

Adding $\Sigma\, V_i Q_i$ to both sides of the above equation we obtain

$$\sum V_i (Q_i + Q_i') = \sum (V_i + V_i') Q_i \qquad (3\text{-}24)$$

By comparing equations 3–21 and 3–24 we may deduce the following theorem: If charges $Q_1, Q_2, \ldots Q_N$ on conductors $1, 2, \ldots N$ produce potentials $V_1, V_2, \ldots V_N$ and charges $Q_1', Q_2', \ldots Q_N'$ produce potentials $V_1', V_2', \ldots V_N'$, then charges $(Q_1 + Q_1')$, $(Q_2 + Q_2')$, $\ldots (Q_N + Q_N')$ produce potentials $(V_1 + V_1')$, $(V_2 + V_2')$, $\ldots (V_N + V_N')$. This is one proof of the principle of superposition.

3.6.3 INDUCED CHARGES ON EARTHED CONDUCTORS

Suppose that in the neighbourhood of a point charge Q_P we have a number N of earthed conductors. Then each one of these conductors will carry an induced charge whose sign is opposite to that of Q_P and whose magnitude is some fraction of Q_P. To find the charge induced on any one of these conductors, conductor 1 for example, we make use of Green's reciprocation theorem. Consider the following two sets of equilibrium conditions:

I. Conductor 1 is raised to some potential V_1, all other conductors are earthed, charge at P is absent.

    charges $Q_P = 0, Q_1, Q_2, \ldots Q_N$

    potentials $V_{P1},\quad V_1, 0, \ldots 0$

II. Conductor 1 is earthed, all other conductors earthed, charge at P is Q.

    charges $Q_P, Q_1', Q_2', \ldots Q_N'$

    potentials $V_P', 0, \quad 0, \ldots 0$

Hence

$$V_{P1} Q_P + V_1 Q_1' = 0$$

and

$$Q_1' = -\frac{Q_P V_{P1}}{V_1} \qquad (3\text{-}25)$$

We note that although V_P' is infinite the product $0 \cdot V_P'$ is always zero. We may therefore calculate the induced charge Q_1' if we know the potential V_{P1} produced at the point P (in the absence of any charge there) when conductor 1 is raised to some potential V_1. It is clear that the charge induced on the ith earthed conductor is given by

$$Q_i' = -\frac{Q_P V_{Pi}}{V_i} \qquad (3\text{-}26)$$

As a simple example consider the case of a point charge Q placed at a distance r from the centre of an earthed conducting sphere of radius a.

We have that

$$V_P = \frac{Q}{4\pi\epsilon_0 r} \text{ and } V = \frac{Q}{4\pi\epsilon_0 a}$$

and

$$Q' = -\frac{Qa}{r} \tag{3-27}$$

A special case of importance is that of a point charge lying between two earthed conductors one of which encloses the other (Fig. 3.10).

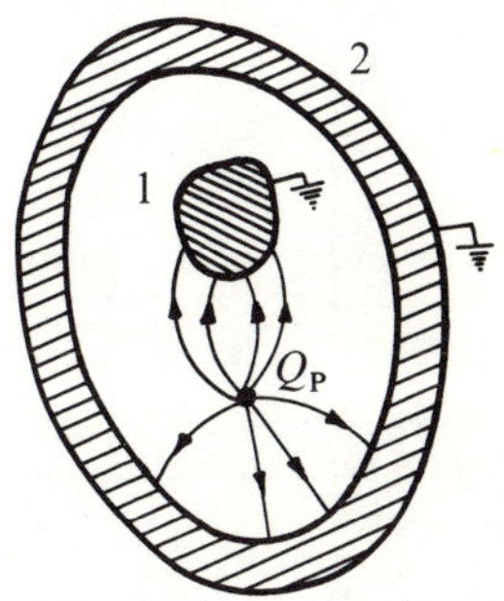

Fig. 3.10 Point charge Q_P between two earthed conductors one of which encloses the other

For such a system consider the following two sets of equilibrium states:

I. Conductors 1 and 2 at potentials V_1 and V_2 respectively, charge at P is absent.

    charges $Q_P = 0, Q_1, Q_2$
    potentials $V_P,\quad\quad V_1, V_2$

II. Conductors 1 and 2 earthed, charge at P is Q_P.

    charges Q_P, Q_1', Q_2'
    potentials $V_P', 0, \quad 0$

By Green's theorem we have

$$Q_P V_P + Q_1' V_1 + Q_2' V_2 = 0 \tag{3-28}$$

Also since all lines of force leaving Q_P terminate either on conductor 1 or conductor 2 we have

$$Q_1' + Q_2' = -Q_P \tag{3-29}$$

Solving equations 3–28 and 3–29 for Q_1' and Q_2' we obtain

$$Q_1' = -\frac{V_P - V_2}{V_1 - V_2} Q_P \qquad (3\text{–}30)$$

$$Q_2' = \frac{V_P - V_1}{V_1 - V_2} Q_P \qquad (3\text{–}31)$$

Example 1. Point charge Q between two earthed infinite plates

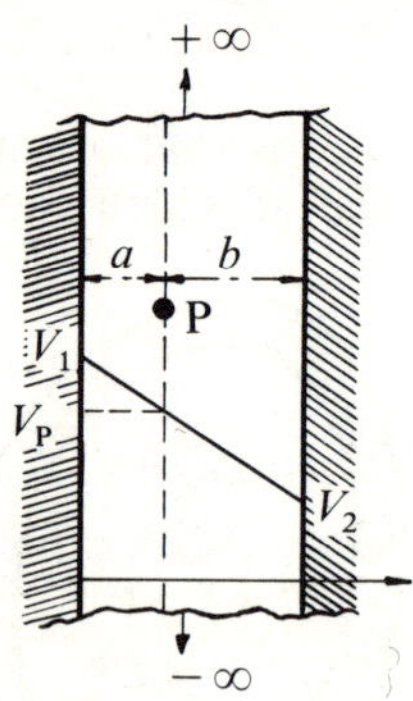

Fig. 3.11 Point P between two infinite parallel plates

(Fig. 3.11). If the potentials of the two plates are V_1 and V_2 $(V_1 > V_2)$, the potential at a point P at a distance a from the first plate and b from the second is given by

$$V_P = -\frac{(V_1 - V_2)}{a+b} a + V_1$$

Hence

$$V_P - V_1 = -\frac{(V_1 - V_2)}{a+b} a$$

$$V_P - V_2 = -\frac{(V_1 - V_2)}{a+b} a + (V_1 - V_2)$$

and

$$Q_1' = -\frac{b}{a+b} Q ; \ Q_2' = -\frac{a}{a+b} Q$$

Example 2. Point charge between two earthed spheres of radii a_1 and a_2 (Fig. 3.12). Let the potential of the inner sphere be V_1 and that of

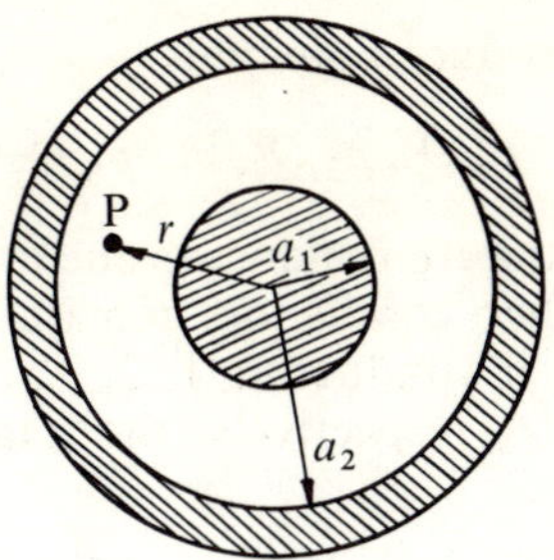

Fig. 3.12 Point P between two concentric spheres

the outer sphere be V_2 ($V_1 > V_2$). We have that

$$V_P - V_1 = V_{P1} = \frac{Q_1}{4\pi\epsilon_0}\left(\frac{a_1 - r}{a_1 r}\right)$$

$$V_P - V_2 = V_{P2} = \frac{Q_1}{4\pi\epsilon_0}\left(\frac{a_2 - r}{a_2 r}\right)$$

$$V_1 - V_2 = V_{12} = \frac{Q_1}{4\pi\epsilon_0}\left(\frac{a_2 - a_1}{a_1 a_2}\right)$$

Hence it follows from equations 3–30 and 3–31 that

$$Q_1' = -\frac{a_1}{r}\frac{(a_2 - r)}{(a_2 - a_1)} Q$$

$$Q_2' = -\frac{a_2}{r}\frac{(r - a_1)}{(a_2 - a_1)} Q$$

3.6.4 POTENTIAL AND CAPACITANCE COEFFICIENTS

Suppose that we have a system of N conductors 1, 2, 3, . . . N, which are fixed in position, isolated from each other and initially uncharged. If we place a unit charge on conductor 1 and leave all the others uncharged, the respective potentials of the N conductors will be, say,

$$p_{11}, p_{21}, p_{31}, \cdots p_{N1}$$

If instead of a unit charge we place a charge Q_1 on the first conductor it is clear that the potentials of the N conductors will now be

$$p_{11}Q_1, p_{21}Q_1, p_{31}Q_1, \cdots p_{N1}Q_1$$

A unit charge placed on the jth conductor will produce potentials

$$p_{1j}, p_{2j}, \cdots p_{Nj}$$

and a charge Q_j will produce potentials

$$p_{1j}Q_j, p_{2j}Q_j, \ldots p_{Nj}Q_j$$

Thus the potential of the ith conductor due to a charge Q_j on the jth conductor, all the other conductors remaining uncharged may be written as $p_{ij}Q_j$. Now if conductors 1, 2, ... N carry simultaneous charges Q_1, Q_2, ... Q_N respectively, the total potential of any one

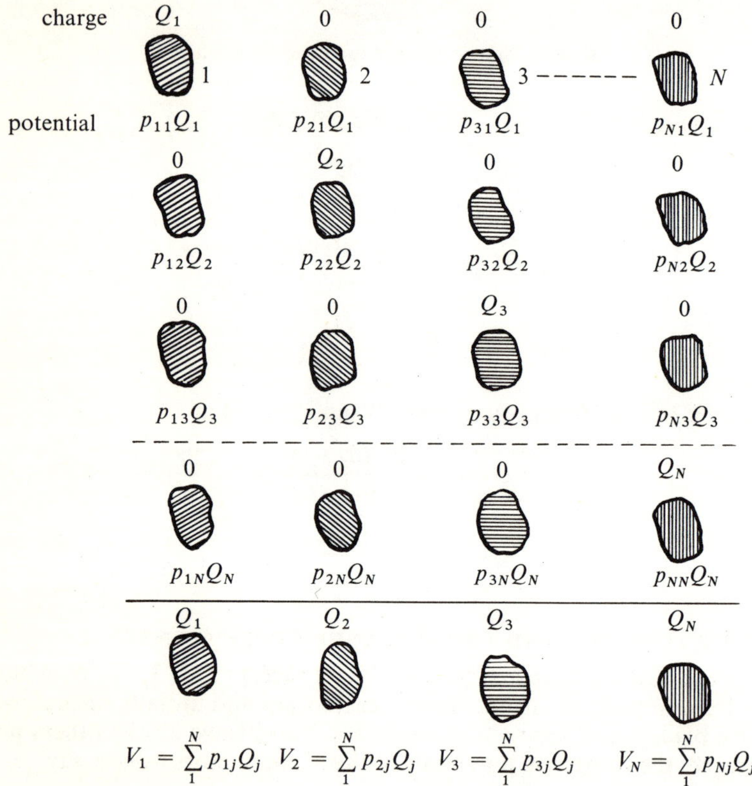

$$V_1 = \sum_1^N p_{1j}Q_j \quad V_2 = \sum_1^N p_{2j}Q_j \quad V_3 = \sum_1^N p_{3j}Q_j \quad V_N = \sum_1^N p_{Nj}Q_j$$

Fig. 3.13 Superposition of potentials for a system of N conductors

conductor of the system can be found by the direct application of the principle of superposition (see Fig. 3.13). The resultant potential of the ith conductor is thus given by

$$V_i = \sum_{j=1}^N p_{ij}Q_j, \quad (i = 1, 2, \ldots N) \tag{3-32}$$

which may be expressed in expanded form as

$$V_1 = p_{11}Q_1 + p_{12}Q_2 + p_{13}Q_3 + \cdots p_{1N}Q_N$$
$$V_2 = p_{21}Q_1 + p_{22}Q_2 + p_{23}Q_3 + \cdots p_{2N}Q_N \qquad (3\text{–}32a)$$
$$\cdots\cdots\cdots\cdots\cdots\cdots\cdots\cdots\cdots\cdots\cdots$$
$$V_N = p_{N1}Q_1 + p_{N2}Q_2 + p_{N3}Q_3 + \cdots p_{NN}Q_N$$

This set of N linear simultaneous equations gives the potentials of the conductors in terms of their charges. The coefficients p_{ij} are known as Maxwell's potential coefficients; p_{ii} are the self-potential coefficients and p_{ij} are the mutual potential coefficients. Both coefficients depend solely on the geometrical configuration of the system of conductors.

We will now use Green's reciprocation theorem to show that the coefficients p_{ij} are symmetrical, that is $p_{ij} = p_{ji}$. Consider the two sets of equilibrium states,

$$\text{I.} \qquad Q_1 = 1, Q_2, Q_3, \ldots Q_N = 0$$

$$V_1 = p_{11}, V_2 = p_{21}, \ldots V_N = p_{N1}$$

$$\text{II.} \qquad Q_1' = 0, Q_2' = 1, Q_3', Q_4', \ldots Q_N' = 0$$

$$V_1' = p_{12}, V_2' = p_{22}, \ldots V_N' = p_{N2}$$

Applying Green's theorem to these two states we obtain that

$$p_{12} = p_{21}$$

It is obvious that this equality can be extended to all other coefficients so that in general

$$p_{ij} = p_{ji} \qquad (3\text{–}33)$$

This means that the potential to which the ith conductor is raised when a unit charge (or a charge Q) is placed on the jth conductor, all other conductors remaining uncharged, is the same as the potential to which the jth conductor is raised when a unit charge (or a charge Q) is placed on the ith conductor, all other conductors remaining uncharged.

Consider the special case of two conductors in which one of them is reduced to a point P. We obtain the important result that the potential to which the uncharged conductor is raised by placing a unit charge at P is the same as the potential produced at P by a unit charge on the conductor. As an example, the potential at any point at a distance r from the centre of a charged conducting sphere of radius a ($r > a$) is $Q/4\pi\epsilon_0 r$. Hence a point charge Q placed at the same distance from the centre will raise the potential of the uncharged sphere to $Q/4\pi\epsilon_0 r$.

We shall now derive the potential coefficients for the system of N conducting spheres discussed in Section 3.3. Changing the variable suffix j to S we rewrite equation 3–10 as

$$V_i = \frac{1}{Ka_i} \sum_{S=1}^{i-1} Q_S + \frac{1}{K} \sum_{S=i}^{N} \frac{Q_S + q_S}{a_S'} - \frac{1}{K} \sum_{S=i}^{N} \frac{q_S}{a_S} \qquad (3\text{–}34)$$

and

$$V_j = \frac{1}{Ka_j} \sum_{S=1}^{j-1} Q_S + \frac{1}{K} \sum_{S=j}^{N} \frac{Q_S + q_S}{a_S'} - \frac{1}{K} \sum_{S=j}^{N} \frac{q_S}{a_S} \qquad (3\text{–}35)$$

Now if we assume the following conditions,

$$Q_S = 0,\ (S \neq i);\ Q_S = 1,\ (S = i);\ q_S = 1,\ (S \geq i+1);$$

$$q_S = 0,\ (S \leq i);\ j > i$$

equations 3–34 and 3–35 become

$$V_i = p_{ii} = \frac{1}{Ka_i'} + \frac{1}{K} \sum_{i+1}^{N} \left(\frac{1}{a_S'} - \frac{1}{a_S} \right) \qquad (3\text{–}36)$$

$$V_j = p_{ij} = \frac{1}{Ka_j} + \frac{1}{K} \sum_{j}^{N} \left(\frac{1}{a_S'} - \frac{1}{a_S} \right) \qquad (3\text{–}37)$$

For two concentric spheres the potential coefficients are

$$p_{11} = \frac{1}{K} \left(\frac{1}{a_1'} + \frac{1}{a_2'} - \frac{1}{a_2} \right) \qquad (3\text{–}38)$$

$$p_{22} = \frac{1}{Ka_2'} \qquad (3\text{–}39)$$

$$p_{12} = \frac{1}{Ka_2'} \qquad (3\text{–}40)$$

This is in agreement with equations 3–4 and 3–5 if we compare them with equation 3–32a.

If the potentials V_1, V_2, ... V_N are known, we may obtain the value of the charges Q_1, Q_2, ... Q_N in terms of these potentials by solving the set of N linear simultaneous equations 3–32a for the charges.

We thus obtain a new set of N linear equations of the form

$$Q_i = \sum_{j=1}^{N} C_{ij} V_j, \quad (i = 1, 2, \ldots N) \qquad (3\text{–}41)$$

which may be expanded to give

$$\begin{aligned}
Q_1 &= C_{11}V_1 + C_{12}V_2 + C_{13}V_3 + \cdots C_{1N}V_N \\
Q_2 &= C_{21}V_1 + C_{22}V_2 + C_{23}V_3 + \cdots C_{2N}V_N \\
&\cdots\cdots\cdots\cdots\cdots\cdots\cdots\cdots\cdots\cdots\cdots\cdots \\
Q_N &= C_{N1}V_1 + C_{N2}V_2 + C_{N3}V_3 + \cdots C_{NN}V_N
\end{aligned} \qquad (3\text{–}41a)$$

The new coefficients C_{ij} are related to the potential coefficients p_{ij} by the relation

$$C_{ij} = \frac{(-1)^{i+j} M_{ij}}{\Delta} \tag{3-42}$$

where Δ is the determinant $|p_{ij}|$ of the set of equations 3–32a.

$$\Delta = \begin{vmatrix} p_{11} & p_{12} & p_{13} & \cdots & p_{1N} \\ p_{21} & p_{22} & p_{23} & \cdots & p_{2N} \\ p_{31} & p_{32} & p_{33} & \cdots & p_{3N} \\ \cdots & \cdots & \cdots & \cdots & \cdots \\ \cdots & \cdots & \cdots & \cdots & \cdots \\ p_{N1} & p_{N2} & p_{N3} & \cdots & p_{NN} \end{vmatrix} \tag{3-43}$$

and M_{ij} is the cofactor of p_{ij} in the determinant Δ and is obtained from it by eliminating the *i*th row and the *j*th column. For example

$$C_{12} = -\frac{1}{\Delta} \begin{vmatrix} p_{21} & p_{23} & \cdots & p_{2N} \\ p_{31} & p_{33} & \cdots & p_{3N} \\ \cdots & \cdots & \cdots & \cdots \\ \cdots & \cdots & \cdots & \cdots \\ p_{N1} & p_{N3} & \cdots & p_{NN} \end{vmatrix}$$

$$C_{13} = \frac{1}{\Delta} \begin{vmatrix} p_{21} & p_{22} & p_{24} & \cdots & p_{2N} \\ p_{31} & p_{32} & p_{34} & \cdots & p_{3N} \\ \cdots & \cdots & \cdots & \cdots & \cdots \\ \cdots & \cdots & \cdots & \cdots & \cdots \\ p_{N1} & p_{N2} & p_{N4} & \cdots & p_{NN} \end{vmatrix}$$

The reader may readily show that the coefficients C_{ij} are also symmetrical so that

$$C_{ij} = C_{ji} \tag{3-44}$$

Like the potential coefficients, the coefficients C_{ij} depend only on the geometrical configuration of the system.

The coefficients C_{jj} are known as the *coefficients of capacitance* or the *self capacitances* of the system. The coefficients C_{ij} $(i \neq j)$ are known as the *coefficients of induction* or the *partial capacitances* of the system.

That the coefficients C_{ij} do represent capacitances of some sort can be seen from the fact that the unit of every one of these coefficients must, according to equation 3–41, be the farad. If we suppose that all the conductors except the *j*th conductor are kept at zero potential (earthed), we have that

$$Q_i = C_{ij} V_j$$

or

$$C_{ij} = \frac{Q_i}{V_j}; \quad (i = 1, 2, \ldots N)$$

If $i = j$,

$$C_{jj} = \frac{Q_j}{V_j}$$

This, according to Section 3.5, is the definition of the capacitance of the jth conductor. The only difference is that in this case C_{jj} represents not the capacitance of the isolated jth conductor but its capacitance in the presence of all the other conductors with these kept at zero potential. Since the algebraic sign of the potential V_j will always be the same as that of the charge Q_j, it follows that *the coefficients C_{jj} are always positive.*

If $i \neq j$ then the coefficient of induction C_{ij} may be defined as the ratio of the charge induced on the ith conductor to the potential of the jth conductor, when all the conductors except the jth are kept at zero potential. For example, with charge Q_1 on conductor 1 and all other conductors grounded, each of the grounded conductors will carry an induced charge given by

$$Q_2' = C_{21}V_1$$
$$Q_3' = C_{31}V_1$$
$$Q_N' = C_{N1}V_1$$

Since the sign of the induced charge will always be opposite to that of the inducing charge, it follows that *the coefficients C_{ij} $(i \neq j)$ are always negative or zero.*

Suppose that in a system of N conductors, conductor j is at unit potential and all others are at zero potential. The corresponding charge on conductor j is C_{jj} and the charges induced on the remaining conductors are C_{ij} $(i \neq j)$. Now if conductor j is not completely surrounded by any other conductor of the system, then some of the lines of force leaving conductor j will terminate on the $N-1$ conductors and some will go to infinity (Fig. 3.14a). Hence not every element of the surface of conductor j will have a corresponding element on one of the other conductors of the system so that

$$C_{jj} > \sum_{i \neq j}^{N} |C_{ij}|$$

If on the other hand one of the conductors of the system, other than the jth conductor, completely surrounds all the other conductors

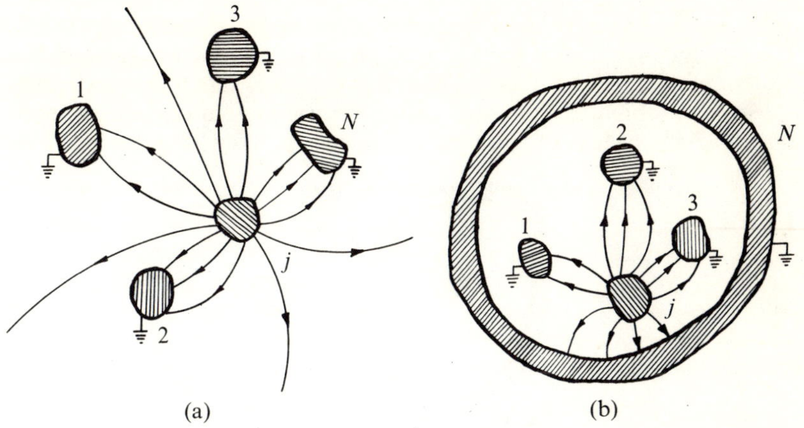

(a)   (b)

Fig. 3.14 System of N conductors with all conductors except the jth grounded. (a) open system (b) closed system

(Fig. 3.14b) then every line of force leaving conductor j will terminate on some other conductor within the system and therefore every element of surface of conductor j will have a corresponding element on one of the other conductors of the system; it follows that

$$C_{jj} = \sum_{i \neq j}^{N} |C_{ij}|$$

Hence it is possible to write then in general

$$C_{jj} \geq \sum_{i \neq j}^{N} |C_{ij}|; \quad (j = 1, 2, \ldots N) \tag{3–44a}$$

If we now admit the existence of a hypothetical conducting sphere of infinite radius whose potential is always zero and which surrounds the entire system of conductors, then we may define an additional coefficient C_{j0} which would account for the charge on any surface element of the jth conductor which does not have a corresponding element on any of the other $N-1$ conductors. In other words it accounts for any lines of force which leave the jth conductor and go to infinity. We may thus state that always

$$C_{jj} = \sum_{i \neq j}^{N} |C_{ij}| + |C_{j0}|; \quad (j = 1, 2, \ldots N) \tag{3–45}$$

or, since the suffixes i and j are interchangeable,

$$C_{ii} = \sum_{j \neq i}^{N} |C_{ij}| + |C_{i0}|; \quad (i = 1, 2, \ldots N) \tag{3–45a}$$

3.6.5 ELECTRIC SCREENING

If a conductor which is kept at zero potential completely surrounds a system of charged conductors, then this system is said to be electrically screened or shielded from any other electric system outside the conductor. Any system outside the conductor is likewise screened from

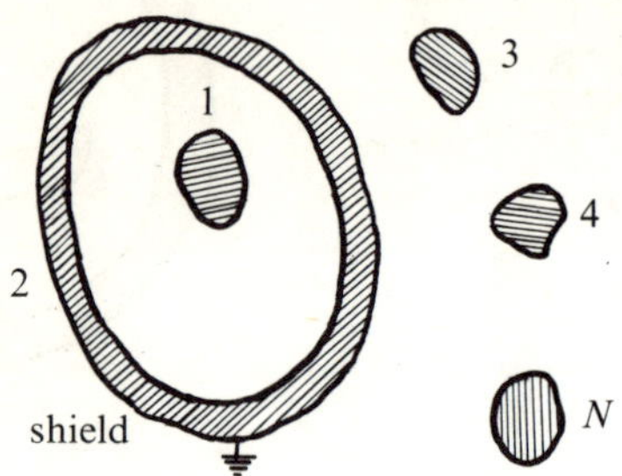

Fig. 3.15 Electrostatic shielding

any inside the conductor. Consider the system of conductors shown in Fig. 3.15. We have that

$$Q_1 = C_{11}V_1 + C_{12}V_2 + C_{13}V_3 + \ldots C_{1N}V_N$$

Suppose that $Q_1 = 0$, then since $V_2 = 0$, $V_1 = 0$ also; hence

$$C_{13}V_3 + \ldots C_{1N}V_N = 0$$

This equation must be satisfied for all values of the potentials $V_3, V_4, \ldots V_N$ and this is only possible if

$$C_{13} = C_{14} = \ldots C_{1N} = 0$$

The charges on the conductors expressed in terms of the potentials now are

$$Q_1 = C_{11}V_1$$
$$Q_2 = C_{21}V_1 + C_{23}V_3 + \ldots C_{2N}V_N$$
$$Q_3 = C_{33}V_3 + \ldots C_{3N}V_N$$
$$Q_N = C_{N3}V_3 + \ldots C_{NN}V_N$$

These equations show quite clearly that the electrical conditions (charge and potential) inside and outside the shield are completely independent of one another.

In practice electrical apparatus and instruments are screened by enclosing them in an earthed metal cage, usually made of a wire grill. The finer the grill mesh the more effective the screening. Very sensitive instruments are completely enclosed in a metal case with an aperture through which readings can be taken. The effectiveness of a grill as a screen may be readily determined in the laboratory by covering an electroscope with a cage having the same mesh size as the grill and observing the deflections of the leaf when charged bodies are held near the cage.

3.6.6 CAPACITORS

Two conductors are said to form a capacitor or a condenser if they carry equal but opposite charges. This means that the whole surface of one conductor may be made to correspond, element by element, to the whole surface of the other conductor so that all lines of force leaving one conductor must necessarily terminate at the surface of the other conductor. The two conductors are known as the plates of the capacitor.

The capacitance of a capacitor is defined as the ratio of the magnitude of the charge on one of the conductors to the potential difference between the conductors

$$C = \frac{Q}{V_1 - V_2} \tag{3-46}$$

If one of the conductors is removed to infinity its potential, V_2 say, becomes zero and C would then be the absolute capacitance of conductor 1, in accordance with equation 3-14.

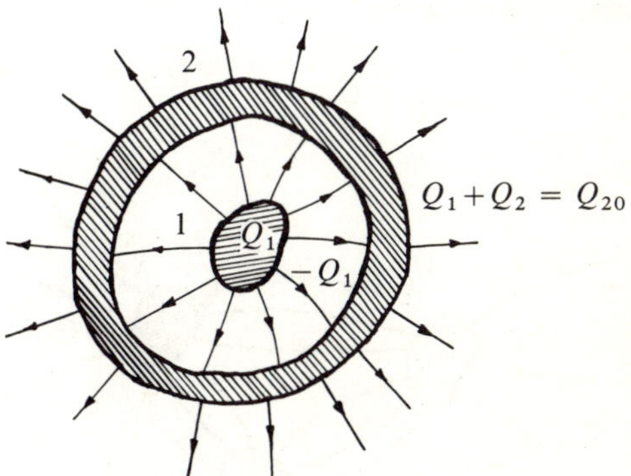

Fig. 3.16 A two-conductor capacitor

Consider two conductors 1 and 2 and let conductor 2 completely surround conductor 1 (Fig. 3.16). Let the conductor potentials be V_1 and V_2 and the corresponding charges Q_1 and Q_2 respectively. We know that the charge on the internal surface of conductor 2 must be equal to $-Q_1$; the capacitance of the condenser formed by these two

conductors is given by equation 3–46. Let us now apply equation 3–14a
to this two-conductor system; we obtain

$$Q_1 = C_{11}V_1 + C_{12}V_2$$
$$Q_2 = C_{21}V_1 + C_{22}V_2$$

From the basic definitions of C_{jj} and C_{ij} it is readily seen that

$$C_{11} = |C_{12}| = |C_{21}| = C$$
$$C_{12} = C_{21} = -C$$

Hence

$$Q_1 = C(V_1 - V_2) \tag{3-47}$$
$$Q_2 = -CV_1 + C_{22}V_2 \tag{3-48}$$

Equation 3–48, which expresses the total charge on conductor 2, may
be written as

$$Q_2 = -C(V_1 - V_2) + (C_{22} - C)V_2 \tag{3-49}$$
$$= \quad Q_{21} \quad + \quad Q_{20}$$

The first term on the right represents the charge $Q_{21} = -Q_1$ induced
on the internal surface of conductor 2, whilst the second term
$Q_{20} = Q_2 + Q_1$ represents the charge on the external surface of conductor 2. Since all lines of force leaving Q_{20} terminate on a charge
$-Q_{20}$ at infinity, we may write

$$C_{20} = C_{22} - C = \frac{Q_{20}}{V_2}$$

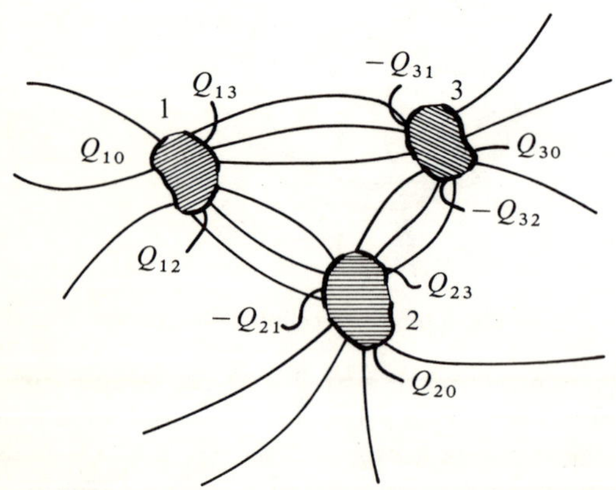

Fig. 3.17 Charge distribution in a system of three conductors at potentials V_1, V_2 and V_3

which is the capacitance of conductor 2 with respect to infinity.

Consider now a system of three conductors in which none of the conductors encloses any of the others (Fig. 3.17). For conductor 1 we have

$$Q_1 = C_{11}V_1 + C_{12}V_2 + C_{13}V_3$$

which may be rewritten as

$$Q_1 = -C_{12}(V_1 - V_2) - C_{13}(V_1 - V_3) + (C_{11} + C_{12} + C_{13})V_1$$

and since the coefficients C_{12} and C_{13} are negative,

$$Q_1 = |C_{12}|(V_1 - V_2) + |C_{13}|(V_1 - V_3) + (C_{11} - |C_{12}| - |C_{13}|)V_1$$
$$ = \qquad Q_{12} \qquad + \qquad Q_{13} \qquad + \qquad Q_{10} \tag{3-50}$$

Similarly

$$Q_2 = |C_{21}|(V_2 - V_1) + |C_{23}|(V_2 - V_3) + (C_{22} - |C_{21}| - |C_{23}|)V_2$$
$$ = \qquad Q_{21} \qquad + \qquad Q_{23} \qquad + \qquad Q_{20} \tag{3-51}$$

$$Q_3 = |C_{31}|(V_3 - V_1) + |C_{32}|(V_3 - V_2) + (C_{33} - |C_{31}| - |C_{32}|)V_3$$
$$ = \qquad Q_{31} \qquad + \qquad Q_{32} \qquad + \qquad Q_{30} \tag{3-52}$$

where

$Q_{12} = -Q_{21} =$ charge on corresponding elements of conductors 1 and 2

$Q_{13} = -Q_{31} =$ charge on corresponding elements of conductors 1 and 3

$Q_{23} = -Q_{32} =$ charge on corresponding elements of conductors 2 and 3

$Q_{10} = (C_{11} - |C_{12}| - |C_{13}|)V_1 = C_{10}V_1 =$ charge on corresponding elements of conductor 1 and infinite sphere

$Q_{20} = (C_{22} - |C_{21}| - |C_{23}|)V_2 = C_{20}V_2 =$ charge on corresponding elements of conductor 2 and infinite sphere

$Q_{30} = (C_{33} - |C_{31}| - |C_{32}|)V_3 = C_{30}V_3 =$ charge on corresponding elements of conductor 3 and infinite sphere

Every two corresponding elements form a capacitor whose capacitance is the ratio of the magnitude of the charge on one of the elements to the potential difference between them; thus

$$|C_{12}| = \frac{Q_{12}}{V_1 - V_2}; \quad |C_{13}| = \frac{Q_{13}}{V_1 - V_3}; \text{ etc.}$$

The above argument is readily extended to a system of N conductors, the charge on the ith conductor may be expressed as

$$Q_i = \sum_{j \neq i}^{N} |C_{ij}|(V_i - V_j) + \left(C_{ii} - \sum_{j \neq i}^{N} |C_{ij}|\right)V_i$$

$$= \sum |C_{ij}|(V_i - V_j) + C_{i0}V_i; \quad (i = 1, 2, \ldots N) \qquad (3\text{–}53)$$

It is therefore now possible to give a physical representation of equation 3–41a by means of a system of capacitors whose capacitances correspond to the coefficients C_{ij} and C_{i0}. The coefficient C_{ij} is the *partial* capacitance formed by two corresponding surface elements carrying equal but opposite charges; one element is on the ith conductor and the other is on the jth conductor. In equation 3–53

$$C_{i0} = C_{ii} - \sum_{j \neq i}^{N} |C_{ij}|$$

or

$$C_{ii} = \sum_{j \neq i}^{N} |C_{ij}| + C_{i0}$$

which is the same as equation 3–45a. The partial capacitance C_{i0} is always associated with those charges on the ith conductor for which the lines of force go to infinity; C_{i0} is known as the *stray capacitance* of the ith conductor. Figure 3.18 shows the partial capacitances for

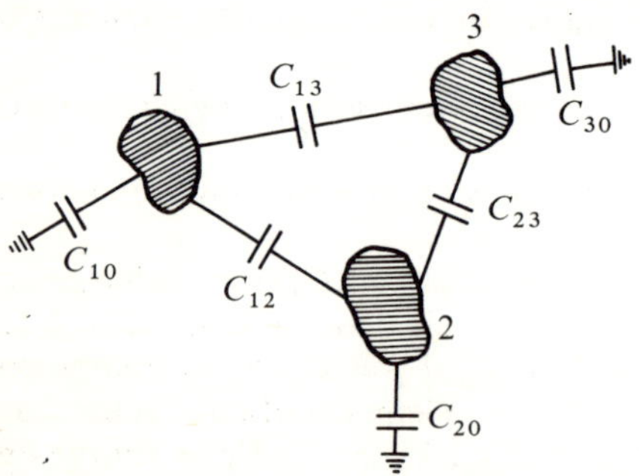

Fig. 3.18 Partial capacitances for a system of three conductors

a system of three conductors; the self capacitance C_{11} of conductor 1 is given by

$$C_{11} = |C_{12}| + |C_{13}| + C_{10}$$

since when conductors 2 and 3 are earthed the capacitors C_{10}, C_{12} and C_{13} are all in parallel whilst capacitors C_{23}, C_{20} and C_{30} are shorted.

3.7 CALCULATION OF THE CAPACITANCE OF SOME IMPORTANT TWO-CONDUCTOR SYSTEMS*

3.7.1 PARALLEL-PLATE CAPACITOR

Consider two infinitely large plates separated by a distance d and whose potentials are V_1 and V_2 ($V_1 > V_2$). The plane surfaces facing one another must necessarily carry equal and opposite charges

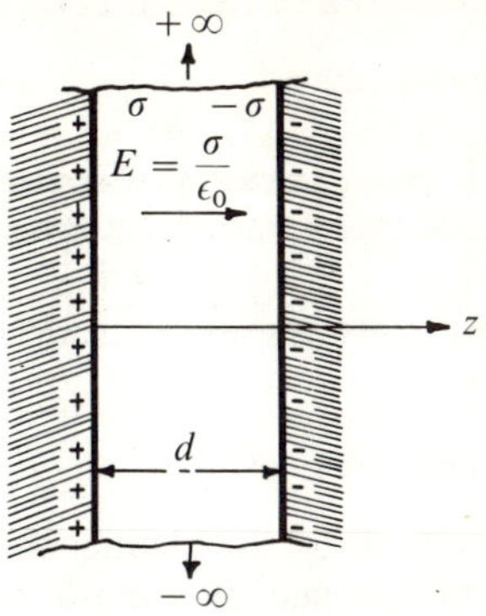

Fig. 3.19 Parallel-plate capacitor

(Fig. 3.19). If $\pm\sigma$ are the surface charge densities on these surfaces, we have by equation 2–58

$$\mathbf{E} = \frac{\sigma}{\epsilon_0}\,\mathbf{a}_z$$

$$V_{12} = V_1 - V_2 = -\int_d^0 \frac{\sigma dz}{\epsilon_0} = \frac{\sigma d}{\epsilon_0}$$

Hence the capacitance per unit area is

$$C' = \frac{\sigma}{V_1 - V_2} = \frac{\epsilon_0}{d}$$

and for an area S of the plates

$$C = \frac{\epsilon_0 S}{d} \quad \text{farad} \tag{3–54}$$

In practice, parallel-plate capacitors have plates of finite size and to obtain a true value of the capacitance the fringing of the field at the plate edges would have to be taken into account. However, if the

* See Section 4.4, for the effect on capacitance of a dielectric medium between capacitor armatures.

separation d is small compared with the smallest linear dimensions of the plates equation 3–54 can be used with negligible error. The field between two finite parallel plates may be kept uniform by using a *guard ring*. This is an additional electrode which surrounds one of the plates and is isolated from it by a narrow gap whose width is much smaller than the plates' separation d; the plate and its guard are maintained at the same potential. By this means the field fringing will be outside the effective area of the condenser plates (Fig. 3.20).

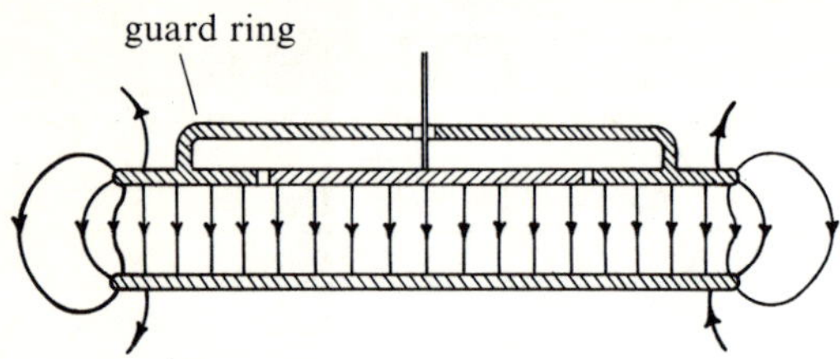

Fig. 3.20 Parallel-plate capacitor with guard ring

We can use equation 3–54 to define the electric constant ϵ_0 as equivalent to the capacitance of a condenser whose plates have an area of $1\,\text{m}^2$ and are separated by $1\,\text{m}$; it is now apparent that the units of ϵ_0 are $\text{F}\,\text{m}^{-1}$ (cf. Chapter 1).

3.7.2 SPHERICAL CAPACITOR

Let the potentials and the charges of the two concentric spherical conductors (Fig. 3.21) be V_1, V_2 and Q_1, Q_2 respectively. The capacitance between the inner sphere and the inner surface of the outer sphere is, making use of equation 3–3,

$$C_{12} = \frac{Q_1}{V_{12}} = \frac{4\pi\epsilon_0 a_1 a_2}{a_2 - a_1} \tag{3–55}$$

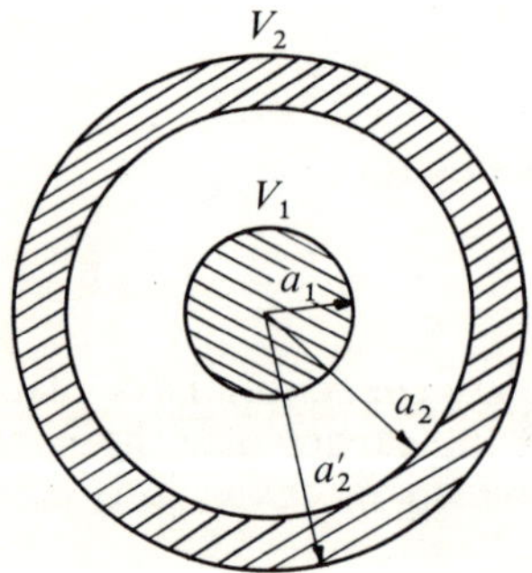

Fig. 3.21 Spherical capacitor

The absolute capacitance of the outer sphere is

$$C_{20} = \frac{Q_2+Q_1}{V_2} = 4\pi\epsilon_0 a_2' \qquad (3\text{-}56)$$

If the inner sphere is earthed the total capacitance of the two conductor system is

$$C = C_{12}+C_{20} = 4\pi\epsilon_0 \left(\frac{a_1 a_2}{a_2-a_1}+a_2'\right) \qquad (3\text{-}57)$$

If the outer sphere is earthed the capacitance C_{20} is effectively shorted and $C = C_{12}$.

We will now solve this problem using potential and capacitance coefficients. The potentials of the conductors are given by

$$V_1 = p_{11}Q_1+p_{12}Q_2$$
$$V_2 = p_{21}Q_1+p_{22}Q_2$$

In Section 3.6.4 we found that the potential coefficients for such a system are

$$p_{11} = \frac{1}{K}\left[\frac{1}{a_1}+\frac{1}{a_2'}-\frac{1}{a_2}\right]; p_{12} = p_{22} = \frac{1}{Ka_2'}$$

where $K = 4\pi\epsilon_0$. To find the capacitance coefficients we use equation 3–42. We have

$$\Delta = \frac{1}{K^2}\left(\frac{a_2-a_1}{a_1 a_2 a_2'}\right)$$

$$C_{11} = \frac{1}{\Delta}p_{22} = \frac{Ka_1 a_2}{a_2-a_1}$$

$$C_{12} = -\frac{1}{\Delta}p_{21} = -\frac{Ka_1 a_2}{a_2-a_1}$$

$$C_{22} = \frac{1}{\Delta}p_{11} = \frac{Ka_1 a_2}{a_2-a_1}+Ka_2'$$

and from equation 3–45 we have

$$C_{20} = C_{22}-|C_{12}| = Ka_2'$$

If we go back to the basic definitions for the capacitance coefficients we see that their above values are in agreement with equations 3–55/57. If the potentials V_1 and V_2 are known then the charges will be given by

$$Q_1 = C_{11}V_1 + C_{12}V_2$$
$$Q_2 = C_{21}V_1 + C_{22}V_2$$

Substituting the above values of the capacitance coefficients we find that

$$Q_1 = \frac{Ka_1a_2}{a_2-a_1}(V_1-V_2) \qquad (3\text{--}58)$$

$$Q_2 = -\frac{Ka_1a_2}{a_2-a_1}(V_1-V_2)+Ka_2'V_2 \qquad (3\text{--}59)$$

3.7.3 COAXIAL CYLINDRICAL CAPACITOR

Consider two infinitely long coaxial conductors, the inner conductor with radius a_1 and the outer conductor with inner radius a_2 (Fig. 3.22).

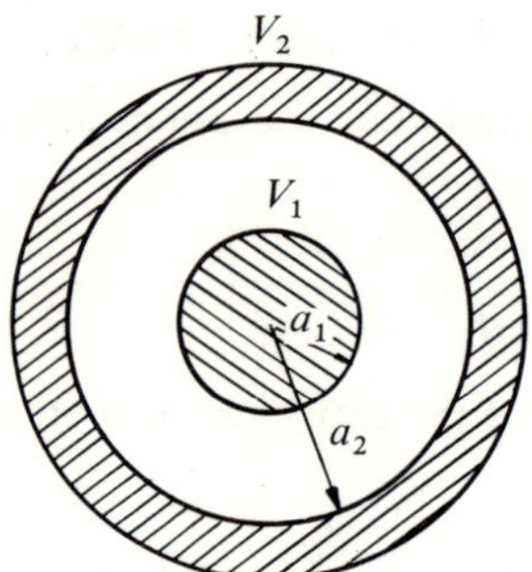

Fig. 3.22 Cross-section through cylindrical capacitor

Let their potentials be V_1 and V_2 respectively and let the charge per unit length on the surface of the inner conductor be Q_1 C m^{-1}. Gauss's law tells us that the field at any point between the two cylinders is radial and is the same as that of an infinite line charge of density Q placed at the centre. The potential difference between the conductors is given by

$$V_1 - V_2 = \frac{Q_1}{2\pi\epsilon_0}\log\frac{a_2}{a_1} \qquad (3\text{--}60)$$

and the capacitance per metre length is

$$C = \frac{Q_1}{V_1-V_2} = \frac{2\pi\epsilon_0}{\log(a_2/a_1)} \text{ F m}^{-1} \qquad (3\text{--}61)$$

For conductors of finite length equation 3–61 is only approximate since there will be fringing of the field at the cylinder ends. The error is negligible whenever the distance between the conductors is small compared with the length of the conductors. As in the case of the parallel plate capacitor one can always eliminate the fringing effect by using guard electrodes (Fig. 3.23).

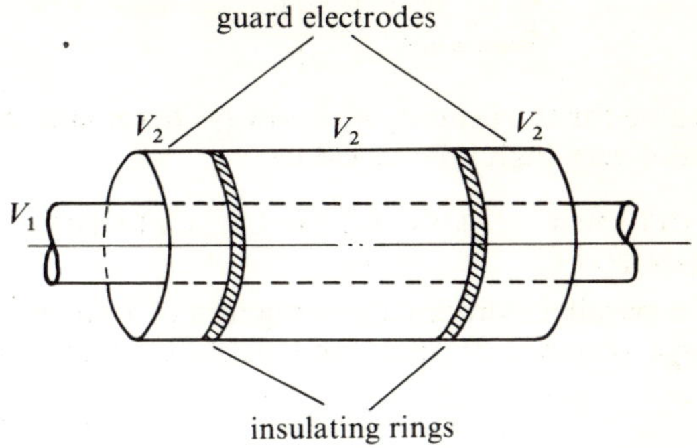

Fig. 3.23 Cylindrical capacitor with guard electrodes

The magnitude of the field intensity at any radius $r < a_2$ is given by

$$E = Q_1/2\pi\epsilon_0 r$$

Substituting for $Q_1/2\pi\epsilon_0$ from equation 3–60 we obtain

$$E = \frac{V_1 - V_2}{r \log (a_2/a_1)}$$

This stress has a maximum value at the surface of the inner conductor

$$E_{\text{max}} = \frac{V_1 - V_2}{a_1 \log (a_2/a_1)}$$

$$= \frac{V_1 - V_2}{a_2} \frac{\mu}{\log \mu} \tag{3–62}$$

where $\mu = a_2/a_1$. For a given value of a_2 we may find the optimum value of μ for which the expression for E_{max} is a minimum. E_{max} has a minimum value when

$$\frac{\mathrm{d} E_{\text{max}}}{\mathrm{d} \mu} = 0 = \frac{\mathrm{d}}{\mathrm{d} \mu} \left[\frac{1}{\mu} \log \mu \right]^{-1}$$

Hence

$$\left[-\frac{1}{\mu^2} \log \mu + \frac{1}{\mu^2} \right] \left[\frac{1}{\mu} \log \mu \right]^{-2} = 0$$

so that

$$\mu = \frac{a_2}{a_1} = e = 2 \cdot 718 \tag{3-63}$$

and

$$(E_{\max})_{\min} = \frac{V_1 - V_2}{a_1} \tag{3-64}$$

The above value for the ratio a_2/a_1 is always taken into consideration in the design of very high voltage cables.

3.7.4 CAPACITANCE OF TWO PARALLEL CYLINDRICAL CONDUCTORS

Consider two parallel cylindrical conductors of radii R_1 and R_2 with their centres a distance D apart and held at potentials V_1 and V_2

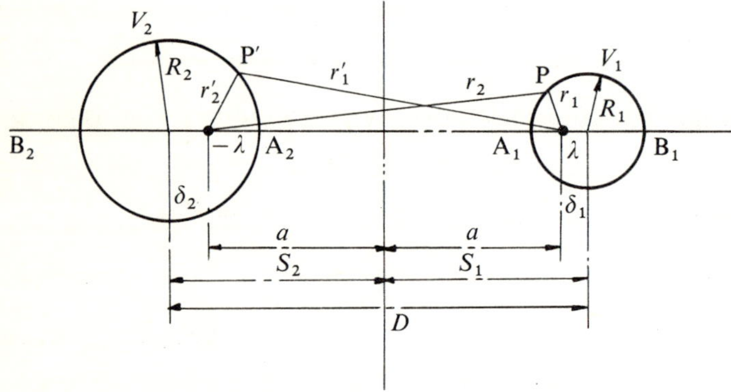

Fig. 3.24 Two parallel cylindrical conductors

respectively (Fig. 3.24). In Chapter 2 we showed that the equipotential surfaces of two infinite line charges $\pm \lambda$ at a distance $2a$ apart are a set of cylinders having their axes parallel to and collinear with the line charges. In our problem, therefore, there exist two infinite line charges such that the cylinders at V_1 and V_2 correspond to two equipotentials of these line charges. With reference to Fig. 3.24 and equation 2-67 we have that

$$V_1 = \frac{\lambda}{2\pi\epsilon_0} \log \frac{r_2}{r_1} + C$$

$$V_2 = \frac{\lambda}{2\pi\epsilon_0} \log \frac{r_2'}{r_1'} + C$$

and

$$V_1 - V_2 = \frac{\lambda}{2\pi\epsilon_0} \log \frac{r_2}{r_1} \frac{r_1'}{r_2'} \tag{3-65}$$

At points A_1 and B_1 we have* (since these are on the same equipotential)

$$\frac{r_2}{r_1} = \frac{D-R_1-\delta_2}{R_1-\delta_1} = \frac{D+R_1-\delta_2}{R_1+\delta_1} = \frac{D-\delta_2}{R_1} = \frac{R_1}{\delta_1} \tag{3-66}$$

and at points A_2 and B_2

$$\frac{r_2'}{r_1'} = \frac{R_2-\delta_2}{D-R_2-\delta_1} = \frac{R_2+\delta_2}{D+R_2-\delta_1} = \frac{R_2}{D-\delta_1} = \frac{\delta_2}{R_2} \tag{3-67}$$

From equations 3-66 and 3-67 we have

$$R_1^2 + \delta_1\delta_2 = \delta_1 D$$
$$R_2^2 + \delta_1\delta_2 = \delta_2 D$$

By multiplying these two equations together we obtain

$$\frac{R_1^2 R_2^2}{(\delta_1\delta_2)^2} - \frac{2R_1 R_2}{\delta_1\delta_2}\left(\frac{D^2-R_1^2-R_2^2}{2R_1 R_2}\right) + 1 = 0 \tag{3-68}$$

Now if we let

$$\frac{r_2}{r_1} \cdot \frac{r_1'}{r_2'} = \frac{R_1 R_2}{\delta_1\delta_2} = u \tag{3-69}$$

and

$$\frac{D^2-R_1^2-R_2^2}{2R_1 R_2} = \beta \tag{3-70}$$

Equation 3-68 becomes

$$u^2 - 2\beta u + 1 = 0$$

and its two roots are

$$u = \beta \pm \sqrt{(\beta^2-1)}$$

Hence equation 3-65 may now be written as

$$V_1 - V_2 = \frac{\lambda}{2\pi\epsilon_0}\log[\beta \pm \sqrt{(\beta^2-1)}] = \pm\frac{\lambda}{2\pi\epsilon_0}\log[\beta + \sqrt{(\beta^2-1)}]$$

$$\tag{3-71}$$

* If $\dfrac{a}{b} = \dfrac{c}{d}$, then $\dfrac{a\pm c}{b\pm d} = \dfrac{a}{b} = \dfrac{c}{d}$

The capacitance per metre between the two conductors is therefore

$$C = \left| \frac{\lambda}{V_1 - V_2} \right| = \frac{2\pi\epsilon_0}{\log\left[\beta + \sqrt{(\beta^2 - 1)}\right]}\ \mathrm{F\,m^{-1}} \qquad (3\text{-}72)$$

where β is given by equation 3-70.

In order to obtain the complete system of equipotentials we must know the value of a. From equations 2-69 and 2-70 and Fig. 3.24 we have that

$$S_1^2 - R_1^2 = S_2^2 - R_2^2 = a^2 \qquad (3\text{-}73)$$

and also that

$$S_1 + S_2 = D$$

By solving for S_1 and S_2 we obtain

$$S_1 = \frac{D^2 - (R_2^2 - R_1^2)}{2D} \qquad (3\text{-}74)$$

$$S_2 = \frac{D^2 + (R_2^2 - R_1^2)}{2D} \qquad (3\text{-}75)$$

If the two cylinders are eccentric with their centres a distance D apart as shown in Fig. 3.25, the equations for $V_1 - V_2$ and for C are the

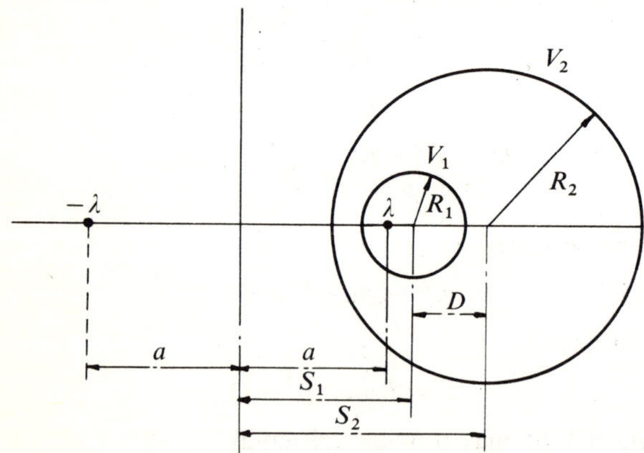

Fig. 3.25 Two eccentric cylindrical conductors

same as those given above with $-\beta$ instead of β. In this case therefore

$$\beta' = -\beta = -\left(\frac{D^2 - R_1^2 - R_2^2}{2R_1 R_2}\right) \qquad (3\text{-}76)$$

and

$$C = \frac{2\pi\epsilon_0}{\log\left[\beta' + \sqrt{(\beta'^2 - 1)}\right]} \tag{3-77}$$

The reader may readily verify that if the cylinders are concentric, that is $D = 0$, equation 3–77 reduces to

$$C = \frac{2\pi\epsilon_0}{\log(R_2/R_1)}$$

which is identical with equation 3–61.

For two eccentric cylinders the corresponding value for S_1 and S_2 are given by

$$S_1 = \frac{R_2^2 - R_1^2 - D^2}{2D} \tag{3-78}$$

$$S_2 = \frac{R_2^2 - R_1^2 + D^2}{2D} \tag{3-79}$$

We shall now consider two important special cases for two parallel cylinders
(a) Both cylinders have equal radii, that is $R_1 = R_2 = R$.
From equation 3–70,

$$\beta = \frac{D^2 - 2R^2}{2R^2} = \frac{D^2}{2R^2} - 1 \tag{3-80}$$

and

$$C = \pi\epsilon_0 \Big/ \log\left\{\frac{D}{2R} + \sqrt{\left[\left(\frac{D}{2R}\right)^2 - 1\right]}\right\} \tag{3-81}$$

When $D \gg R$

$$C \simeq \frac{\pi\epsilon_0}{\log D/R} \tag{3-81a}$$

Turning to the subject of field intensity, it is evident that the maximum potential gradients must lie along the line forming the conductor centres and at the surfaces of the conductors, that is at points A_1 and A_2 (Fig. 3.26). The field at any point between A_1 and A_2 is

$$E = \frac{\lambda}{2\pi\epsilon_0}\left[\frac{1}{r} + \frac{1}{2a - r}\right]$$

$$= \frac{\lambda}{2\pi\epsilon_0}\left[\frac{2a}{r(2a - r)}\right] \tag{3-82}$$

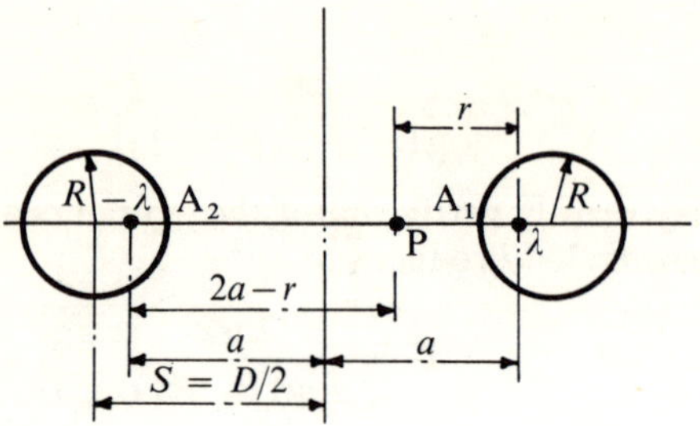

Fig. 3.26 Two parallel cylindrical conductors of equal radii

At the point A_1

$$r = R - (S - a)$$

Since $R_1 = R_2 = R$, we have from equations 3–73 to 3–75

$$S_1 = S_2 = D/2; \quad a = \sqrt{[(D/2)^2 - R^2]}$$

Substituting the values of r, S and a in equation 3–82 gives

$$E_{max} = \frac{\lambda}{2\pi\epsilon_0} \frac{(\tfrac{1}{2}D + R)^{1/2}}{R(\tfrac{1}{2}D - R)^{1/2}}$$

Substituting the value of $\lambda/2\pi\epsilon_0$ from equation 3–71 but with the value for β given by equation 3–80 we obtain

$$E_{max} = \frac{V_1 - V_2}{2R \log\left\{ \dfrac{D}{2R} + \sqrt{\left[\left(\dfrac{D}{2R}\right)^2 - 1 \right]} \right\}} \sqrt{\left(\frac{\dfrac{D}{2R} + 1}{\dfrac{D}{2R} - 1} \right)} \tag{3–83}$$

When $D \gg R$ this reduces to

$$E_{max} = \frac{V_1 - V_2}{2R \log \dfrac{D}{R}} \tag{3–83a}$$

(b) The radius of one of the conductors becomes infinite.

The problem is reduced to that of a cylinder at a distance from an infinite conducting plane at zero potential (Fig. 3.27). We seek the limit approached by β as one of the radii, R_2 say, tends to infinity. We have

$$\beta = \frac{D^2 - R_1^2 - R_2^2}{2R_1 R_2}$$

when $R_2 \gg R_1$, and this may be written as

$$\beta = \frac{D^2 - R_2^2}{2R_1 R_2} = \frac{(D - R_2)(D + R_2)}{2R_1 R_2}$$

Now as $R_2 \to \infty$

$$D - R_2 \to h$$

$$D + R_2 \to 2R_2 + h$$

so that

$$\lim_{R_2 \to \infty} \beta = \frac{h}{R_1}$$

The capacitance between the cylinder and the plane is thus

$$C = \frac{2\pi\epsilon_0}{\log\left\{\dfrac{h}{R_1} + \sqrt{\left[\left(\dfrac{h}{R_1}\right)^2 - 1\right]}\right\}}\, \mathrm{F\,m^{-1}} \qquad (3\text{--}84)$$

for $h \gg R_1$

$$C = \frac{2\pi\epsilon_0}{\log\dfrac{2h}{R_1}} \qquad (3\text{--}84\mathrm{a})$$

The capacitance between a cylinder and an infinite plane is equal to twice the capacitance between two identical cylinders separated by a distance $D = 2h$ (equations 3–81, 3–81a).

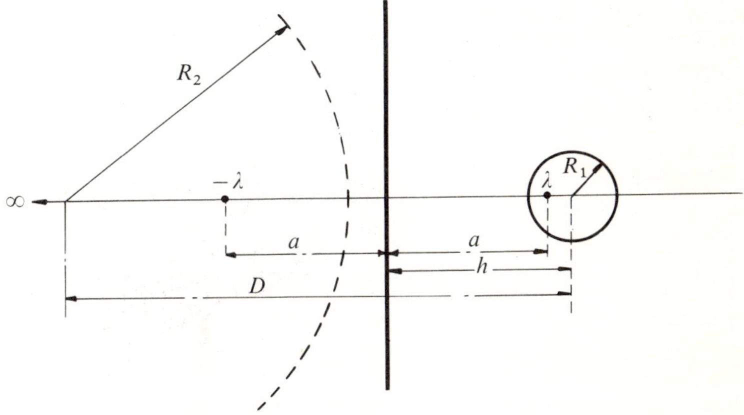

Fig. 3.27 In the limit when R_2 becomes infinite the system reduces to a single conductor opposite an infinite plane of zero potential

3.7.5 CAPACITANCES OF OVERHEAD TRANSMISSION LINES

Equation 3–84 is the general expression for the capacitance per metre-length of a single-conductor overhead transmission line consisting of a conductor of radius R_1 at an average distance h above the earth's surface. The latter may be represented by an infinite conducting plane of zero potential. In overhead lines the distance h is always much greater than the conductor radius so that equation 3–84a may be

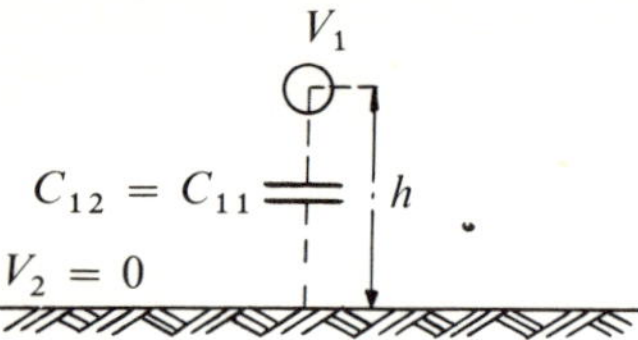

Fig. 3.28 Cylindrical conductor at a distance h above ground

always used without introducing any significant error. For the system shown in Fig. 3.28 we have

$$Q_1 = C_{11}V_1 + C_{12}V_2$$

$$Q_2 = C_{21}V_1 + C_{22}V_2$$

Since $V_2 = 0$, and $Q_1 = -Q_2, |C_{12}| = C_{11}$ and

$$V_1 = \frac{Q_1}{C_{11}} = p_{11}Q_1 = \frac{Q_1}{2\pi\epsilon_0}\log\frac{2h}{R} \tag{3–85}$$

This is the same as the value for the potential of conductor 1 which we would obtain if the earth were replaced by a line conductor 1′, identical with the actual one but carrying an equal and opposite charge, placed as far below the earth's surface as the original con-

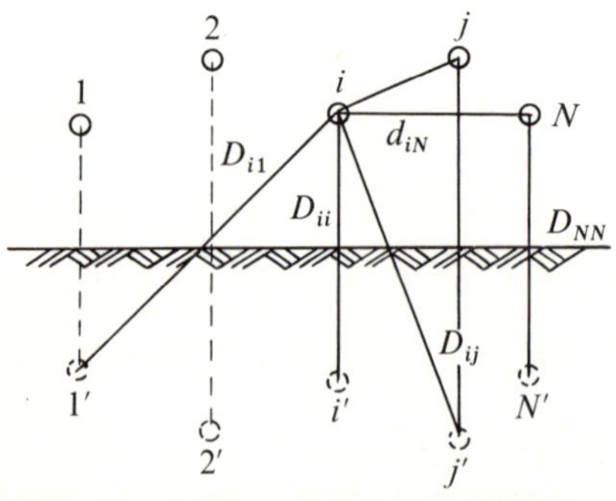

Fig. 3.29 System of N conductors in presence of earth

ductor is above. Conductor 1′ is called the electrical image of line conductor 1 with respect to the earth's surface.

If instead of a single conductor we have a transmission system consisting of N conductors of radii $r_1, r_2, \ldots r_N$ carrying charges $Q_1, Q_2, \ldots Q_N$ and at distances $h_1, h_2, \ldots h_N$ above the earth's surface, the presence of the earth may be taken into consideration by replacing it by a system of N image conductors of radii $r_1, r_2, \ldots r_N$, carrying charges $-Q_1, -Q_2, \ldots -Q_N$ and at distances $h_1, h_2, \ldots h_N$ below the earth's surface. In order to facilitate the writing down of the expressions for the potential coefficients of the system we shall adopt the following notation (Fig. 3.29).

d_{ii} = radius of conductor i or of its image.
d_{ij} = distance between the centres of conductors i and j.
D_{ii} = distance between conductor i and its image.
D_{ij} = distance between conductor i and the image of conductor j.

The reader will readily verify that the self and mutual potential coefficients are given by

$$p_{ii} = \frac{1}{2\pi\epsilon_0} \log \frac{D_{ii}}{d_{ii}}, \quad (i = 1, 2, \ldots N) \tag{3-86}$$

$$p_{ij} = \frac{1}{2\pi\epsilon_0} \log \frac{D_{ij}}{d_{ij}}, \quad (i \neq j = 1, 2, \ldots N) \tag{3-87}$$

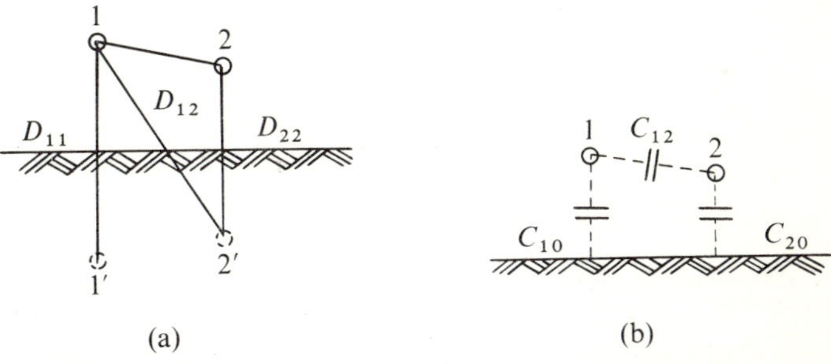

Fig. 3.30 (a) Two-conductor system in presence of earth (b) Equivalent system of capacitances

As a simple example consider the two-conductor circuit shown in Fig. 3.30a. The potentials of the conductors are given by

$$V_1 = p_{11}Q_1 + p_{12}Q_2$$
$$V_2 = p_{21}Q_1 + p_{22}Q_2$$

$\Delta = (p_{11}p_{12} - p_{12}^2)$ and the capacitance coefficients are

$$C_{11} = \frac{1}{\Delta} p_{22} = \frac{1}{\Delta} \frac{1}{2\pi\epsilon_0} \log \frac{D_{22}}{d_{22}}$$

$$C_{22} = \frac{1}{\Delta} p_{11} = \frac{1}{\Delta} \frac{1}{2\pi\epsilon_0} \log \frac{D_{11}}{d_{11}}$$

$$C_{12} = -\frac{1}{\Delta} p_{12} = -\frac{1}{\Delta} \frac{1}{2\pi\epsilon_0} \log \frac{D_{12}}{d_{12}}$$

$$C_{10} = C_{11} - |C_{12}| = \frac{1}{\Delta} \frac{1}{2\pi\epsilon_0} \log \frac{D_{22}d_{12}}{D_{12}d_{22}}$$

$$C_{20} = C_{22} - |C_{12}| = \frac{1}{\Delta} \frac{1}{2\pi\epsilon_0} \log \frac{D_{11}d_{12}}{D_{12}d_{11}}$$

It is clear (see Fig. 3.30b) that the total capacitance C between conductors 1 and 2 consists of the direct capacitance C_{12} between conductors 1 and 2 in parallel with the capacitance between conductors 1 and 2 by way of ground (C_{10} and C_{20} in series). Thus

$$C = C_{12} + \frac{C_{10}C_{20}}{C_{10} + C_{20}}$$

PROBLEMS

Chapter Three

3.1   A spherical drop of water 1 cm in diameter carries a charge of 10 pC. What is the potential at the surface of the drop? If two such drops combine to form a single drop find the potential at the surface of the new drop.

3.2   The spherical electrode of a Van de Graff generator has a diameter of 1 m. It is charged by a moving belt which conveys to the electrode a current of $10\,\mu$A. How long does it take to change the sphere to a potential such that the field intensity at the surface is $30\,\text{kV cm}^{-1}$?

3.3   Consider two concentric spherical conductors; the radius of the inner one is a_1 and the outer one has an inner radius a_2' and an outer radius a_2. If the potentials of the conductors are V_1 and V_2 respectively, find the change on each conductor if (a) both conductors are isolated, (b) the inner conductor is earthed ($V_1 = 0$), (c) the outer conductor is earthed ($V_2 = 0$), and (d) the inner conductor is uncharged.

3.4 A conducting sphere of radius 10 cm and negligible thickness carries a positive charge of $0.2\,\mu$C. A second sphere of radius 6 m and concentric with the first carries a negative charge of $0.02\,\mu$C. Find (a) the potential difference between the two spheres, (b) the field at a point midway between the two spheres, (c) the field at a point 25 cm from the centre and (d) the capacitance of the system.

3.5 Two thin concentric conducting spheres have radii 5 cm and 8 cm. The outer sphere carries a charge of $0.1\,\mu$C and the inner sphere is grounded. Calculate the charge on the inner sphere and the potential difference between the two spheres. What is the capacitance between the two conductors?

3.6 Three isolated concentric spherical conductors in the form of shells of negligible thickness have radii a, b, c ($a > b > c$) and carry charges Q_1, Q_2, Q_3, respectively. Find the potentials of the three shells. Show that if the innermost sphere is connected to earth the potential of the outermost is decreased by (c/a) $(Q_1/a+Q_2/b+Q_3/c)/4\pi\epsilon_0$.

3.7 Two thin concentric conducting spheres have radii a and b ($b > a$). The inner sphere has a charge Q_1 and the outer sphere a charge Q_2. Show that the total charge located within a right circular cone, of apex angle 2θ and vertex at the common centre of the conductors, is $(Q_1 + Q_2)\sin^2(\theta/2)$.

3.8 A point charge Q is placed at a distance r from the centre of two infinite coaxial conducting cylinders of radii a and b ($b > a$). If the two cylinders are earthed show that the charges induced on the inner and outer cylinders are respectively,

$$Q_a = -Q\,\frac{\log{(b/r)}}{\log{(b/a)}}; \quad Q_b = -Q\,\frac{\log{(r/a)}}{\log{(b/a)}}$$

3.9 Three identical conducting spheres are placed at the corners of an equilateral triangle. When the potentials of the spheres are $V, 0, 0$, their charges are Q, Q', Q'. Show that if the potential of each sphere is V', then the charge on each sphere is

$$q = (2Q'+Q)(V'/V)$$

If the charges are Q'', 0, 0, show that the corresponding

potentials are V'', V_0, V_0, where

$$V'' = VQ''(Q+Q')[Q^2+Q'Q-2Q'^2]^{-1}$$
$$V_0 = VQ'Q''[2Q'^2-Q^2-Q'Q]^{-1}$$

3.10 A conducting ellipsoid with major and minor axes 10 cm and 5 cm respectively carries a total charge of 0·1 μC. Find the electric field intensity at the ends of the major and minor axes. What is the capacitance of this conductor?

3.11 A conductor is in the form of an ellipsoid of revolution whose equation is given by

$$\frac{x^2+z^2}{a^2}+\frac{y^2}{b^2} = 1$$

where $2a$ and $2b$ are the lengths of the minor and major axes respectively. If the conductor carries a total charge Q, show that the surface charge density at any point on the conductor surface is given by

$$\sigma = Q/4\pi a^2 b[(x^2+z^2)a^{-4}+y^2 b^{-4}]^{1/2}$$

3.12 Using the result of the previous problem, show that the charge density at any point at a distance r from the centre of a conducting disc of radius a and carrying a total charge Q is

$$\sigma = Q/4\pi a(a^2-r^2)^{1/2}$$

and hence show that the capacitance of the disc is $8\epsilon_0 a$.

3.13 The coefficients of potential of two conductors 1 and 2 are p_{11}, p_{22} and p_{12}. If 1 is insulated and 2 is earthed, show that the capacity of 1 is $[p_{11}-(p_{12}^2/p_{22})]^{-1}$. Find the value of this expression for the case of two conducting concentric spheres of radii a and b ($b > a$) when (i) the inner sphere is earthed and (ii) the outer sphere is earthed.

3.14 Three small conducting spheres, each of radius a, are placed at the corners of an equilateral triangle of side r, with $r \gg a$. Each sphere carries an initial charge Q. If each in turn is now connected to earth and then insulated show that the charge on the last sphere is given by $Q(a/r)^2(3-2ar^{-1})$.

3.15 Two conductors A and B are completely surrounded by a third earthed conductor C. Conductor B is now connected to C and the capacitance between A and earth is found to be 0·02 μF. When A is connected to C the capacitance between B and earth is found to be 0·025 μF. When A and B are connected together (but not to C) the capacitance to earth is found to be 0·03 μF. Find the capacitance between conductors A and B.

3.16 Three identical uncharged spheres of radius a are placed with their centres in line; the central sphere is at a distance r ($r \gg a$) from each of the other two. The central sphere is now given a charge Q and is then connected in turn first to the sphere on its left and then to the one on its right. Show that the charge on the latter sphere is $\frac{1}{8}Q(2r-a)/(r-a)$.

3.17 Two conductors have each a capacitance C with respect to earth and a mutual capacitance kC. Both conductors are initially charged to a potential V and are then isolated. The following operations are now carried out in succession, (i) conductor 1 is earthed and then isolated, (ii) conductor 2 is earthed and then isolated. The above two operations are then repeated a second time. Show that the potential of the first conductor at the end of the four operations is given by $-k^3 V/(1+k)^4$.

3.18 Two parallel very long conductors A and B of diameter 1 cm each are in the same horizontal plane at a distance of 2 m centre to centre (Fig. P3.18). A is at a potential of 10 kV and B is at a

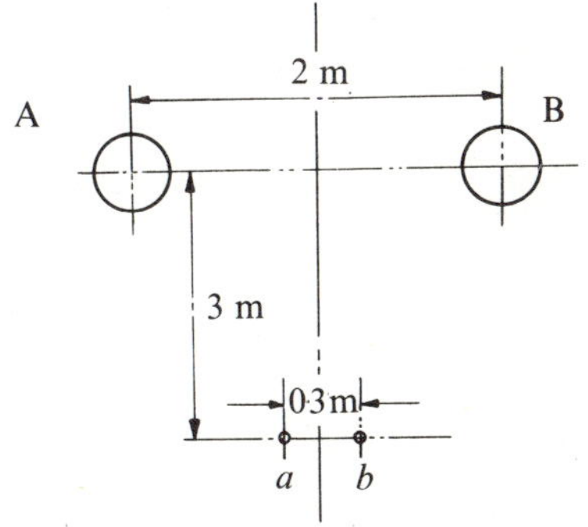

Fig. P. 3.18

potential of -10 kV. Find the potential difference between the two insulated parallel wires a and b each of which has a diameter of 0·2 cm. What is this potential difference if A is kept at 20 kV and B is grounded?

3.19 A transmission line consists of two identical parallel wires each of diameter 1 cm at a distance of 1·5 m centre to centre and at a height of 6 m above the ground. Calculate the capacitance per kilometre between the two wires (a) neglecting the effect of earth, and (b) taking the earth effect into consideration. If one of the wires is at a potential of 10 kV with respect to ground and the other conductor is insulated, find the potential to which the insulated wire is raised for cases (a) and (b).

3.20 In the conductor arrangement described in Problem 3.19, a third grounded conductor of diameter 0·5 cm is placed at a distance of 1 m above the plane of the conductors and is equidistant from both. Find how the presence of the earthed conductor affects the capacitances for cases (a) and (b) of the previous problem.

3.21 Three identical parallel conductors each of diameter 2 cm are in the same horizontal plane with the central conductor at a distance of 3 m from each of the outer conductors. The conductors are at a distance of 10 m above the earth's surface. Find the capacitance of each conductor to neutral and the capacitances between the conductors (a) neglecting the effect of earth, (b) taking the effect of earth into consideration.

3.22 A two-conductor cable consists of two parallel wires each of radius a at a distance $2d$ apart and enclosed symmetrically in an earthed metal sheath of radius R; the remaining space is filled with a dielectric of permittivity ϵ. If the core diameters are sufficiently small, show that the capacitance per unit length between the wires is given by

$$C = \pi\epsilon/\log\frac{2d(R^2-d^2)}{a(R^2+d^2)}$$

CHAPTER FOUR

DIELECTRICS IN THE ELECTRIC FIELD

By a dielectric we mean any medium (gaseous, liquid or solid) which is a good insulator. The term insulator is applied to a medium whenever its primary function is to isolate an electrical system from its surrounding environment. The term dielectric is applied to any insulating medium whenever that medium is used in such a way that its characteristics enter as parameters in the description of an electrical system. However, both terms are often used interchangeably to describe media which are poor conductors of electric current.

An insulating or dielectric medium is a medium in which the electrons are bound to their atoms by such strong forces that, at ordinary temperatures, they are not free to move through the medium under the influence of ordinary applied electric fields. However, when an electric field passes through a dielectric medium it produces important changes in the 'electrical state' of the medium. From the macroscopic point of view these changes are accounted for by the introduction of a quantity known as the electric permittivity of the medium. On the microscopic scale these changes are accounted for by a property of the atoms or molecules of the medium known as the polarizability.

In this chapter we shall study how the basic laws of electrostatics are modified in order to take into account the presence of a dielectric medium. We shall also give an elementary account of the microscopic theory of dielectrics. In order to give a complete picture of the behaviour of dielectrics in an electric field we have found it pertinent to include in this chapter, rather than elsewhere in the book, a discussion of the effect of alternating electric fields on dielectrics.

4.1 DIELECTRIC PERMITTIVITY

It is a fundamental experimental result, first discovered by Cavendish and later rediscovered by Faraday, that the capacitance of a condenser is increased if the space between its plates is filled with a dielectric material. If C_0 is the capacitance of the condenser with the region between the plates evacuated, and C its capacitance when this region is filled with a dielectric then the ratio

$$\frac{C}{C_0} = \epsilon_r \tag{4-1}$$

is found to be independent of the shape or the dimensions of the plates and is solely a characteristic of the particular dielectric medium used. ϵ_r is called the *relative permittivity* or the *dielectric constant* of the medium. In the SI system of units the *permittivity* of the medium is defined as

$$\epsilon = \epsilon_0 \epsilon_r \qquad\qquad (4\text{–}2)$$

where $\epsilon_0 = (36\pi \times 10^9)^{-1}$ farad/metre and is an electric constant. The term dielectric constant is sometimes used for both ϵ and ϵ_r; since this causes confusion the terms permittivity and relative permittivity are to be preferred and we shall use them exclusively. Moreover the term dielectric constant is somewhat misleading since ϵ (or ϵ_r) is constant in only a very limited way. Table 4.1 gives the relative permittivity of a number of gaseous, liquid and solid dielectrics; the values given are for static or low frequency (< 1000 Hz) fields. The relative permittivity of gases is only so very slightly greater than unity that for all practical cases in which the dielectric medium is gaseous the relative permittivity may be taken as equal to unity without introducing any significant error.

The relative permittivity of a medium is a directly measurable quantity which expresses on a macroscopic scale the overall result of the inter-action which occurs on a microscopic scale between an externally applied electric field and the atoms or molecules of the material. This interaction is known as *polarization*.

4.2 MACROSCOPIC CONCEPT OF POLARIZATION

Dielectrics may be broadly divided into non-polar media and polar media. In a non-polar medium an atom or molecule may be represented as a positive nucleus of charge q surrounded by a sym-metrically distributed negative electronic cloud of charge $-q$. In the absence of an applied field the centres of gravity of the positive and negative charges, averaged over time, coincide. When the atom or molecule is acted upon by an electric field the positive and negative charges experience electric forces which tend to move them apart in the direction of the field. The distance moved is very small (10^{-12}–10^{-13} m) because the displacement is limited by strong restoring forces which increase with increasing displacement. The centres of positive and negative charges no longer coincide and the atom or molecule is said to be polarized. Each atom or molecule now forms a very small dipole whose moment is given by $\mathbf{p} = q\,\mathbf{dl}$, where $\mathbf{dl}$ is the distance between the two centres of charge. Dipoles so formed are known as induced dipoles since when the field is removed the charges resume their normal distribution and the dipoles disappear. q will be of the order of electronic charge (10^{-19} C) and $\mathbf{dl}$ of the order of molecular dimensions (10^{-10} m) so that p will be of the order of 10^{-29} C m. In the c.g.s. system of units this is of the order of 10^{-18} e.s.u. which is sometimes taken as a unit of dipole moment known as a debye so that,

$$1 \, \text{C m} = 3 \times 10^{29} \, \text{debye}$$

In considering a dielectric from the macroscopic point of view, we restrict our attention to average values over volumes which are sufficiently small in comparison with the dimensions of the medium under consideration but large enough to contain a sufficient number of atoms or molecules for the purpose of averaging. Thus the sum of the dipole moments in an element of volume Δv is

$$\sum_{i=1}^{N\Delta v} \mathbf{p}_i = N\mathbf{p}\Delta v = \mathbf{P}\Delta v$$

where $\mathbf{p}$ represents the average dipole moment of each atom or molecule and N is the number of atoms or molecules per unit volume. The vector $\mathbf{P}$ is the *dipole moment per unit volume* and is called the electric polarization density or simply the *polarization* of the medium.

In polar dielectrics the molecules, which are normally composed of two or more different atoms, have dipole moments even in the absence of an electric field; that is, the centres of their positive and negative charges do not coincide. Normally these molecular dipoles are randomly oriented throughout the material owing to thermal agitation, so that the average moment over any macroscopic volume element is zero. In the presence of an externally applied field the dipoles are acted upon by torques which tend to orient them in the direction of the field, with the result that each elementary volume now has a net dipole moment given by

$$\sum_{1}^{N\Delta v} \mathbf{p}_i = \mathbf{P}\Delta v$$

Since the polarization $\mathbf{P}$ has the nature of a density and since it may vary from one point (macroscopic) to another, its value at any point should be defined as

$$\mathbf{P} = \lim_{\Delta v \to 0} \frac{1}{\Delta v} \sum \mathbf{p}_i = \lim_{\Delta v \to 0} \frac{\Delta \mathbf{p}}{\Delta v} = \frac{d\mathbf{p}}{dv} \qquad (4\text{--}3)$$

where $\Delta \mathbf{p}$ is the net dipole moment in the volume Δv. The above equation then defines the polarization vector $\mathbf{P}$ at a point.

Consider a polarized dielectric of finite volume v. Assume that the polarization is non-uniform and let its value at any point A (x, y, z) be $\mathbf{P}$. The potential at any point B (x', y', z') outside the dielectric and at

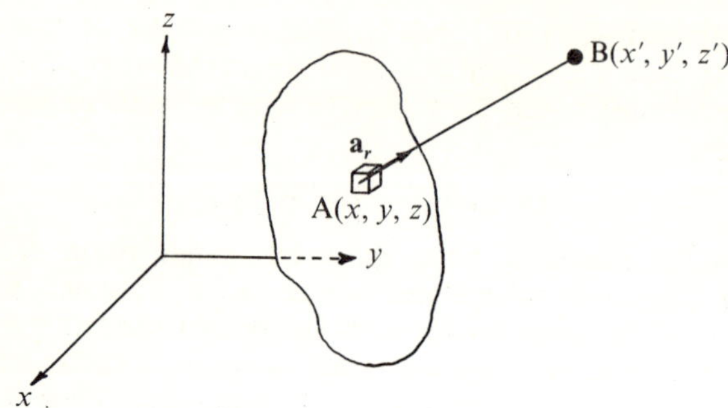

Fig. 4.1 To calculate the potential at a point outside a polarized dielectric of finite volume v

a distance r from A (Fig. 4.1) is given by (using equation 2–81),

$$dV_{\mathrm{B}} = \frac{d\mathbf{p} \cdot \mathbf{a}_r}{4\pi\epsilon_0 r^2} = \frac{\mathbf{P} \cdot \mathbf{a}_r\, dv}{4\pi\epsilon_0 r^2} = -\frac{1}{4\pi\epsilon_0}\, \mathbf{P} \cdot \nabla'\left(\frac{1}{r}\right) dv$$

and the potential due to the whole dielectric by

$$V_{\mathrm{B}} = -\frac{1}{4\pi\epsilon_0} \int_v \mathbf{P} \cdot \nabla'\left(\frac{1}{r}\right) dv \tag{4–4}$$

where the primed operator involves derivatives with respect to the coordinates of B. Because of equation 2–38 this may be rewritten as

$$V_{\mathrm{B}} = \frac{1}{4\pi\epsilon_0} \int_v \mathbf{P} \cdot \nabla\left(\frac{1}{r}\right) dv \tag{4–5}$$

where all derivatives are now to be carried out with respect to the coordinates of point A. By using the vector identity

$$\nabla \cdot \frac{\mathbf{P}}{r} = \frac{1}{r}\nabla \cdot \mathbf{P} + \mathbf{P} \cdot \nabla\left(\frac{1}{r}\right)$$

equation 4–5 may be written as

$$V_{\mathrm{B}} = \frac{1}{4\pi\epsilon_0} \int_v \left[\nabla \cdot \left(\frac{\mathbf{P}}{r}\right) - \frac{\nabla \cdot \mathbf{P}}{r}\right] dv$$

By applying the divergence theorem to the first integral on the right we obtain

$$V_{\mathrm{B}} = \frac{1}{4\pi\epsilon_0} \oint_S \frac{\mathbf{P} \cdot \mathbf{dS}}{r} + \frac{1}{4\pi\epsilon_0} \int_v \frac{(-\nabla \cdot \mathbf{P})\, dv}{r} \tag{4–6}$$

where S is the surface bounding the volume v. This equation may be further written as

$$V_\mathrm{B} = \oint_S \frac{\sigma_p\,\mathrm{d}S}{4\pi\epsilon_0 r} + \int_v \frac{\rho_p\,\mathrm{d}v}{4\pi\epsilon_0 r} \qquad (4\text{--}7)$$

provided that $\sigma_\mathrm{p} = \mathbf{P}\cdot\mathbf{a}_\mathrm{n}$ where $\mathbf{a}_\mathrm{n}$ is the unit vector along the outward normal to the surface S, and $\rho_\mathrm{p} = -\nabla\cdot\mathbf{P}$. Thus the total potential is equivalent to the sum of the potentials produced by a surface charge distribution of density σ_p and a volume charge distribution throughout the interior of the dielectric of density ρ_p. The charge densities σ_p and ρ_p are referred to as 'bound' or 'polarization' charge densities to distinguish them from the free or conduction charge densities. Since the dielectric as a whole must be electrically neutral, it follows that the total polarization charge must be zero, that is:

$$Q_\mathrm{p} = \oint_S \mathbf{P}\cdot\mathbf{dS} + \int_v (-\nabla\cdot\mathbf{P})\,\mathrm{d}v = 0 \qquad (4\text{--}8)$$

In order to obtain a better understanding of the significance of the interpretation of $-\nabla\cdot\mathbf{P}$ and $\mathbf{P}\cdot\mathbf{a}_\mathrm{n}$ as volume and surface charge densities respectively, we will examine separately conditions within the volume of the dielectric and at its surface.

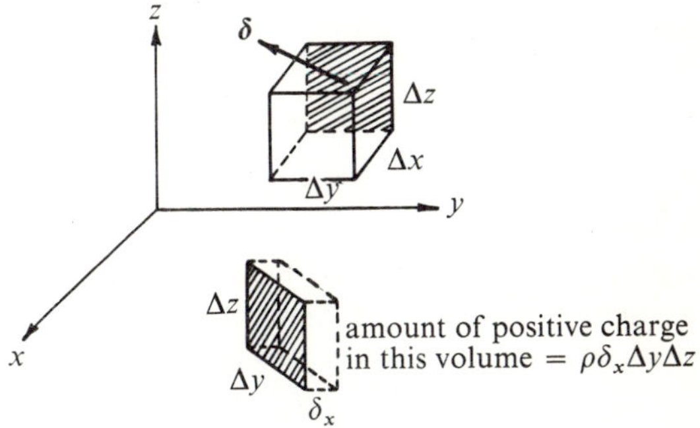

Fig. 4.2 Element of volume of a non-uniformly polarized medium

Consider an element of volume Δv of dimensions Δx, Δy, Δz within the dielectric medium which is assumed to be non-uniformly polarized (Fig. 4.2). The displacement of charge due to polarization causes positive and negative charges to pass through the surfaces bounding the volume Δv. We want to calculate the excess or net charge within Δv which results from this displacement of charge. Let ρ be the density of positive charges (bound) within the medium and ρ' be the density of negative charges. Since the medium is electrically neutral,

$$\rho = -\rho'$$

Let the average displacements of the positive and negative charges be $\boldsymbol{\delta}$ and $\boldsymbol{\delta}'$ respectively. Consider first only the x-component of charge displacement; the amount of positive charge which enters into Δv through the back face is that charge originally contained in the volume $\delta_x \Delta y \Delta z$ which is

$$\rho \delta_x \Delta y \Delta z$$

The amount of positive charge which leaves the front face is, considering only the first two terms of Taylor's expansion,*

$$\rho \delta_x \Delta y \Delta z + \frac{\partial}{\partial x}(\rho \delta_x \Delta y \Delta z)\Delta x$$

The net gain of positive charge is therefore

$$-\frac{\partial}{\partial x}(\rho \delta_x)\Delta x \Delta y \Delta z$$

In a similar way the net charge which enters Δv due to the x-component of the displacement of negative charges is found to be

$$-\frac{\partial}{\partial x}(\rho' \delta'_x)\Delta x \Delta y \Delta z \;=\; \frac{\partial}{\partial x}(\rho \delta'_x)\Delta x \Delta y \Delta z$$

The total gain of charge is therefore

$$-\frac{\partial}{\partial x}\left[\rho(\delta_x - \delta'_x)\right]\Delta x \Delta y \Delta z$$

The quantity $\rho(\delta_x - \delta'_x)$ is the average x-component of the dipole moment per unit volume P_x.

The net gain of charge due to the y- and z-components of charge displacement may be found in a manner analogous to the above. The total gain of charge due to all three displacement components is thus

$$\Delta Q_{\mathrm{p}} \;=\; -\left(\frac{\partial P_x}{\partial x} + \frac{\partial P_y}{\partial y} + \frac{\partial P_z}{\partial z}\right)\Delta x \Delta y \Delta z$$

and the charge density produced by the polarization of the medium is therefore

$$\rho_{\mathrm{p}} \;=\; \lim_{\Delta v \to 0}\frac{\Delta Q_{\mathrm{p}}}{\Delta v} \;=\; -\left(\frac{\partial P_x}{\partial x} + \frac{\partial P_y}{\partial y} + \frac{\partial P_z}{\partial z}\right) \;=\; -\nabla \cdot \mathbf{P} \qquad (4\text{--}9)$$

It is evident that in those regions where the polarization of the medium is uniform the polarization charge density is zero.

---

* Higher terms will eventually go to zero in the limit as the volume element shrinks to a point (see Section 2.15).

Although the polarization inside a dielectric may be uniform it is never so at the boundary between a dielectric and free space or between two different dielectrics. To avoid mathematical difficulties we may consider that near the surface we have a very thin transition layer in which P is a smoothly varying function of position; both P and its derivatives change very rapidly but continuously as we move through the layer from one side of the surface to the other. In this layer it is more convenient to replace the volume charge density by a charge per unit area of surface. Consider the boundary between a dielectric and free space. Let Δv be the volume of a small pillbox-shaped Gaussian surface in which the two surfaces ΔS_1 and ΔS_2 are

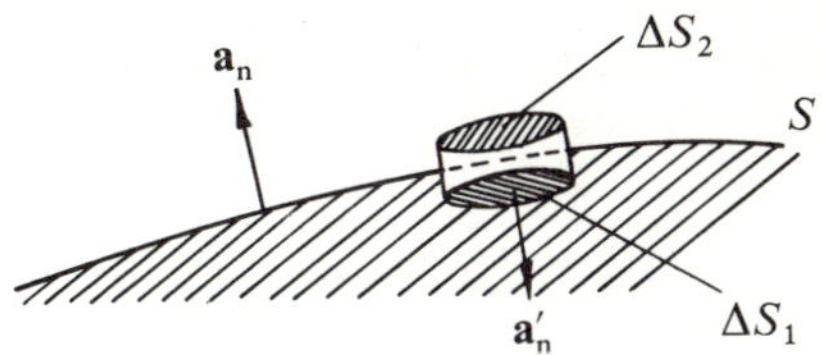

Fig. 4.3 Gaussian pillbox at boundary between a dielectric and free space

parallel to each other and to the surface S (Fig. 4.3). The total charge within Δv is

$$\rho_p \Delta v = \Delta Q_p = -\int_{\Delta v} \nabla \cdot \mathbf{P}\, dv = -\oint_{\Delta S} \mathbf{P} \cdot \mathbf{dS}$$

The last integral is to be evaluated over the whole outer surface of the pillbox. There is no contribution over the area ΔS_2 since P is zero there (free space); the contribution over the curved surface is negligibly small and the only contribution to the integral is that over the area ΔS_1, just inside the dielectric. Hence

$$\Delta Q_p = -\int_{\Delta S_1} \mathbf{P} \cdot \mathbf{dS} = -\mathbf{P} \cdot \Delta \mathbf{S}_1$$

because ΔS_1 is so small. Making use of equation 2–15 we may define a surface polarization charge density as

$$\sigma_p = \lim_{\Delta S \to 0} \frac{\Delta Q_p}{\Delta S_1} = \lim_{\Delta S \to 0} \left(-\frac{1}{\Delta S_1} \mathbf{P} \cdot \Delta \mathbf{S}_1 \right) = -\mathbf{P} \cdot \mathbf{a}'_n$$

where the unit vector $\mathbf{a}'_n$ points along the outward direction normal to the area ΔS_1. If $\mathbf{a}_n$ is the unit vector normal to the surface S then $\mathbf{a}'_n = -\mathbf{a}_n$ and σ_p may be expressed as,

$$\sigma_p = \mathbf{P} \cdot \mathbf{a}_n = P_n \qquad (4\text{--}10)$$

where P_n is the component of $\mathbf{P}$ along the outward direction normal

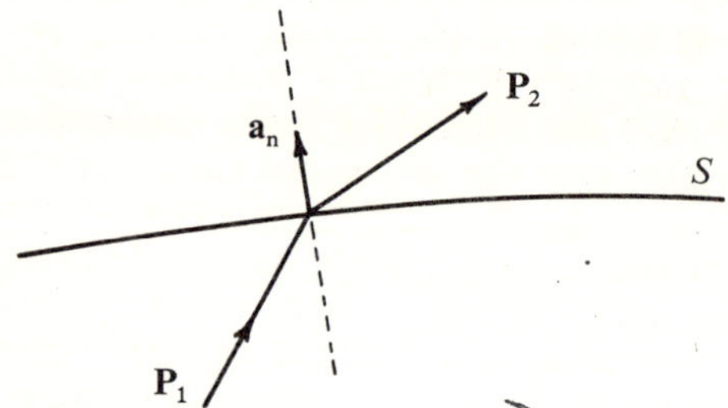

Fig. 4.4 Boundary between two polarized dielectrics

to the surface S. The reader will readily verify that in the more general case where we have two different dielectrics on both sides of the boundary surface, the density of bound charge at the interface is given by

$$\sigma_p = (\mathbf{P}_1 - \mathbf{P}_2) \cdot \mathbf{a}_n = P_{1n} - P_{2n} \tag{4-11}$$

where P_{1n} and P_{2n} are the normal components of the polarizations $\mathbf{P}_1$ and $\mathbf{P}_2$ at points on the opposite sides of the boundary (Fig. 4.4), and $\mathbf{a}_n$ is the unit normal vector pointing from region 1 to region 2.

4.3 GAUSS'S LAW IN A DIELECTRIC AND THE ELECTRIC FLUX DENSITY

In Section 2.8 we showed that Gauss's law for free space had the form

$$\oint_S \epsilon_0 \mathbf{E} \cdot \mathbf{dS} = Q = \int_v \rho \, dv \tag{4-12}$$

where ρ is the density of charges enclosed in the volume v bounded by the surface S. For free space ρ can only represent a density of *free* charges. If the free space is now replaced by a polarized dielectric medium, the total charge within the volume v will consist of the free charges present plus the bound charges created by the polarization of the medium. Equation 4–12 therefore takes the form

$$\oint_S \epsilon_0 \mathbf{E} \cdot \mathbf{dS} = \int_v (\rho + \rho_p) \, dv$$

$$= \int_v \rho \, dv + \int_v -(\nabla \cdot \mathbf{P}) \, dv$$

Using equation 4–8 this may be written as

$$\oint_S \epsilon_0 \mathbf{E} \cdot \mathbf{dS} = \int_v \rho \, dv - \oint_S \mathbf{P} \cdot \mathbf{dS}$$

or

$$\oint_S (\epsilon_0 \mathbf{E} + \mathbf{P}) \cdot \mathbf{dS} = \int_v \rho \, dv \tag{4-13}$$

The surface integral of the sum $(\epsilon_0 \mathbf{E} + \mathbf{P})$ is thus a function only of the free charge enclosed by the Gaussian surface. It is thus possible to define the electric flux density or electric displacement more generally as

$$\mathbf{D} = \epsilon_0 \mathbf{E} + \mathbf{P} \qquad (4\text{--}14)$$

For free space $\mathbf{P} = 0$ and the flux density is

$$\mathbf{D} = \mathbf{D}_0 = \epsilon_0 \mathbf{E}$$

as already defined in Chapter 2, equation 2–39.

We may now state Gauss's law in its most general form: The total flux of the vector $\mathbf{D}$ through any closed surface is equal to the total *free* charge enclosed by that surface. The point form of Gauss's law in a dielectric may be readily derived by applying the divergence theorem to the left-hand side of equation 4–13. This, together with equation 4–14 gives

$$\nabla \cdot \mathbf{D} = \rho \qquad (4\text{--}15)$$

We may rewrite equation 4–14 as

$$\mathbf{D} = \epsilon_0 \left(1 + \frac{\mathbf{P}}{\epsilon_0 \mathbf{E}} \right) \mathbf{E} \qquad (4\text{--}16)$$

If we are to make any use of this equation it is clear that we must know the relationship between the polarization $\mathbf{P}$ and the electric field $\mathbf{E}$. For most dielectrics, at normal temperatures and for fields obtainable in the laboratory, experimental evidence indicates that this relationship is linear; such dielectrics are known as *linear dielectrics*. Most important of the *non-linear* dielectrics are the so-called *ferro-electrics*; in these materials the polarization $\mathbf{P}$ is not uniquely determined for a given value of the applied field, but depends on the previous history of the material so that the *P–E* relationship exhibits hysteresis (Fig. 4.5). We shall not discuss these materials here but it should be pointed out that their treatment from the phenomenological point of view is analogous to that of ferromagnetic materials which will be discussed in Chapter 7.

Irrespective whether their characteristics are linear or non-linear, dielectrics are further classified as *isotropic* and *anisotropic*. If the directions of $\mathbf{P}$ and $\mathbf{E}$ coincide, regardless of the direction of $\mathbf{E}$, the dielectric is said to be isotropic. In anisotropic dielectrics, usually single crystals of symmetry lower than cubic, the orientations of $\mathbf{P}$ and $\mathbf{E}$ do not coincide. From the analytic point of view the simplest dielectrics are those which are linear and isotropic. All gaseous and liquid dielectrics as well as most solid dielectrics of practical importance

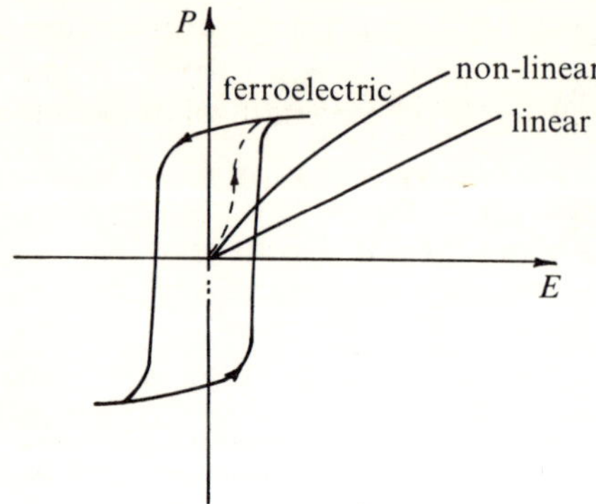

Fig. 4.5 Relationship between P and E for linear, non-linear and ferroelectric isotropic dielectrics

may be treated as linear and isotropic.* For such dielectrics the relationship between **P** and **E** is most conveniently expressed as

$$\mathbf{P} = \epsilon_0 \chi_e \mathbf{E} \qquad (4\text{--}17)$$

where χ_e is a dimensionless constant defined as the *electric susceptibility* of the material. Combining equations 4–16 and 4–17 we have

$$\mathbf{D} = \epsilon_0(1 + \chi_e)\mathbf{E}$$
$$= \epsilon_0\epsilon_r\mathbf{E} = \epsilon\mathbf{E} \qquad (4\text{--}18)$$

where

$$\epsilon_r = 1 + \chi_e \qquad (4\text{--}19)$$

and

$$\epsilon = \epsilon_0\epsilon_r \qquad (4\text{--}20)$$

ϵ_r is the relative permittivity and ϵ is the permittivity of the dielectric. If the medium is homogeneous both χ_e and ϵ will be independent of position. By combining equations 4–17 and 4–19 we obtain

$$\mathbf{P} = \epsilon_0(\epsilon_r - 1)\mathbf{E} \qquad (4\text{--}21)$$

In an anisotropic dielectric each of the components of **P** and **D** may be a function of all three components of **E**. For a linear anisotropic dielectric equations 4–17 and 4–18 are replaced by the more general

---

* Although the majority of solids have a definite crystal structure, the bulk of the material is generally composed of aggregates of randomly oriented crystallites so that on a macroscopic scale the behaviour is isotropic.

expressions,

$$P_1 = \epsilon_0\chi_{e11}E_1 + \epsilon_0\chi_{e12}E_2 + \epsilon_0\chi_{e13}E_3$$
$$P_2 = \epsilon_0\chi_{e21}E_1 + \epsilon_0\chi_{e22}E_2 + \epsilon_0\chi_{e23}E_3 \qquad (4\text{-}22)$$
$$P_3 = \epsilon_0\chi_{e31}E_1 + \epsilon_0\chi_{e32}E_2 + \epsilon_0\chi_{e33}E_3$$

and

$$D_1 = \epsilon_{11}E_1 + \epsilon_{12}E_2 + \epsilon_{13}E_3$$
$$D_2 = \epsilon_{21}E_1 + \epsilon_{22}E_2 + \epsilon_{23}E_3 \qquad (4\text{-}23)$$
$$D_3 = \epsilon_{31}E_1 + \epsilon_{32}E_2 + \epsilon_{33}E_3$$

where the subscripts 1, 2 and 3 denote components in the x, y and z directions respectively. Equations 4–22 and 4–23 may be expressed more compactly as

$$P_i = \epsilon_0\chi_{eij}E_j; \quad (i, j = 1, 2, 3)$$

and

$$D_i = \epsilon_{ij}E_j; \quad (i, j = 1, 2, 3)$$

The coefficients χ_{eij} and ϵ_{ij} may be written out in matrix form as,

$$\chi_{eij} = \begin{bmatrix} \chi_{e11} & \chi_{e12} & \chi_{e13} \\ \chi_{e21} & \chi_{e22} & \chi_{e23} \\ \chi_{e31} & \chi_{e32} & \chi_{e33} \end{bmatrix}; \quad \epsilon_{ij} = \begin{bmatrix} \epsilon_{11} & \epsilon_{12} & \epsilon_{13} \\ \epsilon_{21} & \epsilon_{22} & \epsilon_{23} \\ \epsilon_{31} & \epsilon_{32} & \epsilon_{33} \end{bmatrix}$$

Each matrix represents a tensor of second rank. χ_{eij} is known as the electric susceptibility tensor and ϵ_{ij} as the permittivity tensor. Since it can be shown that $\chi_{eij} = \chi_{eji}$ and $\epsilon_{ij} = \epsilon_{ji}$, the number of independent coefficients is six for each.

For the remainder of this chapter, as well as throughout the rest of the book, we shall assume that all our dielectric media are linear and isotropic.

4.4 BOUNDARY CONDITIONS

We shall first consider the interface between two dielectric media having permittivities ϵ_1 and ϵ_2. We shall assume that at the interface there is a surface density of free charge σ. Consider a small cylindrical pillbox-type surface of negligible thickness and with its two end faces in regions 1 and 2 parallel to the interface (Fig. 4.6). By Gauss's law the flux leaving the top and bottom surfaces must be equal to the free charge enclosed so that

$$(D_{n2} - D_{n1})\Delta S = \sigma\Delta S$$

or

$$D_{n2} - D_{n1} = \sigma \qquad (4\text{-}24)$$

This means that at the boundary there is a discontinuity in the normal component of $\mathbf{D}$; the amount of this discontinuity is equal to the surface density of free charge. Normally there is no free charge on the interface between two dielectrics so that in this case,

$$D_{n2} = D_{n1} \qquad (4\text{-}25)$$

Thus the normal component of $\mathbf{D}$ at an uncharged boundary surface is continuous. Using the relationship $\mathbf{D} = \epsilon\mathbf{E}$ we obtain that

$$\epsilon_1 E_{n1} = \epsilon_2 E_{n2} \qquad (4\text{-}26)$$

Since the electric field is conservative the line integral of $\mathbf{E}$ around a small closed path abcd on both sides of the interface (Fig. 4.6) must

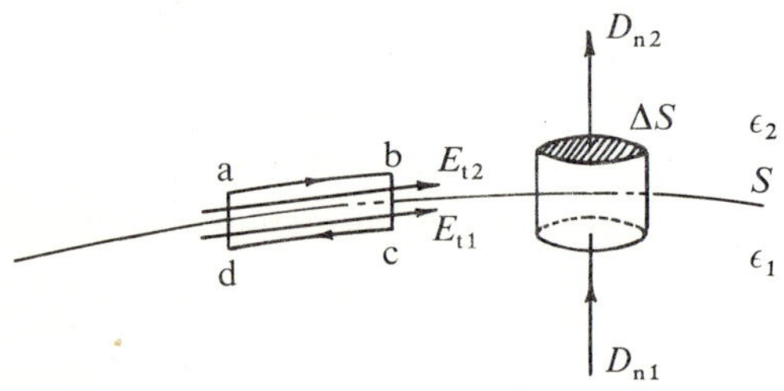

Fig. 4.6 Gaussian pillbox and elementary closed path at boundary between two media

be zero. If we choose the path so that the portions perpendicular to the interface are negligibly short compared with those parallel to the interface we have

$$\oint \mathbf{E} \cdot d\mathbf{l} = \mathbf{E}_2 \cdot \Delta\mathbf{L} + \mathbf{E}_1 \cdot (-\Delta\mathbf{L}) = 0$$

or

$$(E_{t2} - E_{t1})\Delta L = 0$$

and hence

$$E_{t2} = E_{t1} \qquad (4\text{-}27)$$

showing that the tangential component of $\mathbf{E}$ is always continuous across an interface; this result is independent of whether there are or there are no free charges at the interface. From the relationship $\mathbf{D} = \epsilon\mathbf{E}$ we obtain that

$$\epsilon_2 D_{t1} = \epsilon_1 D_{t2} \qquad (4\text{-}28)$$

The boundary conditions (4–24), (4–25) and (4–27) may be expressed in terms of the potential respectively as

$$\epsilon_1 \frac{\partial V_1}{\partial n} - \epsilon_2 \frac{\partial V_2}{\partial n} = \sigma \qquad (4\text{–}29)$$

$$\epsilon_1 \frac{\partial V_1}{\partial n} = \epsilon_2 \frac{\partial V_2}{\partial n} \qquad (4\text{–}30)$$

and

$$V_1 = V_2 \qquad (4\text{–}31)$$

where $\partial/\partial n$ denotes the derivative taken along the normal to the interface. Strictly equation 4–31 must contain an integration constant; this constant should represent a contact potential if it exists. For dielectrics at least it is immeasurably small and the constant is therefore conveniently assumed to be zero. Equations 4–29 or 4–30 and 4–31 are independent.

The boundary conditions given by equations 4–26 and 4–27 may now be combined to show the change in the directions of the vectors **D** and **E** as we move from one medium to the other across the

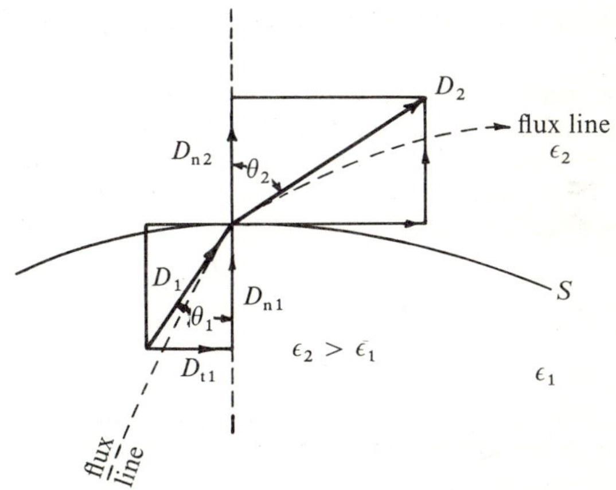

Fig. 4.7 Refraction of flux lines at boundary between two dielectrics

surface. Let $\mathbf{E}_1$ (or $\mathbf{D}_1$) make an angle θ_1 with the normal to the interface and $\mathbf{E}_2$ (or $\mathbf{D}_2$) and angle θ_2 with the same normal (Fig. 4.7); then

$$E_1 \sin \theta_1 = E_2 \sin \theta_2$$
$$\epsilon_1 E_1 \cos \theta_1 = \epsilon_2 E_2 \cos \theta_2$$

Dividing the first equation by the second we get

$$\epsilon_1 \cot \theta_1 = \epsilon_2 \cot \theta_2 \qquad (4\text{-}32)$$

This equation is the law of refraction for both $\mathbf{D}$ and $\mathbf{E}$ at the boundary between two dielectrics. In passing from one medium to another of higher permittivity the lines of force are refracted away from the normal to the interface.

The boundary conditions at the interface between a conductor and a dielectric are quite simple to derive. Since $\mathbf{E}$ and $\mathbf{D}$ are zero inside the conductor it follows that the tangential components of $\mathbf{E}$ and $\mathbf{D}$ on the dielectric side of the interface must both be zero and hence that both $\mathbf{E}$ and $\mathbf{D}$ must be normal to the conductor surface. Thus

$$D = D_{\mathrm{n}} = \sigma \qquad (4\text{-}33)$$

and

$$E = E_{\mathrm{n}} = \sigma/\epsilon \qquad (4\text{-}34)$$

If we compare these equations with equations 3–1 and 3–2 we see that if free space is replaced by a medium of permittivity ϵ then, for a given density of charge on the conductor surface, the flux density remains unchanged whilst the field is reduced by $1/\epsilon_r$. From this we may conclude that flux lines begin and end only at free or true* charges whereas lines of force may begin and end either at true or at bound charges.

We shall conclude this section with two examples of practical interest. Consider a parallel-plate capacitor with a constant potential difference V maintained between the plates which are at a distance d apart (Fig. 4.8a). With vacuum (or air) between the plates the following relationships hold:

$$E = V/d = \sigma/\epsilon_0$$

$$D_0 = \epsilon_0 E = \sigma$$

$$C_0 = \epsilon_0 S/d$$

If the space between the plates is filled with a dielectric of uniform thickness $t = d$ and permittivity ϵ (Fig. 4.8b), we have that

$$E = V/d = \sigma'/\epsilon$$

$$D = \epsilon E = \sigma'$$

* On the surface of a conductor any charges can only be free charges. However, when the conductor surface is in contact with a dielectric, part of these charges are compensated by the surface polarization charges. In a sense, therefore, part of the conductor charges may be said to be 'bound' even though they are fundamentally free. To avoid possible confusion it is best therefore to refer to charges on conductor surfaces as true charges and use the term 'bound charge' exclusively for dielectrics.

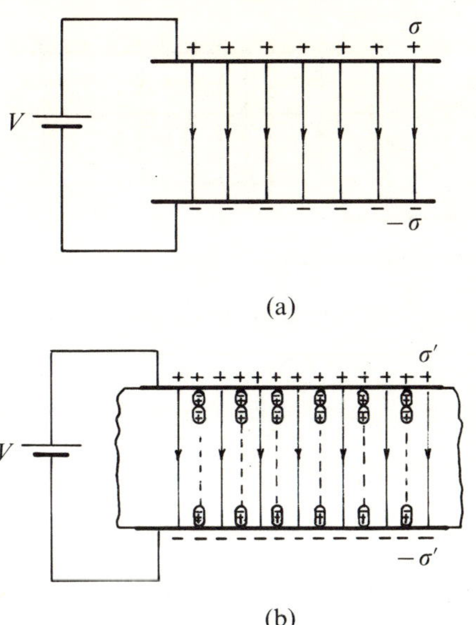

(a)

(b)

Fig. 4.8 Parallel plate capacitor with constant potential difference between plates; (a) free space between plates, (b) dielectric between plates

where σ' now represents the total charge density on the condenser plates and is related to σ by

$$\sigma' = \epsilon_r \sigma$$

It follows that the capacitance of the system is given by

$$C = \frac{\epsilon S}{d} = \epsilon_r C_0$$

which is in agreement with equation 4–1. The density of polarization charges at the top and bottom surfaces of the dielectric is given by

$$\sigma_p = \mp (\sigma' - \sigma)$$

$$= \mp \sigma(\epsilon_r - 1)$$

Let us now assume that the thickness t of the dielectric is less than the distance between the plates d (Fig. 4.9). Since the flux and field lines are normal to the condenser plates and to the dielectric surfaces, the continuity conditions (4–25) and (4–26) give

$$D_0 = D$$

$$\epsilon_0 E_0 = \epsilon E$$

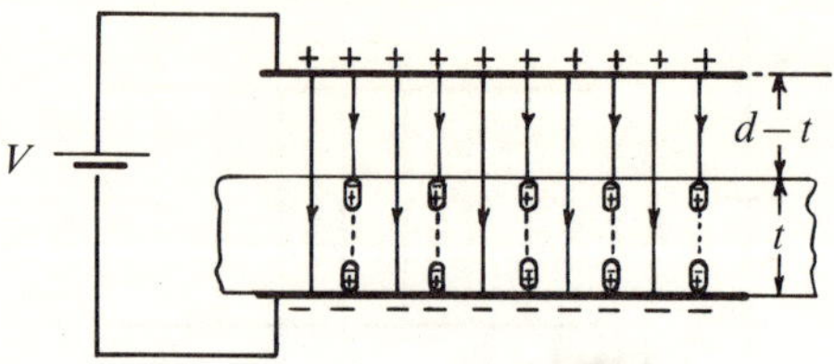

Fig. 4.9 Parallel plate capacitor with dielectric slab of thickness less than plate separation

since also the potential difference between the plates is V we have that

$$Et + E_0(d - t) = V$$

and solving the last two equations for E and E_0 we obtain

$$E = \frac{V}{\epsilon_r d - (\epsilon_r - 1)t}$$

and

$$E_0 = \frac{\epsilon_r V}{\epsilon_r d - (\epsilon_r - 1)t}$$

We thus see that the field in the dielectric is smaller than the average field V/d whereas the field in the free space (or air) region is greater than the average field. The charge density on the two plates is given by

$$\sigma = \pm \frac{\epsilon_r \epsilon_0 V}{\epsilon_r d - (\epsilon_r - 1)t}$$

The capacitance of the system is therefore

$$C = \frac{\sigma S}{V} = \frac{\epsilon_r \epsilon_0 S}{\epsilon_r d - (\epsilon_r - 1)t} \tag{4-35}$$

The reader can easily verify that this is the same as the total capacitance obtained by regarding the system as two condensers C_0 and C_1 in series. This is true because the division between the two media is an equipotential surface.

Consider a single core cable in which the insulation between the core and the sheath consists of two layers having different permittivities. Let ϵ_1 be the permittivity of the inner layer which extends up to a radius r_2 and let ϵ_2 be the permittivity of the outer layer. The dimensions of the cable are shown in Fig. 4.10. If the charge per unit length on the inner conductor is $Q\,\mathrm{C\,m^{-1}}$, the voltage across the inner dielectric is given by

$$V_{12} = \frac{Q}{2\pi\epsilon_1} \log\left(\frac{r_2}{r_1}\right)$$

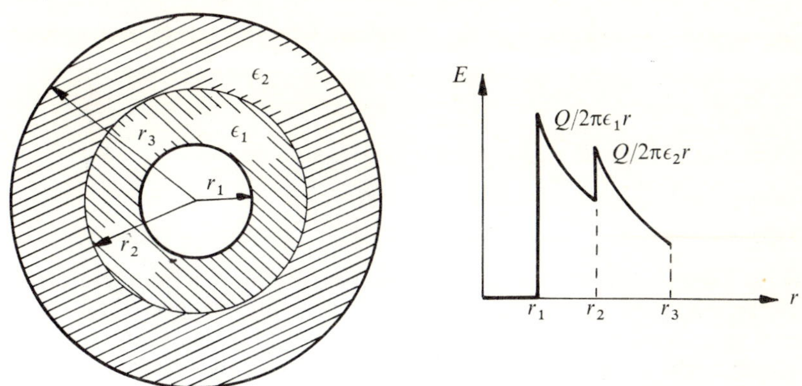

Fig. 4.10 Variation of the radial field intensity in a two-dielectric coaxial cable

and that across the outer dielectric by

$$V_{23} = \frac{Q}{2\pi\epsilon_2} \log\left(\frac{r_3}{r_2}\right)$$

The total voltage between conductor and sheath is therefore given by

$$V_{13} = \frac{Q}{2\pi\epsilon_0}\left[\frac{1}{\epsilon_{r1}}\log\left(\frac{r_2}{r_1}\right) + \frac{1}{\epsilon_{r2}}\log\left(\frac{r_3}{r_2}\right)\right]$$

and the capacitance per metre length is

$$C = \frac{2\pi\epsilon_0\epsilon_{r1}\epsilon_{r2}}{\epsilon_{r2}\log(r_2/r_1) + \epsilon_{r1}\log(r_3/r_1)} \tag{4-36}$$

Since the flux density is radial it is continuous across the boundary. Its value at any point at a distance r from the cable centre is

$$D = \frac{Q}{2\pi r}$$

The electric field intensities in the two dielectric layers are therefore given by

$$E_1 = \frac{Q}{2\pi\epsilon_1 r}; \quad r_1 \leq r < r_2$$

and

$$E_2 = \frac{Q}{2\pi\epsilon_2 r}; \quad r_2 < r \leq r_3$$

The ratio of the maximum to the minimum field intensities is given by

$$\frac{E_{1\,\text{max}}}{E_{2\,\text{min}}} = \frac{\epsilon_2 r_3}{\epsilon_1 r_1}$$

In the case of a single dielectric this ratio would be

$$\frac{E_{1\,\text{max}}}{E_{2\,\text{min}}} = \frac{r_3}{r_1}$$

By choosing ϵ_1 greater than ϵ_2 it is possible to achieve a more uniform field intensity throughout the insulation. This also reduces the maximum stress at the surface of the conductor which is one of the critical factors in the design of high voltage cables. With two dielectrics this maximum stress is given by

$$E_{\text{max}} = \frac{V_{13}}{r_1[\log(r_3/r_1) + \{(\epsilon_1/\epsilon_2) - 1\}\log(r_3/r_2)]} \qquad (4\text{--}37)$$

with only one dielectric $\epsilon_1 = \epsilon_2$ and

$$E_{\text{max}} = \frac{V_{13}}{r_1 \log(r_3/r_1)}$$

In general the value of the radius r_2 must be such that the maximum permissible stress in each medium is not exceeded; this means that

$$\epsilon_1 r_1 E_{1\,\text{max}} = \epsilon_2 r_2 E_{2\,\text{max}}$$

4.5 POISSON'S EQUATION AND COULOMB'S LAW IN A DIELECTRIC

For a linear isotropic dielectric medium we have shown that the relationship between $\mathbf{D}$ and $\mathbf{E}$ is given by $\mathbf{D} = \epsilon\mathbf{E}$. Equation 4–15 may therefore be written as

$$\nabla \cdot \mathbf{D} = \nabla \cdot (\epsilon\mathbf{E}) = \rho$$

Since the field $\mathbf{E}$ is always related to the scalar potential by

$$\mathbf{E} = -\nabla V$$

then

$$\nabla \cdot (\epsilon\nabla V) = -\rho$$

Expanding the left-hand side we obtain

$$\epsilon\nabla^2 V + \nabla V \cdot \nabla\epsilon = -\rho$$

If the medium is homogeneous $\nabla\epsilon = 0$ and

$$\nabla^2 V = -\rho/\epsilon \qquad (4\text{--}38)$$

which is the most general form of Poisson's equation (compare equation 2–120). If there are no free charges in the dielectric this reduces to Laplace's equation $\nabla^2 V = 0$.

Consider a point charge Q_1 in a linear isotropic homogeneous dielectric medium of permittivity ϵ. If we apply Gauss's law over the surface of a sphere of radius r and centre at Q_1 then

$$4\pi r^2 D = Q_1$$

and

$$D = \frac{Q_1}{4\pi r^2}$$

The electric field is therefore given by

$$E = \frac{Q_1}{4\pi\epsilon_0\epsilon_r r^2}$$

The force on a point charge Q_2 at a distance r from Q_1 is

$$F = Q_2 E = \frac{Q_1 Q_2}{4\pi\epsilon_0\epsilon_r r^2} \qquad (4\text{–}39)$$

Equation 4–39 may be regarded as the general form of Coulomb's law. It differs from the corresponding law for free space only by the additional factor $1/\epsilon_r$. It should be born in mind however, that the law is 'general' only in so far as the dielectric medium is linear, isotropic and homogeneous. Coulomb's law may be considered as a universal law of nature only its form for free space (equation 2–2).

4.6 DEPOLARIZING FACTOR

When a dielectric material is introduced in a region of space where the field is $\mathbf{E}_0$ the dipoles (either induced or permanent) align themselves along the field lines and a field, opposing the external field, is established between the positive and negative surface bound charges (positive and negative ends of the dipole chains). This opposing dipole field is called the depolarizing field $\mathbf{E}_{dep}$. It is the difference between the original field and the actual field inside the dielectric,

$$\mathbf{E}_{dep} = \mathbf{E}_0 - \mathbf{E}$$

The *depolarizing factor* is defined as the ratio

$$L = \frac{\epsilon_0 \mathbf{E}_{dep}}{\mathbf{P}} \qquad (4\text{–}40)$$

It is difficult to calculate the exact value of L except for relatively simple geometrical shapes.

For a sphere $L = 1/3$.

For an infinite thin flat plate with its faces perpendicular to the external field, $L = 1$.

For an infinitely long rod with its axis parallel to the external field, $L = 0$.

For an infinitely long circular cylinder with its axis perpendicular to the direction of the external field, $L = 1/2$.

4.7 MICROSCOPIC CONCEPT OF POLARIZATION

By defining the polarization of a dielectric medium as the total dipole moment per unit volume, we have shown in Section 4.3 that it is possible to account for the behaviour of dielectrics in an electric field by the introduction of the electric permittivity. Using this same definition of the polarization we shall show in this section that there exists a relationship between the macroscopic electric permittivity and the microscopic or molecular nature of the dielectric. An electric field polarizes a dielectric by inducing dipoles and (in polar dielectrics) by tending to orient in the field direction randomly distributed dipoles already present in the dielectric. Induced dipoles are created by *electronic* and *ionic polarizations* whilst the alignment of permanent dipoles gives rise to an *orientational polarization*. The total polarization is the sum of these three component polarizations which will now be considered separately.

4.7.1 ELECTRONIC POLARIZATION

An atom may be pictured as consisting of a positive nucleus of charge Ze where Z is the atomic number and e the electronic charge, surrounded by a spherically symmetric electronic cloud of total charge $-Ze$ distributed with uniform density throughout a sphere of radius r_0. In the absence of an electric field the centre of gravity of the negative electron cloud coincides with positive nucleus; the electric moment of the atom is thus zero. An electric field $\mathbf{E}'$ acting on the atom will exert a force ZeE' on the positive nucleus and an equal and opposite force on the electron cloud. This results in a small displacement in the respective centres of charge so that the atom, although still electrically neutral, acquires an induced electric moment p_e. To obtain an estimate of the magnitude of this moment we assume as an approximation that the electric field shifts the electron cloud with respect to the nucleus without changing its spherical shape. Let x be the distance between the nucleus and the displaced centre of the

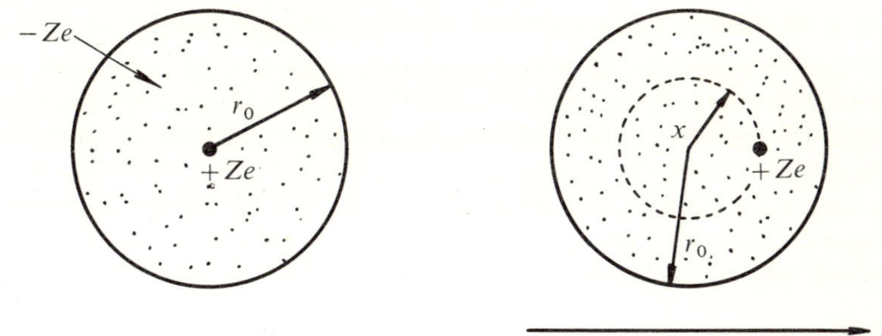

Fig. 4.11 Electronic polarization

electron cloud (Fig. 4.11). This distance may be obtained by equating the force ZeE' due to the field to the oppositely directed Coulomb force between the nucleus and the electron cloud.

By applying Gauss's law it is easy to show that the Coulomb force is that between the nucleus and the negative charge inside the sphere of radius x. This charge is proportional to the volume of this sphere and is equal to $-Zex^3/r_0^3$. Hence the Coulomb force is

$$F_C = -\frac{(Ze)^2 x^3}{4\pi\epsilon_0 x^2 r_0^3} = -\frac{(Ze)^2 x}{4\pi\epsilon_0 r_0^3} \qquad (4\text{--}41)$$

Equating this force to the force due to the field we have

$$ZeE' = -\frac{(Ze)^2 x}{4\pi\epsilon_0 r_0^3}$$

or

$$Zex = 4\pi\epsilon_0 r_0^3 E'$$

The quantity $\mathbf{p}_e = Ze\mathbf{x}$ is by definition the induced dipole moment of the atom. Thus

$$\mathbf{p}_e = 4\pi\epsilon_0 r_0^3 \mathbf{E}'$$
$$= \alpha_e \mathbf{E}' \qquad (4\text{--}42)$$

where the proportionality factor $\alpha_e = 4\pi\epsilon_0 r_0^3$ is called the *electronic polarizability* of the atom and has the dimensions of farad-metre^2, it is of the order of $10^{-40}\,\mathrm{F\,m^2}$. If we have a system of N atoms per unit volume and assume these to be sufficiently far apart so that any inter-action between induced dipoles is negligible*, then the electronic polarization is given by

$$\mathbf{P}_e = N\alpha_e \mathbf{E}' \qquad (4\text{--}43)$$

* This assumption is strictly valid only for monatomic gases such as He, Ar, Ne, etc.

4.7.2 IONIC POLARIZATION

This type of polarization is only found in ionic compounds (such as sodium chloride) whose molecules are formed of atoms having excess charges of opposite polarities. An electric field will tend to shift the relative position of the positive and negative ions of a molecule thus inducing a dipole moment other than that induced by the distortion of the electronic charges around individual atoms. The induced dipole moment may be expressed by an equation similar to equation 4–42,

$$\mathbf{p}_i = \alpha_i \mathbf{E}' \qquad (4\text{–}44)$$

where α_i is the *ionic polarizability* of the molecules. The corresponding ionic polarization is given by

$$\mathbf{P}_i = N\alpha_i \mathbf{E}' \qquad (4\text{–}45)$$

where N is the number of molecules per cubic metre.

4.7.3 ORIENTATIONAL POLARIZATION

This type of polarization is associated with polar dielectrics in which the molecules possess a dipole moment even in the absence of an applied field. Such a moment is not normally observed macroscopically because as a result of thermal agitation the molecules are oriented at random so that the average moment over a physically small volume is zero. In the presence of an external field the dipoles experience a torque tending to orient them in the direction of the field so that the average moment is no longer zero.

The calculation of the orientation polarization was first carried out by Debye by a method analogous to that originally developed by Langevin to explain the behaviour of paramagnetic substances in terms of permanent magnetic moments. To simplify the calculation it is assumed that the permanent electric moments are independent of the temperature and applied field, that they are free to rotate and that the interaction energy between the dipoles is small relative to the thermal energy kT of each dipole, where k is Boltzmann's constant ($1\cdot38 \times 10^{-23}$ joule per degree).

In the absence of a field there is no preferred orientation so that the probability of finding a dipole in any direction is the same. The number of dipoles $N(\theta)$ whose orientations lie between angles θ and $\theta + d\theta$ with an arbitrary direction is proportional to the solid angle subtended by the elemental area $2\pi \sin\theta \, d\theta$ on a sphere of unit radius (Fig. 4.12). When an electric field is applied along the z-direction say, then if the molecules possessed no thermal energy, they would all line up in the direction of the field (position of minimum energy). Because of thermal motion the number of dipoles oriented between θ and $\theta + d\theta$ must be weighted by the Boltzmann factor $\exp(-U/kT)$, where U is the energy of a polar molecule of moment $\mathbf{p}$ in an electric field $\mathbf{E}'$. By equation 2–94, $U = -pE'\cos\theta$ where θ is the angle between the dipole and the field. We thus have that

$$N(\theta) = A2\pi \sin\theta [\exp(pE'\cos\theta/kT)]\,d\theta$$

where A is a proportionality factor determined by the total number of dipole molecules per unit volume.

The average moment p_0 of each molecule in the direction of the field is given by the ratio of the total moment due to orientation in the direction of the field to the total number of molecules:

$$p_0 = \frac{\displaystyle\int_0^\pi A2\pi \sin\theta [\exp(pE'\cos\theta/kT)]p\cos\theta\,d\theta}{\displaystyle\int_0^\pi A2\pi \sin\theta [\exp(pE'\cos\theta/kT)]\,d\theta} \tag{4–46}$$

The factor $p\cos\theta$ appears in the numerator because a dipole of moment p making an angle θ with the field direction contributes to the polarization a component $p\cos\theta$. By letting

$$a = pE'/kT$$

the integrals in equation 4–46 may be readily evaluated to give

$$p_0 = p[\coth a - (1/a)] = pL(a) \tag{4–47}$$

The function $L(a)$ is known as the Langevin function; a plot of this

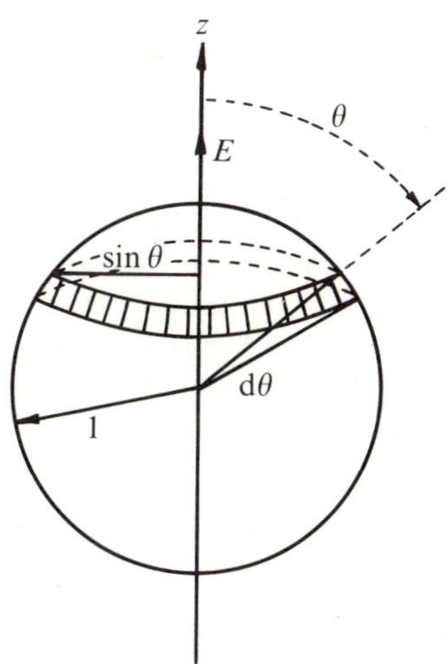

Fig. 4.12 Unit sphere model for calculating orientational polarization

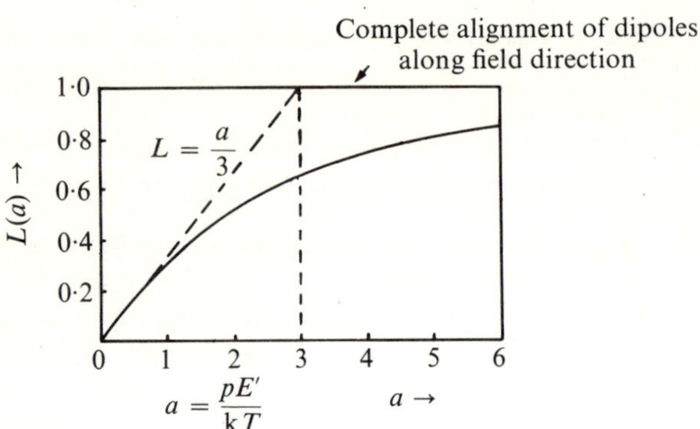

Fig. 4.13 The Langevin function $L(a) = [\coth a - (1/a)]$

function is shown in Fig. 4.13. From the practical point of view even for highly polar molecules and fields of the order of 10^7 volts per metre at room temperature, the value of $a \ll 1$ so that equation 4–47 may be expanded in series; neglecting terms higher than the first order we have

$$p_0 = p\,\frac{a}{3} = \frac{p^2 E'}{3kT} \tag{4-48}$$

$$= \alpha_0 E' \tag{4-49}$$

where $\alpha_0 = p^2/3kT$ is the *orientational polarizability*. The corresponding orientational polarization is

$$\mathbf{P}_0 = N\alpha_0 \mathbf{E}' \tag{4-50}$$

Thus, for a given field, the polarization of a dielectric arising from the orientation of the permanent dipoles is, unlike the electronic and ionic polarizations, strongly dependent on temperature and increases as the temperature decreases.

It should be mentioned here that the above value for α_0 was derived on the assumption that the dipoles are completely free to rotate. This is possible in gases and liquids but only in a few solids. A discussion of the so-called hindered rotation in solids lies outside the scope of this text and interested readers should refer to more specialized books.*

By adding the above three polarizabilities the total polarization may be written as

$$\mathbf{P} = N\left(\alpha_e + \alpha_i + \frac{p^2}{3kT}\right)\mathbf{E}' \tag{4-51}$$

By using the macroscopic definition of the polarization $\mathbf{P}$ given by

* e.g., C. F. J. Bottcher, *Theory of Electric Polarization*, Elsevier, Amsterdam, 1952.

equation 4·21, the above equation becomes

$$\epsilon_0(\epsilon_r - 1)\mathbf{E} = N\left(\alpha_e + \alpha_i + \frac{p^2}{3kT}\right)\mathbf{E}' \qquad (4\text{–}52)$$

Equation 4–52 expresses the relationship between the macroscopic quantity ϵ_r and the microscopic properties α_e, α_i and α_0 of the medium. In this equation $\mathbf{E}$ is the applied field whilst $\mathbf{E}'$ is the electric field acting on each molecule and is usually referred to as the local or internal field. As will be shown in the following section this field is in general different from the externally applied field.

4.8 THE LOCAL FIELD

When the density of molecules is reasonably low, as in a gas, the electrostatic interaction between dipoles can be neglected and ·the external field E can be equated to the local field E'. In solids and liquids, however, the distances between neighbouring molecules are of the same order as molecular dimensions so that the field acting on a given molecule consists partly of the applied field and partly of the field resulting from the mutual interaction of the other molecules, polarized by the external field. To take all interaction forces between molecules into account presents a problem of great mathematical difficulty. In general, however, considerable use can be made of simplified models which give satisfactory and meaningful results as long as the limitations on the validity of these models are recognized.

Interaction forces are usually of two types, short range and long range forces. Forces due to chemical bonds, van der Waals forces and atomic forces of repulsion are all short range forces; forces of interaction between dipoles are of an electrostatic character and predominate at large distances. In considering a simplified model of a dielectric medium only the long range dipolar forces are considered. The interaction of a particular dipole with all others in the medium is further simplified by assuming that beyond short distances from a molecule the medium may be treated as macroscopically continuous. This model was first used by Lorentz for calculating the local field intensity acting on a particular molecule.

Let the reference molecule A be surrounded by a sphere whose radius is large in comparison with intermolecular distances but small compared with the physical dimensions of the dielectric (Fig. 4.14). The field acting at the point A will consist of three components:

(i) a component E due to the free changes on the plates; this is the external field.

(ii) a component E_1 due to the polarization of the dielectric outside

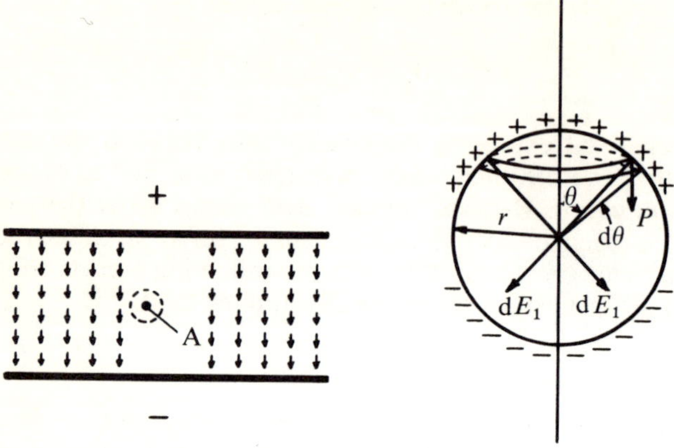

Fig. 4.14 Calculation of the Lorentz field

the sphere assuming that all molecules inside the sphere are removed.

(iii) a component E_2 due to all the molecules inside the sphere.

To calculate E_1 we make use of the definition of the polarization P as a surface charge density (equation 4–10) so that

$$dE_1 = \frac{P\,dS\cos\theta}{4\pi\epsilon_0 r^2}$$

Because of the spherical symmetry of the cavity the component field perpendicular to the external field direction will cancel so that the vertical field at A will be,

$$E_1 = \oint_{sphere} dE_1 \cos\theta = \oint_s \frac{P\cos^2\theta\,dS}{4\pi\epsilon_0 r^2}$$

or

$$E_1 = \int_0^\pi \frac{P\cos^2\theta\,2\pi r\sin\theta\,d\theta}{4\pi\epsilon_0 r^2} = \frac{P}{3\epsilon_0} \tag{4–53}$$

Substituting for P from equation 4–21 we have

$$E_1 = \tfrac{1}{3}(\epsilon_r - 1)E \tag{4–54}$$

In order to calculate the field E_2 accurate information is required on the geometrical configuration and polarizability of the molecules as well as on the interaction energy. Such information, if available, leads in general to great mathematical complications. For our purpose it is sufficient to point out that for solids the field E_2 is zero for crystals having cubic lattices and may be assumed zero for isotropic bodies in general. Thus according to Lorentz the internal field is given by

$$E' = E + \tfrac{1}{3}(\epsilon_r - 1)E$$

or

$$E' = \tfrac{1}{3}(\epsilon_r + 2)E \tag{4-55}$$

Substituting this value of E' into equation 4–52 we obtain

$$\epsilon_0(\epsilon_r - 1)E = \tfrac{1}{3}(\epsilon_r + 2)N\alpha E$$

or

$$\frac{N\alpha}{3\epsilon_0} = \frac{\epsilon_r - 1}{\epsilon_r + 2} \tag{4-56}$$

where $\alpha = \alpha_e + \alpha_i + \alpha_0$. This equation is known as the *Clausius–Mossotti* equation.

4.9 DIELECTRICS IN ALTERNATING FIELDS

When a dielectric material is subjected to an alternating field the orientation of dipoles, and hence the polarization, will tend to reverse every time the polarity of the field changes. As long as the frequency remains low ($< 10^6$ Hz) the polarization follows the alternations of the field without any significant lag and the permittivity is independent of the frequency and has the same magnitude as in a static field. When the frequency is increased the dipoles will no longer be able to rotate sufficiently rapidly so that their oscillations will begin to lag behind those of the field. As the frequency is further raised, the permanent dipoles, if present in the medium, will be completely unable to follow the field and the contribution to the static permittivity from this molecular process, the orientational polarization, ceases. This usually occurs in the radio frequency range (10^6–10^{11} Hz) of the electromagnetic spectrum. At still higher frequencies, usually in the infra-red (10^{11}–10^{14} Hz) the relatively heavy positive and negative ions cannot follow the field variations so that the contribution to the permittivity from the ionic polarization ceases and only the electronic polarization remains.

The above effects lead to a fall in the permittivity of a dielectric with increasing frequency, a phenomenon which is usually referred to as *anomalous dielectric dispersion*. Dispersion arising during the transition from full ionic polarization at radio frequencies to negligible ionic polarization at optical frequencies is usually referred to as *resonance absorption*. Dispersion arising during transition from full orientational polarization at zero or low frequencies to negligible orientational polarization at high radio frequencies is known as *dielectric relaxation*.

4.9.1 RESONANCE ABSORPTION

At optical frequencies the permittivity is almost entirely due to the electronic component of polarization. To determine the dependence of the electronic polarizability on the frequency of the applied field we shall use the classical model of an electron elastically bound to its atom; the results obtained are analogous to those derived from quantum theory.

An electron is elastically bound to its equilibrium position by a restoring force which is proportional to the displacement. Thus from equation 4–41 we have that the restoring force (which is the Coulomb force) is

$$F_C = -\frac{Z^2 e^2 x}{4\pi\epsilon_0 r_0^3} = -ax$$

where a is a constant known as the restoring force constant. The negative sign indicates that the force tends to move the electron towards $x = 0$. The classical differential equation describing the motion of the electron of mass m and charge e in an alternating field $E_0 \cos \omega t$, where E_0 is the amplitude of the field and ω is its angular frequency, is therefore that of a forced harmonic oscillator,

$$m\frac{d^2 x}{dt^2} + ax = e \, \mathrm{Re}(E_0 \, e^{j\omega t}) \qquad (4\text{–}57)$$

where x is the displacement and $\mathrm{Re}(E_0 e^{j\omega t})$ is the real part* of $E_0 e^{j\omega t} = E_0 \cos \omega t$.

The general solution of equation 4–57 consists of two parts, a particular integral and a complementary function. We are only interested here in the particular integral since it gives the displacement x caused by the polarizing field. The particular integral is

$$x = \mathrm{Re}\left(\frac{e}{m}\frac{E_0 e^{j\omega t}}{\omega_0^2 - \omega^2}\right) = X_0 \cos \omega t \qquad (4\text{–}58)$$

where

$$X_0 = \frac{e}{m}\frac{E_0}{\omega_0^2 - \omega^2} \qquad (4\text{–}59)$$

and

$$\omega_0 = (a/m)^{1/2} \qquad (4\text{–}60)$$

ω_0 is the natural or resonance angular frequency of the oscillations. It is clear from equation 4–59 that the amplitude of the oscillations X_0

---

* $e^{j\omega t} = \cos \omega t + j \sin \omega t$

becomes infinitely large when $\omega = \omega_0$. This phenomenon is known as resonance. Oscillations of infinite amplitude do not normally occur in nature because the amplitude is almost always limited or damped by friction forces which constitute a resistance to motion. In the vibrating electron model considered here damping occurs because of the emission of electromagnetic radiation by the oscillating charge and this is an energy consuming process. This process is quite satisfactorily accounted for by introducing a 'radiation resistance force' proportional to the velocity of the electron. The equation of motion of the electron now becomes

$$m \frac{d^2x}{dt^2} + b \frac{dx}{dt} + ax = e\,\mathrm{Re}(E_0 e^{j\omega t}) \tag{4–61}$$

where b is the radiation damping constant. The displacement x is again given by the particular integral of the above equation and is

$$x = \mathrm{Re}\left(\frac{eE_0 e^{j\omega t}}{m(\omega_0^2 - \omega^2) + jb\omega} \right) \tag{4–62}$$

Since the induced dipole moment is defined as $p_e = ex$ we have

$$p_e = \mathrm{Re}\left(\frac{e^2 E_0 e^{j\omega t}}{m(\omega_0^2 - \omega^2) + jb\omega} \right) \tag{4–63}$$

Comparing this equation with equation 4–42 and noting that the coefficient of $E_0 e^{j\omega t}$ is a complex quantity, it follows that

$$p_e = \mathrm{Re}(\hat{\alpha}_e E_0 e^{j\omega t}) \tag{4–64}$$

where

$$\hat{\alpha}_e = \frac{e^2/m}{(\omega_0^2 - \omega^2) + j(b\omega/m)} \tag{4–65}$$

$\hat{\alpha}_e$ is now a complex quantity defined as the complex electronic polarizability. Separating the real and the imaginary components we obtain

$$\hat{\alpha}_e = \frac{e^2}{m} \left[\frac{\omega_0^2 - \omega^2}{(\omega_0^2 - \omega^2)^2 + (b^2\omega^2/m^2)} - j\frac{b\omega/m}{(\omega_0^2 - \omega^2)^2 + (b^2\omega^2/m^2)} \right] \tag{4–66}$$

$$= \alpha_e' - j\alpha_e''$$

The variation of the real and imaginary components of $\hat{\alpha}_e$ with frequency is shown in Fig. 4.15. When ω is much smaller than ω_0 the imaginary component is negligibly small whereas the real component has a static positive value of $e^2/m\omega_0^2$. When ω is much larger than ω_0 the imaginary component is again negligible whilst the real component

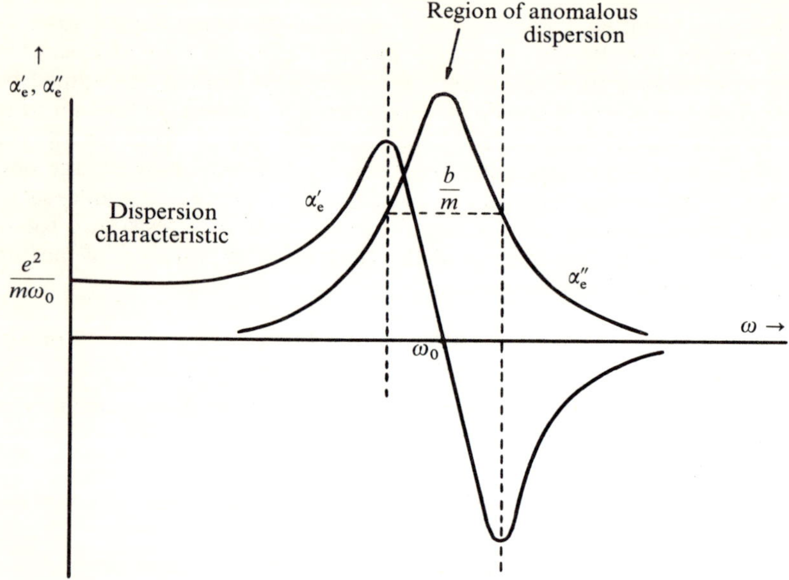

Fig. 4.15 The variations of the real and imaginary components of the electronic polarizability with the angular frequency

approaches zero negatively. When $\omega = \omega_0$ the imaginary component is a maximum whilst the real component is zero. In order to study the behaviour of α'_e and α''_e in the region around the natural frequency ω_0 we introduce the deviation from resonance $\Delta\omega = \omega_0 - \omega$ so that we have

$$\omega_0^2 - \omega^2 \simeq 2\omega_0(\omega_0 - \omega) \simeq 2\omega_0\Delta\omega$$

The real and imaginary components of the polarizability become

$$\alpha'_e = \frac{e^2}{m} \cdot \frac{\Delta\omega/2\omega_0}{(\Delta\omega)^2 + b^2/4m^2} \tag{4-67}$$

$$\alpha''_e = \frac{e^2}{m} \cdot \frac{b/4m\omega_0}{(\Delta\omega)^2 + b^2/4m^2} \tag{4-68}$$

From equations 4–67 and 4–68 it may be readily shown that

(i) α'_e has a maximum value at $\Delta\omega = b/2m$ and a minimum value at $\Delta\omega = -b/2m$.

(ii) α''_e has a maximum value at $\Delta\omega = 0$ and falls to half its maximum value at $\pm\Delta\omega = b/2m$. It is in the interval between the half-value points that α'_e decreases with increasing frequency.

It is desirable at this point to give some approximate estimate of the natural or resonance angular frequency ω_0. For values of $\omega_0 \gg \omega$ it was shown above that the electronic polarizability has the static value of $e^2/m\omega_0^2$. Equating this to the static electronic polarizability as defined by equation 4–42 we have

$$\alpha_e = e^2/m\omega_0^2 = 4\pi\epsilon_0 r_0^3$$

and

$$\omega_0^2 = \frac{e^2}{4\pi\epsilon_0 m r_0^3}$$

with $e = 1{\cdot}6 \times 10^{-19}$ coulomb, $r_0 \simeq 10^{-10}$ metre and $m \simeq 10^{-31}$ kg, the value of ω_0 is approximately 10^{16} radians per second. This lies in the ultra-violet portion of the electromagnetic spectrum and this is why the electronic polarizability is sometimes referred to as the *optical polarizability*.

The dependence of the ionic polarizability on frequency may be derived in a manner analogous to the one used above. The ionic polarizability will again be a complex quantity given by

$$\hat{\alpha}_i = \alpha_i' - j\alpha_i''$$

and α_i' and α_i'' will vary with frequency in a similar manner as do α_e' and α_e''; the only difference is that, because of the relatively larger mass of the ions, the natural frequency of oscillation will be much lower than that of electrons. The natural frequency of oscillation will normally lie in the infra-red region and is of the order of 10^{14} radians per second.

From equation 4–21 it follows that the relative permittivity may be expressed as

$$\epsilon_r = 1 + \frac{\mathbf{P}}{\epsilon_0 \mathbf{E}} \tag{4–69}$$

Also, the polarization $\mathbf{P}$ is given by

$$\mathbf{P} = N\alpha\mathbf{E}'$$

where α now represents the sum of the electronic and ionic polarizabilities. Substituting this value of $\mathbf{P}$ into equation 4–69 and assuming that $\mathbf{E}' = \mathbf{E}$ we obtain

$$\epsilon_r = 1 + \frac{N\alpha}{\epsilon_0} \tag{4–70}$$

Since for fields alternating at high frequencies α is a complex quantity, it follows that at such frequencies the relative permittivity must also be complex; thus

$$\hat{\epsilon}_r = 1 + N(\alpha' - j\alpha'')/\epsilon_0$$

$$= \left(1 + \frac{N\alpha'}{\epsilon_0}\right) - j\frac{N\alpha''}{\epsilon_0}$$

or

$$\hat{\epsilon}_r = \epsilon_r' - j\epsilon_r'' \tag{4-71}$$

For the case in which E' is given by equation 4–50 we have that

$$\frac{\hat{\epsilon}_r - 1}{\hat{\epsilon}_r + 2} = \frac{N\hat{\alpha}}{3\epsilon_0}$$

and the reader may readily show that ϵ_r' and ϵ_r'' will vary with frequency in a manner similar to the variations of α' and α'', the only difference being a shift in the resonance frequency from ω_0^2 to $\omega_0^2 - Ne^2/3\epsilon_0 m$. The variation of ϵ_r' with frequency is termed the *dispersion curve* of the dielectric medium; in those regions where ϵ_r' increases with increasing frequency the dispersion is referred to as *normal dispersion* whereas in the region where ϵ_r' decreases with increasing frequency the dispersion is referred to as *anomalous dispersion*. The variation of ϵ_r'' with frequency is termed the resonance absorption characteristic of the dielectric for, as will be shown in the next section, the imaginary component of the permittivity represents a dielectric loss and hence an absorption of energy.

4.9.2 DIELECTRIC RELAXATION

Relaxation phenomena are associated with the frequency dependence of the orientational polarization and hence with polar dielectric media. In static or slowly varying fields the permanent dipoles align themselves along the field acting upon them and thus contribute fully to the total polarization of the dielectric. At frequencies in the infrared and beyond the variations in the field are too rapid for the dipoles to follow so that their contribution to the polarization, and hence to the permittivity, is negligible. At intermediate frequencies, which usually lie in the radio and microwave frequency range (10^6–10^{11} Hz), the dipoles cannot follow the field variations without a measurable lag because of internal retarding or 'friction' forces.

For frequencies below the infra-red the contributions of the ionic and electronic polarizations to the total polarization P are independent of frequency and may be expressed as

$$P_e + P_i = P_\infty = \epsilon_0(\epsilon_{r\infty} - 1)E \tag{4-72}$$

where $\epsilon_{r\infty}$ is the relative permittivity at frequencies which are too high for the permanent dipoles to follow and arises solely therefore from the electronic and ionic polarizations. Under static conditions

$$P_s = P_o + P_\infty$$

so that the orientational polarization P_o is given by

$$
\begin{aligned}
P_o &= P_s - P_\infty \\
&= \epsilon_0[(\epsilon_{rs}-1)-(\epsilon_{r\infty}-1)]E \\
&= \epsilon_0(\epsilon_{rs}-\epsilon_{r\infty})E
\end{aligned}
\tag{4-73}
$$

where ϵ_{rs} is the relative permittivity under static conditions.

When a static field is suddenly applied to a dielectric the equilibrium value of P_∞ is attained almost instantaneously whilst a finite time is required for P_o to reach its equilibrium value. According to Pellat and Debye P_o approaches its final value exponentially so that the orientational polarization at any instant t after the application of the field is given by

$$P_o(t) = P_o(1-e^{-t/\tau})
\tag{4-74}$$

where τ has the dimensions of time and determines the rate at which the polarization builds up. τ is called the relaxation time of the dielectric medium; it is independent of the frequency but depends upon the temperature. The time derivative of equation 4–74 gives

$$
\begin{aligned}
\frac{dP_o(t)}{dt} &= \frac{1}{\tau}[P_o - P_o(t)] \\[2mm]
&= \frac{1}{\tau}[\epsilon_0(\epsilon_{rs}-\epsilon_{r\infty})E - P_o(t)]
\end{aligned}
$$

In the case of an alternating field, represented symbolically by $\hat{E} = E_0 e^{j\omega t}$, the differential equation becomes

$$\frac{d\hat{P}_o(t)}{dt} = \frac{1}{\tau}[\epsilon_0(\epsilon_{rs}-\epsilon_{r\infty})\hat{E} - \hat{P}_o(t)]
\tag{4-75}$$

The steady state solution of equation 4–75 gives

$$\hat{P}_o(t) = \frac{\epsilon_0(\epsilon_{rs}-\epsilon_{r\infty})\hat{E}}{1+j\omega\tau}
\tag{4-76}$$

and the complex total polarization is therefore

$$
\begin{aligned}
\hat{P}(t) &= \hat{P}_o(t) + \hat{P}_\infty(t) \\[2mm]
&= \frac{\epsilon_0(\epsilon_{rs}-\epsilon_{r\infty})\hat{E}}{1+j\omega\tau} + \epsilon_0(\epsilon_{r\infty}-1)\hat{E}
\end{aligned}
\tag{4-77}
$$

154   DIELECTRICS IN THE ELECTRIC FIELD

The electric flux density may be expressed as

$$\hat{D} = \epsilon_0 \hat{E} + \hat{P}(t)$$

$$= \epsilon_0 \left[\epsilon_{r\infty} + \frac{\epsilon_{rs} - \epsilon_{r\infty}}{1 + j\omega\tau} \right] \hat{E} \tag{4-78}$$

and it follows that

$$\hat{\epsilon}_r = \epsilon_{r\infty} + \frac{\epsilon_{rs} - \epsilon_{r\infty}}{1 + j\omega\tau} \tag{4-79}$$

Separating $\hat{\epsilon}_r$ into a real and an imaginary part we find that

$$\hat{\epsilon}_r = \epsilon'_r - \epsilon''_r$$

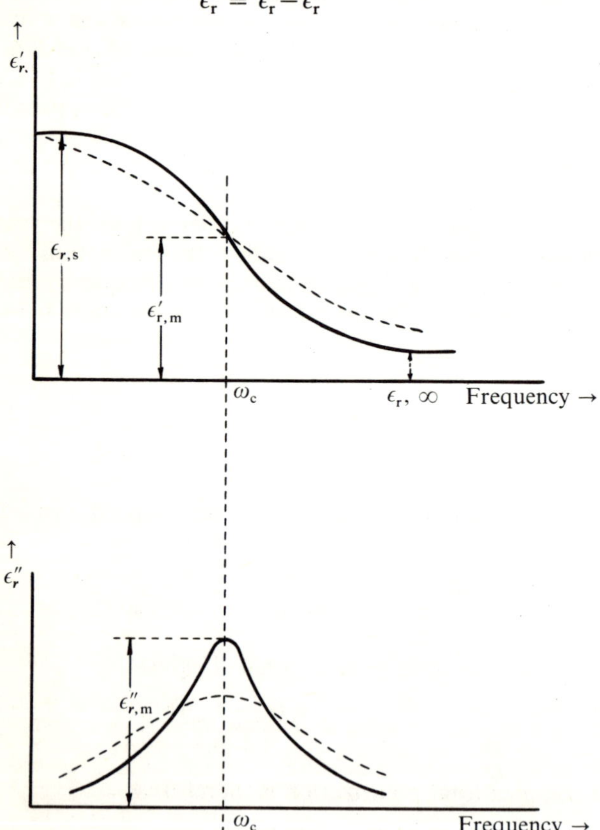

Fig. 4.16 Frequency dependence of ϵ'_r and ϵ''_r as given by Debye equations (solid curves). The broken curves show typical broadening of the characteristics obtained from experimental measurements

where

$$\epsilon_r' = \epsilon_{r\infty} + \frac{\epsilon_{rs} - \epsilon_{r\infty}}{1 + \omega^2\tau^2} \tag{4-80}$$

$$\epsilon_r'' = \frac{(\epsilon_{rs} - \epsilon_{r\infty})\omega\tau}{1 + \omega^2\tau^2} \tag{4-81}$$

Equations 4–80 and 4–81 are known as the Debye relaxation equations. The variations of ϵ_r' and ϵ_r'' with frequency are shown in Fig. 4.16. The decrease of ϵ_r' with increasing frequency represents the anomalous dispersion of the relative permittivity at low and intermediate frequencies whilst the variation of ϵ_r'' with frequency represents the relaxation absorption of the dielectric. At zero frequency ϵ_r' is equal to the static value of the relative permittivity and ϵ_r'' is zero. At sufficiently high frequencies such that $\omega\tau \gg 1$, $\epsilon_r' = \epsilon_{r\infty}$ and ϵ_r'' is negligibly small. The maximum value of ϵ_r'' is reached when $\omega\tau = 1$; the corresponding frequency $f = 1/2\pi\tau$ is called the critical frequency and for this value

$$(\epsilon_r')_{\max} = \tfrac{1}{2}(\epsilon_{rs} + \epsilon_{r\infty}) \tag{4-82}$$

$$(\epsilon_r'')_{\max} = \tfrac{1}{2}(\epsilon_{rs} - \epsilon_{r\infty}) \tag{4-83}$$

Thus by determining the frequency at which ϵ_r'' is a maximum the relaxation time τ may be found. The value of τ will normally lie between 10^{-6} and 10^{-10} seconds.

4.10 DIELECTRIC LOSSES

In discussing the losses which occur in a dielectric when it is subjected to an alternating field we shall first consider the case in which the permittivity of the dielectric may be represented by its static value. This means that the frequency of the applied field is sufficiently low to permit the permanent and induced dipoles to follow the field variations without any measurable lag. It will be assumed that the dielectric is ideal or perfect in the sense that its ohmic conductivity is zero. The relationship between the dielectric displacement density and the sinusoidally varying field intensity is given by,

$$D = \epsilon_0\epsilon_r E_0 \cos \omega t = \mathrm{Re}(\epsilon_0\epsilon_r E_0 e^{j\omega t}) \tag{4-84}$$

The displacement or charging current density is, by definition (see Section 6.6),

$$J_d = \frac{dD}{dt} = \mathrm{Re}(\epsilon_0\epsilon_r j\omega E_0 e^{j\omega t})$$

$$= -\epsilon_0\epsilon_r \omega E_0 \sin \omega t$$

$$= -\epsilon_0\epsilon_r \omega E_0 \cos(\omega t + 90°) = J_{d0} \cos(\omega t + 90°) \tag{4-85}$$

so that the current density leads the field intensity by a temporal phase angle of 90°. The energy absorbed per unit volume of the dielectric (energy density) and per second is given by

$$U = \frac{\omega}{2\pi} \int_0^{2\pi/\omega} J_d E \, dt \tag{4-86}$$

Substituting for J_d and E we obtain

$$U = -\frac{\omega}{2\pi} \int_0^{2\pi/\omega} \epsilon_0 \epsilon_r \omega E_0^2 \sin \omega t \cos \omega t \, dt = 0$$

so that the dielectric does not absorb any energy from the field.

We shall next consider the case in which the frequency of the alternating field to which the ideal dielectric is subjected is sufficiently high so that the polarization is now frequency dependent. It has been shown in Section 4.8 that this frequency dependence leads to a complex permittivity $\hat{\epsilon}_r = \epsilon_r' - j\epsilon_r''$. Equation 4–84 now becomes

$$D = \mathrm{Re}(\epsilon_0 \hat{\epsilon}_r E_0 e^{j\omega t})$$

and the corresponding displacement current density is

$$J_d = \mathrm{Re}\left(\epsilon_0 (\epsilon_r' - j\epsilon_r'') j\omega E_0 e^{j\omega t}\right)$$
$$= -\epsilon_0 \omega E_0 (\epsilon_r' \sin \omega t - \epsilon_r'' \cos \omega t) \tag{4-87}$$

Equation 4–87 shows that the displacement current now consists of two parts. The first part is the same as equation 4–85 and does not lead therefore to any energy absorption. The second part, which is associated with ϵ_r'' is in phase with the applied field and the corresponding energy density is

$$U = \frac{\omega}{2\pi} \int_0^{2\pi/\omega} \epsilon_0 \epsilon_r'' \omega E_0^2 \cos^2 \omega t \, dt$$
$$= \tfrac{1}{2} \omega \epsilon_0 \epsilon_r'' E_0^2 \tag{4-88}$$

This energy appears as heat in the dielectric. Equation 4–88 shows that the energy density absorbed by the dielectric is proportional to ϵ_r'' and consequently ϵ_r'' is known as the *loss factor*; the variation of ϵ_r'' with frequency gives the absorption characteristic or spectrum of the dielectric. In practice it is found convenient to specify the absorption losses in a dielectric (at a given frequency and temperature) by the *loss tangent*. Thus equation 4–87 may be rewritten as

$$J_d = -\epsilon_0 \omega |\hat{\epsilon}_r| E_0 \sin (\omega t - \delta)$$

where

$$\sin \delta = \epsilon_r'' / |\hat{\epsilon}_r|, \quad \cos \delta = \epsilon_r' / |\hat{\epsilon}_r|$$

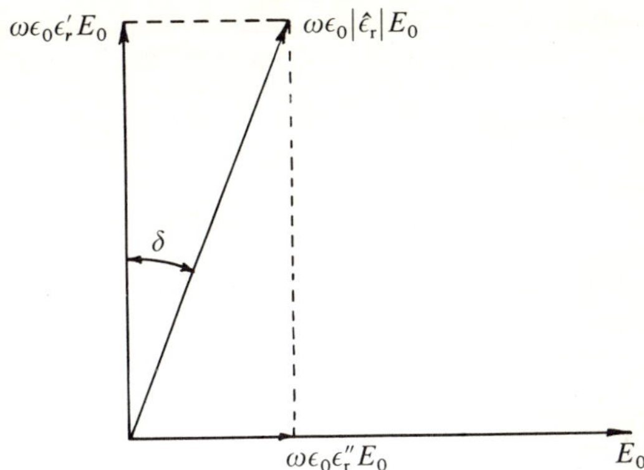

Fig. 4.17 Components of the displacement current and the loss angle for an ideal dielectric ($\gamma = 0$)

and the loss tangent is given by

$$\tan \delta = \frac{\epsilon_r''}{\epsilon_r'} \tag{4-89}$$

δ is the angle between the total displacement current J_d and its lossless component (Fig. 4.17).

A knowledge of the dielectric loss of dielectric materials used for insulation or in the manufacture of condensers is of primary importance in solving high-frequency electrical engineering problems. Excessive losses will not only reduce circuit efficiency but will cause undesirable heating which may lead to electrical or mechanical failure of the dielectric. Table 4.1 lists the loss tangent and ϵ_r' for a number of dielectrics at room temperature and a frequency of 10^6 Hz. Figure 4.18 summarizes the frequency dependence of ϵ_r' and ϵ_r'' for a dielectric having the three most important types of polarization.

Table 4.1 The relative permittivity and loss tangent of some common gaseous, liquid and solid dielectrics. (The values for liquids and solids are for a temperature of 25°C)

		$f < 100\,\text{Hz}$	$f = 10^6\,\text{Hz}$	
	Dielectric medium	ϵ_r	ϵ_r'	$10^4 \tan\delta$
Gases	Vacuum	1		
	Air (760 torr, 0°C)	1·000576		
	Hydrogen ,,	1·000272		
	Carbon dioxide ,,	1·000985		
	Methane ,,	1·00094		
Liquids	Castor oil	4·3		
	Paraffin oil (b.p.)	2·2	2·2	∼1·0
	Transformer oil (BS 138)	2·2	2·2	∼10·0
	Methyl alcohol	31·2	31·0	2000
	Ethyl alcohol	25·8	24·5	900
	n-Hexane	1·9		
	Water	81·0		
Cryogenic Liquids	Helium (b.p. 4·2 K)	1·048		
	Hydrogen (b.p. 20 K)	1·227		
	Nitrogen (b.p. 77 K)	1·433		
	Oxygen (b.p. 90 K)	1·484		
Solids	Amber	2·7	2·65	56
	Beeswax (white)	2·65	2·4	84
	Beeswax (yellow)	2·73	2·5	92
	Diamond	5·5	5·5	
	Epoxy resin	3·6–11·0	3·5–4·0	300
	Germanium	16	16	
	Glass	4·0–7·5	4–7	15–100
	Marble	10–15		
	Mica	6–11	6–11	1·0
	Neoprene	7	5·7	950
	Paper (impregnated with mineral oil)	3·5–4·5		
	Polyethylene	2·25	2·25	1·0
	Polyvinylchloride (PVC)	6·0	3·2–3·6	450–900
	Polystyrene	2·6	2·6	0·7
	PTFE (Teflon)	2·1	2·1	2·0
	Porcelain	5·5	5·5	80
	Quartz (fused)	3·8	3·8	2·0
	Rubber (vulcanized)	3–4	2·7	440
	Silicon	11·5	11·5	
	Steatite	6	6	20

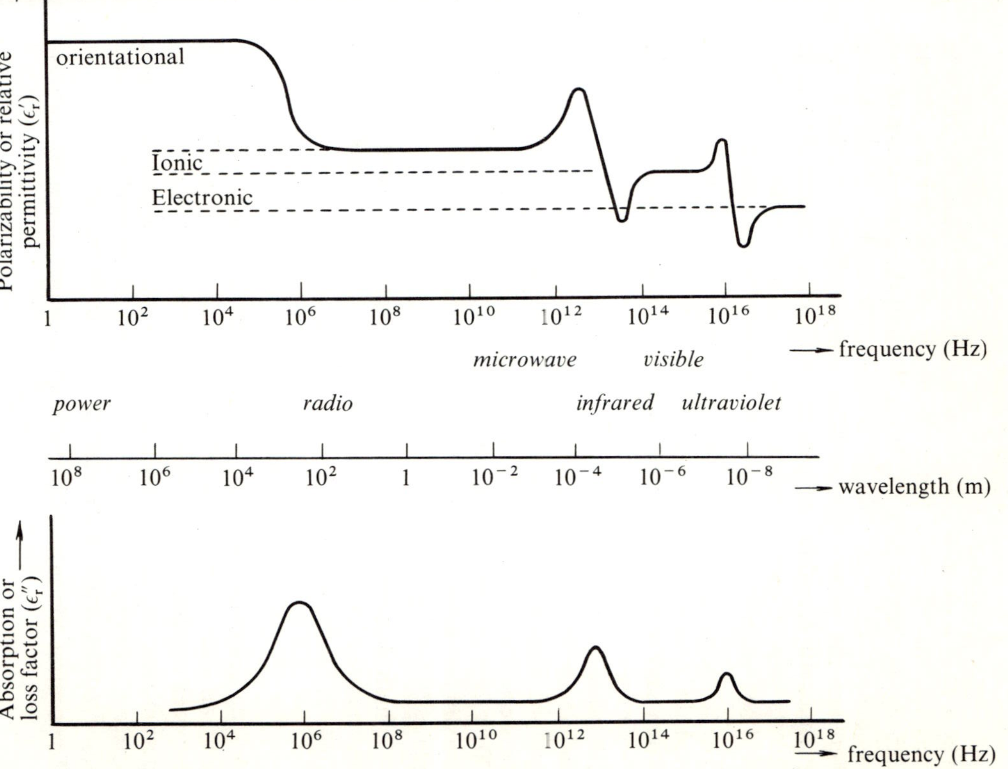

Fig. 4.18 Diagrammatic representation of the variation of the permittivity and loss factor with frequency

PROBLEMS

Chapter Four

4.1   The area of a parallel plate capacitor is $100\,\text{cm}^2$ and the distance between the plates is $5\,\text{mm}$. With air as dielectric the charge density on the plates is $10^{-6}\,\text{C}\,\text{m}^{-2}$. Find the capacitance of the condenser and the voltage between the plates. If the space between the plates is now filled with a dielectric of relative permittivity 4, find (a) the potential difference between the plates, (b) the capacitance of the condenser and (c) the density of free and bound charges on one of the plates. Neglect end effects.

4.2   The distance between the parallel plates of a capacitor is $5\,\text{mm}$ and the area of each plate is $100\,\text{cm}^2$. A dielectric of thickness $3\,\text{mm}$ and same area as the plates is placed between them. A potential difference is now applied between the plates such that the plates carry charges $\pm 10^{-6}\,\text{C}\,\text{m}^{-2}$. Find (a) the potential difference between the plates, (b) the polarization in the dielectric and the density of bound charges at its surfaces, (c) the capacitance between the two plates. If the breakdown voltage of air is $30\,\text{kV/cm}$ and that of the dielectric is $200\,\text{kV/cm}$, what is the maximum potential difference that can be applied between the plates?

4.3   A spherical capacitor consists of a conducting sphere of radius a surrounded by a dielectric of uniform thickness d and permittivity ϵ. Show that the capacitance is given by

$$C = 4\pi\epsilon a(a+d)/(\epsilon_r a+d)$$

4.4   A long coaxial cable consists of an inner conductor of radius a and an outer earthed shield of internal radius b. The space between conductor and shield is filled with a dielectric of relative permittivity ϵ_{r1} up to a radius r ($a \leq r < b$) and the remaining space is occupied by a second dielectric of relative permittivity ϵ_{r2}. Find the capacitance per metre length of the cable.

4.5   The central conductor of a vacuum-insulated concentric cable is

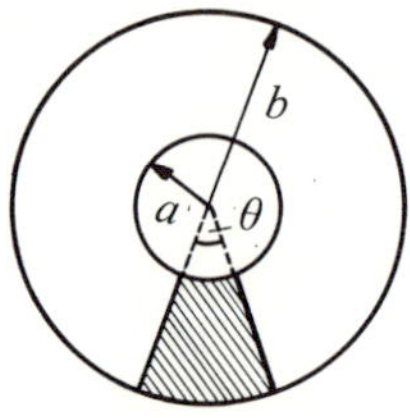

Fig. P. 4.5

supported along its length by a wedge-shaped dielectric as shown in Fig. P4.5. If the permittivity of the dielectric is ϵ find the capacitance of the cable per unit length. The outer conductor is earthed.

4.6 A coaxial cable consists of an inner conductor of radius 0·6 cm and an outer sheath of radius 1·7 cm. The inner conductor is surrounded by a uniform film of air 1 mm thick and the remaining space is filled by a dielectric of relative permittivity 3·5. If the breakdown strength of air is 30 kV/cm find the maximum voltage which can be applied between conductor and sheath. If the air film is replaced by a film of oil of relative permittivity 2·4 and strength 300 kV/cm, find the maximum voltage.

4.7 A homogeneous dielectric of relative permittivity 2 fills the region $z \leq 0$ (region 1). The region $z \geq 0$ (region 2) is free space. If
(a) $\mathbf{E}_1 = 6\mathbf{a}_x + 10\mathbf{a}_y - 6\mathbf{a}_z$, find $\mathbf{E}_2, \theta_2$,
(b) $E_1 = 12, \theta_1 = 27°$, find $\mathbf{E}_2, \theta_2$,
(c) $E_2 = 7, \theta_2 = 50°$, find E_1, θ_1.

4.8 A dielectric sphere of radius a is uniformly polarized with a density P throughout. Show that the potential at a distance $r > a$ from the centre is given by

$$V = \frac{\mathbf{P} \cdot \mathbf{a}_r a^3}{3\epsilon_0 r^2}$$

4.9 A point charge Q is placed at a distance d from a point dipole of moment $\mathbf{p}$. Show that the force on the dipole is $Q\mathbf{p}/2\pi\epsilon_0 d^3$.

4.10 A dielectric material contains $1·5 \times 10^{19}$ polar molecules per cubic metre, each of moment $1·6 \times 10^{-27}$ C m. If all the dipoles are assumed to be aligned in the direction of a uniform field of strength 10^5 V m^{-1}, find the relative permittivity of the dielectric.

4.11 An atom whose polarizability is 10^{-40} F m^2 is at a distance of $0·5 \times 10^{-10}$ m from a proton. Find the dipole moment induced in the atom and the force between the atom and the proton.

4.12 Two identical atoms of polarizability $\alpha = 1·5 \times 10^{-40}$ F m^2 are at a distance $2·5 \times 10^{-10}$ m apart. If a uniform field of strength E V m^{-1} is applied in the direction of the line joining the atom centres, find the ratio between the internal and the external field.

4.13 A solid contains 5×10^{28} atoms per cubic metre and the polarizability of each atom is 2×10^{-40} F m^2. Assuming that the internal field is given by the Lorentz equation find the relationship between the applied field and the internal field.

CHAPTER FIVE

ENERGY AND FORCES IN THE ELECTROSTATIC FIELD

5.1 ENERGY ASSOCIATED WITH A GIVEN CONFIGURATION OF CHARGES

Suppose that a point charge Q_1 is fixed at some point 1 in space. We shall assume that all space is filled with a linear dielectric medium of permittivity ϵ (for free space $\epsilon = \epsilon_0$). If a second point charge Q_2, initially at a very large distance from Q_1 (theoretically, at an infinite distance), is brought to a point 2 at a distance r_{12} from Q_1 the work done is

$$W = Q_2 V_{21}$$

where V_{21} is the potential at point 2 produced by the charge Q_1,

$$V_{21} = \frac{Q_1}{4\pi\epsilon r_{12}}$$

Since

$$V_{12} = \frac{Q_2}{4\pi\epsilon r_{21}}$$

it follows that

$$Q_2 V_{21} = Q_1 V_{12}$$

This relationship is valid for any pair of point charges Q_i and Q_j at a distance r_{ij} apart,

$$Q_i V_{ij} = Q_j V_{ji} \tag{5-1}$$

For the charges Q_1 and Q_2 we may therefore write the energy of the system in the symmetrical form

$$W = \tfrac{1}{2}[Q_1 V_{12} + Q_2 V_{21}]$$

If a third point charge Q_3, initially at a very large distance from both Q_1 and Q_2, is now brought to a point 3 distant r_{13} from Q_1 and r_{23} from Q_2, the additional work required is $Q_3 V_{31} + Q_3 V_{32}$. Hence the total work required to position the three charges is

$$W = Q_2 V_{21} + Q_3 V_{31} + Q_3 V_{32}$$

Because of equation 5–1, this may be rewritten as

$$W = Q_1 V_{12} + Q_1 V_{13} + Q_2 V_{23}$$

and symmetrically as

$$W = \tfrac{1}{2}[Q_1(V_{12}+V_{13})+Q_2(V_{21}+V_{23})+Q_3(V_{31}+V_{32})]$$

The above expression can be generalized to represent the energy of N point charges,

$$W = \frac{1}{2}\sum_{i=1}^{N} Q_i \sum_{j\neq i}^{N} V_{ij}$$

If we let

$$V_i = \sum_{j\neq i}^{N} V_{ij}$$

then

$$W = \tfrac{1}{2}\sum Q_i V_i \text{ joule} \tag{5-2}$$

V_i is the potential at the point i due to all charges except Q_i.

If the number of point charges is very large, we may replace them by a continuous volume charge distribution of free charges of density ρ. In equation 5–2 each charge is now replaced by $\rho\, dv$ and the summation by an integral so that

$$W = \tfrac{1}{2}\int_v \rho V\, dv \tag{5-3}$$

where V is the potential produced by the distribution at some point **P** where the density is ρ (Fig. 5.1).

$$V = \int \frac{\rho'\, dv'}{4\pi\epsilon r} \tag{5-4}$$

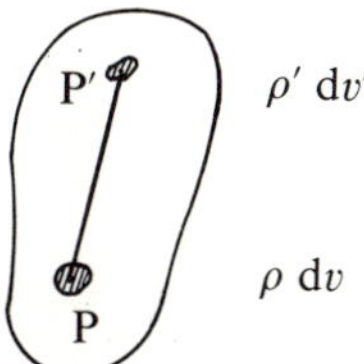

Fig. 5.1 Finite volume distribution of free charge

5.2 CONCEPT OF ENERGY DENSITY IN THE ELECTROSTATIC FIELD

If in equation 5–3 we replace ρ by $\nabla \cdot \mathbf{D}$, the energy W may be expressed as

$$W = \tfrac{1}{2}\int_v V(\nabla \cdot \mathbf{D})\, dv$$

By making use of the vector identity

$$\nabla \cdot (V\mathbf{D}) = V(\nabla \cdot \mathbf{D}) + \mathbf{D} \cdot (\nabla V) \tag{5-5}$$

we obtain

$$W = \tfrac{1}{2} \int_v \nabla \cdot (V\mathbf{D}) \, dv - \tfrac{1}{2} \int_v \mathbf{D} \cdot (\nabla V) \, dv$$

By applying the divergence theorem to the first integral on the right of the above expression we obtain

$$W = \tfrac{1}{2} \oint_S (V\mathbf{D}) \cdot \mathbf{dS} - \tfrac{1}{2} \int_v \mathbf{D} \cdot (\nabla V) \, dv \tag{5-6}$$

where S is the closed surface bounding the volume v.

Let Σ be the surface of a sphere enclosing the charge distribution and

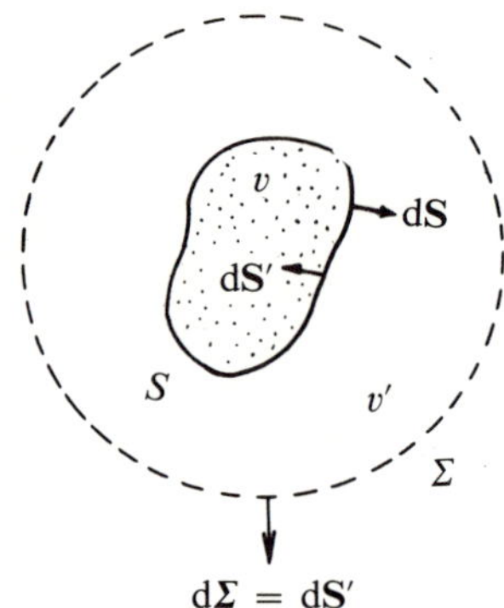

Fig. 5.2 Hypothetical spherical surface enclosing a volume distribution of charge

let v' be the volume bounded by the surfaces S and Σ (Fig. 5.2). Using successively the divergence theorem and the vector identity (5-5) we have that

$$\oint_{S+\Sigma} V\mathbf{D} \cdot \mathbf{dS'} = \int_{v'} \nabla \cdot (V\mathbf{D}) \, dv$$

$$= \int_{v'} V(\nabla \cdot \mathbf{D}) \, dv + \int_{v'} \mathbf{D} \cdot \nabla V \, dv$$

Since the volume v' does not contain any free charges $\nabla \cdot \mathbf{D} = 0$ there and

$$\oint_{S+\Sigma} V\mathbf{D} \cdot \mathbf{dS'} = \int_{v'} \mathbf{D} \cdot \nabla V \, dv \tag{5-7}$$

The surface integral on the left-hand side is the sum of two surface integrals, one over the surface S and the second over the surface Σ. Since the surface S must now be regarded as an external surface with respect to the volume v', it follows that $d\mathbf{S}' = -d\mathbf{S}$ over that surface; also $d\mathbf{S}' = d\mathbf{\Sigma}$ over the surface Σ so that

$$\oint_{S+\Sigma} V\mathbf{D} \cdot d\mathbf{S}' = \oint_{S} V\mathbf{D} \cdot d\mathbf{S}' + \oint_{\Sigma} V\mathbf{D} \cdot d\mathbf{\Sigma}$$

$$= -\oint_{S} V\mathbf{D} \cdot d\mathbf{S} + \oint_{\Sigma} V\mathbf{D} \cdot d\mathbf{\Sigma} \qquad (5\text{--}8)$$

Combining equations 5–7 and 5–8 we obtain

$$\oint_{S} V\mathbf{D} \cdot d\mathbf{S} = \oint_{\Sigma} V\mathbf{D} \cdot d\mathbf{\Sigma} - \int_{v'} \mathbf{D} \cdot \nabla V \, dv$$

By substituting the above equation into equation 5–6 we obtain for the energy

$$W = \tfrac{1}{2} \oint_{\Sigma} V\mathbf{D} \cdot d\mathbf{\Sigma} - \tfrac{1}{2} \int_{v+v'} \mathbf{D} \cdot \nabla V \, dv \qquad (5\text{--}9)$$

If we now allow the surface Σ to expand into a sphere of infinite radius, the surface integral in equation 5–9 will be zero. This is because the potential at sufficiently large distances from the charges will decrease as $1/r$ and hence $\mathbf{D}$ as $1/r^2$, whilst the surface element will increase only as r^2. The volume integral now extends over *all space* and the energy is given by

$$W = -\tfrac{1}{2} \int_{\text{all space}} \mathbf{D} \cdot \nabla V \, dv \qquad (5\text{--}10)$$

Since $\mathbf{E} = -\nabla V$ we have that

$$W = \tfrac{1}{2} \int \mathbf{D} \cdot \mathbf{E} \, dv \qquad (5\text{--}11)$$

If the medium is isotropic $\mathbf{D} = \epsilon\mathbf{E}$ at any point and

$$W = \tfrac{1}{2} \int \epsilon E^2 \, dv \qquad (5\text{--}11\text{a})$$

If in addition the medium is homogeneous, the permittivity may be removed from the integral sign.

The quantity $dW/dv = \tfrac{1}{2}\mathbf{D} \cdot \mathbf{E}$ has the dimensions of energy per unit volume (joules per cubic metre) and one is tempted to interpret it as the density of electrostatic energy stored in the electric field. We should bear in mind, however, that it is possible to give a similar interpretation to the quantity $\tfrac{1}{2}\rho V$ of equation 5–3; in this case the integral does not extend over all space so that the energy will be localized in the finite volume occupied by the charge distribution instead of in all space as implied by equation 5–11. Hence it is not possible to attach any physical significance to the concept of energy density. It must be regarded as a convenient mathematical abstraction just as the electric field itself is.

5.3 ENERGY STORED IN A SYSTEM OF CONDUCTORS

In most problems of practical interest the free charges in a system are those residing on the surfaces of conductors. Consider a system of N

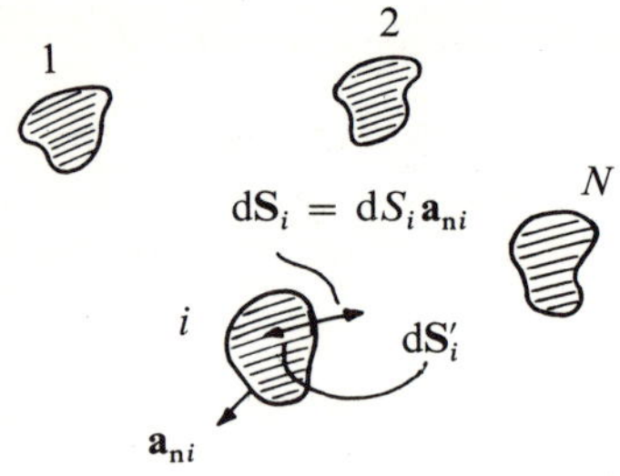

Fig. 5.3 System of N charged conductors

conductors bounded by the surfaces $S_1, S_2, \ldots S_i \ldots S_N$ (Fig. 5.3) and carrying surface charge densities $\sigma_1, \sigma_2, \ldots \sigma_i, \ldots \sigma_N$. Let the dielectric medium in which the conductors lie be linear. The electric energy stored in the space bounded by the conductor surfaces is, from equation 5–10,

$$W = -\tfrac{1}{2} \int_{\text{all space}} \mathbf{D} \cdot \nabla V \, dv$$

By making use of equation 5–5 this may be rewritten as

$$W = \tfrac{1}{2} \int_{\text{all space}} V(\nabla \cdot \mathbf{D}) \, dv - \tfrac{1}{2} \int_{\text{all space}} \nabla \cdot (\nabla \mathbf{D}) \, dv$$

The first integral is zero since $\nabla \cdot \mathbf{D} = 0$ in all the space around the conductors. By applying the divergence theorem to the second integral we obtain

$$W = -\frac{1}{2} \sum_{i=1}^{N} \oint_{S_i} V_i \mathbf{D}_i \cdot d\mathbf{S}_i'$$

where V_i and D_i are to be evaluated at the surface of the ith conductor; V_i is a constant and $\mathbf{D}_i = \sigma_i \mathbf{a}_{ni}$; also $d\mathbf{S}_i' = dS_i \mathbf{a}_{ni}' = -dS \mathbf{a}_{ni}$ and hence

$$W = \frac{1}{2} \sum_{i=1}^{N} V_i \oint_{S_i} \sigma_i \, dS_i$$

$$= \frac{1}{2} \sum_{i=1}^{N} V_i Q_i \tag{5–12}$$

Since the potential and charge of the ith conductor may be expressed by

$$V_i = \sum_{j=1}^{N} p_{ij}Q_j; \quad Q_i = \sum_{j=1}^{N} C_{ij}V_j$$

it follows that the energy of the system may be expressed either in terms of the charges on the conductors or in terms of the conductor potentials. In terms of the charges,

$$W = W_Q = \frac{1}{2}\sum_{i=1}^{N}\sum_{j=1}^{N} p_{ij}Q_iQ_j \tag{5-13}$$

$$= \tfrac{1}{2}(p_{11}Q_1^2 + 2p_{12}Q_1Q_2 + p_{22}Q_2^2 + \cdots) \tag{5-13a}$$

In terms of the potentials

$$W = W_V = \frac{1}{2}\sum_{i}\sum_{j} C_{ij}V_iV_j \tag{5-14}$$

$$= \tfrac{1}{2}(C_{11}V_1^2 + 2C_{12}V_1V_2 + C_{22}V_2^2 + \cdots) \tag{5-14a}$$

Consider two conductors which form the plates of an ideal capacitor. For such a system $Q_1 = -Q_2 = Q$ and

$$W = \tfrac{1}{2}Q^2(p_{11} - 2p_{12} + p_{22})$$

also

$$V_1 = Q(p_{11} - p_{12})$$
$$V_2 = Q(p_{21} - p_{22})$$

so that

$$C = \frac{Q}{V_1 - V_2} = (p_{11} - 2p_{12} + p_{22})^{-1}$$

and the energy is given by

$$W = \frac{1}{2}\frac{Q^2}{C} = \tfrac{1}{2}CV^2 \tag{5-15}$$

where V is the potential difference between the conductors.

We shall now evaluate the energy stored in a capacitor by considering the process of transferring charge from one plate to the other. Suppose that a potential difference V is applied between the plates; when equilibrium is attained the plates will carry an equal and opposite charge $Q = CV$. Assume that the charge Q is transferred in n elements dQ. The energy required to transfer the first charge element is $V\,dQ$; at the end of this process the potential difference between the plates is $V - (dQ/C)$ and the energy required to transfer the second charge element is now $[V-(\mathrm{d}Q/C)]\,\mathrm{d}Q$. The total energy required is therefore,

$$W = V\,\mathrm{d}Q + [V-(\mathrm{d}Q/C)]\,\mathrm{d}Q$$
$$+ [V-2(\mathrm{d}Q/C)]\,\mathrm{d}Q + \cdots [V-(n-1)\,\mathrm{d}Q/C]\,\mathrm{d}Q$$

Since this is an arithmetic progression we have that

$$W = \tfrac{1}{2}n[V\,\mathrm{d}Q + V\,\mathrm{d}Q - (n-1)\,\mathrm{d}Q^2/C]$$
$$= VQ - \tfrac{1}{2}Q^2(1-1/n)/C$$

For an infinite number of steps $n \to \infty$, $\mathrm{d}Q \to 0$ such that $n\,\mathrm{d}Q = Q$ and

$$W = \frac{1}{2}\frac{Q^2}{C} = \tfrac{1}{2}CV^2 \tag{5-15}$$

5.4 DERIVATION OF FORCES AND TORQUES FROM ENERGY CONSIDERATIONS

5.4.1 FORCES ON CONDUCTORS

Consider a system of charged conductors which are in a state of electrostatic equilibrium. In general such a system will not be in mechanical equilibrium unless external mechanical forces are applied which balance the electrostatic (Coulomb) forces. A simple example is the case of two point charges; to keep them at a certain fixed distance apart a mechanical force equal and opposite to the Coulomb force must be applied to each charge.

If the mechanical constraint (force) holding any conductor in its equilibrium position were removed, the electrical force acting on the conductor would tend to move it in such a direction as to either decrease or increase the electric energy of the system, depending upon whether the charges or the potentials of the system remain invariant during the displacement. Let us assume that the ith conductor makes a small linear displacement $\mathrm{d}\eta_i$ under the influence of the electric force acting on it. From equation 5–12 we have that an incremental change in the total electrostatic energy of the system is given by

$$\mathrm{d}W = \frac{1}{2}\sum_i V_i\,\mathrm{d}Q_i + \frac{1}{2}\sum_i Q_i\,\mathrm{d}V_i \tag{5-16}$$

We shall consider two separate cases: (a) the system is isolated, that is to say the charges on the conductors remain constant during the displacement and (b) the conductors are maintained at constant potentials during the displacement; this means that the conductors are connected to a system of external energy sources such as batteries or generators. In both cases we shall assume that during the displacement there is no dissipation of energy.

Case (a) Since the system receives no energy from external sources, the principle of conservation of energy requires that

$$F_i \, d\eta_i + dW = 0 \qquad (5\text{--}17)$$

Because the charge remains constant $dW = dW_Q$ and

$$F_i \, d\eta_i = -dW_Q$$

so that

$$F_i = -\frac{\partial W_Q}{d\eta_i} \qquad (5\text{--}18)$$

where F_i is the electric force in the direction of the coordinate η_i. If instead of being linear the displacement is angular and equal to $d\theta_i$ we have

$$T_i \, d\theta_i = -dW_Q$$

and

$$T_i = -\frac{\partial W_Q}{d\theta_i} \qquad (5\text{--}19)$$

where T_i is the electrical torque. Partial derivatives have been used in equations 5–17 and 5–18 since W_Q will in general be a function of more than one coordinate. The negative sign before both these equations indicates that in this case the electrical forces tend to decrease the energy of the system.

Case (b) The principle of conservation of energy now requires that

$$F_i \, d\eta_i + dW = dW_S \qquad (5\text{--}20)$$

where dW_S represents the energy supplied by the sources which maintain the potentials of the system constant. Since the potentials are constant we have from equation 5–16,

$$dW = dW_V = \tfrac{1}{2} \sum V_i \, dQ_i$$

so that all charges will necessarily change by increments dQ_i. The total energy required to place charges $dQ_1, dQ_2, \ldots dQ_N$ on the conductors which are at constant potentials $V_1, V_2, \ldots V_N$ is

$$dW_S = \sum V_i \, dQ_i$$

Hence equation 5–20 may be written as

$$F_i \, d\eta_i = \sum V_i \, dQ_i - \tfrac{1}{2} \sum V_i \, dQ_i$$
$$= \tfrac{1}{2} \sum V_i \, dQ_i = dW_V$$

and

$$F_i = +\frac{\partial W_V}{\mathrm{d}\eta_i} \tag{5-21}$$

The torque is given by

$$T_i = +\frac{\partial W_V}{\partial \theta_i} \tag{5-22}$$

The energy supplied by the sources is $2\mathrm{d}W_V$. The positive sign before the above two equations indicates that if the potentials of the conductors are kept constant, the electrical forces tend to increase the energy of the system.

In both the above cases the electrical force (and torque) must be the same since it depends only on the initial energy of the system; this energy may be expressed in terms of either charge or potential. Any displacements introduced to calculate the force or torque must be regarded as purely virtual. We may therefore write the force and torque equations as

$$F_i = -\frac{\partial W_Q}{\partial \eta_i} = +\frac{\partial W_V}{\partial \eta_i} \tag{5-23}$$

and

$$T_i = -\frac{\partial W_Q}{\partial \theta_i} = +\frac{\partial W_V}{\partial \theta_i} \tag{5-24}$$

As a simple example consider a parallel-plate capacitor whose plates have an area S and are kept at a distance x apart. Suppose that the plates are initially given equal and opposite charges Q and are then disconnected from the charging source (Fig. 5.4a). The energy stored is

$$W_Q = \tfrac{1}{2}Q^2/C = \tfrac{1}{2}Q^2 x/\epsilon S$$

and

$$\frac{\partial W_Q}{\partial x} = \tfrac{1}{2}Q^2/\epsilon S = \sigma^2 S/2\epsilon$$

so that

$$F_x = -\frac{\sigma^2 S}{2\epsilon} \tag{5-25}$$

If the plates are permanently connected to the poles of a battery so that the potential difference between the plates is V (Fig. 5.4b), the energy

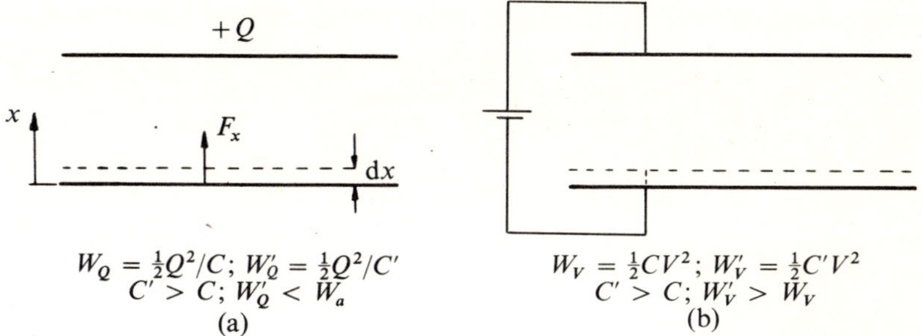

Fig. 5.4 Parallel-plate capacitor whose plates are kept at (a) constant charge, (b) constant potential

stored is

$$W_V = \tfrac{1}{2}CV^2 = V^2\epsilon S/2x$$

and

$$\frac{\partial W_V}{\partial x} = -V^2\epsilon S/2x^2 = -\sigma^2 S/2\epsilon$$

so that

$$F_x = -\frac{\sigma^2 S}{2\epsilon}$$

The negative sign indicates that in both cases the force tends to decrease the separation x. In the first case the energy would decrease whilst in the second the energy would increase.

5.4.2 FORCES ON DIELECTRIC BODIES

Let us assume that a linear isotropic dielectric medium whose permittivity is everywhere ϵ_1 extends over all space. Suppose that due to a certain distribution of free charges the field at any point in the medium is $\mathbf{E}_0$. The energy of the system is given by

$$W_0 = \tfrac{1}{2}\int \mathbf{E}_0 \cdot \mathbf{D}_0 \, dv$$

where $\mathbf{D}_0 = \epsilon_1\mathbf{E}_0$.

Suppose now that in a finite volume v_2 we replace the dielectric of permittivity ϵ_1 with one of permittivity ϵ_2 (Fig. 5.5). If we assume that the charges in the system remain constant, then the potential and field at any point are modified such that

$$\mathbf{D} = \epsilon_1\mathbf{E} \qquad \text{in volume } v_1 \text{ (all space less volume } v_2\text{)}$$
$$\mathbf{D} = \epsilon_2\mathbf{E} \qquad \text{in volume } v_2$$

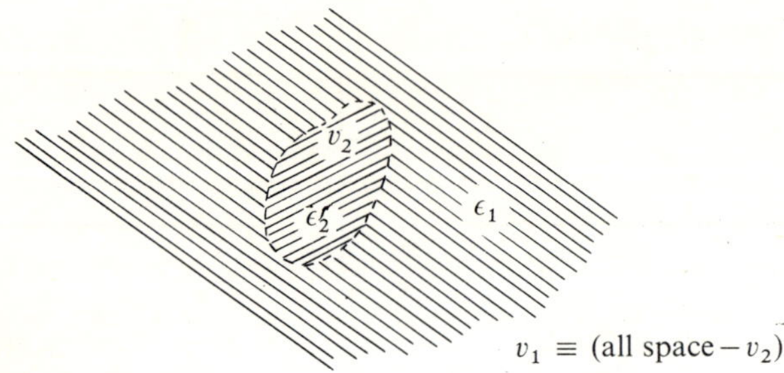

Fig. 5.5 Dielectric volume v_2 and permittivity ϵ_2 in an infinite dielectric medium of permittivity ϵ_1

The energy now has the value

$$W_Q = \tfrac{1}{2} \int_{v_1 + v_2} \mathbf{E} \cdot \mathbf{D} \, dv$$

The change in energy is given by

$$W_Q - W_0 = \tfrac{1}{2} \int_{v_1 + v_2} (\mathbf{E} \cdot \mathbf{D} - \mathbf{E}_0 \cdot \mathbf{D}_0) \, dv$$

$$= \tfrac{1}{2} \int_{v_1 + v_2} (\mathbf{E} + \mathbf{E}_0) \cdot (\mathbf{D} - \mathbf{D}_0) \, dv$$

$$+ \tfrac{1}{2} \int_{v_1 + v_2} (\mathbf{E} \cdot \mathbf{D}_0 - \mathbf{D} \cdot \mathbf{E}_0) \, dv \qquad (5\text{--}26)$$

Since we are dealing with conservative electric fields we may write $\mathbf{E} + \mathbf{E}_0 = -\nabla \phi$ where ϕ is a scalar potential function. The first integral of equation 5–25 may thus be written as

$$I_1 = \tfrac{1}{2} \int_{v_1 + v_2} (\mathbf{E} + \mathbf{E}_0) \cdot (\mathbf{D} - \mathbf{D}_0) \, dv = -\tfrac{1}{2} \int_{v_1 + v_2} \nabla \phi \cdot (\mathbf{D} - \mathbf{D}_0) \, dv$$

By using a vector identity of the same form as equation 5–5, the above equation may be transformed to give

$$I_1 = \tfrac{1}{2} \int_{v_1 + v_2} \phi \nabla \cdot (\mathbf{D} - \mathbf{D}_0) \, dv - \tfrac{1}{2} \int_{v_1 + v_2} \nabla \cdot \phi (\mathbf{D} - \mathbf{D}_0) \, dv$$

Since the source charge density is assumed to remain constant, $\nabla \cdot \mathbf{D} = \rho$ and $\nabla \cdot \mathbf{D}_0 = \rho$ so that the first term is zero. The second term may be transformed into a surface integral and it is left to the reader to show, by arguments similar to those given in Section 5.2, that this integral is also zero. Consequently equation 5–26 becomes

$$W_Q = W_0 + \tfrac{1}{2} \int_{v_1 + v_2} (\mathbf{E} \cdot \mathbf{D}_0 - \mathbf{D} \cdot \mathbf{E}_0) \, dv$$

Within the volume v_1 the function under the integral sign is everywhere zero since in that volume $\mathbf{D}_0 = \epsilon_1\mathbf{E}_0$ and $\mathbf{D} = \epsilon_1\mathbf{E}$; hence

$$W_Q = W_0 + \tfrac{1}{2}\int_{v_2} (\mathbf{E} \cdot \mathbf{D}_0 - \mathbf{D} \cdot \mathbf{E}_0)\,dv$$

within v_2, $\mathbf{E} = \mathbf{E}_2$, $\mathbf{D} = \epsilon_2\mathbf{E}_2$ and $\mathbf{D}_0 = \epsilon_1\mathbf{E}_0$ so that

$$W_Q = W_0 + \tfrac{1}{2}\int_{v_2} (\epsilon_1\mathbf{E}_0 \cdot \mathbf{E}_2 - \epsilon_2\mathbf{E}_2 \cdot \mathbf{E}_0)\,dv$$

$$= W_0 - \tfrac{1}{2}\int_{v_2} (\epsilon_2 - \epsilon_1)\mathbf{E}_2 \cdot \mathbf{E}_0\,dv \tag{5–27}$$

This last result shows that if $\epsilon_2 > \epsilon_1$ there will always be a decrease in the electrostatic energy of the system.

In practice any free charges will in general lie on the surface of conductors. If the potential of these conductors is kept constant when the dielectric of volume v_2 and permittivity ϵ_2 is introduced into the system, the energy supplied by the sources maintaining the conductor potentials constant will again be equal, as in 5.4.1, to twice the negative change in electrostatic energy. In this case the energy stored in the system is given by

$$W_{\mathrm{V}} = W_0 + \tfrac{1}{2}\int_{v_2} (\epsilon_2 - \epsilon_1)\mathbf{E}_2 \cdot \mathbf{E}_0\,dv \tag{5–28}$$

so that if $\epsilon_2 > \epsilon_1$, there will always be an increase in electrostatic energy.

The quantity $W - W_0$ is the amount by which the energy of the system is changed when a dielectric of volume v_2 and permittivity ϵ_2 is introduced into a dielectric medium of permittivity ϵ_1. It is sometimes referred to as the energy of interaction of the dielectric body with the external field $\mathbf{E}_0$.

To calculate the force acting on the dielectric body we assume that it makes a small displacement $d\eta$ and obtain, by arguments similar to those given in Section 5.4.1, that

$$F_\eta = -\frac{\partial W_Q}{\partial \eta} = +\frac{\partial W_{\mathrm{V}}}{\partial \eta} \tag{5–29}$$

where W_Q and W_{V} are now given by equations 5–26 and 5–27 respectively. It is clear that the force will always tend to draw the dielectric medium of higher permittivity into the one of lower permittivity.

Consider the following example. A plane-parallel capacitor has rectangular electrodes each of area lb at a distance d from each other. A dielectric slab of permittivity ϵ, thickness d and base area b is introduced between the electrodes so that a length $(l-x)$ of it remains outside (Fig. 5.6). It is required to find the force which tends to pull the slab

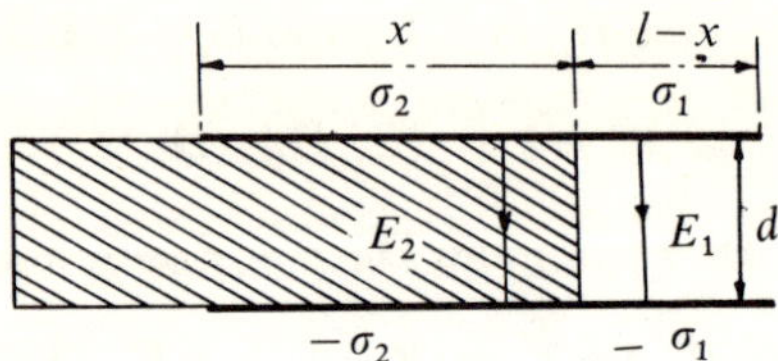

Fig. 5.6 Dielectric slab partially filling the space between two plane-parallel charged electrodes

into the capacitor if the electrodes carry an equal and opposite charge Q. The force is given by

$$F_x = -\frac{\partial W_Q}{\partial x} = \frac{\partial}{\partial x}\frac{1}{2}\int (\epsilon_2-\epsilon_1)\mathbf{E}_2 \cdot \mathbf{E}_0 \, dv$$

In the absence of the dielectric slab the field E_0 is

$$E_0 = \sigma/\epsilon_1 = Q/\epsilon_1 lb$$

With the dielectric slab in position the fields in the two regions 1 and 2 are,

$$E_1 = \sigma_1/\epsilon_1 ; \; E_2 = \sigma_2/\epsilon_2$$

and since these must be equal

$$\sigma_2/\epsilon_2 = \sigma_1/\epsilon_1$$

The total charge is

$$Q = [\sigma_2 x + \sigma_1(l-x)]b$$
$$= \sigma_2[x+\epsilon_1(l-x)/\epsilon_2]b$$

and hence

$$E_2 = Q/[\epsilon_2 x+\epsilon_1(l-x)]b$$

The expression to be differentiated is therefore given by

$$\frac{1}{2}\int (\epsilon_2-\epsilon_1)\mathbf{E}_2 \cdot \mathbf{E}_0 = \frac{1}{2}\frac{Q^2 d}{lb\epsilon_1}\frac{(\epsilon_2-\epsilon_1)x}{\epsilon_2 x+\epsilon_1(l-x)}$$

Differentiating this expression with respect to x and setting $\epsilon_2 = \epsilon = \epsilon_0\epsilon_r$ and $\epsilon_1 = \epsilon_0$ we obtain

$$F_x = \frac{1}{2}\frac{Q^2 d}{b\epsilon_0}\frac{\epsilon_r-1}{[\epsilon_r x+(l-x)]^2} \text{ (newton)}$$

As a second example consider the system shown in Fig. 5.7. It represents a parallel-plate capacitor of plate area S placed in an incompressible liquid of permittivity ϵ and mass density ρ. It is required to find the height to which the liquid will rise when a potential difference V is maintained between the plates. The width of the plates is b and their separation is d.

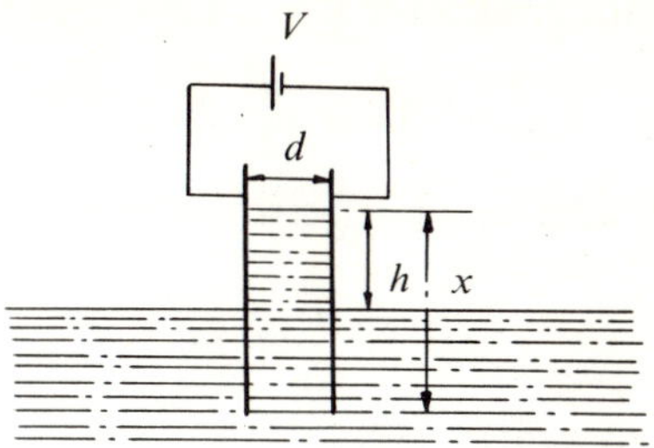

Fig. 5.7 Rise of a dielectric liquid between two parallel plates with a potential difference V between them

Since the potential between the plates is constant, $E_2 = E_0 = E$; also $\epsilon_2 = \epsilon$ and $\epsilon_1 = \epsilon_0$ so that

$$W_V = W_0 + \tfrac{1}{2} \int (\epsilon - \epsilon_0) E^2 \, dv$$

$$= W_0 + \tfrac{1}{2}(\epsilon - \epsilon_0) E^2 xbd$$

Hence

$$F_x = \frac{\partial W_V}{\partial x} = \tfrac{1}{2}(\epsilon - \epsilon_0) E^2 bd$$

At equilibrium this upward force must be equal to the downward gravitational force of the volume hbd of the displaced liquid. This force is

$$F_g = \rho g h b d$$

Hence

$$\rho g h b d = \tfrac{1}{2}(\epsilon - \epsilon_0) E^2 bd$$

and

$$h = \frac{\epsilon - \epsilon_0}{2\rho g} E^2$$

5.5 FORCE AT THE SURFACE OF A CHARGED CONDUCTOR

Let the charge density on the surface of a conductor be $\sigma \, \mathrm{C\,m^{-2}}$ and let the conductor be embedded in a dielectric medium of permittivity ϵ. Suppose that an element dS of the surface is given an incremental displacement dl in the direction of the outward normal. The energy of the system decreases by an amount equivalent to that which was contained in the volume element $dv = dS\,dl$. This decrease in energy is equal to the mechanical work done so that

$$\mathrm{d}F\,\mathrm{d}l = \tfrac{1}{2}(\mathbf{E}\cdot\mathbf{D})\,\mathrm{d}S\,\mathrm{d}l$$

At the surface of the conductor $\mathbf{E} = (\sigma/\epsilon)\mathbf{a}_n$ and $\mathbf{D} = \sigma\mathbf{a}_n$ so that

$$\mathrm{d}F = \frac{\sigma^2}{2\epsilon}\,\mathrm{d}S$$

The force permit area or the electromechanical pressure at the conductor surface is therefore

$$f_S = \frac{\sigma^2}{2\epsilon}\,\mathrm{N\ m}^{-2} \tag{5-30}$$

This force is independent of the sign of the charge and is always directed normally outward from the conductor surface tending to expand it into the dielectric. The force is numerically equal to the energy density just off the surface.

We shall now derive equation 5–29 by a different method. Consider an element of charge $\mathrm{d}Q = \sigma\,\mathrm{d}S$ on the conductor surface. The force acting on this element is

$$\mathbf{dF} = \mathbf{E}'\sigma\,\mathrm{d}S$$

where $\mathbf{E}'$ is equal to the electric field at the surface element produced by all charges on the surface of the conductor minus the field produced by the charge $\mathrm{d}Q$ alone. This latter field must be subtracted because the self-field of a charge $\mathrm{d}Q$ cannot produce a resultant force on that charge. Hence

$$\mathbf{dF} = (\mathbf{E} - \mathbf{E}_S)\sigma\,\mathrm{d}S$$

At the surface of a charged conductor $\mathbf{E} = (\sigma/\epsilon)\mathbf{a}_n$. To a point on $\mathrm{d}S$, the element appears to be an infinite plane so that (see equation 2–49)

$$\mathbf{E}_S = \frac{\sigma}{2\epsilon}\,\mathbf{a}_n$$

and therefore

$$\mathbf{dF} = \frac{\sigma^2\,\mathrm{d}S}{2\epsilon}\,\mathbf{a}_n$$

and

$$\mathbf{F}_S = \frac{\sigma^2}{2\epsilon}\,\mathbf{a}_n \tag{5-30a}$$

As an example of the application of equation 5–30 consider the following problem. A sphere of density $\rho'\ \mathrm{kg\,m}^{-3}$ and radius a metres floats in a dielectric liquid of density $\rho\ \mathrm{kg\,m}^{-3}$ and relative permittivity ϵ_r. It is required to find the total charge Q which the sphere must carry if it

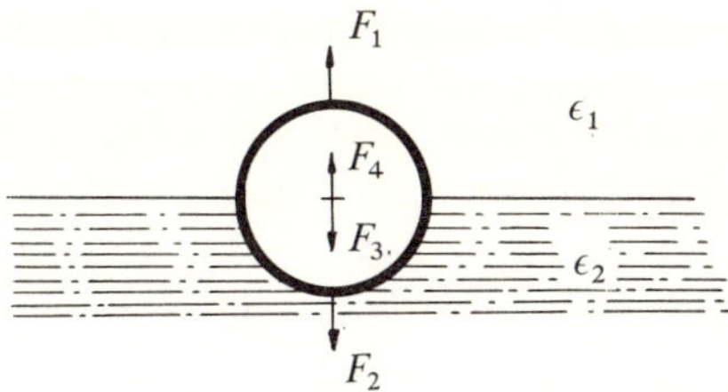

Fig. 5.8 Forces acting on a charged sphere partially submerged in a dielectric liquid

is to float half submerged (Fig. 5.8). The forces acting on the charged sphere are the following:

F_1 = electrical force acting on the non-immersed surface of the sphere
F_2 = electrical force acting on the immersed surface of the sphere
F_3 = weight of sphere
F_4 = buoyant force acting on sphere

At equilibrium

$$F_1 + F_4 = F_2 + F_3$$

that is,

$$\frac{\pi a^2 \sigma_1^2}{2\epsilon_1} + \tfrac{2}{3}\pi a^3 \rho g = \frac{\pi a^2 \sigma_2^2}{2\epsilon_2} + \tfrac{4}{3}\pi a^3 \rho' g$$

where σ_1 and σ_2 are the charge densities on the non-immersed and the immersed surfaces respectively. They are related by the equations

$$2\pi a^2 (\sigma_1 + \sigma_2) = Q$$

and

$$\sigma_1/\epsilon_1 = \sigma_2/\epsilon_2$$

which when solved for σ_1 and σ_2 give

$$\sigma_1 = \frac{Q\epsilon_1}{2\pi a^2 (\epsilon_1 + \epsilon_2)}; \quad \sigma_2 = \frac{Q\epsilon_2}{2\pi a^2 (\epsilon_1 + \epsilon_2)}$$

Substituting these values into the force equation and simplifying we obtain finally

$$Q^2 = \frac{(4\pi)^2 a^5 g (\rho - 2\rho')(\epsilon_1 + \epsilon_2)^2}{3(\epsilon_2 - \epsilon_1)}$$

For $\epsilon_1 = \epsilon_0$ and $\epsilon_2 = \epsilon_0 \epsilon_r$ this becomes

$$Q^2 = \frac{(4\pi)^2 a^5 g (\rho - 2\rho')\epsilon_0 (\epsilon_r + 1)^2}{3(\epsilon_r - 1)}$$

It is evident that the force between two parallel plates of area S carrying equal and opposite charges will be given by equation 5–25. The magnitude of this force is directly proportional to the square of the potential difference between the plates

$$F = \frac{\sigma^2 S}{2\epsilon} = \frac{V^2 \epsilon S}{2x^2}$$

where x is the separation between the plates. The existence of this force is the basis of operation of most electrostatic voltmeters especially those used for the direct measurement of high voltages. As a rule, all electrostatic instruments rely for their operation on the electrostatic forces or torques which exist in a system of charged conductors.*

PROBLEMS

Chapter Five

5.1   In a system of N conductors, the conductors carry charges Q_i at potentials V_i. If an additional insulated and uncharged conductor is introduced into the system, show that the electrostatic energy of the system will decrease.

5.2   A charge Q is distributed with a uniform volume density throughout a sphere of radius a. Show that the electrostatic energy associated with this charge distribution is $3Q^2/20\pi\epsilon_0^2 a$.

5.3   A charge is distributed with a uniform volume density ρ throughout a hemispherical shell of internal radius a and external radius b. Show that the force between this charge distribution and a point charge Q placed at the centre of the hemisphere is $\rho Q(b-a)/4\epsilon_0$.

5.4   A parallel-plate capacitor consists of two identical plates of area S separated by a dielectric slab of equal area, thickness t, and relative permittivity ϵ_r. The upper plate is fixed to the dielectric but the lower plate is not. Show that the lower plate will remain in position if a voltage V is applied between the plates such that

$$V^2 \geq \frac{2t^2 w}{\epsilon_0 \epsilon_r^2 S}$$

where w is the weight of the plate.

* For a comprehensive account of electrostatic instruments the reader should consult *Electrical Measurements and Measuring Instruments* by E. W. Golding and F. C. Widdis, Pitman, London, 1963.

5.5   The plates of a parallel-plate capacitor have an area of $50\,\text{cm}^2$. Find the force required to keep the plates apart at a distance of $0.5\,\text{mm}$ when a potential difference of $1000\,\text{V}$ is applied between the plates. What is the value of this force (a) if the space between the plates is filled with oil of relative permittivity 2.4, and (b) if the capacitor as a whole is totally immersed in oil of the same permittivity.

5.6   Two coaxial cylinders of radii a and b are partially immersed in a dielectric liquid of permittivity ϵ and mass density ρ; the axis of the cylinders is perpendicular to the liquid surface. Show that if a potential difference V is applied between the cylinders the liquid will rise to a height given by,

$$h = \frac{\epsilon_0(\epsilon_\text{r}-1)V^2}{\rho g(b^2-a^2)\log(b/a)}$$

5.7   The half-space between two concentric spherical conductors of radii a and b $(b > a)$ is filled with a dielectric of permittivity ϵ. If a potential difference V is applied between the conductors, show that the electric field at any point between the spheres is given by

$$E = V\frac{ab}{(b-a)r^2}$$

and that the resultant electrostatic force on the inner sphere is

$$F = \tfrac{1}{2}\pi V^2 b^2 \epsilon_0(\epsilon_\text{r}-1)/2(b-a)^2$$

5.8   Two parallel metal plates 1 cm apart are lowered vertically into a dielectric liquid of mass density $0.88\,\text{g cm}^3$ and relative permittivity 2.4. What potential difference must be applied between the plates for the liquid between them to rise a distance of 1 mm? Neglect any edge effects.

5.9   A capacitor consists of two coaxial cylinders of radii a and b $(b > a)$. A potential difference V is applied between the cylinders. Show that the force with which a dielectric tube of permittivity ϵ and radii a and b would be drawn into the capacitor is given by $F = \pi(\epsilon-\epsilon_0)V^2/\log(b/a)$.

5.10 A spherical air capacitor consists of a central sphere of radius a and an outer sphere of inner radius b. The outer sphere consists of two earthed hemispheres. If the inner sphere is raised to a potential V, show that the force needed to separate the two hemispheres is given by $F = \tfrac{1}{2}\pi\epsilon_0 V^2 a^2/(b-a)^2$.

CHAPTER SIX

THE ELECTRIC CURRENT

The motion of electric charges in solids, liquids, gases or in a vacuum constitutes an electric current. The motion of free charges (electrons or ions) in the absence of an external electric field may give rise to a *convection current* or to a *diffusion current*. Convection currents may be generated by the motion of electrostatically charged bodies, by the flow of a gas or a liquid containing a net charge per unit volume or by the motion of charged particles under their own momentum. Diffusion currents are due to the presence of a concentration gradient of charge carriers; according to diffusion theory the carriers will always tend to flow in the direction of decreasing concentration gradient. Diffusion currents are of special importance in the study of semiconductor junctions.

When free charges move under the action of an electric field, the resulting flow of charge constitutes a *conduction current*. Charge flow in metallic conductors and in electrolytic solutions are the commonest examples of such a current. From the standpoint of electromagnetic theory, the conduction current which flows in metallic conductors is the most important type of electric current. The first part of this chapter will therefore be devoted to the study of conduction currents. Later in the chapter we shall discuss a special type of current which is also of great importance in electromagnetic theory. This is the so-called *displacement current* which is to be found wherever there is an electric field which changes with time.

6.1 CONDUCTION CURRENT

As already mentioned in Section 3.1, when an isolated conductor is placed in an electric field the valence or free electrons acquire an average drift velocity in a direction opposite to the field. This migration of electrons produces surface concentrations of charge and continues until a surface charge distribution is built up which reduces the field everywhere inside the conductor to zero and the whole of the conductor becomes an equipotential surface. During this whole process, which is of extremely short duration, the conductor as a whole remains electrically neutral. Now suppose that a permanent difference of potential is maintained between any two points of a conductor, for example by connecting the two points to the poles of a battery. As a result, a field (potential gradient) will exist in the conductor and as before the free electrons will migrate against this field to try and build up an equal and opposite field. In this case, however, there is no build up of surface charges because the positive pole of the battery acts as an electron sink whilst the negative pole acts as an electron source. At the positive pole the electrons are removed as fast as they arrive there whilst at the negative pole the electrons are fed into the conductor at the same rate. This action results in a continuous transport or flow of electrons and constitutes the conduction current. Here again the conductor as a whole remains electrically neutral but it is no longer an equipotential surface.

It is now necessary to give a quantitative definition for an electric current. Since this represents a flow of charge, the current flowing through a given area is defined as the amount of charge which crosses that area per unit time. In order to take into account the possibility of a variation of current with time it is mathematically more accurate to define the current flowing through a surface S as

$$I = \lim_{\Delta t \to 0} \frac{\Delta Q}{\Delta t} - \frac{dQ}{dt} \; [\mathrm{C\,s}^{-1} = \mathrm{A}] \tag{6-1}$$

where ΔQ is the amount of charge which crosses the surface S during the time interval Δt. In SI units the current is measured in amperes, one ampere being equal to a coulomb per second. As already mentioned in Chapter 2 the ampere is the fourth basic unit of the SI system. The positive direction of current flow is *arbitrarily* chosen as that direction in which a positive charge would move in an electric field. The current is a scalar quantity so that although we may speak of the sense in which a current flows through a given surface no definite direction can be ascribed to it.

In order to define the rate of charge flow at some macroscopic point P the concept of *current density* is introduced. Consider a small element

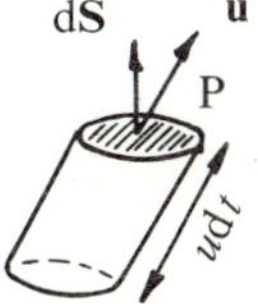

Fig. 6.1 Volume element in the direction of the charge drift velocity **u**

of surface dS surrounding the point P (Fig. 6.1). By choosing dS sufficiently small we may assume that all charges crossing dS have an average drift velocity **u** in the same direction. The total charge dQ which crosses the surface dS in a time dt is clearly that contained in the volume **u** d$t \cdot$ d**S**. If ρ is the volume density of the flowing charge then,

$$dQ = \rho \mathbf{u} \, dt \cdot d\mathbf{S}$$

and

$$\frac{\mathrm{d}Q}{\mathrm{d}t} = \mathrm{d}I = \mathbf{j} \cdot \mathrm{d}\mathbf{S}$$

where

$$\mathbf{j} = \rho\mathbf{u} \, [\mathrm{A \ m}^{-2}] \qquad (6\text{--}2)$$

$\mathbf{j}$ is defined as the current density and is measured in amperes per square metre. It is a vector point function which at any point has the same direction as the velocity of the charges at that point. In a conductor, where the charges are electrons, ρ will be negative and the direction of $\mathbf{j}$ will be opposite to that of the velocity $\mathbf{u}$. Since the electrons move against the field direction, $\mathbf{j}$ and $\mathbf{E}$ will have the same sense in agreement with the definition of a positive current. The total current through an arbitrary area S is given by

$$I = \int_S \mathbf{j} \cdot \mathrm{d}\mathbf{S} \qquad (6\text{--}3)$$

since the current through $\mathrm{d}S$ is the same as that through the area $\mathrm{d}S \cos\theta$ perpendicular to $\mathbf{j}$.

A line of current flow is defined as a curve drawn in a conducting medium such that the direction of the tangent to that curve at any point gives the direction of the current density at that point. A tubular region which is bounded by lines of current flow constitutes a current tube. No current passes through the wall of a current tube so that the total current through any cross-section of a tube is constant. Thus if $\mathrm{d}S_1$ and $\mathrm{d}S_2$ are any two perpendicular cross-sections of a current tube (Fig. 6.2) then

$$j_1 \, \mathrm{d}S_1 = j_2 \, \mathrm{d}S_2 = \mathrm{d}I = \text{constant}$$

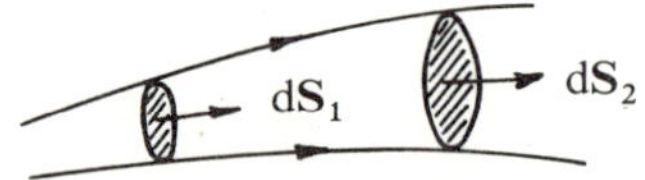

Fig. 6.2 Current tube

If the current is due to more than one type of charge carrier, equation 6–2 may be expressed in a more general form as

$$\mathbf{j} = \sum_i \mathbf{j}_i = \sum_i \rho_i\mathbf{u}_i \qquad (6\text{--}4)$$

where ρ_i and $\mathbf{u}_i$ are the charge density and velocity respectively of the charge carriers of type i; these may be for instance electrons, positive ions, or negative ions.

Since the total density of charge carriers is given by

$$\rho = \sum \rho_i$$

we may very well have the simultaneous conditions

$$\mathbf{j} \neq 0$$

$$\rho = 0$$

These conditions apply in the case of metallic conductors where the conduction current is due solely to the motion of free electrons. Thus at any point inside the metal we have that

$$\rho = \rho_{\text{elec.}} + \rho_{\text{ion}} = 0$$

whereas

$$\mathbf{j} = \rho_{\text{elec.}} \mathbf{u}_{\text{elec.}}$$

In a metal, the number of conduction electrons per unit volume is approximately of the same order of magnitude as the number of atoms per unit volume. This is of the order of 10^{28} m^{-3}. Since the charge per electron is of the order of 10^{-19} C, ρ is of the order of 10^9 C m^{-3}. For a current density of 10^6 A m^{-2} the electron drift velocity $\mathbf{u}$ is of the order of 10^{-3} m s^{-1}. This is very small compared with the thermal random velocity of the electrons which is of the order of 10^6 m s^{-1}. We thus see that although an immense number of electrons take part in the flow of current through a metallic conductor, the electrons merely crawl through the conductor with a velocity of not more than a few millimetres per second.

If the flow of current is confined within a surface layer of vanishingly small thickness we have a so-called *current sheet*. In this case it is convenient to replace the volume current density $\mathbf{j}$ by a surface current density $\mathbf{K}$ measured in amperes per metre.

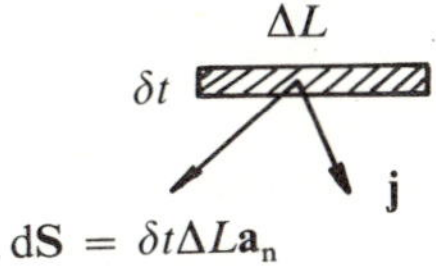

Fig. 6.3 Element of a current sheet

Consider the element of area shown in Fig. 6.3.

$$\Delta I = \mathbf{j} \cdot \Delta \mathbf{S} = \mathbf{j} \cdot \mathbf{a}_n \delta t \, \Delta L$$

We define a surface current density

$$\mathbf{K} = \lim_{\substack{\delta t \to 0 \\ j \to \infty}} \{\mathbf{j} \, \delta t\} \qquad (6\text{--}5)$$

such that the current I remains unchanged and therefore finite. Thus

$$\Delta I = \mathbf{K} \cdot \mathbf{a}_n \Delta L$$

$$\lim_{\Delta L \to 0} \frac{\Delta I}{\Delta L} = \frac{dI}{dL} = \mathbf{K} \cdot \mathbf{a}_n$$

and

$$I = \int \mathbf{K} \cdot \mathbf{a}_n \, dL$$

It is clear that the vector $\mathbf{K}$ can have at most two components. Equation 6–5 requires that the conductivity of the medium (defined below) be infinite since this is the only way in which $\mathbf{j}$ can become infinite.

6.2 OHM'S LAW

If a constant potential difference V_{AB} is maintained between the two ends A and B of a conducting wire which is kept at constant temperature, a current I will flow through it from the higher potential end to the lower. It is found experimentally that the voltage V_{AB} is proportional to the current I, that is

$$V_{AB} = IR \tag{6-6}$$

where the constant of proportionality R is known as the resistance of the wire. Equation 6–6 is the familiar Ohm's law. In SI units, where potential difference is measured in volts and current in amperes, the unit of resistance is the ohm. The resistance of a conductor depends on the material of which the conductor is made, on its shape and size and on its temperature. For a conductor of uniform cross-section, experiment shows that its resistance is directly proportional to its length l and inversely proportional to its cross-sectional area S so that

$$R = \frac{\rho L}{S} \tag{6-7}$$

where ρ is a characteristic property of the material known as the *resistivity* of the material; it has the dimensions of resistance times length and in SI units is expressed in ohm metres. It is generally quite obvious from the context whether ρ is being used to denote resistivity or volume charge density. The reciprocal of ρ is known as the *conductivity* and is denoted by γ; it is measured in reciprocal ohms per metre also written as mhos per metre.

The resistivity is a function of temperature. For all metallic conductors it increases with increasing temperature. If ρ_0 is the resistivity at some base temperature T_0 (usually the ice-point) the variation of ρ with the temperature T may be expressed by the power series

$$\rho = \rho_0[1+\alpha(T-T_0)+\beta(T-T_0)^2+\delta(T-T_0)^3+\cdots]$$

where α, β, δ, etc., are constants which decrease very rapidly in order of magnitude as the powers of $(T-T_0)$ increase. Over a range of a few hundred degrees in the neighbourhood of T_0, only α is significant so that for practical purposes the variation of resistivity with temperature may be expressed as

$$\rho = \rho_0[1+\alpha(T-T_0)] \tag{6-8}$$

α is known as the temperature coefficient of resistance.

Since γ is the reciprocal of ρ, its variation with temperature is given by

$$\gamma = \frac{\gamma_0}{1+\alpha(T-T_0)} \tag{6-9}$$

The conductivity γ (at 20°C) and the temperature coefficient of resistance α of some materials are listed in Table 6-1. Equations 6-8 and 6-9 are not valid for semiconductors and insulators.

Since in field theory one is usually also interested in conditions at some specific point, it is convenient to express Ohm's law in a point form. To do this consider the element of volume of an elementary current tube

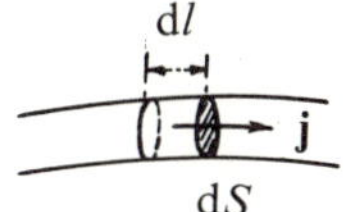

Fig. 6.4 Element of volume within a current tube

as shown in Fig. 6.4. The length dl is so small that the cross-sectional area dS normal to the tube may be considered uniform over that length. The current passing through the tube is

$$dI = j\,dS$$

Let the potential difference across the tube element be $-dV$ (negative because the potential decreases in the direction of current flow). Ohm's law applied to the element gives

$$-dV = j\,dS\,dl/\gamma\,dS$$

or

$$j = -\gamma\frac{dV}{dl}$$

The derivative $-\mathrm{d}V/\mathrm{d}l$ is the component of the electric field along the axis of the tube so that

$$j = \gamma E_l \qquad (6\text{--}10)$$

If the medium is isotropic the resultant electric field intensity at any point is parallel to the current density at that point. Equation 6–10 may therefore be written in vector form as

$$\mathbf{j} = \gamma \mathbf{E} \qquad (6\text{--}11)$$

This equation is the point form of Ohm's law for isotropic media. In such media the lines of current flow coincide with the lines of force. A tube of current will therefore also constitute a tube of force (Section 2.8) and all surfaces intersecting the current lines at right angles are equipotential surfaces.

Materials which obey Ohm's law are said to be linear or ohmic materials. This means that the conductivity γ is independent of the field. All metals and metal alloys are linear for current densities as high as 10^{10}–10^{11} $\mathrm{A\,m}^{-2}$. Experiments with higher current densities are rendered very difficult by the intense heating of the material. Electrolytic solutions, gases and insulators may under certain experimental conditions behave as linear media.

Single crystals are, in general, anisotropic in their conducting properties, that is, the direction of $\mathbf{j}$ and $\mathbf{E}$ no longer coincide. In this case the conductivity is represented by an electrical conductivity tensor γ_{ij} and equation 6–11 is replaced by

$$j_i = \gamma_{ij}E_j, \; (i, j = 1, 2, 3) \qquad (6\text{--}12)$$

which defines a set of equations analogous to those given by equation 4–22.

Throughout this book we shall be concerned with conduction currents flowing in linear isotropic conductors to which equation 6–11 is applicable.

6.3 JOULE'S LAW

If a potential difference $V = V_A - V_B$ is maintained by some external source connected between the ends of a conductor of resistance R, a current $I = V/R$ will flow through the conductor. This means that every second a charge of I coulombs is moved from a potential V_A to a potential V_B. The energy per second or power expended during this process is

$$P = I(V_A - V_B) = IV = I^2R \; [\text{watt}] \qquad (6\text{--}13)$$

This electric energy is irreversibly converted into heat. In SI units P is expressed in joules per second or watts. The heat energy dissipated in the conductor in a time t seconds is

$$W = Pt = I^2 Rt \qquad (6\text{-}14)$$

Equation 6–14 is known as Joule's law.

If we apply equation 6–13 to an element of volume dv of a current tube length dl and perpendicular cross-sectional area dS we have that

$$dP = dI\,dV = j\,dS\mathbf{E} \cdot \mathbf{dl}$$

but since $\mathbf{j}$ and $\mathbf{dl}$ have the same direction and $dS\,dl = dv$, then

$$dP = \mathbf{j} \cdot \mathbf{E}\,dv$$

or

$$\frac{dP}{dv} = \mathbf{j} \cdot \mathbf{E}\,[\mathrm{W\,m^{-3}}] \qquad (6\text{-}15)$$

in an isotropic medium $\mathbf{j}$ and $\mathbf{E}$ are also in the same direction so that

$$\frac{dP}{dv} = jE = \gamma E^2 \qquad (6\text{-}16)$$

which is the power dissipated per unit volume of a linear isotropic medium. The total power dissipated is given by

$$P = \int \gamma E^2\,dv \qquad (6\text{-}17)$$

6.4 EQUATION OF CONTINUITY OF CURRENT

Consider an arbitrary volume v bounded by a closed surface S. Let the current density at any point on this surface be $\mathbf{j}$; the integral

$$\oint_S \mathbf{j} \cdot \mathbf{dS}$$

will give the net electric current leaving the surface, that is the difference between the inward and outward rate of flow of charge. If there is no rate of change of charge within the closed surface then the charge flowing into the volume is the same as that flowing out so that,

$$\oint_S \mathbf{j} \cdot \mathbf{dS} = 0 \qquad (6\text{-}18)$$

By transforming the surface integral into a volume integral we have

$$\int_v \nabla \cdot \mathbf{j}\,dv = 0$$

or

$$\nabla \cdot \mathbf{j} = 0 \qquad (6\text{-}19)$$

This last equation is the point form of equation 6–18 and states that the lines of steady current flow close upon themselves—that is, they have neither sources nor sinks. A vector function whose divergence is zero is said to be *solenoidal*.

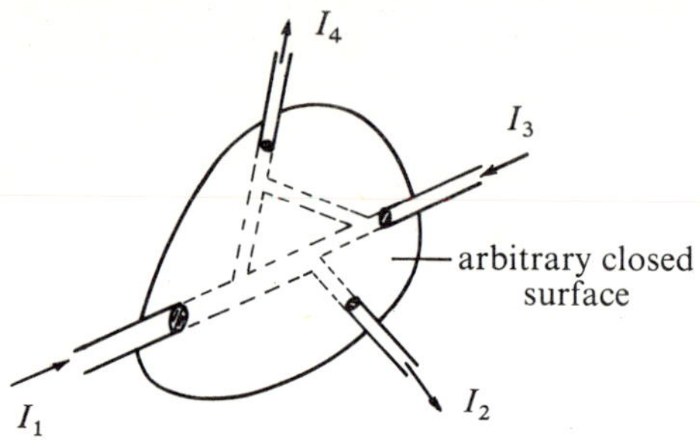

Fig. 6.5 Closed surface enclosing an arbitrary number of conductor junctions

Equation 6–18 is in effect a general statement of the familiar Kirchhoff current law as applied to circuit analysis. Thus let the closed surface enclose an arbitrary number of conductor junctions (Fig. 6.5). It is obvious that the integrand in equation 6–18 is zero except at the intersection of the surface with a conductor. The integral over this intersection gives the current carried by the conductor either out of the surface (positive) or into the surface (negative). The resultant value of the integral in this case gives $I = 0$.

Suppose now that the integral $\oint_S \mathbf{j} \cdot \mathbf{dS}$ is not zero. In this case the principle of conservation of charge requires that the net rate of flow of charge through the surface be equal to the rate of change of charge within the surface. A net flow of current *out* of the surface S must therefore be equal to the rate of *decrease* of the total charge Q within the volume bounded by S; that is,

$$\oint \mathbf{j} \cdot \mathbf{dS} = -\frac{dQ}{dt} = -\frac{\partial}{\partial t} \int_v \rho \, dv$$

$$= -\int_v \frac{\partial \rho}{\partial t} \, dv$$

and

$$\int_v \nabla \cdot \mathbf{j} \, dv = -\int_v \frac{\partial \rho}{\partial t} \, dv$$

Since this relation is true whatever the volume over which we integrate, the integrands must be equal, so that

$$\nabla \cdot \mathbf{j} = -\frac{\partial \rho}{\partial t}$$

or

$$\nabla \cdot \mathbf{j} + \frac{\partial \rho}{\partial t} = 0 \qquad (6\text{–}20)$$

This is the *equation of continuity* and it may be regarded as the mathematical expression of the principle of conservation of charge. The first term accounts for the flow of free charges (conduction) into a small volume whilst the second term is the rate of increase of the local free charge density in that volume.

For steady current flow $\partial \rho / \partial t = 0$ and equation 6–20 reduces to equation 6–19. By combining this latter equation with Ohm's law we obtain

$$\nabla \cdot \gamma \mathbf{E} = 0$$

and since $\mathbf{E} = -\nabla V$ we have, for a homogeneous and isotropic medium, that

$$\nabla^2 V = 0 \qquad (6\text{–}21)$$

Hence in a stationary electric conduction field the potential V satisfies Laplace's equation. This means that problems involving distributions of steady currents in a homogeneous conducting medium may be solved in the same way as problems involving static field distributions in homogeneous insulating media. If we assume that the electrodes at which the current enters and leaves the conducting medium are perfectly conducting (infinite conductivity) their surfaces will be equipotentials. The potential distribution in the medium is the same as in a capacitor whose plates are formed by the two conductors. The electric field lines, which are everywhere orthogonal to lines of constant V, are identical to lines of current flow. This fact forms the basis of operation of the electrolytic tank analogue which is used to obtain experimentally the potential and field distributions for problems which are too difficult or impossible to solve analytically.

The electrolytic tank (Fig. 6.6), which is a conduction-field analogue, consists of a Bakelite or Perspex tank containing tap water or some other convenient weak electrolyte—e.g., copper sulphate—into which are placed the electrodes of the desired shape. The electrical conductivity of these electrodes must be very high compared with that of the electrolyte. The electrodes are connected to external current sources and sinks and the potential at any point between the electrodes is measured by means of a thin-wire travelling probe inserted into the

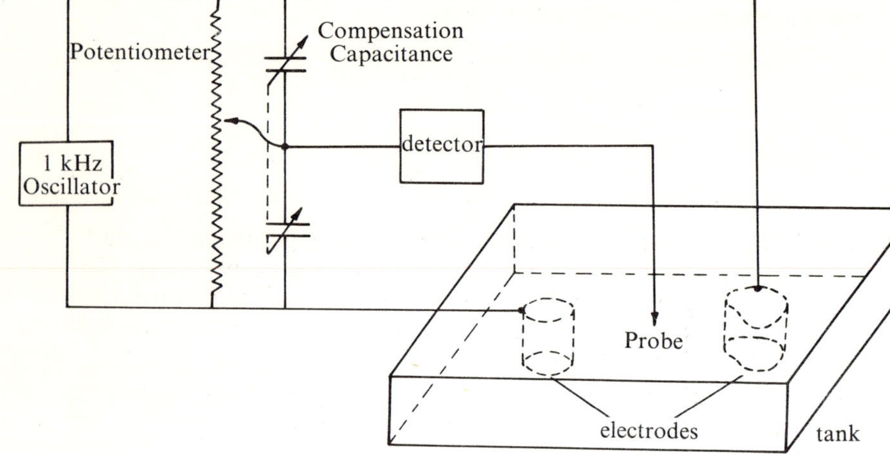

Fig. 6.6 Electrolytic tank

liquid and connected to a convenient measurement circuit. In order to overcome polarization effects a.c. sources are used at a frequency between 500 Hz and 1500 Hz and stainless steel or etched and graphited (to prevent tarnishing and improve conductivity at junction of electrode and electrolyte) brass or copper is recommended for use as electrodes.

The simplest form of the measurement circuit is that in which the probe is connected via a detector to a calibrated resistance potential divider, which together with the resistance through the liquid between the probe and the two electrodes form a Wheatstone bridge. Adjustable capacitors are generally added to compensate for any polarization capacitance at the electrodes. In recent years sophisticated bridges and detectors as well as automatic plotting systems have been developed for increasing the accuracy of electrolytic tank analogues.*

To obtain the equipotentials the potentiometer is set to, say, 10% and the position of the probe is varied until the detector indicates zero potential. The position of the probe is plotted and the process repeated until the number of null points obtained is sufficient to give an equipotential line. The potentiometer is then set to a different value, say, 20%, and a new equipotential is obtained. By repeating the process at 10% (or smaller) intervals a family of equipotentials is obtained. To reproduce the position of the probe accurately a conventional pantograph arrangement may be used or the tank may be fitted with a transparent plotting table which allows the position of the probe to be

* D. Vitkovitch, *Field Analysis: Experimental and Computational Methods*, Van Nostrand, London, 1966.

accurately indicated by means of a pair of illuminated cross-wires located directly above it.

A device similar to the electrolytic tank, but much simpler and cheaper, is the analogue plotter which uses conducting (Teledeltos) paper instead of the liquid electrolyte.* The current is fed into the conducting paper via electrodes formed by painting conducting silver paint on to the required electrode cross-sections drawn onto the paper. A probe, power supply, voltage divider and a null detector serve to locate the equipotentials in a manner similar to that described for the electrolytic tank. However, as no electrolyte or polarization effects are involved, such analogues can be operated with either direct or alternating current.

6.5 RELATIONSHIP BETWEEN RESISTANCE AND CAPACITANCE

The analogy between conduction and electrostatic problems leads to a simple relation between the resistance and capacitance between two electrodes. The basic relations for electrostatics and for conduction are,

electrostatics

$$\mathbf{D} = \epsilon\mathbf{E}$$

$$\oint \mathbf{D} \cdot \mathbf{dS} = Q$$

$$\int_1^2 \mathbf{E} \cdot \mathbf{dl} = V_2 - V_1$$

conduction

$$\mathbf{j} = \gamma\mathbf{E}$$

$$\oint \mathbf{j} \cdot \mathbf{dS} = I$$

$$\int_1^2 \mathbf{E} \cdot \mathbf{dl} = V_2 - V_1$$

where Q is the total charge on each of the electrodes in the electrostatic case and I is the total current flowing *between* the electrodes in the conduction case. $V_2 - V_1$ is the potential difference maintained between the electrodes. By definition

$$C = \frac{Q}{V} = \frac{\oint \mathbf{D} \cdot \mathbf{dS}}{\int \mathbf{E} \cdot \mathbf{dl}} = \frac{\epsilon \oint \mathbf{E} \cdot \mathbf{dS}}{\int \mathbf{E} \cdot \mathbf{dl}}$$

and

$$R = \frac{V}{I} = \frac{\int \mathbf{E} \cdot \mathbf{dl}}{\oint \mathbf{j} \cdot \mathbf{dS}} = \frac{\int \mathbf{E} \cdot \mathbf{dl}}{\gamma \oint \mathbf{E} \cdot \mathbf{dS}}$$

* C. W. Park and J. J. Barale, 'Analog Field Plotter; Description and use', *Electrical Engineering*, September 1961, pp. 699–701.

R. Croxford, 'Analogue of Cylindrically Symmetric Fields using Teledeltos Paper', *Proc. Instn elec. Engrs*, 1967, **114**, pp. 664–5.

P. H. G. Allen, 'Field Mapping with Conducting Paper', *Physics Education*, 1968, **3**, pp. 266–72.

The product of the above two expressions yields

$$RC = \epsilon/\gamma \tag{6-22}$$

Consider as an example two coaxial circular cylinders of radii a_1 and $a_2 (a_2 > a_1)$. If the space between the cylinders is filled with a material of conductivity γ the resistance between the cylinders per unit of axial length is readily obtained by combining equation 6–22 with equation 3–61. This gives

$$R = \frac{\log a_2/a_1}{2\pi\gamma}$$

The resistance of the various geometrical configurations for which the capacitance was derived in Chapter 3 may be readily obtained in a similar manner. For cases in which the capacitance between two conductors is not amenable to calculation, the dielectric medium (of permittivity ϵ) between the conductors may be replaced by an electrolyte of known conductivity γ and the resistance of the system measured. The capacitance may then be calculated from equation 6–22.

6.6 BOUNDARY CONDITIONS FOR STEADY CURRENTS

Consider the boundary between two media whose permittivities and conductivities are ϵ_1, γ_1 and ϵ_2, γ_2 respectively (Fig. 6.7). If we apply

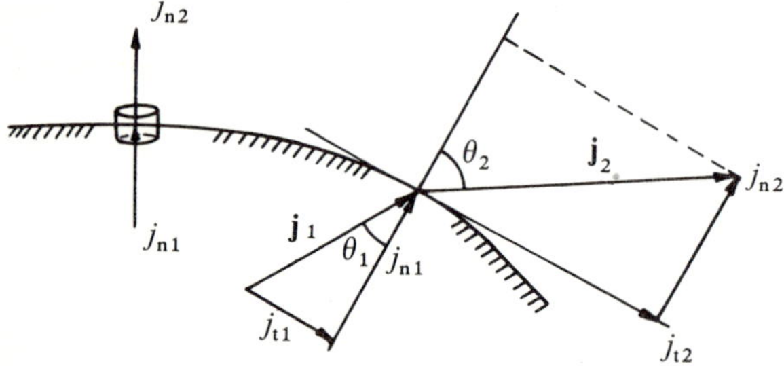

Fig. 6.7 Boundary between two media of different permittivities and conductivities

equation 6–18 to a pillbox surface of negligible thickness enclosing the interface, we obtain a relation between the normal components of the current densities on both sides of the boundary. This is

$$j_{n1} = j_{n2} = j_n \tag{6-23}$$

Since the tangential component of the electric field intensity is continuous across the boundary (see Section 4.4) we have that,

$$E_{t1} = E_{t2}$$

and therefore

$$j_{t1}/\gamma_1 = j_{t2}/\gamma_2 \qquad (6\text{--}24)$$

It follows that

$$\tan\theta_1 = j_{t1}/j_n;\ \tan\theta_2 = j_{t2}/j_n$$

and hence

$$\frac{\tan\theta_1}{\tan\theta_2} = \frac{j_{t1}}{j_{t2}} = \frac{\gamma_1}{\gamma_2}.$$

and

$$\gamma_1\cot\theta_1 = \gamma_2\cot\theta_2 \qquad (6\text{--}25)$$

If region 2 is a perfect insulator, i.e. $\gamma_2 = 0$, then

$$\mathbf{j}_2 = 0 \qquad \text{and} \qquad j_{n1} = j_{n2} = 0$$

so that at the interface between a conductor and a perfect insulator there can only be a tangential flow of current in the conductor. If $\gamma_1 \gg \gamma_2$ then for all values of $\theta_1 \neq 90°$, θ_2 is extremely small. For example if $\gamma_1/\gamma_2 = 500$, for $\theta_1 = 88°$ we find that $\theta_2 = 3°$. Thus the current lines in medium 2 are perpendicular to the interface which is therefore an equipotential surface, j_{t2} and hence E_{t2} being negligibly small. A practical case is that in which medium 1 is a conductor and medium 2 is an electrolyte; the ratio γ_1/γ_2 is of the order of 10^6 so that the conductor surface is, to a very good approximation, an equipotential. As mentioned in Section 6.4 this fact is made use of in the electrolytic tank analogue.

From equation 4–23 we have that,

$$\epsilon_2 E_{n2} - \epsilon_1 E_{n1} = \sigma$$

which may be re-expressed as,

$$(\epsilon_2/\gamma_2)j_{n2} - (\epsilon_1/\gamma_1)j_{n1} = \sigma$$

and since $j_{n1} = j_{n2}$, it follows that

$$\left(\frac{\epsilon_2}{\gamma_2} - \frac{\epsilon_1}{\gamma_1}\right)\mathbf{j}\cdot\mathbf{a}_n = \sigma \qquad (6\text{--}26)$$

where $\mathbf{a}_n$ is the unit normal to the interface directed from medium 1 to medium 2. Equation 6–26 shows that in general a surface charge will appear on the boundary between two media of finite conductivities across which a current is flowing. If both media are metallic conductors, $\epsilon_1 = \epsilon_2 = \epsilon_0$ and equation 6–26 becomes

$$\epsilon_0[(1/\gamma_2) - (1/\gamma_1)]\mathbf{j}\cdot\mathbf{a}_n = \sigma$$

6.7 DISPLACEMENT CURRENT

The relationship between the electric displacement **D** and the volume charge density ρ may be expressed by the point form of Gauss's law as

$$\nabla \cdot \mathbf{D} = \rho$$

Hence it follows that,

$$\nabla \cdot \left(\frac{\partial \mathbf{D}}{\partial t}\right) = \frac{\partial \rho}{\partial t} \tag{6-27}$$

By combining equation 6–27 with the equation of continuity (6–20) we obtain,

$$\nabla \cdot \left(\mathbf{j} + \frac{\partial \mathbf{D}}{\partial t}\right) = 0 \tag{6-28}$$

which may be expressed in integral form as

$$\oint_s \left(\mathbf{j} + \frac{\partial \mathbf{D}}{\partial t}\right) \cdot \mathbf{dS} = 0 \tag{6-29}$$

The term $\partial \mathbf{D}/\partial t$ has the dimensions of current density (A m^{-2}). Since it results from the variation of the electric displacement density (and hence of the electric field) it is called the *displacement current density*. By definition therefore,

$$\mathbf{j}_d = \frac{\partial \mathbf{D}}{\partial t} \tag{6-30}$$

In section 4–3 we defined the displacement **D** as

$$\mathbf{D} = \epsilon_0 \mathbf{E} + \mathbf{P}$$

The displacement current density is thus

$$\mathbf{j}_d = \epsilon_0 \frac{\partial \mathbf{E}}{\partial t} + \frac{\partial \mathbf{P}}{\partial t} \tag{6-31}$$

The second term on the right-hand side expresses the variation of the polarization of the medium with time and is called the *polarization current density*. It represents an actual displacement of elastically bound charges and as such it is confined to dielectric media in which the polarization varies with time. It may be regarded as a special type of conduction current which can flow only under a changing electric field.

The first term on the right-hand side of equation 6–31 represents a vacuum displacement current and is associated with a time-varying field in free space. It is really a fictitious current in the sense that it does not arise from the motion or displacement of any charge, either free or bound. A plausible physical explanation for the existence of this current is not possible.

It is clear from equation 6–28 that the addition of the displacement current density to the conduction current density makes the total current density a solenoidal quantity. It is thus possible to state that a current will always flow in a closed circuit provided that the time variation of the electric displacement is taken into account either as an additional contribution or as an alternative to the conduction current. A classical example is that of a battery charging a capacitor (Fig. 6.8). The current which enters and leaves the closed surface S

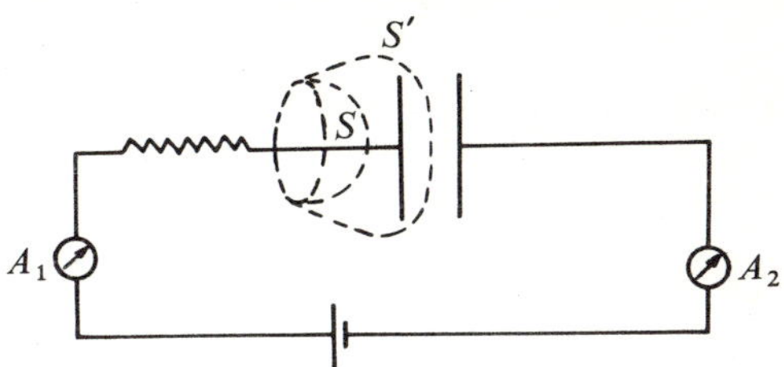

Fig. 6.8 Capacitor charging currents. At any instant during charging A_1 and A_2 indicate the same current

is a conduction current. On the other hand the current which enters the closed surface S' is a conduction current whilst that which leaves it is a displacement current.

The concept of displacement current was first introduced by Maxwell and is considered as his greatest theoretical contribution to electromagnetic theory in general and to the theory of the propagation of electromagnetic waves in particular. Maxwell's assumption of the existence of a displacement current enabled him to formulate a self-consistent set of electromagnetic field equations which lead to the classical wave equations of the electromagnetic field; these give, as far as we know, an accurate description of the propagation of electromagnetic waves and hence justify Maxwell's hypothesis. Experimental proof of the propagation of electromagnetic waves in accordance with Maxwell's predictions was first given by Hertz in his series of epoch-making experiments (1886–88). We shall make further use of the concept of displacement current in Chapter 7.

6.8 RATIO OF DISPLACEMENT CURRENT DENSITY TO CONDUCTION CURRENT DENSITY

The total current in a medium of permittivity ϵ and conductivity γ is given by,

$$\mathbf{J} = \mathbf{j} + \mathbf{j}_d$$

$$= \gamma\mathbf{E} + \frac{\partial \mathbf{D}}{\partial t}$$

and the ratio of displacement to conduction current densities is

$$\frac{\mathbf{j}_d}{\mathbf{j}} = \left(\epsilon_0 \frac{\partial \mathbf{E}}{\partial t} + \frac{\partial \mathbf{P}}{\partial t}\right)\bigg/ \gamma \mathbf{E} \qquad (6\text{--}32)$$

In metallic conductors the term $\partial \mathbf{P}/\partial t$ does not exist so that equation 6–32 becomes

$$\frac{\mathbf{j}_d}{\mathbf{j}} = \frac{\epsilon_0 \partial \mathbf{E}/\partial t}{\gamma \mathbf{E}} \qquad (6\text{--}33)$$

For a periodic field varying sinusoidally with an angular frequency ω such that $E = E_0 \cos \omega t$, the ratio of the amplitudes of the two current densities is

$$\frac{j_d}{j} = \frac{\omega \epsilon_0}{\gamma} \qquad (6\text{--}34)$$

For a conductor such as copper $\gamma = 5\cdot 8 \times 10^7 \, \text{ohm}^{-1}\text{m}^{-1}$ and the ratio ϵ_0/γ is of the order of 10^{-19}. We thus see that even for frequencies of the order of 10^{16} Hz (vacuum ultra-violet and soft X-rays) the displacement current is only about 1% of the conduction current. It is therefore entirely justifiable to neglect the displacement current in metallic conductors up to frequencies well above the microwave region.

For a dielectric the ratio $\mathbf{j}_d/\mathbf{j}$ is given by equation 6–32. If the field varies sinusoidally, then the displacement current density is given by equation 4–87 and the ratio of the amplitudes of the current densities is

$$\frac{j_d}{j} = \frac{\omega \epsilon_0 |\hat{\epsilon}_r|}{\gamma} \qquad (6\text{--}35)$$

where $|\hat{\epsilon}_r| = (\epsilon_r'^2 + \epsilon_r''^2)^{1/2}$. For an excellent insulator such as quartz, for which $\gamma = 10^{-17}\,\text{ohm}^{-1}\text{m}^{-1}$, $|\hat{\epsilon}_r| \simeq 3\cdot 8$, the ratio $\epsilon_0 |\hat{\epsilon}_r|/\gamma$ is of the order of 10^6 so that the conduction current is negligibly small compared with the displacement current.

From the practical point of view the frequency is, as a rule, the important factor which determines whether a medium behaves as a conductor or as a dielectric. As an example, for sea water ($\gamma = 4\,\text{ohm}^{-1}\text{m}^{-1}$, $|\hat{\epsilon}_r| \leq 80$) the ratio $\epsilon_0 |\hat{\epsilon}_r|/\gamma$ is of the order of 10^{-10} so that for frequencies below 10^7 Hz it acts as a conductor whilst above 10^{12} Hz it acts as a dielectric.

6.9 CHARGE RELAXATION

In this section we shall show that in a homogeneous medium of finite conductivity in which there are no sources of free charge, a steady state free-charge density cannot be maintained.

Since $\mathbf{j} = \gamma\mathbf{E}$, we may rewrite the continuity equation 6–20 as

$$\nabla \cdot (\gamma\mathbf{E}) = -\partial\rho/\partial t$$

Expanding the left-hand side we obtain

$$\gamma\nabla \cdot \mathbf{E} + \mathbf{E} \cdot \nabla\gamma = -\partial\rho/\partial t \tag{6–36}$$

We also know that the flux density is related to the charge density by

$$\nabla \cdot \mathbf{D} = \nabla \cdot (\epsilon\mathbf{E}) = \rho \tag{6–37}$$

If we assume that the medium is homogeneous, so that both γ and ϵ are independent of space coordinates, equations 6–36 and 6–37 reduce to

$$\gamma\nabla \cdot \mathbf{E} = -\partial\rho/\partial t \tag{6–38}$$

and

$$\nabla \cdot \mathbf{E} = \rho/\epsilon \tag{6–39}$$

By combining these two equations we obtain the differential equation

$$\frac{\partial\rho}{\partial t} + \frac{\gamma}{\epsilon}\rho = 0 \tag{6–40}$$

whose solution is readily seen to be

$$\rho = \rho_0\exp(-\gamma t/\epsilon) \tag{6–41}$$

where ρ_0 is the volume density of free charge at the time $t = 0$. The time required for this charge to decrease to $1/e$ of its value is

$$\tau = \frac{\epsilon}{\gamma}\ \text{second} \tag{6–42}$$

τ is called the *relaxation time*. For a perfectly non-conducting medium such as absolute vacuum $\gamma = 0$ and τ is infinite. However, since all physical media have a finite conductivity, τ will always have a finite value and there can be no steady state charges in the bulk of the medium. Because the physical size of any medium is finite, the principle of conservation of charge requires that any free charges initially distributed throughout a volume relax to the surfaces that bound that volume. For a good conductor such as silver for which $\gamma = 6\cdot17 \times 10^7\ \text{ohm}^{-1}\,\text{m}^{-1}$ and $\epsilon = \epsilon_0$, the relaxation time is $1\cdot43 \times 10^{-19}\ \text{s}$; all metallic conductors have a relaxation time of this same order of magnitude. Since this is immeasurably small, it is safe to conclude that the volume density of free charge within a conductor is zero at all times and that any free charge will always be concentrated on its surface. This conclusion is true irrespective of whether a conduction current is flowing through the conductor or not, because even with a current flowing the equation $\rho_{\text{elec.}} + \rho_{\text{ion}} = 0$ is always satisfied. Even at the highest frequencies attainable with macroscopic oscillators ($\simeq 10^{10}$ Hz) there can be no accumulation of charge within the volume of a conductor because the relaxation time is very much smaller than the period.

For poor conductors (such as sea water) and electrolytes, the relaxation time is of the order of 10^{-10} s. For insulators τ may be measured in minutes, hours, or even days depending upon the value of the conductivity which, as shown in Table 6.1, may vary over quite a large range.

Table 6.1 The conductivity and temperature coefficient of resistance of some common materials

Material	Conductivity at 20°C (ohm^{-1} m^{-1})	Temperature coefficient (α) per °C
Aluminium	$3 \cdot 54 \times 10^7$	$0 \cdot 004$
Copper	$5 \cdot 8 \times 10^7$	$0 \cdot 0043$
Gold	$4 \cdot 1 \times 10^7$	$0 \cdot 004$
Iron (99·98% pure)	10^7	$0 \cdot 006$
Lead	$4 \cdot 55 \times 10^6$	$0 \cdot 0042$
Mercury	10^6	$0 \cdot 0010$
Nickel	$1 \cdot 28 \times 10^7$	$0 \cdot 0068$
Platinum	$0 \cdot 94 \times 10^7$	$0 \cdot 004$
Silver	$6 \cdot 13 \times 10^7$	$0 \cdot 0041$
Sodium	$2 \cdot 14 \times 10^7$	$0 \cdot 0055$
Tin	$0 \cdot 93 \times 10^7$	$0 \cdot 0046$
Tungsten	$1 \cdot 84 \times 10^7$	$0 \cdot 0048$
Zinc	$1 \cdot 68 \times 10^7$	$0 \cdot 0042$
Constantan (55Cu 45Ni)	$2 \cdot 0 \times 10^6$	$\pm 0 \cdot 00004$
Nichrome (80Ni 20Cr)	10^6	$0 \cdot 00008$
Manganin (84Cu 4Ni 12Mn)	$2 \cdot 3 \times 10^6$	$\pm 0 \cdot 00001$
Carbon (graphite)	$\sim 10^5$	
Germanium (pure)	2	
Silicon (pure)	$3 \cdot 8 \times 10^{-4}$	
Water (distilled)	10^{-2}–10^{-5}	
Sea water	~ 4	
NaCl solution (saturated)	$22 \cdot 6$	
NaCl solution (5%)	7	
Amber	5×10^{-14}	
Glass	10^{-9}–10^{-12}	
Mica	10^{-11}–10^{-15}	
Polystyrene	10^{-15}–10^{-19}	
PTFE (Teflon)	10^{-15}–10^{-18}	
Quartz (fused)	10^{-17}	
Rubber (Hard)	10^{-13}–10^{-16}	
Sulphur	10^{-14}–10^{-15}	
Mineral oil	10^{-11}–10^{-15}	

PROBLEMS

Chapter Six

6.1 Figure P6.1 shows a toroidal sector of rectangular cross-section and uniform conductivity γ. Show that the resistance between the two rectangular ends is given by

$$R = \theta/\gamma t \log{(b/a)}$$

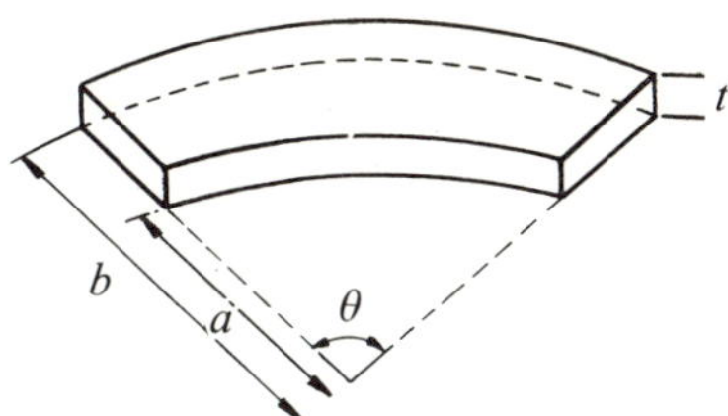

Fig. P. 6.1

6.2 A metal hemisphere of negligible resistivity and radius a is embedded in the ground with its flat part level with the ground surface. If the conductivity of the ground is γ, show that the resistance between the hemisphere and ground is $1/2\pi\gamma a$. If a current I flows from the sphere to ground and a man is standing at a distance r from the centre of the hemisphere, show that if the man takes a step of length l towards the ground electrode, then the potential difference between his feet is given by

$$V = \frac{I}{2\pi\gamma} \cdot \frac{l}{r(r-1)}$$

Find V if $I = 1000\,\text{A}$, $l = 0{\cdot}8\,\text{m}$, $r = 2\,\text{m}$, $\gamma = 10^{-2}\,\text{ohm}^{-1}\,\text{m}^{-1}$.

6.3 Two spherical electrodes of diameter a and negligible resistivity are embedded in a medium of infinite extent and conductivity γ; the separation d between their centres is large compared with a. If a current enters one sphere and leaves at the other, show that the resistance between the spheres is given by

$$R = (d-a)/2\pi\gamma d\,a$$

6.4 A spherical electrode of radius a and negligible resistivity is embedded in an infinite medium of conductivity γ and permittivity ϵ. If a current I leaves the electrode, show, by using equation 6–26 or otherwise, that the total charge on the electrode surface is $\epsilon I/\gamma$.

6.5 A thin spherical shell of radius a and thickness t has a conductivity γ. Current enters and leaves the shell by two small circular electrodes of radius $b(\ll t)$ and negligible resistivity placed at diametrically opposite points of the shell. Show that the resistance of the shell between the electrodes is given by

$$R = \frac{1}{\gamma \pi t} \log(2a/b)$$

6.6 Figure P6.6 shows the sector of a hollow conducting sphere of internal radius a and external radius b, and conductivity γ. Show that

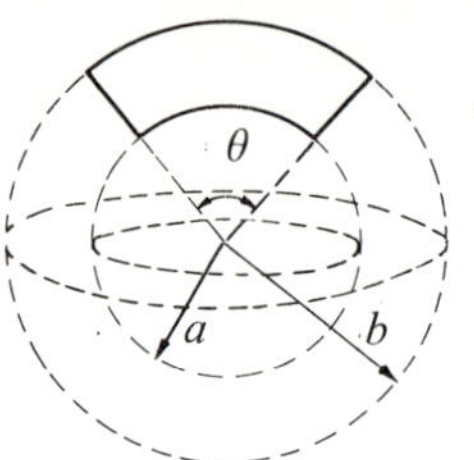

Fig. P. 6.6

the resistance between the two spherical ends of the sector is given by

$$R = \frac{1}{2\pi\gamma(1-\cos\theta)} \cdot \frac{b-a}{ab}$$

6.7 A Faraday disc of thickness t has a radius b and a central hole of radius a. If the conductivity of the disc material is γ, show that the resistance between the hole and the rim of the disc is given by

$$R = \frac{1}{2\pi\gamma t} \log\left(\frac{b}{a}\right)$$

6.8 The current density at a point P lying in region 1 on the boundary between two regions 1 and 2 makes an angle $\theta_1 = 88°$ with the normal at that point. Find the angle between the current density vector and the normal at the point immediately opposite P in region 2 if $\gamma_1 = 500\gamma_2$ and if $\gamma_1 = 50\gamma_2$. What practical use can be made of this result?

6.9 Show that the ratio of the magnitudes of displacement and con duction current densities for harmonically varying fields is $\gamma/\omega\epsilon$. At what frequency is this ratio unity for
(a) copper, $\gamma = 5\cdot8 \times 10^7$ ohm^{-1} m^{-1}, $\epsilon_r = 1$
(b) sea water, $\gamma = 4 \times 10^{-3}$ ohm^{-1} m^{-1}, $\epsilon_r = 81$
(c) marble, $\gamma = 10^{-8}$ ohm^{-1} m^{-1}, $\epsilon_r = 8$

CHAPTER SEVEN

THE MAGNETIC FIELD OF STEADY CURRENTS IN FREE SPACE

Up to the end of the eighteenth century the phenomena of electricity and magnetism (effects produced by bodies possessing the particular property of magnetism or magnetizability) were considered to be quite distinct from one another, although observations that steel objects became magnetized after a lightning discharge had led some philosophers to suspect some kind of relationship between the two. For many years experimental attempts to show this failed and it was only in 1820, when Oersted demonstrated that the flow of a steady current in a wire produces a marked deflection of a magnetic compass needle placed in the vicinity of the wire, that a definite link was established between electricity and magnetism.

Oersted's discovery laid the foundations for the development of the science of electromagnetism. His experiments were repeated and extended and before long Ampère, Biot and Savart had worked out a complete and perfect quantitative formulation of the magnetic action of steady electric currents.*

In this chapter we shall study the quantitative relationships which exist between a steady magnetic field and its direct current source, for free-space conditions. The effects of material media will be discussed in the next chapter.

In electrostatics our starting point was the law of force between point charges. Here we shall take as our starting point the law of force between current elements.

7.1 THE CURRENT ELEMENT

It will be recalled that in electrostatics we always used the concept of point charge as the basis for our calculations; whenever we had a charge distribution our point charge took the form of a differential element of charge, such as $\rho \, dv$, $\sigma \, dS$ or $\lambda \, dL$, and the resultant electric quantity (potential or field) at any point due to the entire distribution was obtained by integrating the corresponding differential point-charge expression over the whole distribution.

In dealing with electric current it is again convenient to introduce the

* For a complete chronological account of the developments which eventually led to a complete formulation of the classical theory of electromagnetism, see *A History of the Theories of Aether and Electricity*, Vol. I, by E. T. Whittaker (Nelson, 1951).

concept of a 'differential current element'. The definition of such an element is based on that of the current density **j** as the rate of charge flow at some macroscopic point; a *volume current element* is thus defined as

$$\text{current element} = \mathbf{j}\,dv \tag{7-1}$$

where dv is an element of volume. In order to fix the boundaries of dv, let **dl** be an element of length in the same direction as **j**; then an element of area **dS** can always be defined by

$$dv = \mathbf{dl} \cdot \mathbf{dS}$$

Hence we have that

$$\mathbf{j}\,dv = \mathbf{j}\,\mathbf{dl} \cdot \mathbf{dS} \tag{7-2}$$

and since **dl** and **j** are collinear it follows that

$$\mathbf{j}\,dv = (\mathbf{j} \cdot \mathbf{dS})\,\mathbf{dl} = dI\,\mathbf{dl} \tag{7-3}$$

where dI is the current through the volume element. Equation 7–3 defines a *linear current element*.

Whenever the current is constrained to flow in thin wires (which is the most common case in practice) it is convenient to integrate the basic current element, equation 7–2, over the wire cross-section; for these cases the current element is defined as

$$\mathbf{dl} \int \mathbf{j} \cdot \mathbf{dS} = I\,\mathbf{dl} \tag{7-4}$$

If the current is confined to flow along sheets of negligible thickness

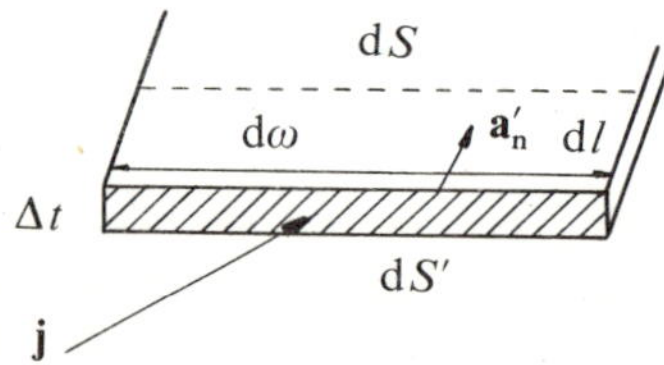

Fig. 7.1 Element of current sheet

the differential current element may be expressed in terms of the surface current density **K** as (see Fig. 7.1)

$$\mathbf{j}\,dv = \mathbf{j}(\mathbf{dS}' \cdot \mathbf{dl}) = \mathbf{j}\Delta t(d\omega \mathbf{a}'_n \cdot \mathbf{dl}) = \mathbf{K}\,dS \tag{7-5}$$

It should be noted that the current element, whether defined by equations 7–1, 7–4 or 7–5, is a **vector** point function. The dimensions of a current element are ampere metre.

7.2 FORCE BETWEEN TWO CURRENT ELEMENTS. THE BIOT-SAVART LAW

Consider two distinct linear current elements $I_1\,\mathrm{dl}_1$ and $I_2\,\mathrm{dl}_2$ located in space at points 1 and 2 respectively (Fig. 7.2a). If $\mathbf{r}_{12}$ is the vector distance from element 1 to element 2 (noting $|r_{12}| = r$) then the force

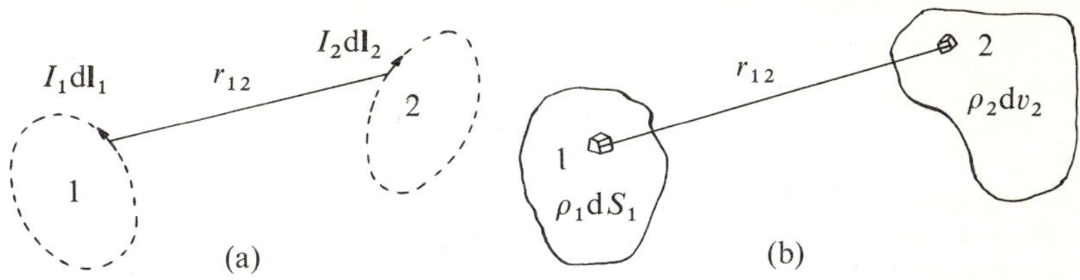

Fig. 7.2 (a) Two distinct current elements (b) Two distinct volume elements of charge

on current element $I\,\mathrm{dl}_2$ due to element $I\,\mathrm{dl}_1$ is given by the second-order differential

$$\mathrm{d}^2\mathbf{F}_2 = C\,\frac{I_2\,\mathrm{dl}_2 \times I_1\,\mathrm{dl}_1 \times \mathbf{r}_{12}}{r^3} \qquad (7\text{-}6)$$

In SI units the constant of proportionality has the value $\mu_0/4\pi$, where μ_0 is a magnetic constant whose value is defined as

$$\mu_0 = 4\pi \times 10^{-7} \text{ henry metre}^{-1} \qquad (7\text{-}7)$$

The units of μ_0 are newton ampere^{-2} which, as will be shown later, are equivalent to henry metre^{-1}. Equation 7–6 may thus be written as

$$\mathrm{d}^2\mathbf{F}_2 = \frac{\mu_0}{4\pi}\,\frac{I_2\,\mathrm{dl}_2 \times I_1\,\mathrm{dl}_1 \times \mathbf{r}_{12}}{r^3} \text{ newton} \qquad (7\text{-}8)$$

The direction of the force $\mathrm{d}^2\mathbf{F}_2$ is always perpendicular to the directions of dl_2 and of the normal to the plane containing dl_1 and $\mathbf{r}_{12}$; the sense of the force will depend upon the sense of the vector elements dl_1 and dl_2. We may rewrite equation 7–8 as

$$\mathrm{d}^2\mathbf{F}_2 = I_2\,\mathrm{dl}_2 \times \left(\frac{\mu_0}{4\pi}\,\frac{I_1\,\mathrm{dl}_1 \times \mathbf{r}_{12}}{r^3} \right) \qquad (7\text{-}9)$$

The quantity between the brackets depends solely on the current element $I\,\mathrm{dl}_1$ at point 1 and on its position relative to the current element $I_2\,\mathrm{dl}_2$ at point 2 so that we may write

$$\mathrm{d}^2\mathbf{F}_2 = I_2\,\mathrm{dl}_2 \times \left(\begin{array}{l} \text{action at point 2 produced by} \\ \text{current at point 1} \end{array} \right) \qquad (7\text{-}10)$$

Equation 7–9 is the law which expresses the force between two current elements. The reader will readily see that this law is in many respects analogous to Coulomb's law which expresses the force between two static point charges or charge elements. Thus if ρ_1 and ρ_2 are the charge densities of two distinct static charge distributions 1 and 2 (Fig. 7.2b), the force exerted by an element of charge of the first distribution on an element of charge of the second distribution is given by

$$d^2\mathbf{F}_2 = \rho_2\, dv_2 \left(\frac{1}{4\pi\epsilon_0} \frac{\rho_1\, dv_1 \mathbf{r}_{12}}{r^3} \right)$$

In this expression the quantity in the brackets may be referred to as the action at point 2 produced by the point-charge element at point 1. By definition we know that this 'action' is the differential field intensity $d\mathbf{E}_1$ produced at 2 by the charge element at 1.

By analogy with electrostatics, the 'action at point 2 . . .' referred to in equation 7–10 is represented by the differential quantity $d\mathbf{B}$ defined by

$$d\mathbf{B}_1 = \frac{\mu_0}{4\pi} \frac{I_1\, d\mathbf{l}_1 \times \mathbf{r}_{12}}{r^3} \text{ tesla} \tag{7–11}$$

The vector $\mathbf{B}$ is called the magnetic flux density or magnetic induction. A more appropriate name would be magnetic field intensity, but unfortunately the former two nomenclatures are so widespread that we have not attempted to change it. The SI units of magnetic induction are

$$\text{newton ampere}^{-1} \text{ metre}^{-1} = \text{henry ampere metre}^{-2}$$

$$= \text{weber metre}^2 = \text{tesla}$$

Equation 7–11 is the *Biot–Savart law* for free space. This law expresses the magnetic flux density at some field point 2 produced by a current element located at a source point 1 (Fig. 7.3). The direction of the flux

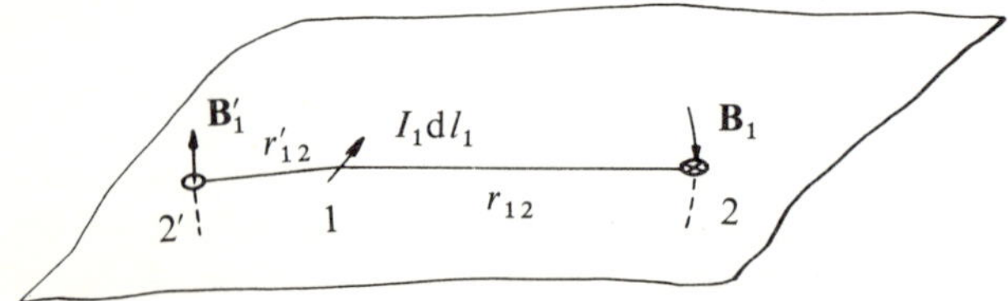

Fig. 7.3 Quantities involved in the Biot–Savart law

density is normal to the plane containing $I_1\, d\mathbf{l}_1$ and $\mathbf{r}_{12}$. Both sense and direction are best remembered by means of the *right-hand rule*. If the element is grasped in the right hand with the thumb pointing in the direction of the current, the fingers which encircle the element will point in the direction of the magnetic flux density.

Although the differential form of the Biot–Savart law given above is convenient for carrying out mathematical computations, it is apparent that it is void of any physical significance since it is impossible to isolate a current element $I\,d\mathbf{l}$ from the rest of the circuit of which it necessarily is a part. On the other hand the integration of equation 7–11 over the entire circuit of which $I_1\,d\mathbf{l}_1$ is a part does have a physical sense; it gives the magnetic flux density $\mathbf{B}_1$ produced by this circuit at the point 2:

$$\mathbf{B}_1 \;=\; \frac{\mu_0}{4\pi}\oint_1 \frac{I_1\,d\mathbf{l}_1 \times \mathbf{r}_{12}}{r^3} \qquad (7\text{–}12)$$

It is this integral form of the Biot–Savart law that can be verified experimentally—e.g., using current balances.

Integrating equation 7–9 over circuit 1 we obtain the first-order differential force

$$d\mathbf{F}_2 \;=\; I_2\,d\mathbf{l}_2 \times \frac{\mu_0}{4\pi}\oint_1 \frac{I_1\,d\mathbf{l}_1 \times \mathbf{r}_{12}}{r^3}$$

which, using equation 7–12 gives

$$d\mathbf{F}_2 \;=\; I_2\,d\mathbf{l}_2 \times \mathbf{B}_1 \qquad (7\text{–}13)$$

This is the force experienced by a current element $I_2\,d\mathbf{l}_2$ placed in an external magnetic field whose magnetic flux density at $d\mathbf{l}_2$ is $\mathbf{B}_1$. The force is perpendicular to the plane containing the current element and $\mathbf{B}_1$; its direction is that of the advance of a right-handed screw turned from $d\mathbf{l}_2$ to $\mathbf{B}_1$ through the smaller angle between these two vectors (Fig. 7.4). Equation 7–13 defines the magnetic induction $\mathbf{B}$ for any medium as the force per ampere metre (current-length) normal to $\mathbf{B}$.

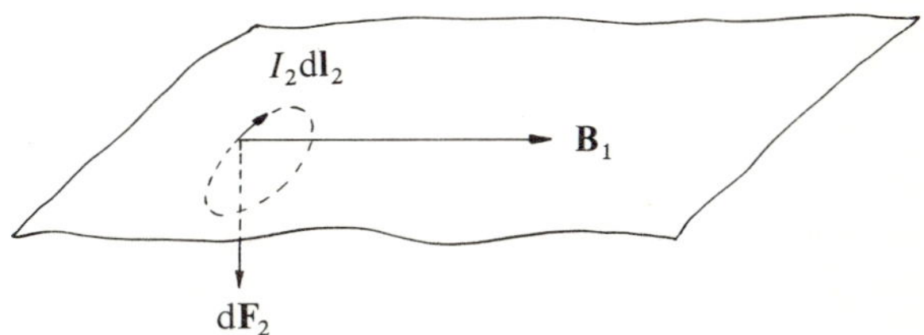

Fig. 7.4 Force experienced by a current element placed in an external magnetic field

The total force which the current circuit 1 (of which $I_1\,d\mathbf{l}_1$ is an element) exerts on the current circuit 2 (of which $I_2\,d\mathbf{l}_2$ is an element)

may now be expressed by

$$\mathbf{F}_2 = \oint_2 d\mathbf{F}_2 = \oint_2 I_2 \, d\mathbf{l}_2 \times \mathbf{B}_1$$

substituting for $\mathbf{B}_1$ from equation 7–12 we obtain

$$\mathbf{F}_2 = \frac{\mu_0}{4\pi} I_1 I_2 \oint_1 \oint_2 \frac{d\mathbf{l}_2 \times (d\mathbf{l}_1 \times \mathbf{r}_{12})}{r^3} \tag{7–14}$$

This equation is sometimes called *Grassmann's formula*.

The total force which circuit 2 exerts on circuit 1 is given by

$$\mathbf{F}_1 = \frac{\mu_0}{4\pi} I_1 I_2 \oint_1 \oint_2 \frac{d\mathbf{l}_1 \times (d\mathbf{l}_2 \times \mathbf{r}_{21})}{r^3} \tag{7–15}$$

Since the integrands in equations 7–14 and 7–15 are asymmetrical it would appear that the forces $\mathbf{F}_2$ and $\mathbf{F}_1$ do not obey Newton's third law that action and reaction should be equal and opposite. That these forces do not in fact violate Newton's law may be demonstrated as follows.

The vector triple product in the integrand of equation 7–14 may be expanded to give

$$d\mathbf{l}_2 \times (d\mathbf{l}_1 \times \mathbf{r}_{12}) = (d\mathbf{l}_2 \cdot \mathbf{r}_{12}) \, d\mathbf{l}_1 - (d\mathbf{l}_1 \cdot d\mathbf{l}_2) \mathbf{r}_{12}$$

so that

$$\mathbf{F}_2 = \frac{\mu_0}{4\pi} I_1 I_2 \oint_1 \oint_2 \left\{ \frac{(d\mathbf{l}_2 \cdot \mathbf{r}_{12}) \, d\mathbf{l}_1}{r^3} - \frac{(d\mathbf{l}_1 \cdot d\mathbf{l}_2)\mathbf{r}_{12}}{r^3} \right\}$$

The first of these integrals may be rewritten as

$$\oint_1 \oint_2 \frac{(d\mathbf{l}_2 \cdot \mathbf{r}_{12}) \, d\mathbf{l}_1}{r^3} = -\oint_1 \oint_2 d\mathbf{l}_1 \, d\mathbf{l}_2 \cdot \nabla \left(\frac{1}{r} \right)$$

Because the integrand $d\mathbf{l}_2 \cdot \nabla(1/r)$ is an exact differential, its integral around any closed path is zero. Using this result we have that

$$\mathbf{F}_2 = -\frac{\mu_0}{4\pi} I_1 I_2 \oint_1 \oint_2 \frac{(d\mathbf{l}_1 \cdot d\mathbf{l}_2)\mathbf{r}_{12}}{r^3} \tag{7–16}$$

which is symmetric with respect to the two circuits. Since $\mathbf{r}_{12} = -\mathbf{r}_{21}$ it follows that $\mathbf{F}_2 = -\mathbf{F}_1$ in agreement with Newton's third law.*

---

* For two isolated non-parallel current elements Newton's third law is apparently violated. It has been shown that this anomaly is removed if the momentum of the electromagnetic field of the current elements is taken into consideration. The interested reader should consult the article by L. Page and N. Adams, 'Action and Reaction between Moving Charges', *Am. J. Phys.*, **13**, 141–7 (1945).

The foregoing equations for force and flux density are applicable to those cases in which the currents flow along filamentary circuits, that is when the conductors are of negligible thickness in comparison to the other dimensions of the problem. Although this is true in most practical situations, there are cases in which the current must be considered as spread out over a definite area. For such cases we must go back to our original definition of a current element as $\mathbf{j}\,\mathrm{d}v$. The force and flux density expressions are then given by

$$\mathbf{F}_2 = \int_{v_2} (\mathbf{j}_2 \times \mathbf{B}_1)\,\mathrm{d}v_2 = -\int_{v_1} (\mathbf{j}_1 \times \mathbf{B}_2)\,\mathrm{d}v_1 \qquad (7\text{–}17)$$

and

$$\mathbf{B}_1 = \frac{\mu_0}{4\pi} \int_{v_1} \frac{\mathbf{j}_1 \times \mathbf{r}_{12}}{r^3}\,\mathrm{d}v_1 \qquad (7\text{–}18)$$

For a surface current of density $\mathbf{K}$ the corresponding expressions are

$$\mathbf{F}_2 = \int_{S_2} (\mathbf{K}_2 \times \mathbf{B}_1)\,\mathrm{d}S_2 \qquad (7\text{–}19)$$

and

$$\mathbf{B}_1 = \frac{\mu_0}{4\pi} \int_{S_1} \frac{\mathbf{K}_1 \times \mathbf{r}_{12}}{r^3}\,\mathrm{d}S_1 \qquad (7\text{–}20)$$

7.2.1 FORCE ON A MOVING POINT CHARGE

The force on a current element $\mathbf{j}\,\mathrm{d}v$ placed in a magnetic field of flux density $\mathbf{B}$ is from equation 7–17,

$$\mathrm{d}\mathbf{F} = \mathbf{j} \times \mathbf{B}\,\mathrm{d}v$$

Since by definition $\mathbf{j} = \rho\mathbf{u}$ (see Section 6.1), then

$$\mathrm{d}\mathbf{F} = \rho\mathbf{u} \times \mathbf{B}\,\mathrm{d}v$$

but $\rho\,\mathrm{d}v$ is the total charge contained in the volume element $\mathrm{d}v$. Regarding this as a point charge Q we may drop the differentials and write

$$\mathbf{F} = Q\mathbf{u} \times \mathbf{B} \qquad (7\text{–}21)$$

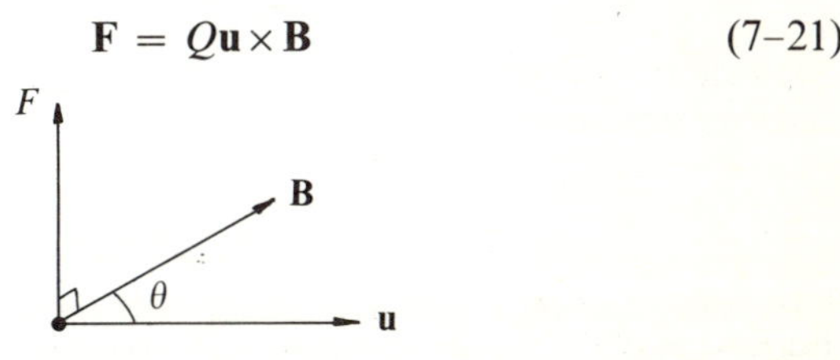

Fig. 7.5 Direction of force of a charged particle moving with velocity $\mathbf{u}$ in a magnetic field $\mathbf{B}$

This is the force in newtons experienced by a charge of magnitude Q coulomb which is moving with a velocity $\mathbf{u}$ metre second^{-1} through a magnetic field of flux density $\mathbf{B}$ tesla. It is clear that $\mathbf{F}$ is perpendicular to both $\mathbf{u}$ and $\mathbf{B}$ (Fig. 7.5). For a single electron the force is

$$\mathbf{F} = -e\mathbf{u} \times \mathbf{B} \tag{7-22}$$

Since equation 7–21 can be verified experimentally, it can be used instead of equation 7–12 or 7–13 to define the vector $\mathbf{B}$. The direction of $\mathbf{B}$ is defined as the direction in which a moving charge experiences no magnetic force. The sense of $\mathbf{B}$ is defined such that, for a positive charge, the three vectors $\mathbf{F}$, $\mathbf{u}$ and $\mathbf{B}$ form, in that order, a right-handed system.

If a charge $Q[\mathrm{C}]$ moves with a velocity $\mathbf{u}[\mathrm{ms}^{-1}]$ in a region where there is an electric field $\mathbf{E}[\mathrm{V\,m}^{-1}]$ and a magnetic field $\mathbf{B}[\mathrm{T}]$, then the total force $[\mathrm{N}]$ acting on Q is given by

$$\mathbf{F_L} = Q(\mathbf{E} + \mathbf{u} \times \mathbf{B}) \tag{7-23}$$

This is known as the *Lorentz force*.

The reader's attention should be drawn to the following two points:

(a) Excluding gravitational forces, the only force which a charged particle at rest can experience is an electric force. This force is therefore an accelerating force.

(b) Whereas a free charge moving under the action of an electric force will gain kinetic energy from the field, the kinetic energy of a free charge moving in a stationary magnetic field remains constant. This is because the magnetic force is always normal to the velocity and hence cannot transfer any energy to the moving charge. The magnetic force is thus a purely deflecting force.

7.2.2 MAGNETIC FIELD OF A MOVING POINT CHARGE*

The magnetic flux density at a distance $\mathbf{r}$ from a current element $\mathbf{j}\,\mathrm{d}v$ is (from equation 7–18)

$$\mathrm{d}\mathbf{B} = \frac{\mu_0}{4\pi} \frac{\mathbf{j} \times \mathbf{r}\,\mathrm{d}v}{r^3}$$

Making the substitutions $\mathbf{j} = \rho\mathbf{u}$, $\rho\,\mathrm{d}v = Q$ and dropping the differentials we obtain

$$\mathbf{B} = \frac{\mu_0}{4\pi} Q \frac{\mathbf{u} \times \mathbf{r}}{r^3} \text{ tesla} \tag{7-24}$$

---

* The equivalence of a moving electrified body and an electric current as sources of magnetic field was first demonstrated experimentally by Rowland in 1876. See E. T. Whittaker, *A History of the Theories of Aether and Electricity*, Vol. I (Nelson, 1951), p. 305.

This is the magnetic flux density at a point distant $\mathbf{r}$ (Fig. 7.6) from an electric point charge Q coulomb moving with a velocity $\mathbf{u}$ metre second^{-1}. Equation 7–24 is accurate only when the velocity u is small compared with the velocity of light; in a conductor $u \simeq 10^{-3}\,\mathrm{m\,s^{-1}}$.

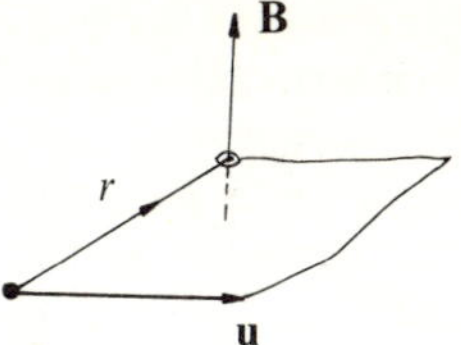

Fig. 7.6 Flux density of a point charge moving with velocity $\mathbf{u}$

7.3 SOLENOIDAL NATURE OF THE MAGNETIC FLUX DENSITY

The magnetic flux density at a field point P produced by a closed circuit carrying a current I is given by the integral form of the Biot–Savart law.

$$\mathbf{B} = \frac{\mu_0}{4\pi} \oint_C \frac{I\,d\mathbf{l} \times \mathbf{r}}{r^3}$$

where $\mathbf{r}$ is the vector distance from the current element to the point P. The divergence of $\mathbf{B}$ at P is given by

$$\nabla \cdot \mathbf{B} = \frac{\mu_0}{4\pi} \nabla \cdot \oint_C \frac{I\,d\mathbf{l} \times \mathbf{r}}{r^3} = \frac{\mu_0}{4\pi} \oint_C \nabla \cdot \left(\frac{I\,d\mathbf{l} \times \mathbf{r}}{r^3} \right) \qquad (7\text{--}25)$$

The interchange of the integration and differentiation operations does not affect the end result because the integration involves the coordinates of $d\mathbf{l}$ only whilst the differentiation involves the coordinates of the point P only. Making use of the vector identity

$$\nabla \cdot (\mathbf{a} \times \mathbf{b}) = \mathbf{b} \cdot \nabla \times \mathbf{a} - \mathbf{a} \cdot \nabla \times \mathbf{b}$$

the integrand of equation 7–25 may be rewritten as,

$$\nabla \cdot \left(\frac{I\,d\mathbf{l} \times \mathbf{r}}{r^3} \right) = \frac{\mathbf{r}}{r^3} \cdot \nabla_\mathrm{P} \times I\,d\mathbf{l} - I\,d\mathbf{l} \cdot \nabla_\mathrm{P} \times \left(\frac{\mathbf{r}}{r^3} \right)$$

where the subscript P indicates that all derivatives are to be carried out with respect to the coordinates of the point P. Since $I\,d\mathbf{l}$ is independent of the coordinates of P, the first term is zero. The second term may be written as

$$I\,d\mathbf{l} \cdot \nabla_\mathrm{P} \times \left(\frac{\mathbf{r}}{r^3} \right) = -I\,d\mathbf{l} \cdot \nabla_\mathrm{P} \times \left[\nabla_\mathrm{P} \left(\frac{1}{r} \right) \right] = 0$$

since the curl of the gradient of any function is always zero. It follows that

$$\nabla \cdot \mathbf{B} = 0 \qquad (7\text{–}26)$$

Thus the magnetic flux density vector $\mathbf{B}$ is solenoidal. This is an important and fundamental result which is also completely general in the sense that it holds in free space as well as in a magnetizable medium and even when the currents which produce $\mathbf{B}$ are time dependent. The physical significance of equation 7–26 is that the lines of magnetic flux density have neither sources nor sinks so that free magnetic charges or poles do not exist. A conclusion which is commonly drawn from the fact that the divergence of $\mathbf{B}$ is zero, is that the lines of B are always continuous and form closed loops. However, as was first pointed out by Slepian*, this property is by no means general. The reason behind this is that the very concept of field lines is a purely topographical one, that is we use such lines to delineate graphically the direction and strength of a field at the various points in space. The tangent to a line at some point represents the field direction at that point; the strength of the field is represented by the density of the lines. Hence the divergenceless character of a field may be represented by a set of lines which start and end *with equal density* on opposite sides of an arbitrary surface, but which need not necessarily form closed loops. The following passage, quoted from Slepian's paper, is particularly elucidative: 'Since it is only the density and direction of the lines of force of the assemblage (of force lines) which are related to the electric vector or magnetic field, it should be quite evident that the individuality and continuity of individual lines of force of the assemblage are entirely irrelevant properties, so far as any observable electromagnetic phenomenon is concerned, postulating as we do that the vector fields are entirely sufficient for describing such phenomena. As we prolong the lines of force, we may at any point end all the lines and start new ones *with the same density*, without in any way impairing the correspondence between the assemblage of lines of force and the vector field. The assemblage of broken lines serves just as well as the assemblage of continuous lines of force.'

Examples of current configurations whose magnetic flux lines cannot be represented by closed curves are given in a paper by McDonald†. The lines of $\mathbf{B}$ produced by all the current circuits discussed in this chapter can and will be represented by closed curves. This is perfectly justifiable as long as we make sure that the use of such closed curves

* J. Slepian, 'Lines of Force in Electric and Magnetic Fields', *Am. J. Phys.*, **19**, 87–90 (1951).
† K. L. McDonald, 'Topology of Steady Current Magnetic Fields', *Am. J. Phys.*, **22**, 586–96 (1954).

does not invalidate the concept of line density as a representation of the magnitude of the field.

By analogy with the equation for the electric lines of force (equation 2–7), it is evident that the equation for the lines of magnetic induction (**B** everywhere tangent to these lines) is given by

$$\frac{\mathrm{d}x}{B_x} = \frac{\mathrm{d}y}{B_y} = \frac{\mathrm{d}z}{B_z}$$

7.4 MAGNETIC FLUX

The flux of the vector **B** through any surface is called the magnetic flux. It is defined by

$$\Phi = \int_S \mathbf{B} \cdot \mathrm{d}\mathbf{S} \tag{7–27}$$

The SI unit of magnetic flux is the weber. If the surface S is closed, then making use of the divergence theorem we have that

$$\Phi = \oint_S \mathbf{B} \cdot \mathrm{d}\mathbf{S} = \int_v \nabla \cdot \mathbf{B}\,\mathrm{d}v = 0$$

so that the total flux crossing any *closed* surface is always zero. That is as much flux enters the surface as leaves it. It follows that the magnetic flux passing through any open surface bounded by the same closed contour C is constant. This flux is said to link the contour C and is often referred to as the *flux linkage*.

In Section 2.8 we defined a tube of electric flux. In the same way we may define a tube of magnetic flux. It is an imaginary tube with its walls everywhere parallel to the **B**-lines and hence with a constant magnetic flux over any cross-section.

7.5 MAGNETIC FLUX DENSITY OF SOME SIMPLE CURRENT DISTRIBUTIONS

7.5.1 STRAIGHT WIRE SEGMENT

Suppose that the segment is of length $2l$ and carries a uniform steady current I. To find the magnetic flux density at some point P external to the segment, we will assume that the wire has a negligibly small diameter or that the distance of the point P from the wire is much greater than the wire diameter. Since the segment has cylindrical symmetry about the z-axis (Fig. 7.7) it is more convenient to assign the cylindrical coordinates z and ρ to the point P.

Applying the Biot–Savart law we have that at P

$$\mathbf{B} = \frac{\mu_0}{4\pi} \int_{-l}^{l} \frac{I\,\mathrm{d}\zeta\, \mathbf{a}_z \times \mathbf{a}_r}{r^2} = \frac{\mu_0}{4\pi} \int_{-l}^{l} \frac{I\,\mathrm{d}\zeta\, \sin\theta}{r^2}\, \mathbf{a}_\phi$$

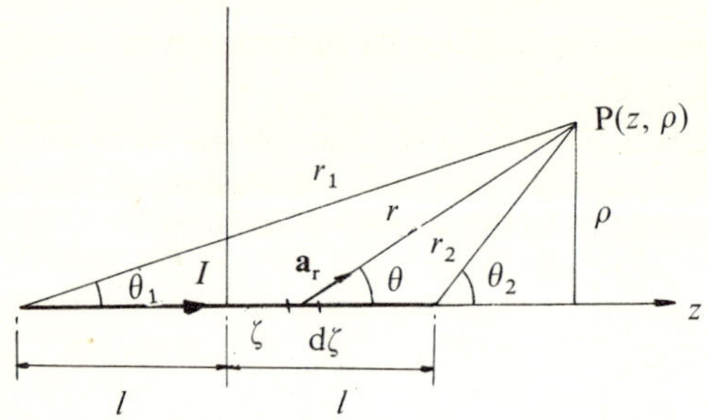

Fig. 7.7 To calculate the flux density produced at any point by a wire segment

From the geometry of the figure we have that,

$$\cot \theta = \frac{z - \xi}{\rho} \quad \text{and} \quad \operatorname{cosec} \theta = \frac{r}{\rho}$$

Differentiating the first expression and making use of the second we obtain,

$$-\operatorname{cosec}^2 \theta \, d\theta = -\frac{r^2 \, d\theta}{\rho^2} = -\frac{d\zeta}{\zeta}$$

so that

$$d\zeta = \frac{r^2 \, d\theta}{\rho}$$

Substituting this value of $d\zeta$ into the expression for **B** gives

$$\mathbf{B} = \frac{\mu_0}{4\pi} I \int_{\theta_1}^{\theta_2} \frac{\sin \theta \, d\theta}{\rho} \, \mathbf{a}_\phi$$

$$= \frac{\mu_0}{4\pi} \frac{I}{\rho} (\cos \theta_1 - \cos \theta_2) \mathbf{a}_\phi \tag{7-28}$$

We thus conclude that the lines of **B** form circular loops whose centres lie on the axis of the conductor. When a circuit consists of several straight segments, the resultant magnetic flux density at any point is the vector sum of the flux density produced at that point by each individual segment.

For an *infinitely long line* the angles θ_1 and θ_2 in equation 7–28 approach 0 and π respectively and the flux density at any point is given by

$$\mathbf{B} = \pm \frac{\mu_0 I}{2\pi\rho} \mathbf{a}_\phi \tag{7-29}$$

where the algebraic signs have been introduced to show that the sense of $\mathbf{B}$ is related to that of I by the right-hand rule. We may now use equation 7–25 in conjunction with equation 7–13 to determine the force between two parallel infinitely long current-carrying wires at a

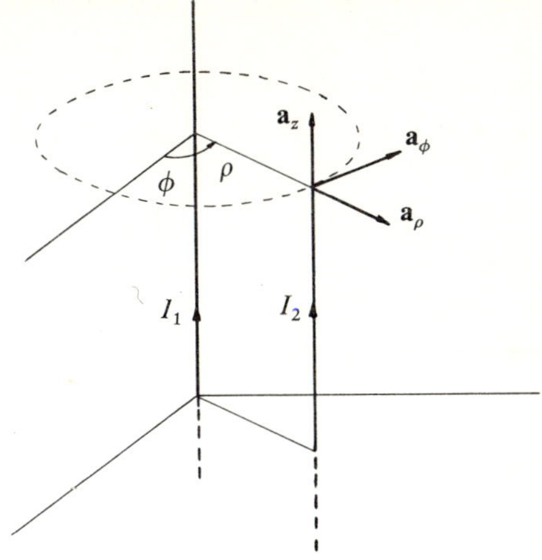

Fig. 7.8 To determine the force between two parallel infinitely long current-carrying wires

distance ρ metres apart. Let the currents flowing in the wires be I_1 and I_2 (Fig. 7.8). The force on a current element $I\,dl_2$ of conductor 2 is

$$d\mathbf{F}_2 = I_2\,d\mathbf{l}_2 \times \mathbf{B}_1$$

where,

$$\mathbf{B}_1 = \frac{\mu_0 I_1}{2\pi\rho}\,\mathbf{a}_\phi$$

Setting $d\mathbf{l}_2 = dl\mathbf{a}_z$ we obtain

$$d\mathbf{F}_2 = \frac{\mu_0 I_1 I_2}{2\pi\rho}\,dl\mathbf{a}_z \times \mathbf{a}_\phi$$

The force per unit length experienced by wire 2 is therefore

$$\frac{d\mathbf{F}_2}{dl} = \mathbf{f}_2 = -\frac{\mu_0 I_1 I_2}{2\pi\rho}\,\mathbf{a}_\rho \quad [\mathrm{N\,m^{-1}}] \tag{7–30a}$$

and that experienced by wire 1 is

$$\frac{d\mathbf{F}_1}{dl} = \mathbf{f}_1 = -\mathbf{f}_2 = \frac{\mu_0 I_1 I_2}{2\pi\rho}\,\mathbf{a}_\rho \quad [\mathrm{N\,m^{-1}}] \tag{7–30b}$$

For the directions of current flow shown in Fig. 7.8, the directions of each of these forces is in agreement with the right-hand screw rule. We thus see that if the currents flow in the same direction the force between the wires is attractive, whilst if they flow in opposite directions the force is repulsive. These two cases are pictorially represented in Fig. 7.9.

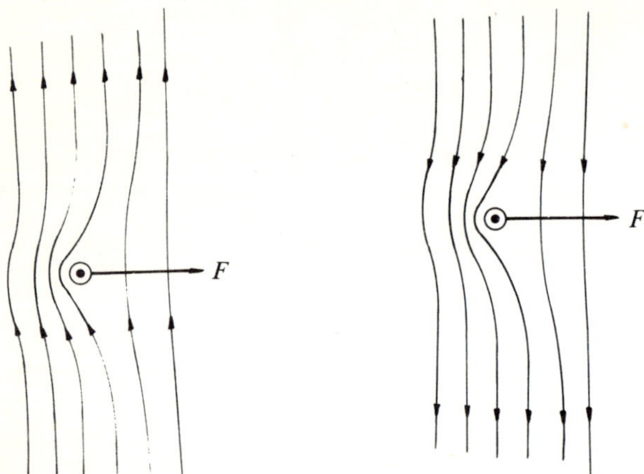

Fig. 7.9 Direction of force on current-carrying wire relative to the direction of current and external magnetic field

From the formal point of view equation 7–30 serves as a basis for the experimental definition of the unit of current in the SI system of units, namely the ampere: 'Two infinitely long parallel wires one metre apart will exert a force (of attraction or repulsion) of 2×10^{-7} newton per metre on each other when each is carrying a current of one ampere.'

Since it is practically impossible to have infinitely long wires, the establishment of the ampere as the unit of current by the above method cannot be carried out in the laboratory with a sufficient degree of accuracy. In practice therefore this is carried out by means of a Raleigh-type current balance* which measures the force between circular coils.

7.5.2 SEMI-INFINITE AND INFINITE CURRENT SHEETS

Consider a current I flowing through an infinitely long conducting strip of width L and whose uniform thickness is negligibly small compared with L. Suppose that the strip lies in the xy-plane and let the current flow be in the positive x-direction (Fig. 7.10). Because of the assumed thinness of the film it is more convenient to use the concept

* A detailed description of such a balance may be found in *Electrical Measurements* by F. A. Laws (McGraw-Hill, 2 edn, 1938).

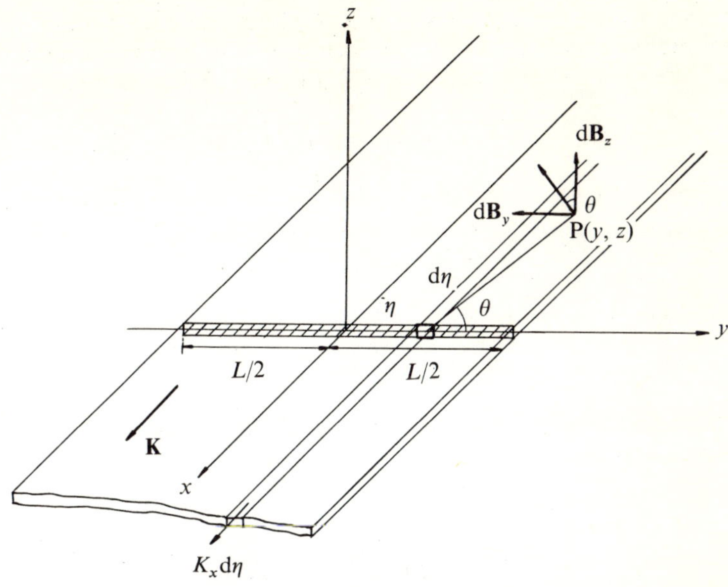

Fig. 7.10 Semi-infinite current sheet

of surface current density already defined in equation 6–5. In the present case then $\mathbf{K} = K_x\mathbf{a}_x$ and $I = K_x L$. To find the flux density at any point P (x, y, z) we subdivide the sheet into a number of filaments. The field at P produced by any of these filaments is equivalent to that of an infinitely long thin wire similarly located. It is evident that no filament can produce an x-component of the flux density.

Since the current sheet is semi-infinite the point P can always be chosen so as to lie in the zy-plane. From Fig. 7.10 and equation 7–29 the y- and z-components of the magnetic flux density at P are given by

$$B_y = \frac{\mu_0 K_x}{2\pi} \int_{-1/2L}^{1/2L} \frac{z\,d\eta}{z^2+(y-\eta)^2} = -\frac{\mu_0 K_x}{2\pi}\left[\tan^{-1}\frac{y-\eta}{z}\right]_{-1/2L}^{1/2L}$$

$$= -\frac{\mu_0 K_x}{2\pi}\left[\tan^{-1}\frac{y+\tfrac{1}{2}L}{z} - \tan^{-1}\frac{y-\tfrac{1}{2}L}{z}\right] \tag{7–31a}$$

and

$$B_z = \frac{\mu_0 K_x}{2\pi} \int_{-1/2L}^{1/2L} \frac{(y-\eta)\,d\eta}{z^2+(y-\eta)^2} = -\frac{\mu_0 K_x}{4\pi}\left\{\log[z^2+(y-\eta)^2]\right\}_{-1/2L}^{1/2L}$$

$$= \frac{\mu_0 K_x}{4\pi}\log\frac{z^2+(y+\tfrac{1}{2}L)^2}{z^2+(y-\tfrac{1}{2}L)^2} \tag{7–31b}$$

Let us now find the flux density at a point very near the surface of the semi-infinite sheet. To do this we first let $y \to 0$, bearing in mind that $\tan^{-1}(-\alpha) = -\tan\alpha$; equations 7–31a and 7–31b reduce to

$$B_y = -\frac{\mu_0 K_x}{\pi} \tan^{-1}\left(\frac{L}{2z}\right)$$

$$B_z = 0$$

For $z \to 0^+$,

$$B_y = -\frac{\mu_0 K_x}{2}$$

for $z \to 0^-$,

$$B_y = +\frac{\mu_0 K_x}{2}$$

If the current sheet becomes infinite in the y- as well as in the x-direction, that is $\pm\frac{1}{2}L \to \pm\infty$, we have that

$$B_z = 0$$

$$B_y = -\frac{\mu_0 K_x}{2\pi}\left[\tan^{-1}\frac{y-\eta}{z}\right]_{-\infty}^{+\infty}$$

For positive values of z and finite values of y

$$B_y = -\frac{\mu_0 K_x}{2\pi}\left[\tan^{-1}(-\infty) - \tan^{-1}(+\infty)\right]$$

$$= -\frac{\mu_0 K_x}{2\pi}\left[\left(-\frac{\pi}{2}\right) - \left(\frac{\pi}{2}\right)\right]$$

$$= -\frac{\mu_0 K_x}{2}$$

For negative finite values of z and finite values of y we have that

$$B_y = +\tfrac{1}{2}\mu_0 K_x$$

We may now combine these last two results in the single vector equation

$$\mathbf{B} = \tfrac{1}{2}\mu_0 \mathbf{K} \times \mathbf{a}_n \qquad (7\text{–}32)$$

where $\mathbf{a}_n$ is the unit normal pointing away from the current sheet.

Thus the magnitude of the magnetic flux density produced by an infinite current sheet at a point is independent of the distance of the point from the sheet whilst the direction of the flux density is parallel to the sheet and perpendicular to the direction of current flow (Fig. 7.11).

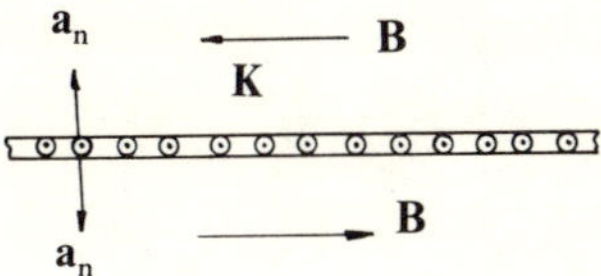

Fig. 7.11 Flux density of an infinite current sheet

In passing from one side of a current sheet to the other the component of the flux density which is parallel to the sheet and normal to the current density suffers a discontinuity and changes abruptly by $\mu_0 K$. We have shown that this is true for a semi-infinite and an infinite current sheet, but it can be shown to be true also for a finite current sheet. Taking this for granted, a little thought will show that this result is also true if the current sheet consists of any curved surface. The electrostatic counterpart of this problem was given in Section 2.9 where it was shown that the normal component of the electric field intensity suffers a discontinuity σ/ϵ_0 when crossing from one side of a charged surface to the other, where σ is the surface charge density.

Just as two infinite parallel planes carrying uniform charge densities of opposite polarities produce a uniform field between them, and zero everywhere else, so two infinite current sheets carrying equal and opposite surface current densities will produce a uniform magnetic flux density $\mu_0 K$ between the sheets and zero outside (Fig. 7.12).

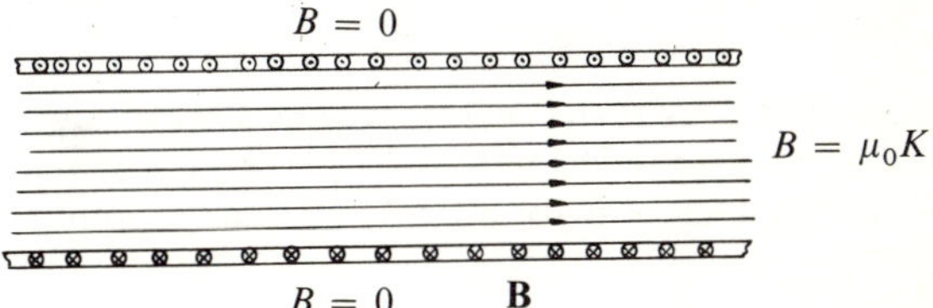

Fig. 7.12 Uniform magnetic field between two infinite current sheets carrying equal and opposite surface current densities

7.5.3 FLUX DENSITY AT A POINT ON THE AXIS OF A CIRCULAR CURRENT LOOP

Consider a wire of negligible thickness in the form of a plane circular ring of radius a and carrying a current I in a positive direction with respect to the z-axis (Fig. 7.13). Let P be a point on the axis of the ring at a distance z from its plane. By symmetry it is clear that the resultant flux density at P must be directed along the axis. The axial component due to two symmetrically located current elements is,

$$dB_z = 2 \frac{\mu_0}{4\pi} \frac{Ia\,d\phi}{r^2} \sin\alpha = \frac{\mu_0}{2\pi} \frac{Ia^2}{(z^2+a^2)^{3/2}} d\phi$$

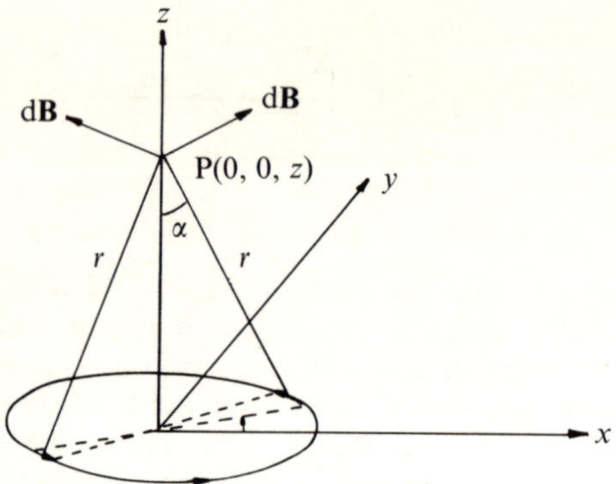

Fig. 7.13 To determine the flux density at a point on the axis of a plane circular current loop

The resultant flux density is

$$B_z = \int_{\phi=0}^{\pi} \mathrm{d}B_z = \mu_0 \frac{Ia^2}{2(z^2+a^2)^{3/2}} = \mu_0 \frac{I}{2a} \sin^3 \alpha \qquad (7\text{--}33a)$$

Since B_z is a function of z^2 it follows that for both positive and negative values of z the sense of B_z is always along the positive z-direction. At the centre of the loop,

$$B_z = \frac{\mu_0 I}{2a} \qquad (7\text{--}33b)$$

If the loop consists of N turns of wire wound very close together, then the current I in equations 7–33a and 7–33b should be replaced by NI. Figure 7.14 shows lines of **B** for a plane circular current loop.

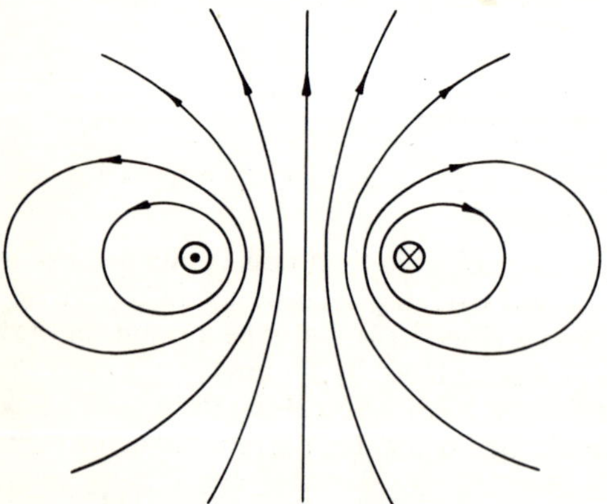

Fig. 7.14 Flux lines for a plane circular current loop

7.5.4 FLUX DENSITY AT A POINT ON THE AXIS OF A SOLENOID

A solenoid consists of a helical coil of wire wound in the form of a long cylinder. Suppose that such a coil has a radius a and consists of N turns of wire carrying a current I and that the wire diameter is small compared with the coil diameter. If the spacing between turns is sufficiently small, which is usually the case, the N turns may be considered uniformly distributed over the length L of the coil with a linear density $n = N/L$. A section of length dz (Fig. 7.15) is then equivalent to a

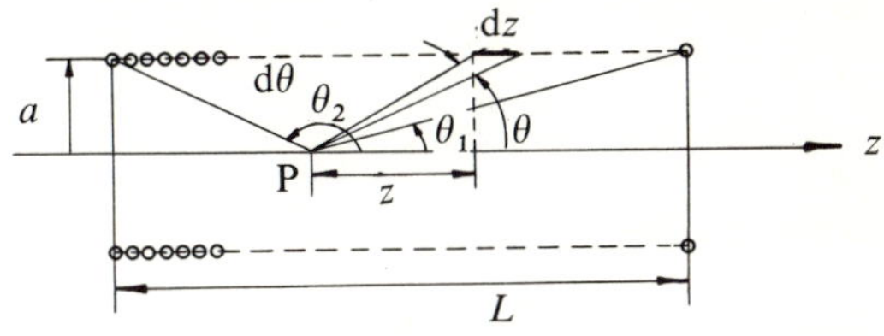

Fig. 7.15 To determine the flux density at a point on the axis of a solenoid

circular loop of radius a carrying a current $In\,dz$. The flux density due to this loop at a point P on the axis of the solenoid is, by equation 7–33a

$$dB_z = \mu_0 \frac{Ina^2\,dz}{2(z^2+a^2)^{3/2}}$$

From the geometry of the figure it is clear that:

$$\sin\theta = \frac{a}{(z^2+a^2)^{1/2}} \quad \text{and} \quad \tan\theta = \frac{a}{z}$$

Differentiating the first expression and making use of the second we obtain,

$$dz = -\frac{(z^2+a^2)^{3/2}}{a^2}\sin\theta\,d\theta$$

Substituting this value of dz in the expression for dB_z gives

$$dB_z = -\frac{\mu_0 In}{2}\sin\theta\,d\theta$$

Integrating this differential from $\theta = \theta_2$ to $\theta = \theta_1$—i.e., in the direction of increasing z—one obtains

$$B_z = \frac{\mu_0 In}{2}[\cos\theta_1 - \cos\theta_2] \tag{7–34a}$$

If either $\theta_1 = 90°$ or $\theta_2 = 90°$, that is at axial points at either end of the solenoid, the flux density is given by

$$B_z = \frac{\mu_0 In}{2} \frac{L}{(a^2 + L^2)^{1/2}} \qquad (7\text{–}34\text{b})$$

For $L \gg a$, this reduces to

$$B_z = \frac{\mu_0 In}{2} \qquad (7\text{–}34\text{c})$$

For a point at the centre of the solenoid, $\theta_1 = \theta$ and $\theta_2 = 180° - \theta$ so that

$$B_z = \frac{\mu_0 In}{2} \frac{L}{(a^2 + L^2/4)^{1/2}} \qquad (7\text{–}34\text{d})$$

and for $L \gg a$ this becomes

$$B_z = \mu_0 In \qquad (7\text{–}34\text{e})$$

If the solenoid is infinitely long then $\theta_1 = 0$, $\theta_2 = 180°$ and

$$B_{z\infty} = \mu_0 In \qquad (7\text{–}35)$$

anywhere along the axis. It will be shown in Section 7.7 that this is also the value of the flux density at *any* point inside an infinitely long

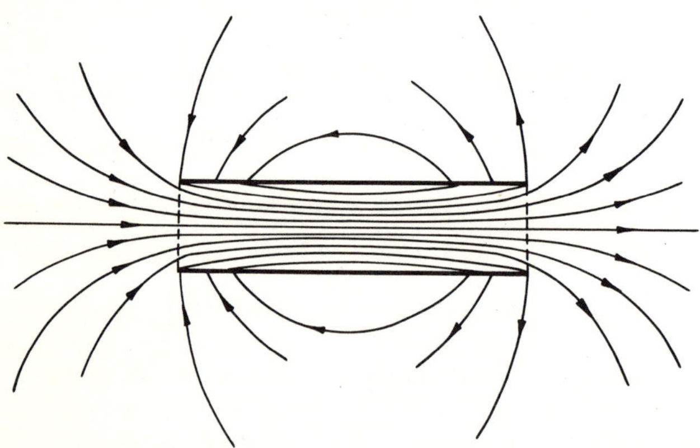

Fig. 7.16 Magnetic flux lines inside and around a solenoid

solenoid. Figure 7.16 shows the **B**-lines inside and around a solenoid and Fig. 7.17 shows the variation of the normalized flux density $(B_z/B_{z\infty})$ along the axis of a solenoid for which $L/a = 2, 4, 8$. It is seen that, away from the solenoid ends, the field remains very nearly constant.

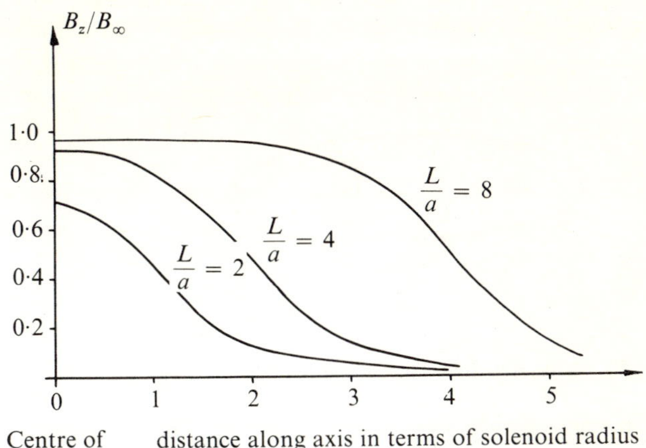

Fig. 7.17 Variation of the normalized flux density along the axis of a solenoid for different length to radius ratios

7.6 AMPÈRE'S CIRCUITAL LAW

Consider a current I flowing around a closed wire loop. We already know that the magnetic flux density at any point P is given by

$$\mathbf{B} = \frac{\mu_0 I}{4\pi} \oint_C \frac{d\mathbf{l}_1 \times \mathbf{r}}{r^3}$$

where the integration is to be carried out over the complete current loop and hence with respect to the coordinates of the element $d\mathbf{l}_1$. Suppose now that the point P is displaced a distance $d\mathbf{l}_2$, the direction of $d\mathbf{l}_2$ being arbitrary, and consider the dot product $\mathbf{B} \cdot d\mathbf{l}_2$,

$$\mathbf{B} \cdot d\mathbf{l}_2 = \frac{\mu_0 I}{4\pi} \left\{ \oint_C \frac{d\mathbf{l}_1 \times \mathbf{r}}{r^3} \right\} \cdot d\mathbf{l}_2 = \frac{\mu_0 I}{4\pi} \oint_C \frac{(d\mathbf{l}_1 \times \mathbf{r}) \cdot d\mathbf{l}_2}{r^3}$$

where $d\mathbf{l}_2$ can be taken inside the integral sign since its coordinates remain constant during the integration. It is clear that the position of the point P relative to the current loop remains unaltered whether we displace the point P itself a distance $d\mathbf{l}_2$ or whether we keep P fixed and displace every point on the current loop, that is the entire loop, by a distance $d\mathbf{l}_2' = -d\mathbf{l}_2$ (Fig. 7.18). It follows that

$$\mathbf{B} \cdot d\mathbf{l}_2 = -\frac{\mu_0 I}{4\pi} \oint_C \frac{(d\mathbf{l}_1 \times \mathbf{r}) \cdot d\mathbf{l}_2'}{r^3}$$

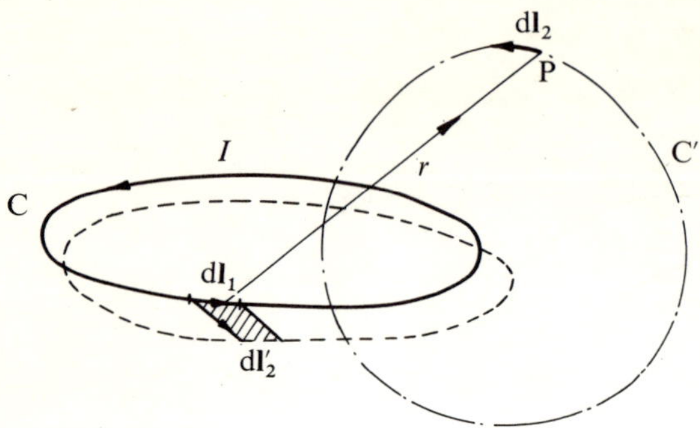

Fig. 7.18 To prove Ampère's circuital law

The scalar triple product in the integrand may be rewritten as

$$(dl_1 \times \mathbf{r}) \cdot dl_2' = -\mathbf{r} \cdot (dl_1 \times dl_2')$$

so that

$$\mathbf{B} \cdot dl_2 = \frac{\mu_0 I}{4\pi} \oint_C \frac{\mathbf{r} \cdot (dl_1 \times dl_2')}{r^3}$$

The vector product $dl_1 \times dl_2'$ represents the area swept by the element dl_1 when it is displaced a distance dl_2'. We therefore have that

$$\mathbf{B} \cdot dl_2 = \frac{\mu_0 I}{4\pi} \oint_C \frac{\mathbf{r} \cdot d\mathbf{S}}{r^3} = \frac{\mu_0 I}{4\pi} \oint \frac{dS \cos \theta}{r^2}$$

The integrand is the projection of the area dS onto a plane normal to $\mathbf{r}$ divided by r^2; this, by definition, is the solid angle $d^2\Omega$ (second-order differential) subtended by the area dS at the point P. The integral is therefore the solid angle subtended at P by the element of area swept out by the whole of the circuit C when it is displaced by dl_2'. Hence

$$\mathbf{B} \cdot dl_2 = \frac{\mu_0 I}{4\pi} d\Omega$$

Suppose now that the point P is given successive displacements such that in the end it describes any closed path C'; the value of the integral of $\mathbf{B} \cdot dl_2$ around this path is the same as that which would be obtained if the point were kept fixed whilst the entire circuit C were taken around the path C'. In so doing the path C sweeps out a closed surface S. If the path C' links the circuit C the point P will lie inside S and the solid angle subtended at P by the swept surface will be 4π. If on the other hand C' does not link C the point P will lie outside S and the solid angle subtended at P by the swept surface will be zero. These results may be expressed mathematically as

$$\oint \mathbf{B} \cdot d\mathbf{l} = \mu_0 I \qquad (7\text{--}36\text{a})$$

if C' links C, and therefore a current I, and

$$\oint \mathbf{B} \cdot d\mathbf{l} = 0 \qquad (7\text{--}36\text{b})$$

if C' does not link C, that is, if it does not link any current. The above two equations constitute *Ampère's circuital law*. In words it states that the line integral of the magnetic flux density around any closed path is proportional to the total current encircled by that path.

The sign of the right-hand side of equation 7–36a depends upon the direction in which the path C' is traversed. By convention it is positive if the direction of advance of a right-hand screw, turned in the direction in which C' is traversed, coincides with the direction of current flow (Fig. 7.19).

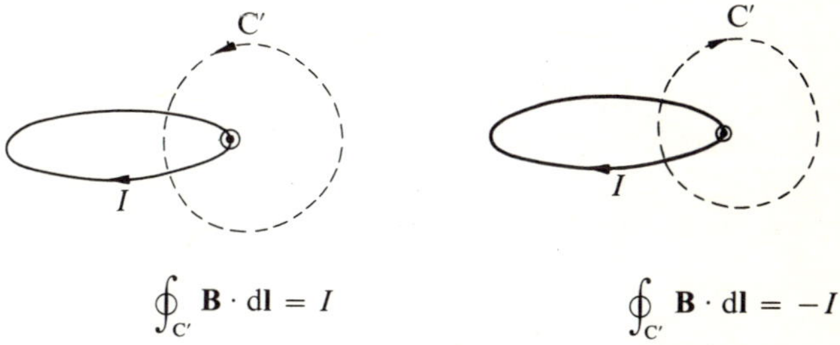

$$\oint_{C'} \mathbf{B} \cdot d\mathbf{l} = I \qquad\qquad \oint_{C'} \mathbf{B} \cdot d\mathbf{l} = -I$$

Fig. 7.19 Convention for the sign of the integral in equation 7–36a

If the path C' encircles more than one circuit then Ampère's law takes the form

$$\oint \mathbf{B} \cdot d\mathbf{l} = \mu_0 \sum_{1}^{n} I_i \qquad (7\text{--}37)$$

where I_i is the current flowing in the ith circuit and may be either positive or negative (cf. above rule). In particular, if the path encircles a current loop consisting of N turns, then,

$$\oint \mathbf{B} \cdot d\mathbf{l} = \mu_0 N I \qquad (7\text{--}38)$$

When a continuous distribution of current, defined by a current density $\mathbf{j}$, passes through the contour C' Ampère's law is given by the more general form

$$\oint_{C'} \mathbf{B} \cdot d\mathbf{l} = \mu_0 \int_{S} \mathbf{j} \cdot d\mathbf{S} \qquad (7\text{--}39)$$

where dS is the element of area of a surface S bounded by the contour C'. A little thought will show that S can be any surface bounded by C'. The orientation of dS is given by $\mathbf{dS} = dS\mathbf{a_n}$ where the unit normal $\mathbf{a_n}$ points in the direction of advance of a right-handed screw turned in the direction in which C' is traversed.

7.7 EXAMPLES ON THE APPLICATION OF AMPÈRE'S CIRCUITAL LAW

7.7.1 MAGNETIC FLUX DENSITY DUE TO AN INFINITELY LONG WIRE OF FINITE RADIUS

Let the radius of the wire be R and suppose that the wire is of a non-magnetic material and carries a current I uniformly distributed over its cross-section such that the current density at any point inside the wire is $j = I/\pi R^2$. Because of the symmetry of the problem, we can say that the magnetic flux density at any point can only have a component B_ϕ which will be constant at all points of a circular path

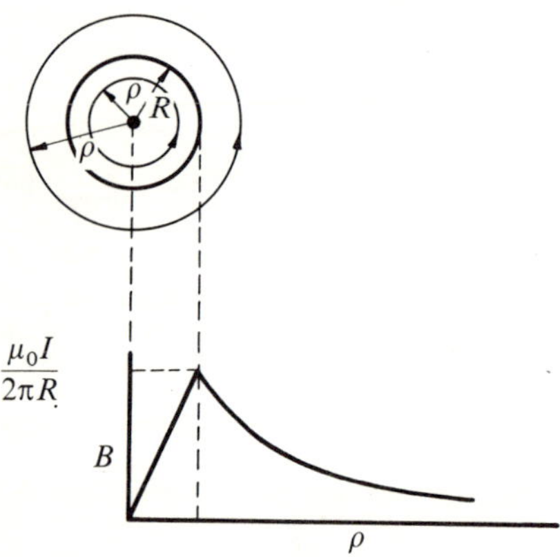

Fig. 7.20 Variation of the flux density inside and outside an infinitely long current-carrying conductor

concentric with the wire (Fig. 7.20). We may thus use Ampère's law around such a path of radius ρ. If $\rho \leq R$ we have, from equation 7–39, that

$$\oint B_\phi \rho \, d\phi = \mu_0 \int_0^{2\pi} \int_0^{\rho} j\rho \, d\phi \, d\rho$$

so

$$\int_0^{2\pi} B_\phi \rho \, d\phi = \frac{\mu_0 I}{\pi R^2} \frac{2\pi \rho^2}{2}$$

and we have that

$$B_\phi = \frac{\mu_0 I \rho}{2\pi R^2} \quad (\rho \leq R) \tag{7-40a}$$

If $\rho \geq R$ then applying equation 7–36a it is readily seen that

$$B_\phi 2\pi\rho = \mu_0 I$$

so

$$B_\phi = \frac{\mu_0 I}{2\pi\rho} \quad (\rho \geq R) \tag{7-40b}$$

This expression is the same as equation 7–29 which was obtained by the use of the Biot–Savart law. It is apparent that the use of Ampère's law greatly simplifies its derivation. Figure 7.20 shows the variation of B as a function of the radial distance from the centre of the wire.

7.7.2 MAGNETIC FLUX DENSITY IN A COAXIAL LINE

Consider a coaxial transmission line consisting of two infinitely long concentric cylinders as shown in Fig. 7.21. The inner conductor has a radius a and carries a current I; the outer conductor has an inner radius b, a thickness t and carries the return current $-I$. Assume that in both conductors the current density is uniform. It follows from Ampère's law that the magnetic flux density in the region $\rho \leq b$ is the same as that given above for the single wire, namely

$$B_\phi = \frac{\mu_0 I \rho}{2\pi a^2} \quad (\rho \leq a)$$

$$B_\phi = \frac{\mu_0 I}{2\pi\rho} \quad (a \leq \rho \leq b)$$

In the region $b \leq \rho \leq b+t$ the flux density is given by

$$B_\phi = \frac{\mu_0 I}{2\pi\rho}\left[1 - \frac{\rho^2 - b^2}{(b+t)^2 - b^2}\right] \tag{7-41}$$

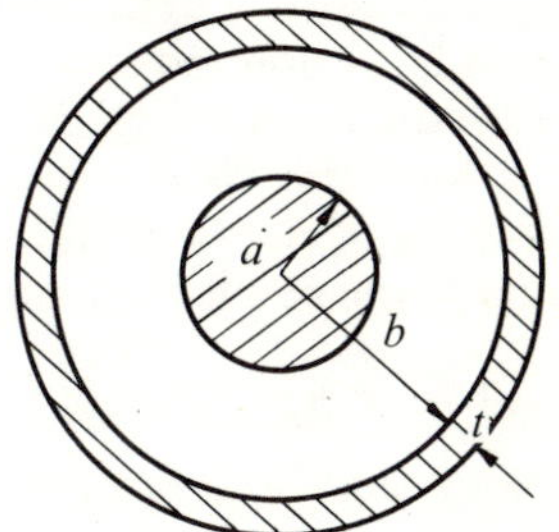

Fig. 7.21 Cross-section of a coaxial transmission line

7.7.3 MAGNETIC FLUX DENSITY AT ANY POINT INSIDE AN INFINITELY LONG SOLENOID

Because the solenoid is infinitely long, it is obvious that the field everywhere in its interior must be parallel to the axis and independent of the axial coordinate z (Fig. 7.22). Consider the line integral of $\mathbf{B}$

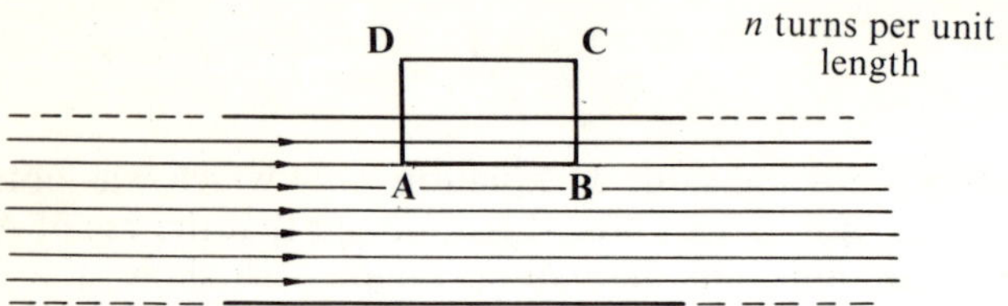

Fig. 7.22 To determine the flux density at any point inside an infinitely long solenoid

over the closed path ABCD, where $AB = CD = l$. The current enclosed by this path is nIl where n is the number of turns per unit length. Because the field outside the solenoid is zero we must have that

$$\oint \mathbf{B} \cdot \mathrm{d}\mathbf{l} = B_z l = \mu_0 n I l$$

or

$$B_z = \mu_0 n I$$

which is the same as equation 7–35. Since the segment AB need not lie on the solenoid axis itself, it follows that equation 7–35 gives the magnetic flux density at *any* point inside the solenoid.

7.7.4 MAGNETIC FLUX DENSITY INSIDE A TOROID

We have just shown that $\mathbf{B}$ at any point inside an infinitely long solenoid is constant and given by $B_z = \mu_0 n I$. In practice no solenoid can be made infinitely long. However, the flux density produced by such a hypothetical solenoid may be achieved, to a good degree of approximation, if a long solenoid is bent into the form of a closed circle whose mean radius is large compared with the turn radius of the solenoid. Such a coil is called a toroid (Fig. 7.23). From the symmetry of the configuration it is seen that $\mathbf{B}$ is constant at all points on a circular path of radius r $(r_1 \leq r \leq r_2)$. If the toroid has N turns and carries a current I, then by Ampère's law we must have that

$$\oint \mathbf{B} \cdot \mathrm{d}\mathbf{l} = \mu_0 N I$$

Hence

$$B_i 2\pi r = \mu_0 N I$$

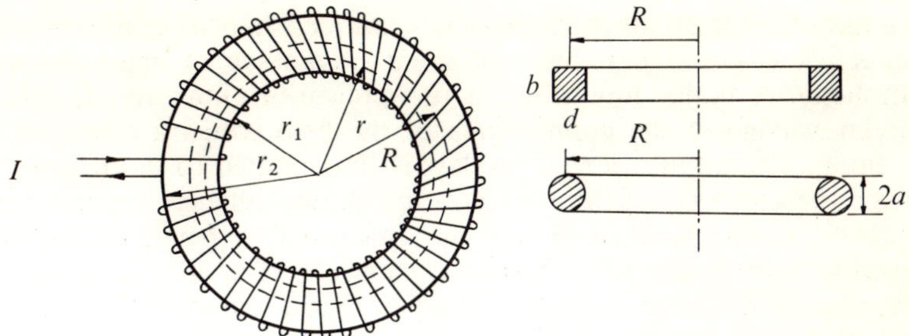

Fig. 7.23 Toroids with rectangular or circular cross-sections

and

$$B_l = \frac{\mu_0 NI}{2\pi r} \quad (r_1 \leq r \leq r_2) \tag{7-42}$$

This flux density is not uniform over the cross-section of the toroid. In many practical applications, however, in which the radius R of the coil axis is large compared with the cross-sectional dimensions of the coil itself, the flux density at any point inside the toroid is, to a good approximation, given by

$$B_l = \frac{\mu_0 NI}{L} = \mu_0 nI$$

where $L = 2\pi R$ is the mean length of the toroid. This is the same expression as that for an infinitely long solenoid.

In practice the cross-section of a toroid may be either rectangular or circular. If it is rectangular the magnetic flux inside the toroid is given by

$$\Phi = \frac{\mu_0 NIb}{2\pi} \log \frac{R + \tfrac{1}{2}d}{R - \tfrac{1}{2}d} \quad \text{weber} \tag{7-43}$$

where b is the breadth of the coil, R is the radius of the coil axis and d is the radial depth of the coil. If the cross-section is circular the flux is given by

$$\Phi = \mu_0 NI[R - \sqrt{(R^2 - a^2)}] \quad \text{weber} \tag{7-44}$$

where a is the radius of the coil cross-section. Both the above expressions are exact and may be derived using equation 7–42 as the starting point. When the cross-sectional dimensions of the coil are small compared with R the flux is given simply by

$$\Phi = \mu_0 nIS \quad \text{weber} \tag{7-45}$$

where S is the cross-sectional area of the coil.

We have seen from all the foregoing examples that the application of Ampère's circuital law has the great advantage of mathematical simplicity. A major limitation to its application, however, is that a foreknowledge of the components of the field which are present is essential. The application of the law is thus limited to more cases in which the geometry of the configuration shows sufficient symmetry to enable the direction of **B** and its functional dependence on the position coordinates to be intuitively known. Ampère's law may be regarded as the magnetic counterpart of Gauss's law in electrostatics and like the latter it is one of the corner-stones of electromagnetic theory.

7.8 POINT FORM OF AMPÈRE'S LAW

Let the magnetic flux density at any point P in space be **B**. With reference to a set of cartesian axes **B** may be expressed as

$$\mathbf{B} = B_x\mathbf{a}_x + B_y\mathbf{a}_y + B_z\mathbf{a}_z$$

Let us now apply Ampère's circuital law to a small rectangular contour 12341 parallel to the yz-plane and having the point P as centre (Fig. 7.24). Because the contour is small, the value of the flux density

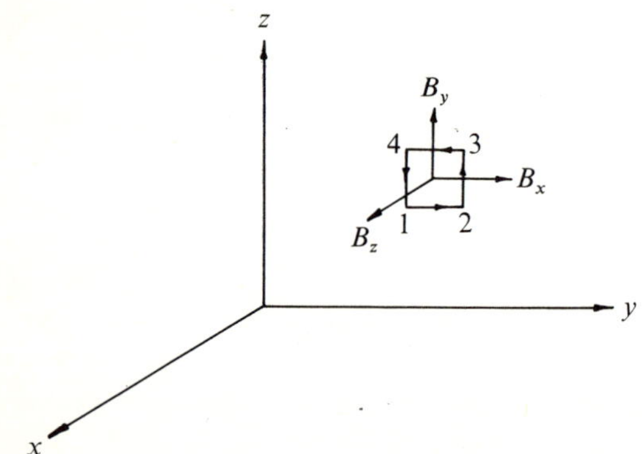

Fig. 7.24 To determine the point form of Ampère's circuital law

along the sides of the rectangle may be obtained by using Taylor's expansion (see footnote on p. 126). Thus

$$\text{along 12}\quad B_{y12} = B_y - \frac{\partial B_y}{\partial z}\frac{\Delta z}{2}$$

$$\text{along 23}\quad B_{z23} = B_z + \frac{\partial B_z}{\partial y}\frac{\Delta y}{2}$$

$$\text{along 34} \quad B_{y34} = B_y + \frac{\partial B_y}{\partial z}\frac{\Delta z}{2}$$

$$\text{along 41} \quad B_{z41} = B_z - \frac{\partial B_z}{\partial y}\frac{\Delta y}{2}$$

The integral around the closed contour is therefore given by

$$\oint_{1}^{4}{}_{2}^{3} \mathbf{B} \cdot \mathbf{dl} = B_{y12}\Delta y + B_{z23}\Delta z - B_{y34}\Delta y - B_{z41}\Delta z$$

$$= \left(\frac{\partial B_z}{\partial y} - \frac{\partial B_y}{\partial z}\right)\Delta y\Delta z$$

By Ampère's law this integral must be equal to μ_0 times the current enclosed by the path 12341. Hence

$$\left(\frac{\partial B_z}{\partial y} - \frac{\partial B_y}{\partial z}\right)\Delta y\Delta z = \mu_0 j_x \Delta y\Delta z$$

and

$$\lim_{\substack{\Delta y \to 0 \\ \Delta z \to 0}} \frac{\oint \mathbf{B} \cdot \mathbf{dl}}{\Delta y\Delta z} = \left(\frac{\partial B_z}{\partial y} - \frac{\partial B_y}{\partial z}\right) = \mu_0 j_x$$

If the higher terms of the Taylor series were carried through we would have obtained an expression of the form

$$\frac{\oint \mathbf{B} \cdot \mathbf{dl}}{\Delta y\Delta z} + \epsilon_{yz}$$

where ϵ_{yz} tends to zero as both Δy and Δz tend to zero simultaneously. The term $(\partial B_z/\partial y - \partial B_y/\partial z)$ is by definition the x-component of a vector which we shall call the '*curl or rotation of* $\mathbf{B}$', written curl $\mathbf{B}$ or rot $\mathbf{B}$. Thus

$$(\text{curl } \mathbf{B})_x = \frac{\partial B_z}{\partial y} - \frac{\partial B_y}{\partial z} \tag{7–46a}$$

If we carry out similar calculations for closed paths having the point P as centre and parallel to the zx- and to the xy-planes we obtain the respective expressions

$$(\text{curl } \mathbf{B})_y = \frac{\partial B_x}{\partial z} - \frac{\partial B_z}{\partial x} = \mu_0 j_y \tag{7–46b}$$

$$(\text{curl } \mathbf{B})_z = \frac{\partial B_y}{\partial x} - \frac{\partial B_x}{\partial y} = \mu_0 j_z \tag{7–46c}$$

which may also be obtained by the cyclic permutation of x, y, z, in equation 7–46a. The sum of the above three expressions defines the curl of **B** in cartesian coordinates. Thus

$$\text{curl } \mathbf{B} = \left(\frac{\partial B_z}{\partial y} - \frac{\partial B_y}{\partial z}\right)\mathbf{a}_x + \left(\frac{\partial B_x}{\partial z} - \frac{\partial B_z}{\partial x}\right)\mathbf{a}_y + \left(\frac{\partial B_y}{\partial x} - \frac{\partial B_x}{\partial y}\right)\mathbf{a}_z$$

which may be written more concisely as

$$\text{curl } \mathbf{B} = \begin{vmatrix} a_x & a_y & a_z \\ \dfrac{\partial}{\partial x} & \dfrac{\partial}{\partial y} & \dfrac{\partial}{\partial z} \\ B_x & B_y & B_z \end{vmatrix} = \nabla \times \mathbf{B} \tag{7–47}$$

Since each component of the curl is equal to μ_0 times the corresponding component of the current density we may write

$$\nabla \times \mathbf{B} = \mu_0 \mathbf{j} \tag{7–48}$$

That is, the curl of the magnetic flux density at any point is proportional to the current density at that point. This is the point form of Ampère's law; it is the magnetic counterpart of the point form of Gauss's law. Just as the divergence operation $(\nabla \cdot)$ applied to the electric flux density **D** at any point tells us whether any of the sources (charge density) of this field exist at that point or not, so the curl operation $(\nabla \times)$ applied to the magnetic flux density field **B** at any point tells us whether or not any of the sources (current density) of this flux exist at that point.

Any vector field whose curl is zero everywhere is said to be an *irrotational* or a vortex-free field.

Although in the above analysis we have confined ourselves to cartesian coordinates and the particular vector **B**, a more general definition of the component in the direction of a unit vector **a** of the curl of *any* continuous and differentiable vector point function **F** at a point P is

$$\mathbf{a} \cdot \text{curl } \mathbf{F} = (\text{curl } \mathbf{F})_a = \lim_{S \to 0} \frac{\oint_C \mathbf{F} \cdot d\mathbf{l}}{S} \tag{7–49}$$

where C is a simple closed curve surrounding P and located in the plane which contains P and which is perpendicular to **a**, S is the area enclosed by C and the integration round C is performed in a sense which would advance a right-hand screw in the direction of **a**. Equation 7–49 is independent of the choice of coordinates and the form of the curl in any coordinate system can be derived explicitly by expressing the line and surface elements in terms of the coordinates of the chosen system.

The expressions for curl $\mathbf{B}$ in cylindrical and spherical coordinates are given below for convenience.

Cylindrical
$$\mathbf{V} \times \mathbf{B} = \left(\frac{1}{r}\frac{\partial B_z}{\partial \phi} - \frac{\partial B_\phi}{\partial z}\right)\mathbf{a}_r + \left(\frac{\partial B_r}{\partial z} - \frac{\partial B_z}{\partial r}\right)\mathbf{a}_\phi$$

$$+ \frac{1}{r}\left[\frac{\partial(rB_\phi)}{\partial r} - \frac{\partial B_r}{\partial \phi}\right]\mathbf{a}_z \qquad (7\text{--}50)$$

Spherical
$$\mathbf{V} \times \mathbf{B} = \frac{1}{r\sin\theta}\left[\frac{\partial(B_\phi \sin\theta)}{\partial \theta} - \frac{\partial B_\theta}{\partial \phi}\right]\mathbf{a}_r + \frac{1}{r}\left[\frac{1}{\sin\theta}\frac{\partial B_r}{\partial \phi}\right.$$

$$\left. - \frac{\partial(rB_\phi)}{\partial r}\right]\mathbf{a}_\theta + \frac{1}{r}\left[\frac{\partial(rB_\theta)}{\partial r} - \frac{\partial B_r}{\partial \theta}\right]\mathbf{a}_\phi \qquad (7\text{--}51)$$

7.8.1 SURFACE CURL

When the function $\mathbf{F}$ is discontinuous across a boundary surface the

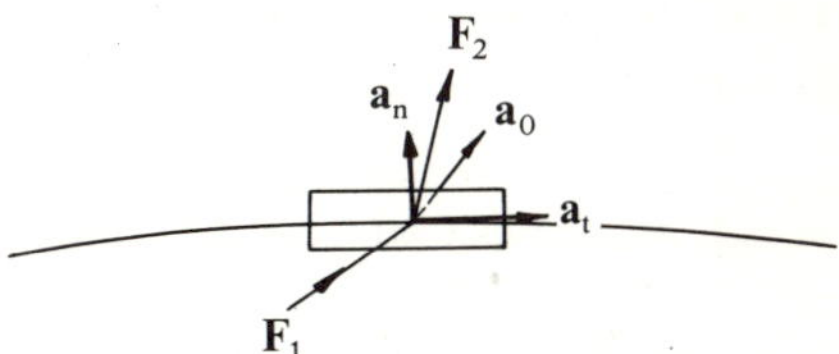

Fig. 7.25 To determine the surface curl at a boundary surface

limit in equation 7–49 has no significance since (Fig. 7.25),

$$\lim_{\substack{\Delta L \to 0 \\ \Delta h \to 0}}\left[\frac{(\mathbf{F}_2 - \mathbf{F}_1)\cdot\mathbf{a}_t\Delta L}{\Delta L \Delta h}\right]$$

will always be infinite. We may however define a surface curl

$$\mathbf{a}_0 \cdot (^S\mathbf{V}\times\mathbf{F}) = \lim_{\Delta L \to 0}\left[\frac{1}{\Delta L}\oint \mathbf{F}\cdot d\mathbf{l}\right]$$

$$= (\mathbf{F}_2 - \mathbf{F}_1)\cdot\mathbf{a}_t$$

and since $\mathbf{a}_t = \mathbf{a}_0 \times \mathbf{a}_n$, the surface curl of $\mathbf{F}$ is given by

$$^S\mathbf{V}\times\mathbf{F} = \mathbf{a}_n \times (\mathbf{F}_2 - \mathbf{F}_1) \qquad (7\text{--}52)$$

As a simple example to illustrate the curl operation as given by the point form of Ampère's law, consider the infinitely long cylindrical conductor of radius R and carrying a uniformly distributed current I.

We already know that the magnetic flux density at any point is given by

$$B_\phi = \frac{\mu_0 I}{2\pi\rho}; \qquad \rho \geq R$$

and

$$B_\phi = \frac{\mu_0 I}{2\pi R^2}\rho; \quad \rho \leq R$$

If we use the expression for curl $\mathbf{B}$ in cylindrical coordinates we see that for a point outside the conductor

$$\nabla \times \mathbf{B} = 0$$

At a point inside the conductor

$$\nabla \times \mathbf{B} = \left[\frac{1}{\rho}\frac{\partial}{\partial\rho}(\rho B_\phi)\right]\mathbf{a}_z = \frac{1}{\rho}\frac{\mu_0 I}{2\pi R^2}\frac{\partial}{\partial\rho}(\rho)^2 = \frac{\mu_0 I\mathbf{a}_z}{\pi R^2} = \mu_0\mathbf{j}$$

7.9 STOKES'S THEOREM

From Ampère's circuital law we have that

$$\oint_C \mathbf{B} \cdot d\mathbf{l} = \mu_0 I = \mu_0 \int_S \mathbf{j} \cdot d\mathbf{S}$$

but since $\nabla \times \mathbf{B} = \mu_0\mathbf{j}$, it follows that

$$\oint_C \mathbf{B} \cdot d\mathbf{l} = \int_S \nabla \times \mathbf{B} \cdot d\mathbf{S} \tag{7–53}$$

where the direction of $d\mathbf{S}$ is related to the direction in which C is traversed by the dextral rule. Equation 7–53 is known as *Stokes's theorem*. Although we have derived it for the magnetic flux density vector, the theorem is valid for any vector function. To show this consider any smooth closed contour C spanned by a smooth simple surface of area S and of arbitrary shape. Let C be traced through in a

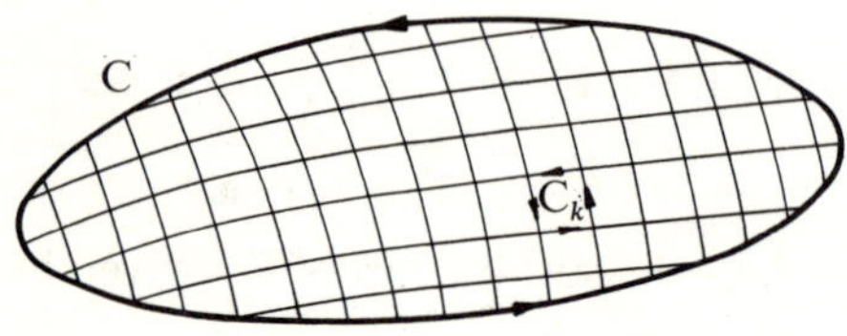

Fig. 7.26 Subdivision of surface bounded by closed contour C into elemental areas

right-handed sense (Fig. 7.26). We can subdivide the surface S into N elemental areas ΔS_k with perimeters C_k. For the perimeter C_k we have that

$$(\text{curl}\,\mathbf{F})_a = \frac{\oint_{C_k} \mathbf{F} \cdot \mathbf{dl}}{\Delta S_k} + \delta_k$$

where δ_k tends to zero as $\Delta S_k \to 0$, and where $(\text{curl}\,\mathbf{F})_a$ is the normal component of curl $\mathbf{F}$ at some interior point of ΔS_k. The above expression may be rewritten as

$$\oint_{C_k} \mathbf{F} \cdot \mathbf{dl} = (\nabla \times \mathbf{F})_a \Delta S_k - \delta_k \Delta S_k$$

$$= (\nabla \times \mathbf{F}) \cdot \Delta \mathbf{S}_k - \delta_k \Delta S_k$$

If we sum over all N elements of area we obtain

$$\sum_{k=1}^{N} \oint_{C_k} \mathbf{F} \cdot \mathbf{dl} = \sum (\nabla \times \mathbf{F}) \cdot \Delta \mathbf{S}_k - \sum \delta_k \Delta S_k$$

Now the sum of the line integrals round C_1, C_2, . . . C_k, . . ., all in the same sense, is the line integral round the bounding contour C. This is because any part of a contour common to two adjacent areas within S is always traversed in opposite senses, so that the integration over the common part of the boundaries are in opposite directions and cancel one another. In the limiting case when $N \to \infty$ and $\Delta S_k \to 0$, the above summation becomes:

$$\oint_C \mathbf{F} \cdot \mathbf{dl} = \int_S \nabla \times \mathbf{F} \cdot \mathbf{dS} \qquad (7\text{--}53\text{a})$$

which is Stokes's theorem; it relates the line integral of a vector to the surface integral of the curl of the vector. A useful modified form of Stokes's theorem can be derived by applying the theorem to the vector $\phi\mathbf{N}$, where ϕ is a scalar function and $\mathbf{N}$ is an arbitrary *constant* vector. Thus

$$\oint_C \phi\mathbf{N} \cdot \mathbf{dl} = \int_S \nabla \times (\phi\mathbf{N}) \cdot \mathbf{dS} = \int_S (\phi\nabla \times \mathbf{N} + \nabla\phi \times \mathbf{N}) \cdot \mathbf{dS}$$

$$= \int_S (\nabla\phi \times \mathbf{N}) \cdot \mathbf{dS} = \int_S \mathbf{N} \cdot \mathbf{dS} \times \nabla\phi \qquad (7\text{--}53\text{b})$$

Since $\mathbf{N}$ is constant we may take it out of the integral sign on both sides so that

$$\mathbf{N} \cdot \oint_C \phi\,\mathbf{dl} = \mathbf{N} \cdot \int_S \mathbf{dS} \times \nabla\phi$$

and

$$\oint_C \phi\,\mathbf{dl} = \int_S \mathbf{dS} \times \nabla\phi \qquad (7\text{--}53\text{c})$$

Applying Stokes's theorem to the line integral of the electric field around a closed path we have

$$\oint_C \mathbf{E} \cdot d\mathbf{l} = \int_S \nabla \times \mathbf{E} \cdot d\mathbf{S}$$

and since the *electrostatic* field $\mathbf{E}$ is conservative, its integral around any closed path is zero, hence

$$\int_S \nabla \times \mathbf{E} \cdot d\mathbf{S} = 0 \tag{7-54}$$

The only condition which can satisfy this equation is that the integrand be zero; thus

$$\nabla \times \mathbf{E} = 0 \tag{7-55}$$

7.10 THE MAGNETIC VECTOR POTENTIAL

The magnetic flux density $\mathbf{B}$ produced at a point P by a current distribution as in Fig. 7.27 is given by the Biot–Savart law

$$\mathbf{B_P} = \frac{\mu_0}{4\pi} \int_v \frac{\mathbf{j} \times \mathbf{r_{QP}}}{r^3} \, dv$$

Consider now the curl of the vector $(\mathbf{j}/r)$ evaluated at the field point P,

$$\nabla_P \times \left(\frac{\mathbf{j}}{r}\right) = \nabla_P \left(\frac{1}{r}\right) \times \mathbf{j} + \frac{1}{r} \nabla \times \mathbf{j}$$

The second term is zero since $\mathbf{j}$ is not a function of the coordinates of the point P. Also,

$$\nabla_P \left(\frac{1}{r}\right) \times \mathbf{j} = -\frac{\mathbf{a_{QP}}}{r^2} \times \mathbf{j}$$

so that,

$$\nabla_P \times \left(\frac{\mathbf{j}}{r}\right) = \frac{\mathbf{j} \times \mathbf{r_{QP}}}{r^3}$$

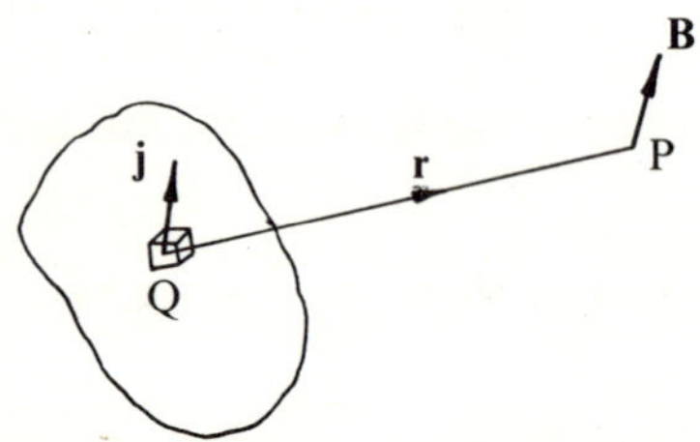

Fig. 7.27 To determine the magnetic vector potential outside a volume distribution of current

The flux density may therefore be written in the form

$$\mathbf{B}_{\mathrm{P}} = \nabla_{\mathrm{P}} \times \frac{\mu_0}{4\pi} \int_v \frac{\mathbf{j}}{r}\, \mathrm{d}v$$

where the curl has been taken outside the integral sign since the differentiations involved must be carried out with respect to the coordinates of the point P, which are independent of the coordinates of the point Q with respect to which the integrations are to be carried out. The vector,

$$\mathbf{A} = \frac{\mu_0}{4\pi} \int_v \frac{\mathbf{j}}{r}\, \mathrm{d}v \quad [\mathrm{Wb\,m}^{-1}] \qquad (7\text{–}56)$$

is called the magnetic vector potential and in the SI system of units is measured in weber per metre. It is a vector point function whose curl at any point gives the flux density $\mathbf{B}$ at that point, that is

$$\mathbf{B} = \nabla \times \mathbf{A} \qquad (7\text{–}57)$$

If we compare equation 7–56 with equation 2–21 we note that the functional relationship between $\mathbf{A}$ and $\mathbf{j}$ is identical to that between the electrostatic scalar potential V and the volume charge density ρ. Because $\mathbf{j}$ is a vector, $\mathbf{A}$ will also be a vector quantity and hence the use of the term *vector* potential. It should be stressed here that the definition of $\mathbf{A}$ as a 'potential' is due solely to the mathematical similarity between $\mathbf{A}$ and V mentioned above; unlike the scalar potential V, $\mathbf{A}$ has no physical significance. It is obvious that equations 7–56 and 7–57 replace the Biot–Savart law. Whether it is easier to calculate $\mathbf{B}$ directly from the Biot–Savart law or indirectly as the curl of $\mathbf{A}$ depends really on the geometry of the problem. The greatest usefulness of the vector potential, however, lies in the study of time varying fields, inductance problems and electromagnetic radiation.

We shall now show that for direct current the divergence of the magnetic vector potential, as given by equation 7–56 is zero. We have

$$\nabla_{\mathrm{P}} \cdot \mathbf{A} = \frac{\mu_0}{4\pi} \int_v \nabla_{\mathrm{P}} \cdot \left(\frac{\mathbf{j}}{r}\right) \mathrm{d}v$$

Using the identity

$$\nabla \cdot \left(\frac{\mathbf{j}}{r}\right) = \mathbf{j} \cdot \nabla \left(\frac{1}{r}\right) + \frac{1}{r} \nabla \cdot \mathbf{j}$$

we obtain

$$\nabla_{\mathrm{P}} \cdot A = \frac{\mu_0}{4\pi} \int_v \left[\mathbf{j} \cdot \nabla_{\mathrm{P}} \left(\frac{1}{r}\right) + \frac{1}{r} \nabla_{\mathrm{P}} \cdot \mathbf{j} \right] \mathrm{d}v$$

The second part of the integral is zero because $\mathbf{j}$ is not a function of the coordinates of the field point P. Also since by equation 2–38

$$\nabla_{\mathrm{P}}\left(\frac{1}{r}\right) = -\nabla_{\mathrm{Q}}\left(\frac{1}{r}\right)$$

we have that

$$\nabla \cdot \mathbf{A} = \frac{\mu_0}{4\pi}\int_v\left[-\mathbf{j}\cdot\nabla_{\mathrm{Q}}\left(\frac{1}{r}\right)\right]dv$$

and making use of the above identity for $\nabla \cdot (\mathbf{j}/r)$ we obtain that

$$\nabla_{\mathrm{P}} \cdot \mathbf{A} = \frac{\mu_0}{4\pi}\int_v\left[\frac{1}{r}\nabla_{\mathrm{Q}}\cdot\mathbf{j}-\nabla_{\mathrm{Q}}\cdot\left(\frac{\mathbf{j}}{r}\right)\right]dv$$

Since we are dealing with steady currents, $\nabla_{\mathrm{Q}} \cdot \mathbf{j} = 0$ (equation 6–19). The second integral may be transformed into a surface integral by using the divergence theorem; we thus have that

$$\nabla_{\mathrm{P}} \cdot \mathbf{A} = -\frac{\mu_0}{4\pi}\oint_S\frac{\mathbf{j}\cdot d\mathbf{S}}{r}$$

As the original volume integral was to be integrated over all the volume through which a current flows, it follows that we may always choose the surface S to be that of a sphere of very large radius such that $\mathbf{j} = 0$ everywhere on S. We may therefore state that for direct currents the condition

$$\nabla \cdot \mathbf{A} = 0 \tag{7–58}$$

is always fulfilled.

For those cases in which the current is constrained to flow along a thin wire, it is more convenient to replace the current element $\mathbf{j}\,dv$ by the element $I\,d\mathbf{l}$. The volume integral in equation 7–56 is then replaced by a closed line integral around the entire circuit so that

$$\mathbf{A} = \frac{\mu_0}{4\pi}I\oint_C\frac{d\mathbf{l}}{r} \tag{7–59}$$

In the case of a surface current of density $\mathbf{K}$ the expression for $\mathbf{A}$ becomes

$$\mathbf{A} = \frac{\mu_0}{4\pi}\int_S\frac{\mathbf{K}\,dS}{r} \tag{7–60}$$

The line integral of the magnetic vector potential around any *closed* path is equal to the total magnetic flux enclosed by that path. This is easily demonstrated by using equation 7–57, Stokes's theorem and equation 7–27; thus

$$\oint_C \mathbf{A} \cdot d\mathbf{l} = \int_S \nabla \times \mathbf{A} \cdot d\mathbf{S} = \int_S \mathbf{B} \cdot d\mathbf{S} = \Phi \qquad (7\text{–}61)$$

An alternative method of deriving equation 7–56 is as follows. Since div $\mathbf{B} = 0$ and the divergence of any vector which is itself the curl of another vector is always zero, we may define a vector $\mathbf{A}$ such that

$$\mathbf{B} = \nabla \times \mathbf{A} \qquad (7\text{–}57)$$

The point form of Ampère's law tells us that

$$\nabla \times \mathbf{B} = \nabla \times (\nabla \times \mathbf{A}) = \mu_0 \mathbf{j}$$

Now the curl of the curl of $\mathbf{A}$ is given by the vector identity

$$\nabla \times (\nabla \times \mathbf{A}) = \nabla(\nabla \cdot \mathbf{A}) - \nabla^2 \mathbf{A}$$

This identity is correct only for the cartesian form of the operator ∇. Hence for any cartesian component A_i ($i = x,\ y$ or z) of the vector $\mathbf{A}$ we have that

$$\nabla_i(\nabla \cdot \mathbf{A}) - \nabla^2 A_i = \mu_0 j_i \qquad (7\text{–}62)$$

Equation 7–57 does not completely specify the vector function $\mathbf{A}$ because any other vector function $\mathbf{A}'$ given by

$$\mathbf{A}' = \mathbf{A} + \nabla f \qquad (7\text{–}63)$$

where f is any scalar function, will produce the same magnetic flux density as $\mathbf{A}$, since $\nabla \times \nabla f = 0$. For the vector function $\mathbf{A}$ to be uniquely defined (to within an additive constant only) both its curl and its divergence must be specified (Helmholtz's theorem). The choice of the divergence of $\mathbf{A}$ is completely arbitrary. For the case of direct currents it is convenient to set $\nabla \cdot \mathbf{A} = 0$. Equation 7–62 becomes

$$\nabla^2 A_i = -\mu_0 j_i$$

so that the cartesian components of $\mathbf{A}$ satisfy the three scalar equations

$$\nabla^2 A_x = -\mu_0 j_x$$
$$\nabla^2 A_y = -\mu_0 j_y \qquad (7\text{–}64)$$
$$\nabla^2 A_z = -\mu_0 j_z$$

Each of these equations has the same form as Poisson's equation for the electrostatic potential. The solutions must therefore be of the same form as equation 2–27 so that,

$$A_x = \frac{\mu_0}{4\pi} \int_v \frac{j_x\, dv}{r}; \quad A_y = \frac{\mu_0}{4\pi} \int_v \frac{j_y\, dv}{r}; \quad A_z = \frac{\mu_0}{4\pi} \int_v \frac{j_z\, dv}{r}$$

The vector sum of these components gives,

$$A = \frac{\mu_0}{4\pi} \int_v \frac{j\,dv}{r} \tag{7-56}$$

The reader may readily verify that by taking the curl of A one obtains B as given by the Biot–Savart law. Since $\nabla \cdot A$ has been chosen as zero it follows that $\nabla \cdot j$ must also be zero (steady currents).

Although by this second method we have arrived at the same result for the magnetic vector potential as that obtained by the first method, it is clear that this has only been possible because of the choice $\nabla \cdot A = 0$; the derivation of equation 7–56 is thus somewhat forced, since the choice of $\nabla \cdot A$ is completely arbitrary. In the first method on the other hand the expression for A and the fact that $\nabla \cdot A = 0$ are direct consequences of the experimentally demonstrable law of Biot and Savart.

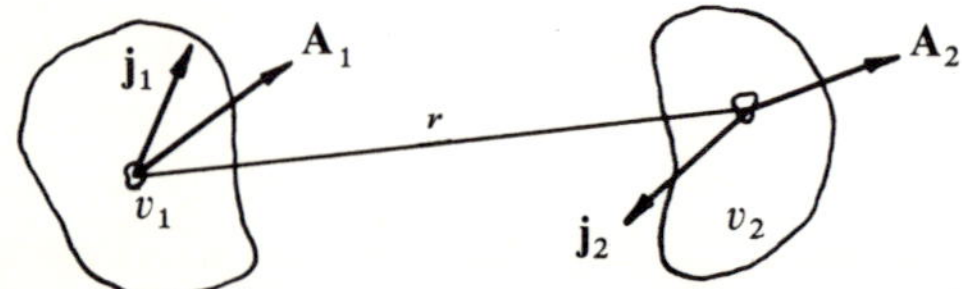

Fig. 7.28 Two distinct volume distributions of current

If we have two systems of distributed currents, j_1 and j_2 throughout volumes v_1 and v_2 respectively (Fig. 7–28), then if A_1 and A_2 are the magnetic vector potentials produced by each distribution, we have that

$$\int_{v_1} (j_1 \cdot A_2)\,dv_1 = \int_{v_2} (j_2 \cdot A_1)\,dv_2 \tag{7-65}$$

This equality is readily verified since

$$A_1 = \frac{\mu_0}{4\pi} \int_{v_1} \frac{j_1}{r}\,dv_1\,; \quad A_2 = \frac{\mu_0}{4\pi} \int_{v_2} \frac{j_2}{r}\,dv_2$$

7.10.1 MAGNETIC VECTOR POTENTIAL OF INFINITELY LONG TRANSMISSION LINES

The magnetic flux density due to an infinitely long cylindrical conductor of non-magnetic material carrying an axial current I uniformly distributed over the cross-section has already been shown to be

$$B = \frac{\mu_0 I \rho}{2\pi R^2}\, a_\phi \cdot \rho \le R$$

$$B = \frac{\mu_0 I}{2\pi \rho}\, a_\phi \quad \cdot \rho \ge R$$

where R is the conductor radius. Since the only component of the current density is the one along the z-axis it is clear that the vector potential will also have a single component in the same direction. Thus

$$\mathbf{A} = A_z \mathbf{a}_z$$

From equation 7–57 we have that

$$\mathbf{B} = \nabla \times \mathbf{A} = -\frac{\partial A_z}{\partial \rho} \mathbf{a}_\phi$$

When $0 \leq \rho \leq R$ we have that

$$-\frac{\partial A_z}{\partial \rho} = \frac{\mu_0 I \rho}{2\pi R^2}$$

and

$$A_z = -\frac{\mu_0 I}{4\pi} \frac{\rho^2}{R^2} + C$$

When $\rho \geq R$ we have that

$$-\frac{\partial A_z}{\partial \rho} = \frac{\mu_0 I}{2\pi \rho}$$

and

$$A_z = -\frac{\mu_0 I}{2\pi} \log \rho + C_1$$

Since the potential must be continuous at $\rho = R$ it follows that

$$-\frac{\mu_0 I}{2\pi} \log R + C_1 = -\frac{\mu_0 I}{4\pi} + C$$

so that

$$C_1 = \frac{\mu_0 I}{2\pi} \log R - \frac{\mu_0 I}{4\pi} + C$$

and hence

$$A_z = -\frac{\mu_0 I}{2\pi} \log \frac{\rho}{R} - \frac{\mu_0 I}{4\pi} + C, \quad (\rho \geq R) \tag{7-66a}$$

$$A_z = -\frac{\mu_0 I}{4\pi} \frac{\rho^2}{R^2} + C, \quad (0 \leq \rho \leq R) \tag{7-66b}$$

C is an arbitrary constant whose value will depend on the choice of the zero reference level of the magnetic potential. If the zero level is chosen at the centre of the wire then $C = 0$ and the vector potential A_z will be everywhere oppositely directed to the current in the wire. The equation of the lines of induction is

$$\frac{\mathrm{d}x}{B_x} = \frac{\mathrm{d}y}{B_y}$$

And since

$$B_x = \frac{\partial A_z}{\partial y}, \quad B_y = -\frac{\partial A_z}{\partial x}$$

it follows that

$$\frac{\partial A_z}{\partial x}\,\mathrm{d}x + \frac{\partial A_z}{\partial y}\,\mathrm{d}y = 0$$

$$\mathrm{d}[A_z] = 0$$

or

$$A_z = \text{constant}$$

so that the magnetic flux lines are concentric circles about the current.

Consider next the two parallel current-carrying conductors shown in Fig. 7.29. The magnetic potential at any point may be obtained by

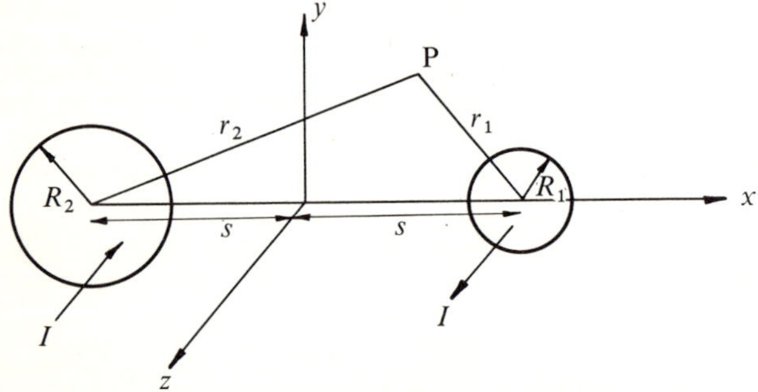

Fig. 7.29 Two infinitely long parallel current-carrying conductors

superimposing the vector potentials produced by each of the conductors at that point. For a point outside both conductors and for the current directions shown in the figure we have that

$$A_z = \frac{\mu_0 I}{2\pi}\log\frac{r_2}{R_2} + \frac{\mu_0 I}{4\pi} + C_1 - \frac{\mu_0 I}{2\pi}\log\frac{r_1}{R_1} - \frac{\mu_0 I}{4\pi} + C_2$$

$$= \frac{\mu_0 I}{2\pi}\left[\log\frac{r_2}{r_1} + \log\frac{R_1}{R_2}\right] + C_1 + C_2$$

If we choose the plane passing midway between the two conductor centres as zero reference then

$$0 = \frac{\mu_0 I}{2\pi} \log \frac{R_1}{R_2} + C_1 + C_2$$

and

$$A_z = \frac{\mu_0 I}{2\pi} \log \frac{r_2}{r_1}, \quad (r_1 > R_1, r_2 > R_2) \tag{7-67}$$

This equation has the same form as that for the electrostatic potential of two equal but opposite line charges located at the wire centres and carrying a uniform linear charge density $\lambda = \pm \epsilon_0 \mu_0 I$ (equation 2–59). In the region outside both wires, therefore, the surfaces of constant potential are a family of cylinders, we may thus write

$$A_z = kV$$

where k is a constant $(k = \epsilon_0 \mu_0 I/\lambda)$ and V is the electrostatic potential. The magnetic flux density is given by

$$\mathbf{B} = \nabla \times \mathbf{A} = \frac{\partial A_z}{\partial y} \mathbf{a}_x - \frac{\partial A_z}{\partial x} \mathbf{a}_y = -\mathbf{a}_z \times \operatorname{grad} A_z$$

$$= -\mathbf{a}_z \times \operatorname{grad} kV$$

$$\mathbf{B} = k\mathbf{a}_z \times \mathbf{E} \tag{7-68}$$

Thus for points outside both conductors the magnetic and electric field lines are mutually orthogonal. The reader may readily verify that the cartesian components of $\mathbf{B}$ are

$$B_x = -\frac{\mu_0 I}{2\pi} y(r_1^{-2} - r_2^{-2})$$

$$B_y = \frac{\mu_0 I}{2\pi} [(y+s)r_2^{-2} - (y-s)r_1^{-2}] \tag{7-69}$$

$$B_z = 0$$

The equation of the induction lines is again given by $A_z = $ constant. It will be noted that for a comparable arrangement of linear charges and currents the electrostatic equipotentials correspond to lines of magnetic induction.

For a field point inside conductor 1 and therefore external to conductor 2 the vector potential is given by the sum of equations 7–66a and 7–66b,

$$A_z = \frac{\mu_0 I}{2\pi} \log \frac{r_2}{R_2} + \frac{\mu_0 I}{4\pi} \left[1 - \frac{r_1^2}{R_1^2} \right] + C_1 + C_2$$

For the same zero reference,

$$A_z = \frac{\mu_0 I}{2\pi}\log\frac{r_2}{R_2} + \frac{\mu_0 I}{4\pi}\left[1 - \frac{r_1^2}{R_1^2}\right], \quad (r_1 < R_1, r_2 > R_2) \qquad (7\text{-}70)$$

Inside conductor 1 therefore the magnetic equipotential surfaces are non-circular cylinders; the equation of their cross-section is obtained by setting equation 7–70 equal to a constant.

7.11 MAGNETIC MOMENT OF A CURRENT LOOP

The *vector* moment about a fulcrum O of a vector **t** acting at a point P is, by definition, $\mathbf{r}\times\mathbf{t}$ where **r** is the radius vector OP. If d**l** is an infinitesimal line element, then the line integral

$$\mathbf{G}_0 = \tfrac{1}{2}\int_L \mathbf{r}\times d\mathbf{l} \qquad (7\text{-}71)$$

defines the vector moment of the path segment L about the fulcrum O (Fig. 7.30). The reason for the introduction of the factor 1/2 will

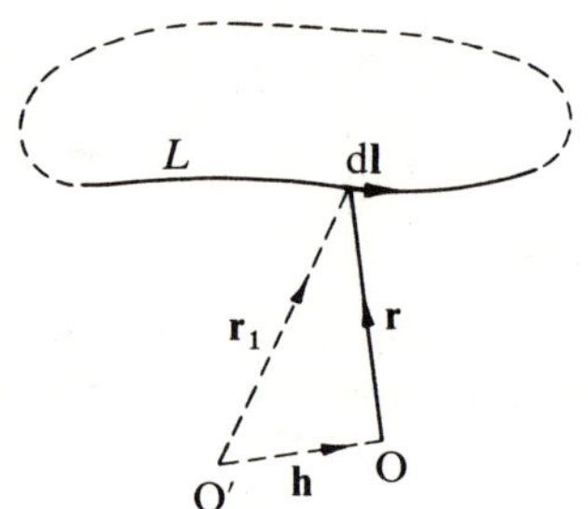

Fig. 7.30 Vector moment of a path segment

become apparent presently. Suppose now that we choose a new fulcrum at some point O′; the vector moment about O′ is

$$\mathbf{G}_{O'} = \tfrac{1}{2}\int_L \mathbf{r}'\times d\mathbf{l} = \tfrac{1}{2}\int_L (\mathbf{h}+\mathbf{r})\times d\mathbf{l}$$

$$= \tfrac{1}{2}\int_L \mathbf{h}\times d\mathbf{l} + \tfrac{1}{2}\int_L \mathbf{r}\times d\mathbf{l}$$

Since **h** is a constant it may be taken out of the integral sign so that

$$\mathbf{G}_{O'} = \tfrac{1}{2}\mathbf{h}\times\int_L d\mathbf{l} + \mathbf{G}_0$$

The first integral represents the vector summation (resultant) of an infinite number of elemental vectors. If the integral is taken around a *closed* path its value will be zero since the vector diagram is closed (zero resultant). In this case therefore

$$\tfrac{1}{2}\mathbf{h} \times \oint_G \mathrm{d}\mathbf{l} = 0$$

and

$$\mathbf{G}_{O'} = \mathbf{G}_0 = \tfrac{1}{2}\oint \mathbf{r} \times \mathrm{d}\mathbf{l} \qquad (7\text{--}72)$$

so that the vector moment of a closed path is independent of the position of the fulcrum and depends solely on the form of the path. Now the quantity $\tfrac{1}{2}(\mathbf{r} \times \mathrm{d}\mathbf{l})$ represents the vector area of the infinitesimal triangle whose sides are $\mathbf{r}$ and $\mathrm{d}\mathbf{l}$ and the integral around the closed path represents therefore the vector area defined by the point O and the contour. Thus

$$\tfrac{1}{2}\oint_C \mathbf{r} \times \mathrm{d}\mathbf{l} = \int_S \mathrm{d}\mathbf{S} \qquad (7\text{--}73)$$

Since the line integral has the same value irrespective of the choice of the origin O, it follows that the surface integral may be taken over any regular surface spanning the contour C. As an illustrative example

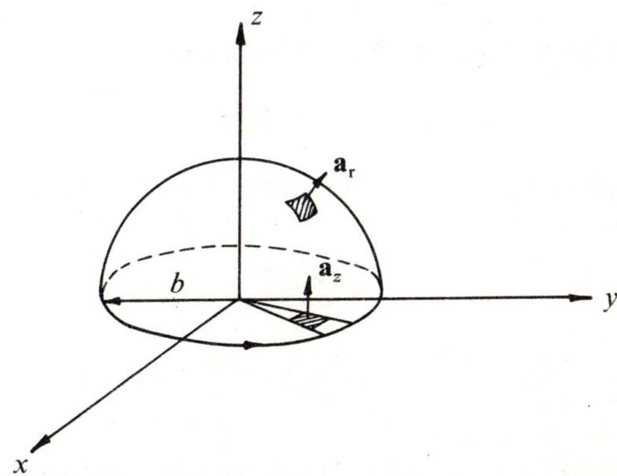

Fig. 7.31 To determine the vector moment of a plane circular current loop about the origin

consider the plane circular loop of radius b shown in Fig. 7.31. The vector moment about the origin is

$$\tfrac{1}{2}\oint \mathbf{r} \times \mathrm{d}\mathbf{l} = \tfrac{1}{2}\int_0^{2\pi} b\mathbf{a}_r \times (b\,\mathrm{d}\phi\mathbf{a}_\phi) = \pi b^2 \mathbf{a}_z$$

$$\int \mathrm{d}\mathbf{S} = \int_0^{2\pi}\int_0^b r\,\mathrm{d}\phi\,\mathrm{d}r\mathbf{a}_z = \pi b^2 \mathbf{a}_z$$

Suppose now that instead of integrating over the plane surface bounded by the circle we choose to integrate over a hemispherical surface bounded by the circle; we have

$$\int d\mathbf{S} = \iint b^2 \sin\theta \, d\theta \, d\phi \mathbf{a_r}$$

Since the unit vector $\mathbf{a_r}$ is a function of position (θ and ϕ) we must express it as a function of the fixed unit directional vectors $\mathbf{a_x}$, $\mathbf{a_y}$, $\mathbf{a_z}$. This is given by equation 1–42. Carrying out the double integration after substituting for this value of $\mathbf{a_r}$ we find that both the $\mathbf{a_x}$ and $\mathbf{a_y}$ terms are zero and that

$$\int d\mathbf{S} = \int_0^{\pi/2} \int_0^{2\pi} b^2 \sin\theta \cos\theta \, d\theta \, d\phi \mathbf{a_z} = \pi b^2 \mathbf{a_z}$$

If we have a current I flowing around a closed filamentary loop, we replace the line element $d\mathbf{l}$ in equations 7–6a and 7–73 by the current element $I\,d\mathbf{l}$ and obtain the quantity

$$\mathbf{m} = \tfrac{1}{2}I \oint_C \mathbf{r} \times d\mathbf{l} = I \int_S d\mathbf{S} = I \int_S dS\mathbf{a_n} \qquad (7\text{–}74a)$$

which is by definition the *magnetic moment* of the current loop. If the loop consists actually of N closely spaced turns of thin wire, then the current I must be replaced by NI in which case

$$\mathbf{m} = NI \int_S dS\mathbf{a_n} \qquad (7\text{–}74b)$$

Since current always flows in closed paths, it is clear that $\mathbf{m}$ will depend solely upon the form of the contour C. Figure 7.32 shows the right-

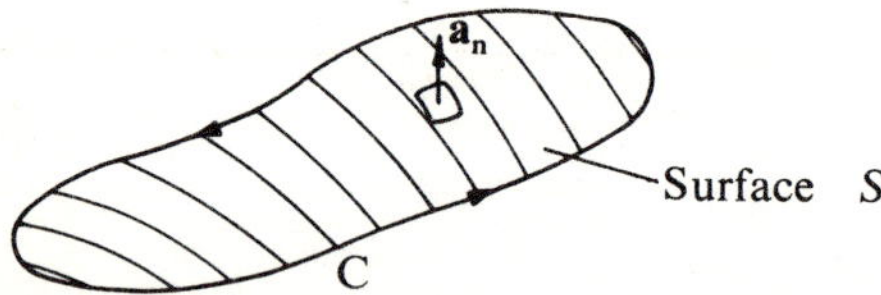

Fig. 7.32 Right-handed relationship between the unit normal vector $\mathbf{a_n}$ and the direction of current flow

handed relationship between the unit normal vector $\mathbf{a_n}$ and the direction of current flow. For a plane contour C, the most convenient choice of S is the part of the plane bounded by C, over this surface $\mathbf{a_n}$ is constant and

$$\mathbf{m} = IS\mathbf{a_n} \qquad (7\text{–}74c)$$

so that the magnitude of the moment is equal to the current multiplied by the loop area whilst the direction is that of the unit normal vector $\mathbf{a_n}$.

7.12 VECTOR POTENTIAL OF A CURRENT LOOP AND MAGNETIC DIPOLE MOMENT

Consider a filamentary current loop carrying a current I (Fig. 7.33a). The magnetic vector potential at any point P is given by

$$\mathbf{A} = \frac{\mu_0 I}{4\pi} \oint_C \frac{d\mathbf{l}}{r}$$

where r is the distance from $d\mathbf{l}$ to P. Using the modified form of Stokes's theorem as given by equation 7–53c we may rewrite $\mathbf{A}$ as

$$\mathbf{A} = \frac{\mu_0 I}{4\pi} \int_S \mathbf{a}_n \times \nabla_Q \left(\frac{1}{r}\right) dS$$

where S is an arbitrary surface bounded by the contour C and r is now

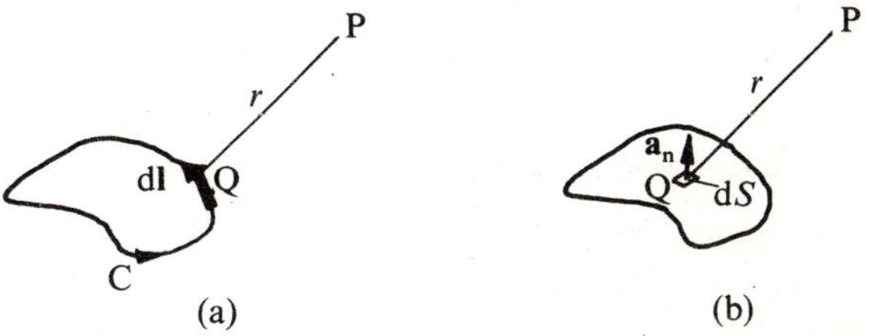

Fig. 7.33 To determine the vector potential at any point P due to a filamentary current loop

the distance from dS to P (Fig. 7.33b). The integrations must be carried out with respect to the coordinates of the point Q. Since

$$\nabla_Q(1/r) = -\nabla_P(1/r) = r^{-2}\mathbf{a}_r$$

we have that

$$\mathbf{A} = \frac{\mu_0 I}{4\pi} \int_S \frac{\mathbf{a}_n \times \mathbf{a}_r}{r^2} dS \qquad (7\text{–}75)$$

where the direction of $\mathbf{a}_n$ is related to the sense in which C is traversed (taken to coincide with the direction of current flow) by the right-hand rule. Suppose now that the current loop is vanishingly small; in other words suppose that the distance r is large compared with the largest possible linear dimension of the loop. Both $\mathbf{a}_r$ and $1/r^2$ in equation 7–75 may then be assumed constant for all points 'within' the loop, and taken out of the integral sign. Thus

$$\mathbf{A} = \frac{\mu_0 I}{4\pi} \left(\int d\mathbf{S}\right) \times \frac{\mathbf{a}_r}{r^2}$$

combining this expression with equation 7–74a we obtain

$$\mathbf{A} = \frac{\mu_0}{4\pi}\,\mathbf{m} \times \nabla_Q\left(\frac{1}{r}\right)$$

$$= -\frac{\mu_0}{4\pi}\,\mathbf{m} \times \nabla_P\left(\frac{1}{r}\right) \tag{7–76}$$

Taking the centre of the infinitesimal loop as the origin of a system of

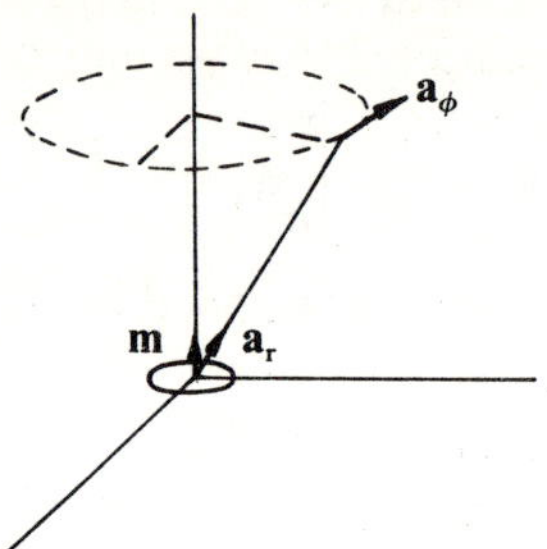

Fig. 7.34 To determine the magnetic potential of an infinitesimal current loop

spherical coordinates (Fig. 7.34), it is clear that $\mathbf{A}$ has only a ϕ-component $(\mathbf{a}_n \times \mathbf{a}_r = \mathbf{a}_\phi \sin\theta)$,

$$\mathbf{A} = A_\phi\mathbf{a}_\phi = \frac{\mu_0}{4\pi}\frac{m\sin\theta}{r^2}\,\mathbf{a}_\phi \tag{7–77}$$

Equation 7–76 may be obtained by a more general approach to the problem using an analysis similar to that given in Section 2.12. Suppose that the entire current loop can be contained within a sphere of finite radius a, such that $r \gg a$, and let the origin be chosen at the

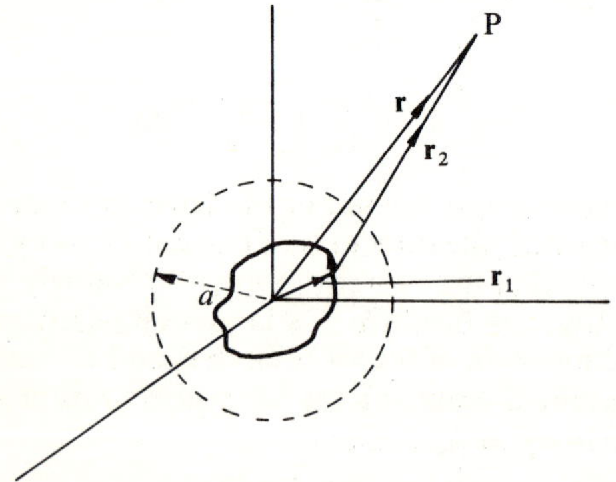

Fig. 7.35 To determine the magnetic vector potential of an arbitrary current loop in terms of magnetic multipoles

centre of the sphere (Fig. 7.35). The magnetic vector potential at P is

$$\mathbf{A} = \frac{\mu_0 I}{4\pi} \oint \frac{d\mathbf{l}}{r_2}$$

We have already shown in Section 2.12 that

$$\frac{1}{r_2} = |\mathbf{r} - \mathbf{r}_1|^{-1} = \frac{1}{r}\left(1 + \frac{\mathbf{r} \cdot \mathbf{r}_1}{r^2} + \cdots\right)$$

Substituting this expression into the equation for $\mathbf{A}$ and noting that r is constant we have

$$\mathbf{A} = \frac{\mu_0 I}{4\pi}\left[\frac{1}{r}\oint d\mathbf{l} + \frac{1}{r^3}\oint (\mathbf{r} \cdot \mathbf{r}_1)\, d\mathbf{l} + \cdots\right] \qquad (7\text{-}78)$$

The first term is zero so that the dominant term of the expansion is the second one. Making use of the expansion

$$\mathbf{A} \times (\mathbf{B} \times \mathbf{C}) = (\mathbf{A} \cdot \mathbf{C})\mathbf{B} - (\mathbf{A} \cdot \mathbf{B})\mathbf{C}$$

and noting (Fig. 7.35) that $d\mathbf{r}_1 = d\mathbf{l}$, the integrand of the second term may be rewritten as

$$(\mathbf{r} \cdot \mathbf{r}_1)\, d\mathbf{r}_1 = (\mathbf{r}_1 \times d\mathbf{r}_1) \times \mathbf{r} + (\mathbf{r} \cdot d\mathbf{r}_1)\mathbf{r}_1 \qquad (i)$$

The total differential of $(\mathbf{r} \cdot \mathbf{r}_1)\mathbf{r}_1$ is

$$d[(\mathbf{r} \cdot \mathbf{r}_1)\mathbf{r}_1] = (\mathbf{r} \cdot \mathbf{r}_1)\, d\mathbf{r}_1 + \mathbf{r}_1\, d(\mathbf{r} \cdot \mathbf{r}_1)$$

$$= (\mathbf{r} \cdot \mathbf{r}_1)\, d\mathbf{r}_1 + \mathbf{r}_1(\mathbf{r} \cdot d\mathbf{r}_1) \qquad (ii)$$

since $\mathbf{r}$ is constant. Combining (i) and (ii) and replacing $d\mathbf{r}_1$ by $d\mathbf{l}$ we obtain

$$(\mathbf{r} \cdot \mathbf{r}_1)\, d\mathbf{l} = \tfrac{1}{2}(\mathbf{r}_1 \times d\mathbf{l}) \times \mathbf{r} + \tfrac{1}{2}d[(\mathbf{r} \cdot \mathbf{r}_1)\mathbf{r}_1]$$

Integration of the second term around a closed path gives zero since the initial and final points are identical. Hence

$$\mathbf{A} = \frac{\mu_0 I}{4\pi}\left[\oint \tfrac{1}{2}(\mathbf{r}_1 \times d\mathbf{l})\right] \times \mathbf{r}r^{-3}$$

$$= -\frac{\mu_0 I}{4\pi}\left(\int d\mathbf{S}\right) \times \nabla_P\left(\frac{1}{r}\right) = -\frac{\mu_0}{4\pi}\mathbf{m} \times \nabla_P\left(\frac{1}{r}\right) \qquad (7\text{-}76)$$

The vector potential as given by equation 7-76 is known as the *magnetic dipole potential* of the current loop. The higher terms in equation 7-78 represent the vector potentials of magnetic multipoles of higher order; they remain negligible as long as $r \gg a$. If, however, the circuit contour is allowed to shrink to infinitesimal dimensions about the origin, the potential at any point (except the origin) will be given by the dipole term alone. The reason for using the term 'dipole' will become clear from what follows.

The magnetic flux density at the field point P is

$$\mathbf{B} = \nabla_P \mathbf{A} = -\frac{\mu_0}{4\pi} \nabla_P \times \left[\mathbf{m} \times \nabla_P \left(\frac{1}{r} \right) \right]$$

Since $\mathbf{m}$ is constant as far as the point P is concerned, the identity

$$\nabla_P \times \left(\frac{\mathbf{m}}{r} \right) = \frac{1}{r} \nabla_P \times \mathbf{m} + \nabla_P \left(\frac{1}{r} \right) \times \mathbf{m}$$

reduces to

$$\nabla_P \times (\mathbf{m}/r) = -\left[\mathbf{m} \times \nabla_P(1/r) \right]$$

so that

$$\mathbf{B} = \frac{\mu_0}{4\pi} \nabla_P \times \nabla_P \times \left(\frac{\mathbf{m}}{r} \right)$$

$$= \frac{\mu_0}{4\pi} \left[\nabla_P \left(\nabla_P \cdot \frac{\mathbf{m}}{r} \right) - \nabla_P^2 \left(\frac{\mathbf{m}}{r} \right) \right]$$

since $\nabla^2(1/r) = 0$ and $\mathbf{m}$ is constant we have that

$$\mathbf{B} = \frac{\mu_0}{4\pi} \nabla_P \left(\nabla_P \cdot \frac{\mathbf{m}}{r} \right)$$

Also

$$\nabla_P \cdot \frac{\mathbf{m}}{r} = \frac{1}{r} \nabla_P \cdot \mathbf{m} + \mathbf{m} \cdot \nabla_P \left(\frac{1}{r} \right)$$

$$= \mathbf{m} \cdot \nabla_P(1/r)$$

Hence finally,

$$\mathbf{B} = \frac{\mu_0}{4\pi} \nabla_P \left[\mathbf{m} \cdot \nabla_P \left(\frac{1}{r} \right) \right] \tag{7-79a}$$

The reader can readily show that we may also express the magnetic flux density in the alternate forms,

$$\mathbf{B} = \frac{\mu_0}{4\pi} \frac{m}{r^3} (2 \cos \theta \mathbf{a}_r + \sin \theta \mathbf{a}_\theta) \tag{7-79b}$$

and

$$\mathbf{B} = \frac{\mu_0}{4\pi} \left[\frac{3(\mathbf{m} \cdot \mathbf{r})\mathbf{r}}{r^5} - \frac{\mathbf{m}}{r^3} \right] \tag{7-79c}$$

If we now compare these last three equations with equations 2–82, 2–83 and 2–84 respectively, we see that each pair has exactly the same mathematical form so that the magnetic flux lines produced by a vanishingly small current loop are exactly of the same form as the field lines produced by an ideal or point dipole.

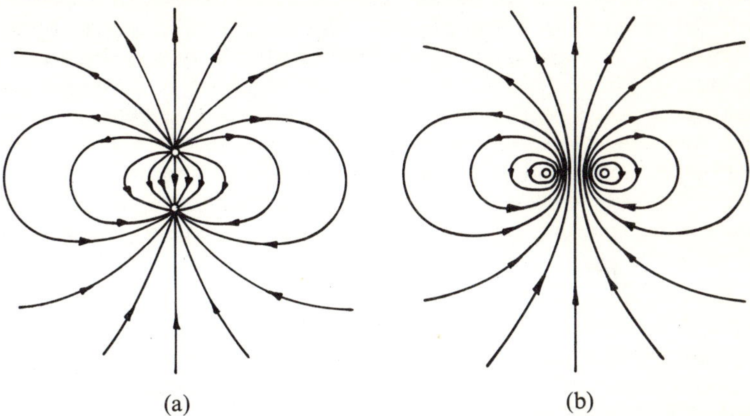

(a)   (b)

Fig. 7.36 (a) E lines of two equal and opposite point charges (b) B lines of a plane circular current loop

Figure 7–36a shows the E lines of two equal and opposite point charges (dipole) and Fig. 7.36b shows the B lines due to a current loop. At distances which are large compared with the dimensions (charge separation, size of loop) of the sources, the fields are alike (dipole field). The fields close to the sources are, however, entirely different. The E lines start and end at the positive and negative charges ($\nabla \cdot \mathbf{E} \neq 0$) whereas the B lines form closed circuits linking the loop ($\nabla \cdot \mathbf{B} = 0$). For a point dipole and a vanishingly small current loop we can therefore say that the E and B fields are identical on a macroscopic scale but are basically different on a microscopic scale.

Because of the close correspondence which exists between the magnetic behaviour of an elementary current loop and the electric behaviour of a charge dipole, it is possible to regard $\mathbf{m}$ as a magnetic dipole moment. This means that a small current loop can be considered as equivalent to a magnetic dipole consisting of two hypothetical isolated magnetic poles m (sometimes called magnetic masses) separated by a distance $\mathbf{d}$ such that $\mathbf{m} = m\mathbf{d}$. The direction of the magnetic dipole is related to its equivalent current loop by the dextral rule; that is, a right-handed screw turned in the direction of the current would advance from the negative end or south pole to the positive end or north pole of the dipole.

The reader should at this point compare the way in which the respective concepts of electric and magnetic dipoles have been arrived at. The concept of an electric dipole is based primarily on the physical independent existence of two electric charges of opposite polarities. Since isolated magnetic poles or charges are not known to exist*— that is, no phenomenon has been found which permits us to define an isolated magnetic pole (monopole)—the concept of magnetic dipole is completely artificial and has been introduced merely because the distant magnetic field of a small current loop† is mathematically analogous to the electric field of a point dipole. It is well to bear in mind that this analogy is made explicit by the definition of the moment of the current loop as $\mathbf{m} = I \int d\mathbf{S}$ and that *magnetism must be regarded as a by-product of moving electric charge*, be it the current circulating in a coil or the spinning or orbiting electrons in magnetic materials (see Chapter 8).

7.13 THE MAGNETIC SCALAR POTENTIAL

It will be recalled that the electrostatic field is a conservative field. This fact is represented mathematically in integral and in point form by the respective expressions,

$$\oint \mathbf{E} \cdot d\mathbf{l} = 0$$

and

$$\nabla \times \mathbf{E} = 0$$

The last equation enables us to express $\mathbf{E}$ as the gradient of a scalar potential function V, since the curl of the gradient of any function is always zero. In effect, in Sections 2.6 and 2.7 we showed, by physical considerations, that

$$-\int_{B}^{A} \mathbf{E} \cdot d\mathbf{l} = V_{AB} = V_A - V_B$$

and

$$\mathbf{E} = -\nabla V$$

Consider now the integral and point forms of Ampère's circuital law as given by

$$\oint \mathbf{B} \cdot d\mathbf{l} = 0 \tag{7-80}$$

---

* Although the existence of monopoles has been predicted by P. A. M. Dirac's quantum theory, all attempts to find such poles have so far failed.

† At a large distance from any magnetic field source—e.g., single current loop, coil, permanent magnet etc.—all magnetic fields are dipole fields; this of course is not so for electric fields.

and

$$\nabla \times \mathbf{B} = 0 \qquad (7\text{--}81)$$

Equation 7–80 holds for all paths of integration which do not enclose any current whilst equation 7–81 holds at all points where the current density is zero. We may rewrite equation 7–81 as

$$\nabla \times (-\nabla V_m) = 0$$

so that

$$\mathbf{B} = -\nabla V_m \qquad (7\text{--}82)$$

where V_m is a scalar function. By analogy with electrostatics the function V_m is called the *magnetic scalar potential*. The SI unit for this potential is weber metre^{-1}. If equation 7–80 holds then the line integral of $\mathbf{B}$ between any two points is independent of the path and, again by analogy with electrostatics, we may write

$$V_{mAB} = -\int_B^A \mathbf{B} \cdot d\mathbf{l} = V_{mA} - V_{mB} \qquad (7\text{--}83)$$

It must be stressed here that the analogy with electrostatics is purely mathematical. V_m has no physical significance in the sense that V has. The reason for this is that isolated magnetic poles do not exist (whereas isolated electric point charges do) so that it would be meaningless to define V_m as a potential energy per unit pole.

The equation $V_m = $ constant defines the magnetic equipotential surfaces. Such surfaces are everywhere perpendicular to the vector $\mathbf{B}$ (equation 7–82). Moreover, since V_m can only be defined for regions where the current is zero, it follows that the equipotential surfaces are open and terminate at the surfaces of the conductors (current-bearing regions).

Since $\nabla \cdot \mathbf{B} = 0$ always, it follows that in any region in which the current density is zero

$$\nabla^2 V_m = 0, \quad (j = 0) \qquad (7\text{--}84)$$

so that V_m in that region is a solution of Laplace's equation.

If we rewrite equation 7–79a as

$$\mathbf{B} = -\nabla_P \left[\frac{\mu_0}{4\pi} \mathbf{m} \cdot \nabla_Q \left(\frac{1}{r} \right) \right]$$

we immediately see that the magnetic scalar potential of a magnetic dipole may be defined by

$$V_m = \frac{\mu_0}{4\pi} \mathbf{m} \cdot \nabla_Q \left(\frac{1}{r} \right) = -\frac{\mu_0}{4\pi} \mathbf{m} \cdot \nabla_P \left(\frac{1}{r} \right) \qquad (7\text{--}85)$$

V_m is positive at all points on the 'north' side of the loop and negative at all points on the 'south' side of the loop.

Consider now any closed circuit C around which a current I is flowing. Let S be the area of an arbitrary surface bounded by C and suppose that this area is divided up into a multitude of elementary areas as shown in Fig. 7.37. Each element of area dS may be regarded as

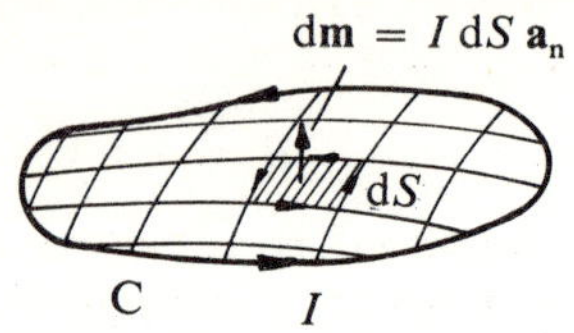

Fig. 7.37 Equivalence of current loop and magnetic shell

bounded by a small current loop carrying a current I in the same sense as the current in the external circuit C. Since the edge of any interior element is common to two adjacent loops, it is clear that the summation of currents in all element edges will leave only the original current I flowing around C. By choosing the elements dS infinitesimally small, each element may be considered as a magnetic point dipole of moment $\mathbf{dm} = I\,dS\mathbf{a_n}$ and the whole surface S as a magnetic dipole sheet or magnetic shell of strength (magnetic moment per unit area) $I\mathbf{a_n}$ (or $NI\mathbf{a_n}$ if loop C has N turns). The magnetic scalar potential at any point P outside the shell is given by (see Fig. 7.38)

$$V_m = \frac{\mu_0}{4\pi} \int_S I\,dS\mathbf{a_n} \cdot \nabla_Q\left(\frac{1}{r}\right)$$

$$= \frac{\mu_0}{4\pi} \int_S \frac{I\,dS\mathbf{a_n} \cdot \mathbf{a_r}}{r^2} = \frac{\mu_0 I}{4\pi} \int \frac{dS\cos\theta}{r^2}$$

Now $dS\cos\theta/r^2$ is the solid angle $d\Omega$ subtended by the element dS at the point P. $d\Omega$ is positive if dS is external with respect to point P— i.e., the normal points away from point of observation P and is negative if dS is internal with respect to P. Since the potential must be positive on the north side of the loop and negative on the south side it is obvious that

$$V_m = -\frac{\mu_0 I}{4\pi} \int d\Omega = -\frac{\mu_0 I}{4\pi}\Omega \qquad (7\text{–}86a)$$

where Ω is the solid angle subtended at P by the circuit C. It is sometimes convenient to rewrite equation 7–86a as

$$V_m = \pm\frac{\mu_0 I}{4\pi}|\Omega| \qquad (7\text{–}86b)$$

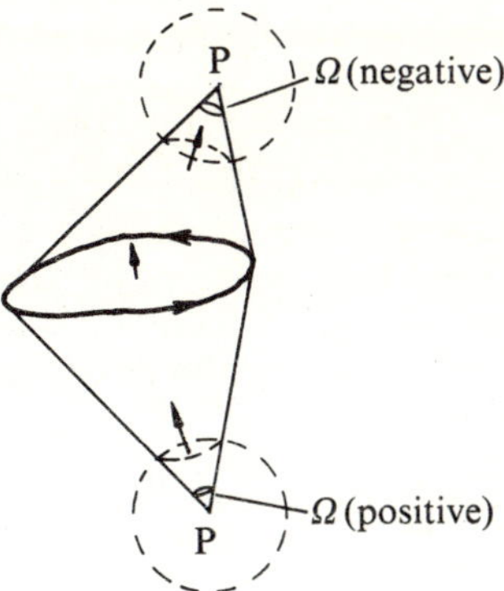

Fig. 7.38 Solid angle subtended by circuit at a point P lying on the positive or negative side of the circuit

in which the choice of the sign depends upon whether the point P lies on the positive or on the negative side of the loop. Equation 7–86a is analogous to equation 2–100 for the potential of an electric double layer of uniform strength.

Since the solid angle subtended by a surface at a point is discontinuous by 4π as the point crosses the surface, it follows that both Ω and V_m are not single-valued functions of position. To show this explicitly, let P_1 and P_2 be two points that are just above and just below the

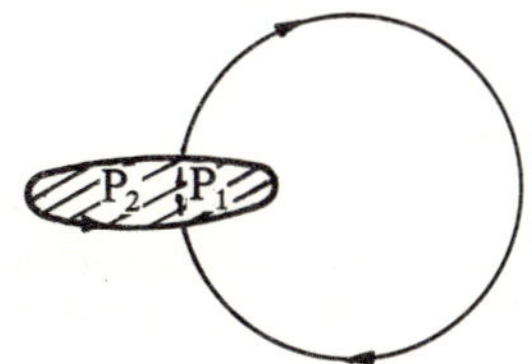

Fig. 7.39 Points P_1 and P_2 just above and just below the magnetic shell enclosed by a current loop

magnetic shell enclosed by a plane current loop C (Fig. 7.39). We have that

$$\text{at } \mathbf{P_1}, \quad \Omega_1 = -2\pi, \quad V_{m1} = \tfrac{1}{2}\mu_0 I$$

$$\text{at } \mathbf{P_2}, \quad \Omega_2 = +2\pi, \quad V_{m2} = -\tfrac{1}{2}\mu_0 I$$

Therefore in going *from* point P_2 *to* P_1 the magnetic potential changes by $\mu_0 I$. In general the potential at some point P will be given by

$$V_m = \frac{\mu_0 I}{4\pi}(-\Omega + 4\pi n) \tag{7–87}$$

where Ω is the principal value at P and n is an integer whose value is equal to the number of times we have encircled the current loop to return to the point P. n is positive if the coil is encircled in the same direction as the lines of B and negative if in the opposite direction.

The fact that the magnetic potential changes discontinuously by $\mu_0 I$ when we encircle a current loop once, may be regarded as a confirmation of Ampère's circuital law. Taking the line integral of **B** from point P_1 round to point P_2—i.e., in the direction of **B**—we have from equations 7–82 and 7–86 that

$$\int_{P_1}^{P_2} \mathbf{B} \cdot \mathbf{dl} = \frac{\mu_0 I}{4\pi} \int_{\Omega_1}^{\Omega_2} \nabla\Omega \cdot \mathbf{dl}$$

Since Ω varies regularly over the whole integration interval we may rewrite the above equation as

$$\int_{P_1}^{P_2} \mathbf{B} \cdot \mathbf{dl} = \frac{\mu_0 I}{4\pi} [\Omega_2 - \Omega_1]$$

Keeping the points P_1 and P_2 on either side of the shell we may nevertheless assume that they are infinitesimally close together; we then have $\Omega_1 = -2\pi$, $\Omega_2 = +2\pi$ and the path from P_1 round to P_2 may be considered closed so that

$$\oint \mathbf{B} \cdot \mathbf{dl} = \frac{\mu_0 I}{4\pi} [2\pi - (-2\pi)] = \mu_0 I$$

which is Ampère's circuital law.

Equation 7–81 made it possible for us to define **B**, in a region where the current density is zero, as the gradient of a magnetic scalar potential V_m. We have shown that V_m is not a single-valued function (equation 7–87); the magnetic flux density on the other hand will always be single-valued because the addition of a constant value to Ω will not affect its gradient. We now ask: is it possible to make V_m single-valued? The answer is yes, by considering equation 7–80. This equation requires that the region throughout which V_m is defined must be so

Fig. 7.40 Possible barrier surfaces or cuts for a long single conductor and for a current loop

limited that it is not possible to find a closed line integral of **B** which links any current. This requirement may be met by introducing hypothetical barrier surfaces or 'cuts' through which no integration path is allowed to pass. A barrier surface must always include the surfaces of all conductors present. Figure 7.40 shows possible barrier surfaces in the case of a single long conductor and of a current loop. All the dotted paths satisfy equation 7–80, whilst the regions enclosed by these paths and the regions between the paths and the barrier surface satisfy equation 7–81. The values of V_m at points lying just above and just below any barrier differ by $\mu_0 I$.

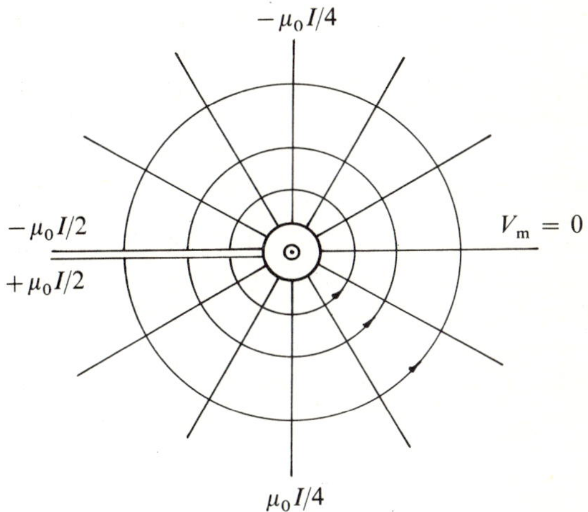

Fig. 7.41 Scalar magnetic potential lines (radial) and magnetic flux lines for an infinitely long cylindrical conductor carrying a current I

As an example consider an infinitely long cylindrical conductor carrying a current I (Fig. 7.41). The magnetic flux density at any point outside the conductor is,

$$\mathbf{B} = \frac{\mu_0 I}{2\pi\rho}\,\mathbf{a}_\phi$$

Applying equation 7–79 we have

$$-\nabla V_m = -\frac{1}{\rho}\frac{\partial V_m}{\partial \phi}\,\mathbf{a}_\phi = \frac{\mu_0 I}{2\pi\rho}\,\mathbf{a}_\phi$$

Hence

$$\frac{\partial V_m}{\partial \phi} = -\frac{\mu_0 I}{2\pi}$$

and

$$V_{\mathrm{m}} = -\frac{\mu_0 I}{2\pi}\,\phi$$

The constant of integration has been chosen as zero, that is $V_{\mathrm{m}} = 0$ when $\phi = 0$. Clearly V_{m} will decrease linearly with increasing ϕ so that the potential at any point will be multi-valued. In order to make V_{m} single-valued, we may choose as our barrier surface the plane $\phi = \pi$ so that the conductor cannot be encircled; the potential is then single-valued and is given by

$$V_{\mathrm{m}} = -\frac{\mu_0 I}{4\pi}\,\phi, \quad (-\pi \le \phi \le \pi) \tag{7-88}$$

Note that from one side of the barrier to the other the potential changes discontinuously by $\mu_0 I$. The scalar magnetic potential lines are radial and the magnetic flux lines are concentric circles about the current. If we compare this problem with that of an infinite line of electrostatic charge we see that the flux and the equipotential lines have simply interchanged places (see also equation 7–68). This is true for any comparable arrangement of linear currents and charges (principle of duality) and Figs. 7.42 and 7.43 illustrate this for linear currents or charges.

As an example of the use of the magnetic scalar potential we will recalculate, using equation 7–86b, the magnetic flux density on the

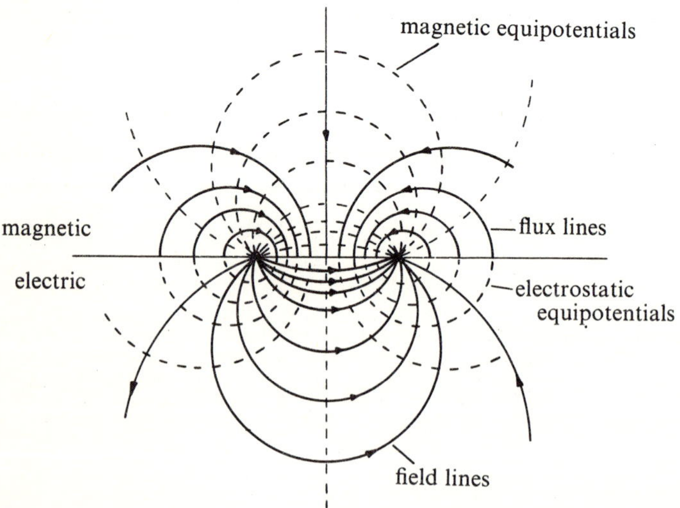

Fig. 7.42 The principle of duality for line currents and line charges of opposite polarities

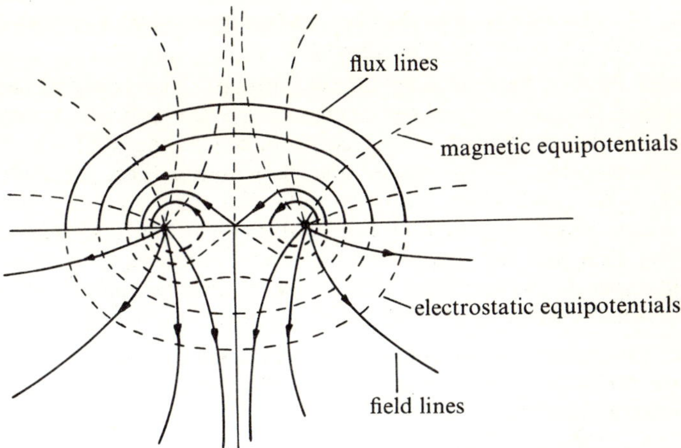

Fig. 7.43 The principle of duality for line currents and line charges of similar polarities

axis of a plane circular coil of radius a (Fig. 7.13). The solid angle subtended by the coil at P is

$$|\Omega| = 2\pi(1-\cos\alpha) = 2\pi(1-z/r)$$

Hence

$$V_{\mathrm{m}} = \pm\tfrac{1}{2}\mu_0 I(1-z/r)$$

The sign of V_{m} will depend upon the direction of flow of current around the loop relative to the point P. For the direction shown in the figure it is positive (Ω is negative). Hence

$$\mathbf{B} = -\nabla V_{\mathrm{m}} = \tfrac{1}{2}\mu_0 I\nabla(z/r)$$

and

$$B_z = \frac{\mu_0 I a^2}{2(z^2+a^2)^{3/2}}$$

which is in agreement with equation 7–33a.

What we have shown so far in this chapter is essentially that given a certain distribution of current, we have four methods by which we can determine the magnetic flux density $\mathbf{B}$ at any point in space and hence the configuration of the magnetic field. These methods are

(a) by using Ampère's circuital law,
(b) directly from the Biot–Savart law,
(c) by calculating the magnetic vector potential $\mathbf{A}$ and then determining $\mathbf{B}$ from the relationship $\mathbf{B} = \nabla \times \mathbf{A}$,

(d) by calculating the scalar vector potential V_m and then determining $\mathbf{B}$ from the relationship $\mathbf{B} = -\nabla V_m$.

Although the first method is by far the simplest, it can only be used in cases where the geometry of the configuration has sufficient symmetry so that the determination of the components of the field which are present is possible. No particular one of the other three methods can be said to be easier than the other two. All three methods involve integrations which may or may not be difficult to perform depending upon the data and the requirements of the problem. The choice of method depends on the nature of the current distribution as well as on the individual's judgement. The following points are worth noting:

(a) The determination of V_m involves a scalar integral which, in simple problems, is usually easier to evaluate than the vector integral required to determine $\mathbf{A}$.

(b) V_m is multi-valued unless an artificial surface barrier is used to render it single-valued. On the other hand $\mathbf{A}$ is everywhere single-valued.

(c) V_m cannot be found inside current-carrying regions whereas $\mathbf{A}$ exists both inside and outside such regions.

(d) V_m and $\mathbf{A}$ are two convenient independent mathematical representations for the magnetic field but neither representation has any physical significance.

7.14 FORCE AND TORQUE ON A CURRENT LOOP

Consider a filamentary loop of arbitrary shape carrying a current I and placed in a region of uniform magnetic flux density $\mathbf{B}$. The force acting on any element of the loop is by equation 7–13

$$\mathrm{d}\mathbf{F} = I\,\mathrm{d}\mathbf{l} \times \mathbf{B}$$

and the total force is therefore

$$\mathbf{F} = I \oint \mathrm{d}\mathbf{l} \times \mathbf{B}$$

Since $\mathbf{B}$ is uniform, and therefore independent of position, it can be removed from under the integral sign and we have that

$$\mathbf{F} = -I\mathbf{B} \times \oint \mathrm{d}\mathbf{l} = 0$$

so that the resultant force on a current loop in a *uniform* magnetic field is zero. This result is valid for volume distributions of current as well; since the current is stationary, the current distribution may be resolved into filaments all of which close upon themselves within the volume. As the force on each closed filament contour is zero, the resultant force will be zero. More generally, therefore, we may state that when any closed circuit carrying a stationary current is placed in an external uniform magnetic field, the resultant force exerted on the circuit by this field is zero.

In order to find the torque acting on a current loop placed in a uniform magnetic field, we will make use of the equivalence of an elementary current loop and a magnetic dipole. By analogy with electrostatics we may write the torque on the loop as (see equation 2–97),

$$\mathrm{d}T = |\mathbf{dm} \times \mathbf{B}| = I|\mathbf{dS} \times \mathbf{B}| = I\,\mathrm{d}SB\sin\theta \qquad (7\text{–}89)$$

This torque tends to align the axis of the loop with the direction of the external field $\mathbf{B}$, such that the sense of the field produced by the loop itself and the sense of $\mathbf{B}$ coincide ($\theta = 0$ in Fig. 7.44). For a current

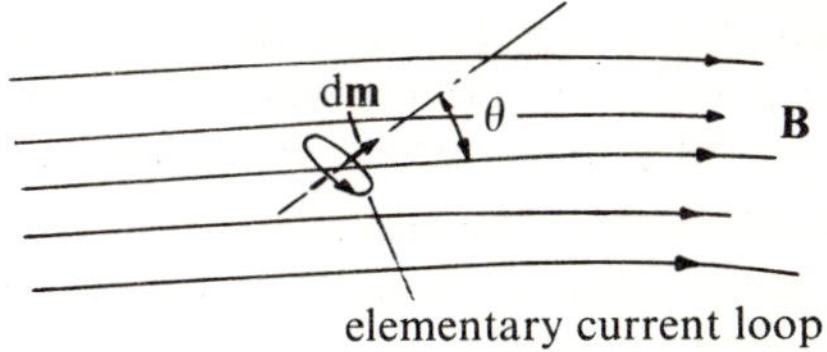

Fig. 7.44 Elementary current loop in an external magnetic field

loop of non-infinitesimal dimensions we have already shown that, by using any surface bounded by the circuit formed by the loop, the loop may be replaced by an infinitely large number of elementary loops or dipoles. The torque on the loop may be obtained therefore by integrating equation 7–89 over the surface S (which it will be recalled, may be arbitrarily chosen) spanning the loop. This gives

$$\mathbf{T} = \int_{S} I\,\mathbf{dS} \times \mathbf{B} = \left(\int_{S} I\,\mathbf{dS}\right) \times \mathbf{B}$$

since $\mathbf{B}$ is uniform. Now the magnetic moment of the current loop is given by

$$\mathbf{m} = \int_{S} I\,\mathbf{dS}$$

so that

$$\mathbf{T} = \mathbf{m} \times \mathbf{B} \qquad (7\text{–}90)$$

A further discussion of forces and torques will be found in Chapter 10.

7.15 SELF- AND MUTUAL INDUCTANCE*

The magnetic flux density $\mathbf{B}$ at any point in space is, according to the Biot–Savart law, directly proportional to the current which is the

* Alternative definitions of self- and mutual inductance are given in Chapters 9 and 10.

source of **B**. It follows that the magnetic flux Φ enclosed by any circuit C_1 is also proportional to the current producing that flux. If the current is flowing in the circuit itself then the flux linking it may be written as

$$\Phi = LI \tag{7-91}$$

where the constant of proportionality L is called the *self-inductance* of the circuit. Thus

$$L = \Phi/I \quad \text{(weber ampere}^{-1} = \text{henry)} \tag{7-92a}$$

If the circuit consists of N closely wound turns such that the same flux links all the turns then

$$L = N\Phi/I \tag{7-92b}$$

The self-inductance of a circuit may therefore be defined as the magnetic flux which links the circuit when a current of one ampere flows through it. The SI unit of inductance is the *henry*.

If the current is not flowing in the circuit itself, but in some neighbouring circuit C_2 say, then the flux linking circuit C_1 is given by

$$\Phi_{12} = M_{12}I_2 \tag{7-93}$$

and

$$M_{12} = \frac{\Phi_{12}}{I_2} \quad \text{(henry)} \tag{7-94a}$$

Where Φ_{12} is the flux linking circuit 1 and due to the current I_2 in circuit 2 (Fig. 7.45). The constant of proportionality M_{12} is defined as

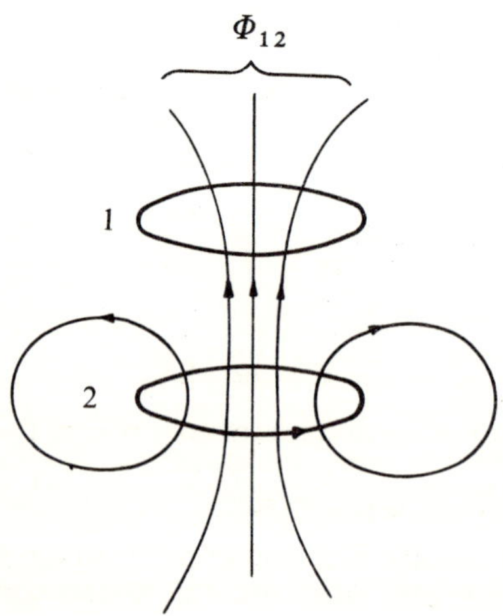

Fig. 7.45 Flux linking circuit 1 due to current in circuit 2

the *mutual inductance* between circuit 1 and circuit 2. If we interchange the role of the two circuits we have that

$$M_{21} = \frac{\Phi_{21}}{I_1} \qquad (7\text{-}94b)$$

It will be shown below that $M_{12} = M_{21} = M$; this simply means that the flux linking circuit 1 due to a unit current in circuit 2 is equal to the flux linking circuit 2 due to a unit current in circuit 1. This equality, however, is only true if the magnetic permeability μ is constant throughout space. Unless otherwise stated we shall assume that our medium is free space.

For the case in which circuit 1 has N_1 closely wound turns and circuit 2 has N_2 closely wound turns, it is evident that

$$N_1\Phi_{12} = M_{12}I_2$$
$$N_2\Phi_{21} = M_{21}I_1$$

and

$$M = \frac{N_1\Phi_{12}}{I_2} = \frac{N_2\Phi_{21}}{I_1} \qquad (7\text{-}94c)$$

The mutual inductance is considered positive if the currents I_1 and I_2 have the same direction; that is, if the direction of the mutual flux is the same as that of the self flux. The self-inductance on the other hand is always positive.

7.15.1 INDUCTANCE FORMULAS

We shall initially assume that all circuits are filamentary; that is, the cross-sectional dimensions of the conductors forming the circuits are negligibly small in comparison with all other dimensions. We shall then discuss the more general case in which the circuit conductors have a finite area, which is of course the actual case.

(a) Filamentary circuits. Consider two circuits 1 and 2 which carry currents I_1 and I_2 respectively and which are in a non-magnetic medium. The total flux Φ_{12} which links circuit 1 due to the current I_2 in circuit 2 is

$$\Phi_{12} = \int_{S_1} \mathbf{B}_2 \cdot d\mathbf{S}$$

where S_1 is any surface bounded by the contour C_1. Making use of equation 7–61 we may write

$$\Phi_{12} = \oint_{C_1} \mathbf{A}_2 \cdot d\mathbf{l}_1 \qquad (7\text{-}95)$$

where $\mathbf{A}_2$ is the magnetic vector potential at the element $d\mathbf{l}_1$. According to equation 7–59 this potential is given by

$$\mathbf{A}_2 = \frac{\mu_0 I_2}{4\pi} \oint_{C_2} \frac{d\mathbf{l}_2}{r_{12}} \tag{7-96}$$

where the integral is taken around circuit 2. Thus we may write

$$\Phi_{12} = \frac{\mu_0 I_2}{4\pi} \oint_{C_1} \oint_{C_2} \frac{d\mathbf{l}_1 \cdot d\mathbf{l}_2}{r_{12}} \tag{7-97}$$

Since the double integral is completely symmetrical with respect to circuits 1 and 2, it follows that the flux Φ_{21} which links circuit 2 due to a current I_1 in circuit 1 is given by

$$\Phi_{21} = \frac{\mu_0 I_1}{4\pi} \oint_{C_1} \oint_{C_2} \frac{d\mathbf{l}_1 \cdot d\mathbf{l}_2}{r_{12}} \tag{7-98}$$

From equations 7–94, 7–97 and 7–98 it follows therefore that

$$M_{12} = M_{21} = M = \frac{\mu_0}{4\pi} \oint_{C_1} \oint_{C_2} \frac{d\mathbf{l}_1 \cdot d\mathbf{l}_2}{r_{12}} \tag{7-99a}$$

This equation is known as Neumann's formula for the mutual inductance; it is valid for free space or for a medium whose permeability is constant and the same everywhere throughout the space in which the circuits are placed. The algebraic sign of M will depend upon the direction of integration around C_1 and C_2; equation 7–99 will automatically give the appropriate sign if the sense of the line elements $d\mathbf{l}_1$ and $d\mathbf{l}_2$ is taken in the direction of the respective currents I_1 and I_2. If circuits 1 and 2 have N_1 and N_2 filamentary closely wound turns respectively then

$$M = \frac{\mu_0}{4\pi} N_1 N_2 \oint_{C_1} \oint_{C_2} \frac{d\mathbf{l}_1 \cdot d\mathbf{l}_2}{r_{12}} \tag{7-99b}$$

Although it is in general difficult to calculate the mutual inductance by Neumann's formula, the formula does show explicitly that for filamentary circuits the mutual inductance is a purely geometrical quantity; that is, it depends only on the shapes, dimensions and relative positions of the circuits of the system. Neumann's formula also shows explicitly the reciprocal property of mutual inductance.

If we have an isolated circuit carrying a current I, then it is reasonable to assume, in the first instance, that the flux which links the circuit due to its proper current will be given by an equation analogous to equation 7–97; that is,

$$\Phi = \frac{\mu_0 I}{4\pi} \oint_{C_1} \oint_{C_2} \frac{d\mathbf{l}_1 \cdot d\mathbf{l}_2}{r_{12}} \tag{7-100}$$

where r_{12} is now the distance between two elements $\mathbf{dl}_1$ and $\mathbf{dl}_2$ on the same circuit. However, $1/r_{12}$ becomes infinite when the two elements coincide and hence the integral becomes infinite. The reason for this is due to our initial mathematical assumption that the circuits we were considering are filamentary ones and hence of negligible cross-sectional area. Since in practice all circuits have a finite cross-section, they will always have a finite self-flux linkage and hence a finite self-inductance.

(b) Non-filamentary circuits. When the conductors of the circuits have finite cross-sections, the main problem which arises is how to calculate the flux which links a given circuit, for in this case we have no precise contour C and hence no precise surface S for calculating the flux. To overcome this difficulty we suppose that each circuit is divided into a very large number of parallel elementary current tubes such that the current in the ith tube is

$$\Delta I_i = \mathbf{j}_i \cdot \Delta \mathbf{S}_i$$

and each tube is then considered as a filamentary circuit C_i. To calculate the mutual inductance of two non-filamentary circuits A and B carrying currents I_A and I_B respectively, we divide each circuit

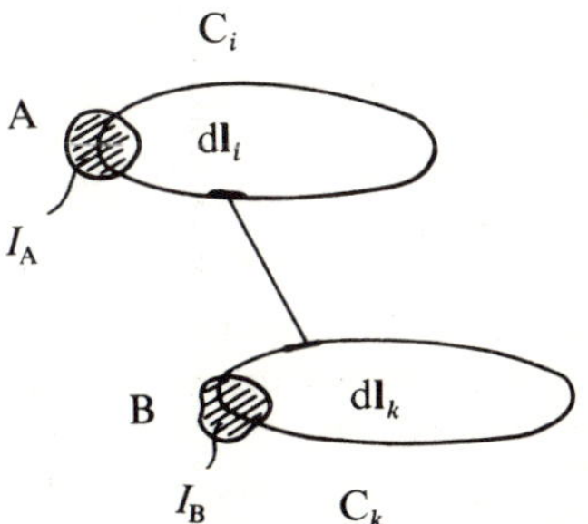

Fig. 7.46 Two non-filamentary current loops

into filamentary current tubes (Fig. 7.46). The flux linking circuit C_i of loop A due to the current ΔI_k in circuit C_k of loop B is given by

$$\Phi_{ik} = \frac{\mu_0}{4\pi} \Delta I_k \oint_{C_i} \oint_{C_k} \frac{\mathbf{dl}_i \cdot \mathbf{dl}_k}{r_{ik}}$$

The flux linking C_i due to all current-tube elements in B is

$$\Phi_{iB} = \frac{\mu_0}{4\pi} \sum_{k=1}^{N} \Delta I_k \oint_{C_i} \oint_{C_k} \frac{\mathbf{dl}_i \cdot \mathbf{dl}_k}{r_{ik}} \tag{7-101}$$

The optimal value of the flux Φ_{AB} linking all current-tube elements in A is given by the weighted average of the flux Φ_{iB}; thus

$$\sum_i \Phi_{iB}\Delta I_i = \Phi_{AB} \sum_i \Delta I_i = \Phi_{AB}I_A$$

and

$$\Phi_{AB} = \frac{1}{I_A} \sum \Phi_{iB}\Delta I_i$$

Substituting for Φ_{iB} from equation 7–101 we have that

$$\Phi_{AB} = \frac{\mu_0}{4\pi} \frac{1}{I_A} \sum_i^N \Delta I_i \sum_k^N \Delta I_k \oint_{C_i} \oint_{C_k} \frac{dl_i \cdot dl_k}{r_{ik}}$$

As the number of current-tubes N tends to infinity the summations over all tubes may be substituted by integrals over the cross-sections of the conductors. Thus

$$\Phi_{AB} = \frac{\mu_0}{4\pi} \frac{1}{I_A} \int_{S_A} j_A \cdot dS \int_{S_B} j_B \cdot dS \oint_{C_i} \oint_{C_k} \frac{dl_i \cdot dl_k}{r_{ik}} \qquad (7\text{–}102)$$

and since by equation 7–3 we have that

$$j \cdot dS\, dl = j\, dv$$

we may rewrite equation 7–102 as

$$\Phi_{AB} = \frac{\mu_0}{4\pi} \frac{1}{I_A} \int_{v_A} \int_{v_B} \frac{j_A \cdot j_B}{r} dv_A\, dv_B \qquad (7\text{–}103)$$

where j_A is the current density at some point inside conductor A, j_B is the current density at some point inside conductor B, r is the distance between the two points and dv_A and dv_B are elements of volume of conductors A and B respectively. The mutual inductance between the two circuits is given by

$$M = \frac{\Phi_{AB}}{I_B} = \frac{\mu_0}{4\pi} \frac{1}{I_A I_B} \int_{v_A} \int_{v_B} \frac{j_A \cdot j_B}{r} dv_A\, dv_B \qquad (7\text{–}104)$$

The self-flux linking a circuit whose conductor has a finite cross-section and through which a current I is flowing is still given by an equation of the form of equation 7–103; thus

$$\Phi = \frac{\mu_0}{4\pi} \frac{1}{I} \int_v \int_v \frac{j_1 \cdot j_2}{r} dv_1\, dv_2 \qquad (7\text{–}105)$$

where the subscripts 1 and 2 refer to any two distinct elements of the same circuit, r is the distance between these elements and v is the volume of the conductor. The self-inductance of the circuit is thus given by

$$L = \frac{\mu_0}{4\pi}\frac{1}{I^2}\int_v\int_v \frac{\mathbf{j}_1\cdot\mathbf{j}_2}{r}\,dv_1\,dv_2 \qquad (7\text{--}106)$$

The integral in this case is finite even though $1/r$ by itself becomes infinite. This is because integrals of the type

$$\int_v \frac{f(x',y',z')\,dv}{r}$$

where $f(x',y',z')$ is a function which is everywhere finite within v, are always finite. This can be easily shown. For a fixed point P inside v

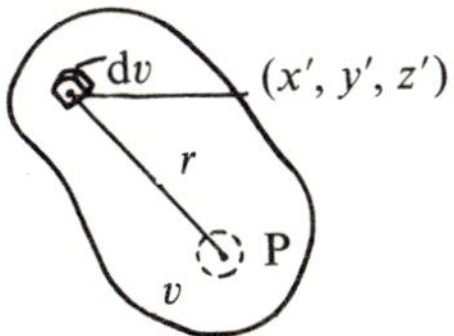

Fig. 7.47 The point P is surrounded by an infinitesimally small sphere of radius δ

(Fig. 7.47) we have a discontinuity at $r = 0$. We exclude this discontinuity by surrounding the point P by an infinitesimally small sphere of radius δ. Outside the sphere the integrand is everywhere finite. If the maximum value of f inside the sphere is f_m then clearly inside the sphere

$$\int \frac{f}{r}\,dv \le f_m \int_0^S 4\pi r^2\,dr/r$$

$$\le f_m 2\pi\delta^2$$

This tends to zero as $\delta \to 0$ so that the integral is infinite. In our case the function f is represented either by $\mathbf{j}_1$ or by $\mathbf{j}_2$. Making use of the magnetic vector potential as defined by equation 7–56, the self-inductance may be expressed as

$$L = \frac{1}{I^2}\int_v \mathbf{A}\cdot\mathbf{j}\,dv \qquad (7\text{--}107)$$

The formal equations 7–104 and 7–106 bring forth the important result that the inductances depend on the *current distribution* inside the conductors and since this distribution depends primarily upon frequency, the inductances—especially self-inductance—will be frequency dependent. However, it is clear from these equations that both

M and *L* are independent of the *magnitude* of the currents flowing since the respective expressions involve the ratios $\mathbf{j}_A/I_A$ and $\mathbf{j}_B/I_B$. Whenever the current is uniformly distributed over the cross-section of the conductors, that is for direct currents or alternating currents of low frequency, it is apparent that inductance is a purely geometric quantity. In Chapter 10 we shall present an alternative definition of *L* in terms of the magnetic energy stored in the field instead of in terms of flux linkage per ampere as above.

It is clear from equations 7–104 and 7–106 that the units for the magnetic constant μ_0 may be expressed as henry metre^{-1} (see Section 7.2).

In general the calculation of the mutual and self-inductance of circuits by means of equations 7–93, 7–104 and 7–106 involves, for all but the simplest of geometries, integrations which are very complex and often not possible to perform except by numerical methods. Whenever possible and whenever a high degree of accuracy is not essential it is much easier to use equations 7–92 and 7–94 to determine *M* and *L* respectively. We shall now give a few examples on the calculation of the mutual and self-inductance of a number of circuit configurations of practical importance.

7.15.2 MUTUAL INDUCTANCE OF TWO COAXIAL CIRCULAR FILAMENTS

Let the radii of the two filaments be a_1 and a_2 and let the planes of the coils be parallel and at a distance *z* apart (Fig. 7.48). By Neumann's formula the mutual inductance is given by

$$M = \frac{\mu_0}{4\pi} \oint_{C_1} \oint_{C_2} \frac{\mathrm{d}\mathbf{l}_1 \, \mathrm{d}\mathbf{l}_2 \cos\theta}{r_{12}}$$

where θ is the angle between the vector elements $\mathrm{d}\mathbf{l}_1$ and $\mathrm{d}\mathbf{l}_2$. Since the elements lie in parallel planes it is apparent from Fig. 7.48 that $\theta = \phi_1$.

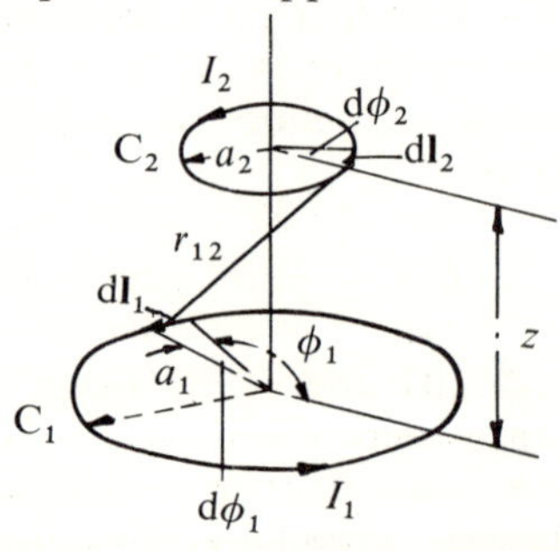

Fig. 7.48 To determine the mutual inductance between two coaxial circular filaments

Also from the figure we see that

$$dl_1 = a_1 d\phi_1; \quad dl_2 = a_2 d\phi_2$$

and

$$r_{12} = [z^2 + a_1^2 + a_2^2 - 2a_1 a_2 \cos\theta]^{1/2}$$

thus

$$M = \frac{\mu_0}{4\pi} a_1 a_2 \int_0^{2\pi} \int_0^{2\pi} \frac{\cos\phi_1}{r_{12}(\phi_1)} d\phi_1 d\phi_2$$

$$= \tfrac{1}{2}\mu_0 a_1 a_2 \int_0^{2\pi} \frac{\cos\phi_1}{r_{12}(\phi_1)} d\phi_1 \tag{7-108}$$

The evaluation of the above integral is not particularly simple in spite of the simple geometry of the configuration. We give here the result only which is

$$M = \mu_0 (a_1 a_2)^{1/2} \left[\left(\frac{2}{k} - k\right) F - \frac{2E}{k} \right] \tag{7-109}$$

where

$$k = \frac{4a_1 a_2}{z^2 + (a_1 + a_2)^2}$$

and F and E are the complete elliptic integrals of the first and second kinds respectively.* If the separation of the loops z is large compared with the radii a_1 and a_2, the mutual inductance is given to a first approximation by

$$M = \tfrac{1}{2}\mu_0 \pi (a_1 a_2)^2 / z^3 \tag{7-110a}$$

This result may be obtained either by using the binomial expansion of $1/r_{12}$ in equation 7–108, or more simply by using equation 7–94 and calculating the flux using equation 7–33a and assuming (since $z \gg a_1, a_2$) that the flux density due to coil 1 has a constant value over the area of coil 2. If one of the loops is able to rotate about an axis perpendicular to $O_1 O_2$ then equation 7–110a is replaced by

$$M = \tfrac{1}{2}\mu_0 \pi \frac{a_1^2 a_2^2}{z^3} \cos\theta \tag{7-110b}$$

In this case the coefficient of mutual inductance will be negative for values of θ between $\pi/2$ and π.

* See H. B. Dwight, *Tables of Integrals and Other Mathematical Data*, p. 180 (Macmillan, New York, 1961).

The expression between square brackets in equation 7–109 tends to infinity as $k \to 1$; this happens when $z \to 0$ and $a_1 \to a_2$. In this case the Neumann formula as given by equation 7–99 is no longer valid because the circuits may no longer be considered as filamentary and the finite diameter of the conductors must be taken into consideration by using equation 7–104 to calculate M.

7.15.3 MUTUAL INDUCTANCE BETWEEN A COIL AND AN INFINITE SOLENOID OR TOROID

The mutual inductance between an infinitely long solenoid and a coil of N_2 turns coaxial with the solenoid (Fig. 7.49) may be easily

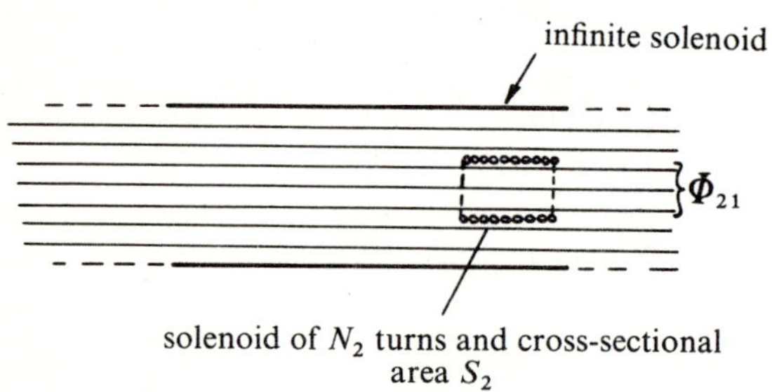

Fig. 7.49 To determine the mutual inductance between two solenoids

calculated using equation 7–91. From equation 7–35 we have that

$$\Phi_{12} = \mu_0 n I_1 S_2$$

where S_2 is the cross-sectional area of the coil. Hence

$$M = \mu_0 n N_2 S_2 \tag{7–111}$$

If instead of an infinite solenoid we have a toroidal coil (with non-magnetic core) linked by a secondary coil of N_2 turns—e.g., a search coil—the mutual inductance between the two circuits is given by

$$M = N_2 \Phi_{21}/I_1$$

where Φ_{21} is given by equations 7–43 or 7–44 or 7–45 depending upon the geometry of the toroid.

These two cases demonstrate clearly the usefulness of the reciprocal property of M; it would be much more difficult to calculate the flux of a current in the search coil which links the infinite solenoid or the toroid.

7.15.4 SELF-INDUCTANCE OF SOLENOIDS AND TOROIDS

The self-inductance per unit length of an infinitely long cylindrical solenoid having n closely wound turns per unit length and a cross-sectional radius a is given by (see equation 7–35) the self-inductance per unit length (L_u)

$$L_u = n\Phi/I = 4\pi^2 a^2 n^2 \times 10^{-7} \quad \text{henry metre}^{-1} \qquad (7\text{–}112)$$

For coils of finite length l, the above formula is multiplied by a correction factor $K(<1)$, known as the Nagaoka factor, to take into account the falling off of the field near the ends of the solenoid. The inductance is then given by

$$L = 4\pi^2 a^2 l n^2 K \times 10^{-7}$$
$$= 2\pi^2 a N^2 (2a/l) K \times 10^{-7} \quad \text{henry} \qquad (7\text{–}113)$$

where N is the total number of turns; the factor K is given in tables* as a function of $l/2a$.

The self-inductance of a toroid of rectangular cross-section is given by (see equation 7–43)

$$L = \frac{\mu_0 N^2 b}{2\pi} \log \frac{2R+d}{2R-d} \quad \text{henry} \qquad (7\text{–}114)$$

and of a toroid of circular cross-section by (see equation 7–44)

$$L = \mu_0 N^2 [R - \sqrt{(R^2 - r^2)}] \quad \text{henry} \qquad (7\text{–}115)$$

7.15.5 SELF-INDUCTANCE OF A LONG TWO-WIRE TRANSMISSION LINE

The line consists of two identical parallel cylindrical conductors each having a radius R and the conductor axes are at a distance D apart (Fig. 7.50). The conductors carry the same current I in opposite

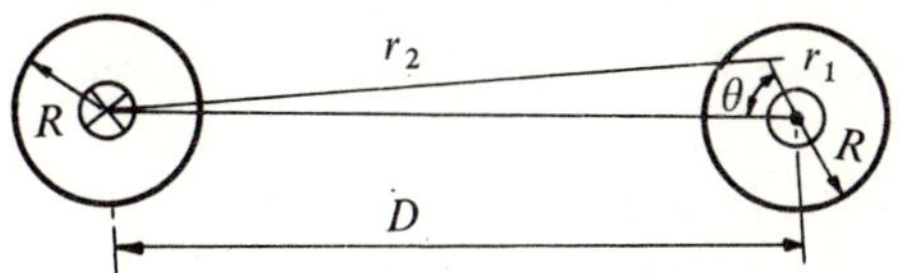

Fig. 7.50 Two-conductor transmission line

directions and we shall assume that this current is uniformly distributed over the conductor cross-section. We shall use equation

* See F. W. Grover, *Inductance Calculations*, p. 144 (Dover, New York, 1946).

7-104 to determine the self-inductance of this system. Thus

$$L = \frac{1}{I^2}\left[\int_{v_1} \mathbf{A}_1 \cdot \mathbf{j}_1 \, dv_1 + \int_{v_2} \mathbf{A}_2 \cdot \mathbf{j}_2 \, dv_2\right]$$

Since at all points inside any of the conductors $\mathbf{A}$ and $\mathbf{j}$ have the same direction (see Section 7.10) we have that

$$\mathbf{A}_1 = -\mathbf{A}_2 = \mathbf{A}$$

$$\mathbf{j}_1 = -\mathbf{j}_2 = \mathbf{j}$$

$$\mathbf{A}_1 \cdot \mathbf{j}_1 = \mathbf{A}_2 \cdot \mathbf{j}_2 = Aj$$

so that

$$L = \frac{2}{I^2}\int_v Aj \, dv$$

where $dv = dS \, dl$; if we let $dl = 1$, then the inductance per unit length of the two-wire line is given by

$$L_u = \frac{2}{I^2}\int_S Aj \, dS = \frac{2}{IS}\int_S A \, dS$$

since the current density is uniform. The magnetic vector potential at any point inside conductor 1 is by equation 7-70

$$A = \frac{\mu_0 I}{4\pi}\left[1 - \frac{r_1^2}{R^2} + 2\log\frac{r_2}{R}\right]$$

so that

$$L_u = \frac{\mu_0}{2\pi S}\int_0^R\int_0^{2\pi}\left[1 - \frac{r_1^2}{R^2} - 2\log R + 2\log r_2\right] r_1 \, dr_1 \, d\theta$$

$$= \frac{\mu_0}{2\pi S}\left\{[\tfrac{1}{2}\pi R^2 - 2\pi R^2 \log R] + \int_0^R\int_0^{2\pi} r_1 \log r_2^2 \, dr_1 \, d\theta\right\}$$

Let us denote the double integral term by $\mathcal{I}$. From Fig. 7.50 we see that

$$r_2^2 = r_1^2 + D^2 - 2r_1 D \cos\theta$$

$$= D^2\left[1 + \left(\frac{r_1}{D}\right)^2 - 2\left(\frac{r_1}{D}\right)\cos\theta\right]$$

so that

$$\mathcal{I} = \int_0^R\int_0^{2\pi}\left\{\log\left[1 + \left(\frac{r_1}{D}\right)^2 - 2\left(\frac{r_1}{D}\right)\cos\theta\right] + \log D^2\right\} r_1 \, dr_1 \, d\theta$$

The first integral is zero* so that

$$\mathscr{I} = 2\pi R^2 \log D$$

and

$$L_u = \frac{\mu_0}{2\pi S}\left[\tfrac{1}{2}\pi R^2 - 2\pi R^2 \log R + 2\pi R^2 \log D\right]$$

and since $S = \pi R^2$ we find that

$$L_u = \frac{\mu_0}{\pi}\left[\frac{1}{4} + \log\frac{D}{R}\right] \quad \text{henry metre}^{-1} \tag{7–116}$$

The self-inductance per unit length per conductor is,

$$L_{cu} = \frac{\mu_0}{8\pi} + \frac{\mu_0}{2\pi}\log\frac{D}{R} \quad \text{henry metre}^{-1} \tag{7–117}$$

The constant term $\mu_0/8\pi$ represents the internal inductance (due to internal flux linkages) of the conductor; for non-magnetic conductors it has the constant value of 0·05 microhenries per metre independent of the conductor radius. The logarithmic term represents the external inductance of the conductor (due to external flux linkages). It is apparent that L_u and L_{cu} tend to infinity as R tends to zero. That the first and second terms of equation 7–117 represent the internal and external inductance respectively will be explicitly demonstrated in Chapter 10.

Equation 7–116 is exact provided that the current is assumed to be uniformly distributed over the cross-section of the conductors. This condition is satisfied for direct and slowly alternating currents ($<100\,\text{Hz}$). At high frequencies the current is confined to a thin layer along the surface of the conductor (skin effect) so that the conductor no longer has an internal inductance. In this case

$$L_{cu} = \frac{\mu_0}{2\pi}\log\frac{D}{R} \quad \text{henry metre}^{-1} \tag{7–118}$$

which is also valid for thin-walled tubular conductors. Equation 7–118, however, is a good approximation only if $D \gg R$ because the skin current is not uniformly distributed around the circumference of the wire. In fact the variation of the current density around the circumference is analogous to the distribution of charges in the corresponding electrostatic problem (Section 3.7) and the exact expression for the

* H. B. Dwight, *Tables of Integrals and Other Mathematical Data*, 865.73 (Macmillan, New York, 1961).

inductance per unit length is given by*

$$L_u = \frac{\mu_0}{2\pi} \log\left\{ \frac{D}{2R} + \left[\left(\frac{D}{2R} \right)^2 - 1 \right]^{1/2} \right\} \quad \text{henry metre}^{-1} \qquad (7\text{–}119)$$

7.16 THE COEFFICIENT OF COUPLING

The coefficient of coupling K of two coils is defined as the ratio of the mutual inductance to the square root of the product of the self-inductances. Thus,

$$K = \frac{M}{\sqrt{(L_1 L_2)}} \qquad (7\text{–}120)$$

Consider two coils 1 and 2 having N_1 and N_2 closely wound turns respectively. The self-inductances of these coils are

$$L_1 = N_1 \Phi_1 / I_1 ; \quad L_2 = N_2 \Phi_2 / I_2$$

and the mutual inductance between the two coils is

$$M = N_1 \Phi_{12} / I_2 = N_2 \Phi_{21} / I_1$$

We therefore have that

$$K^2 = \frac{\Phi_{12} \Phi_{21}}{\Phi_2 \Phi_1}$$

The ratio Φ_{12}/Φ_2 is the fraction of the flux of coil 2 that links coil 1 and the ratio Φ_{21}/Φ_1 is the fraction of the flux of coil 1 that links coil 2. Since neither of these ratios can be greater than unity, it follows that

$$|K| \leq 1 \qquad (7\text{–}121)$$

and the mutual inductance between the two coils will be always equal to or less than the square root of the product of the self-inductances of the coils; thus

$$M \leq \sqrt{(L_1 L_2)} \qquad (7\text{–}122)$$

or

$$M = K\sqrt{(L_1 L_2)}$$

The condition $K = 1$ represents the case (never quite attained in

---

* It is a fact that if the frequency is so high that the current may be considered as confined to the conductor surfaces, then for any pair of parallel straight conductors of uniform but otherwise arbitrary cross-section and in free space, the product of inductance per unit length and capacitance per unit length is equal to $\epsilon_0 \mu_0 = 1/c^2$, where c is the velocity of light in free space. This relation is useful because exact capacitance formulae are usually known for most simple geometries.

practice) in which all the magnetic flux set up by any one of the coils links the other coil (maximum mutual inductance possible). The condition $K = 0$ represents the case in which none of the magnetic flux set up by one of the coils links the other ($M = 0$). If $K < 0\cdot5$ we say that the coils are loosely coupled; if $K > 0\cdot7$ we say that the coils are tightly coupled.

PROBLEMS

Chapter Seven

7.1   An infinitely long copper conductor is placed in a uniform horizontal magnetic field of strength 10^{-3} tesla; the field is perpendicular to the conductor axis. What must the current density in the conductor be in order to float the conductor in the earth's gravitational field? Density of copper $8\cdot9\,\mathrm{g\,cm^{-3}}$.

7.2   A square loop of side l carries a current I. Find the force acting on any one side of the loop and derive an expression for the vector potential $\mathbf{A}$ and the flux density $\mathbf{B}$ at points on the axis perpendicular to the plane of the loop and passing through its centre. Hence show that the flux density at the centre of the loop is $\mu_0 2\sqrt{(2)}I/\pi l$.

7.3   In the above problem find $\mathbf{B}$ at the centre of the loop by using the equation $\mathbf{B} = -\nabla V_{\mathrm{m}}$.

7.4   A current loop is in the form of a plane regular polygon of n sides inscribed in a circle of radius R. If the loop carries a current I show that the flux density at the centre of the loop is given by

$$B = \frac{\mu_0 In}{2\pi R} \tan \frac{\pi}{n}$$

From this expression derive B at the centre of a circular current loop of radius R.

7.5   Show that the integral from $-\infty$ to $+\infty$ of the flux density on the axis of a circular current loop carrying a current I is exactly $\mu_0 I$.

7.6   A three-phase transmission line consists of three very long parallel conductors lying in a horizontal plane with the outer conductors at a distance D from the central one. At a given instant the current in the central conductor is I and that in each of the outer conductors is $-\sqrt{(3)}I/2$. Find the force on each of the conductors.

7.7 Show that the flux inside a toroid of rectangular cross-section is given by equation 7–43.

7.8 Show that the flux inside a toroid of circular cross-section is given by equation 7–44.

7.9 Use any convenient argument or method to show that the field everywhere outside an infinitely long solenoid is zero.

7.10 Use equation 7–86 to find the magnetic flux density on the axis of (a) a short solenoid and (b) a long solenoid.

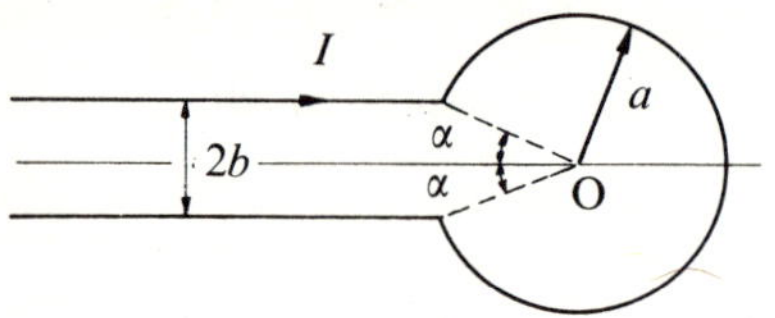

Fig. P. 7.11

7.11 For the loop shown in Fig. P7.11 show that the magnetic flux density at O is given by

$$B = \frac{\mu_0 I}{2\pi a}\left[(\pi - a) + (1 - \cos\alpha)/\sin\alpha\right]$$

7.12 Show that the magnetic flux density at some point P in the plane of a circular loop of radius a is given by

$$B = \frac{\mu_0 a I}{4\pi(a^2 - h^2)} E_{(h/a,\ 2\pi)}$$

where I is the current in the loop, h is the distance of the point P from the centre of the loop and $E_{(h/a,\ 2\pi)}$ is an elliptic integral of the second kind.

7.13 Two infinitely long line currents are located at $x = \pm a$, $y = 0$ with their axes parallel to the z-axis. If the currents are $\pm I$ derive the equation for the flux lines.

7.14 Two identical plane circular coils (Helmholtz coils) of radius a carry a current I in the same sense and are placed coaxially at a distance $2d$ apart. Plot the variation of the flux density along the common axis in the region between the two coils. Find the relationship between a and d for which the gradient of the axial field midway between the two coils is a minimum (maximum field uniformity).

7.15 An infinitely long conductor has a circular cross-section of radius R. The conductor has a cylindrical hole of radius r with its axis parallel to that of the conductor and at a distance d from its centre. If the conductor carries a uniform current density $\mathbf{j}$ show that the magnetic flux density inside the hole is uniform and given by $\mathbf{B} = \frac{1}{2}\mu_0\mathbf{j} \times \mathbf{d}$, where $\mathbf{d}$ is the vector distance from the centre of the conductor to the centre of the hole.

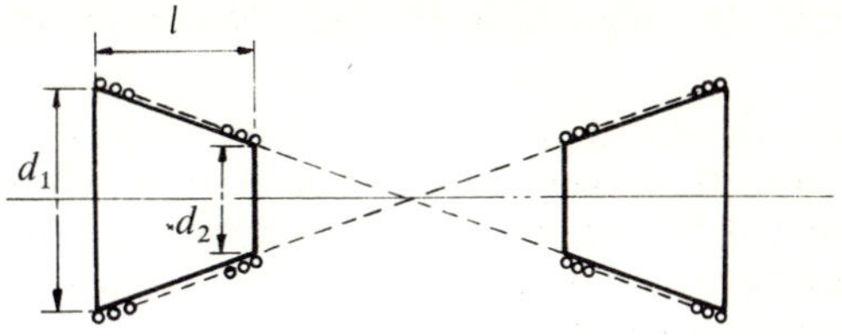

Fig. P. 7.16

7.16 Each of two identical coils consists of N turns uniformly wound on the frustum of a right circular cone (Fig. P7.16). If the coils carry a current I in the same direction, find the magnetic flux density at the common apex of the cones.

7.17 A thin wire is bent into the form of a circular helix of angle θ, radius a and N complete turns. If a current I flows through the wire show that the axial magnetic flux density at the centre of the helix is given by

$$B = \mu_0 NI/2a(1 + \pi^2 N^2 \tan^2 \theta)^{1/2}$$

7.18 An electron moves in a circular orbit with an angular velocity ω. Find the magnetic moment of the loop and the magnetic flux density at its centre.

7.19 A thin disc of radius a carries a uniform surface charge density and rotates about its axis with angular velocity ω. Show that the magnetic dipole moment of the disc is $\frac{1}{4}\sigma\omega\pi a^4$.

7.20 A thin disc of radius a and thickness t carries a uniform volume charge density ρ. If the disc rotates about its axis with an angular velocity $\boldsymbol{\omega}$, show that the magnetic flux density at the centre of the disc is $\frac{1}{2}\mu_0\rho t\omega a$.

7.21 A charge Q is uniformly distributed over the surface of a sphere of radius a. If the sphere is rotated about one of its diameters with an angular velocity ω, show that the rotating sphere is equivalent to a magnetic dipole at the centre of moment $\frac{1}{3}a^2\omega Q$. Show that the magnetic scalar potential at points inside and outside the rotating sphere are given by

$$-\mu_0\omega Qr \cos \theta/6\pi a \quad \text{and} \quad \mu_0\omega Qa^2 \cos \theta/12\pi r^2$$

respectively, where r is the distance from the centre of the sphere to the field point and θ is the angle between r and the axis of rotation.

7.22 An electric charge Q is uniformly distributed throughout the volume of a sphere of radius a. If the sphere rotates about one of its diameters with an angular velocity ω, find the magnetic moment of the sphere and the flux density at the centre of the sphere.

7.23 A filamentary planar rectangular loop of sides a and b ($b > a$) is placed with its longer sides parallel to an infinitely long thin conductor; the distance between the near side of the loop and the conductor is c. Show that the mutual inductance between the two circuits is given by

$$M = \frac{\mu_0 b}{2\pi} \log\left[1 + (a/c)\right]$$

7.24 Two magnetic dipoles $\mathbf{m}_1$ and $\mathbf{m}_2$ are at a distance r apart. Find the torque experienced by each dipole due to the field of the other and show that the force and couple on $\mathbf{m}_1$ are equal and opposite to the force and couple on $\mathbf{m}_2$.

7.25 Show that the kinetic energy of a charged particle moving in a magnetic field of flux density $\mathbf{B}$ is constant. If the charged particle enters a uniform magnetic field with a velocity $\mathbf{u}$ perpendicular to the field, show that the path of the particle is a helix.

7.26 An ion beam of radius R consists of positively charged particles of charge Q C and mass m kg, moving with a velocity u m s^{-1} parallel to the beam axis. Assuming that the current density is constant throughout the cross-section of the beam, find the radial acceleration of an ion at the periphery of the beam, taking into account the electric and magnetic fields produced by the beam. Is the assumption of uniform current density justifiable?

7.27 Two long coaxial cylinders of radii a and b ($b > a$) are placed with their axis parallel to a uniform magnetic field of density B. If a voltage V is now applied between the two cylinders show that a particle of charge q and mass m, initially at rest on the surface of the inner cylinder, will reach the anode at grazing incidence if

$$V = \frac{qB^2 b^2}{8m}\left(1 - \frac{a^2}{b^2}\right)^2$$

CHAPTER EIGHT

THE MAGNETIC FIELD IN THE PRESENCE OF MATERIAL BODIES; PERMANENT MAGNETS

In the preceding chapter we studied the magnetic fields produced by current carriers residing in vacuum. In the present chapter we shall see how the previous results may be modified to take into account the presence of material media. Although we are primarily interested in the behaviour of material media in a magnetic field from the macroscopic point of view, we have found it useful to include, later on in the chapter, an atomic or microscopic interpretation of the behaviour of a medium in a magnetic field. The models used are classical but they are sufficient to give the reader a reasonable amount of insight into the physical mechanisms which underlie most of the observed magnetic phenomena in matter.

It is well known that any material medium, whether gas, liquid or solid, is composed ultimately of atoms. These atoms have an electronic structure in which the electrons may be regarded as moving in planar orbits about a central nucleus; the electrons also rotate about their own axes (electron spin). Both the orbital and spin motions of the electrons constitute current loops of atomic dimensions; these loops are generally referred to as *amperian currents*. Each loop has a magnetic moment associated with it (see Section 7.11) and may appropriately be described, at all distances which are large compared to the largest dimension of the loop, as a magnetic dipole. It should be mentioned here that the only experimental evidence for the existence of amperian currents comes from the magnetic moments to which they give rise. The resultant magnetic moment of an atom is obtained by the vector combination of the orbital and spin magnetic moments. (The nucleus also has a magnetic moment but for most materials its contribution is negligibly small compared with the contributions of the orbital and spin moments.) The rules governing this combination are derived from quantum mechanics (Hund's rules) and will not be discussed here. It must be borne in mind, however, that the response of a medium to an external applied magnetic field is essentially determined by the magnitude of the resultant magnetic moment.

When the atoms or molecules of a medium have a resultant magnetic moment, the application of an external magnetic field tends to align these moments in the field direction such that the field is intensified

(Section 7.14). This effect, if very small, is known as *paramagnetism*; if the effect is very large it is known as *ferromagnetism*. The application of a magnetic field to any atom or molecule, which may or may not have a resultant magnetic moment, will always induce a component of magnetic moment in opposition to the direction of the field. This effect is known as *diamagnetism*.

A detailed account of dia-, para- and ferromagnetism, treated from the classical atomic point of view, will be found in Section 8.8.

8.1 THE MAGNETIZATION VECTOR M

In the analysis of material media from the macroscopic point of view, the existence of amperian currents and the magnetic moments (dipoles) associated with them are accounted for by introducing the concept of a magnetic moment per unit volume **M**, analogous to the polarization **P** in dielectrics. To do this, we restrict our attention to average values over volumes which are sufficiently small in comparison with the dimensions of the medium under consideration, but large enough to contain a sufficient number of atoms or molecules for the purpose of averaging. Thus the vector sum of the magnetic moments in an element of volume Δv is

$$\sum_{i=1}^{N\Delta v} \mathbf{m}_i = N\mathbf{m}\,\Delta v = \mathbf{M}\,\Delta v \qquad (8\text{--}1)$$

where **m** represents the average magnetic moment of each atomic current loop and N is the number of such loops (magnetic dipoles) per unit volume. The vector **M** is the magnetic moment per unit volume and is called the magnetic polarization density or simply the *magnetization* of the medium. The SI units of **M** are $A\,m^{-1}$. The macroscopic mathematical definition of **M** at a point is

$$\mathbf{M} = \lim_{\Delta v \to 0} \frac{1}{\Delta v} \sum \mathbf{m}_i = \lim_{\Delta v \to 0} \frac{\Delta \mathbf{m}}{\Delta v} = \frac{d\mathbf{m}}{dv} \qquad (8\text{--}2)$$

where $\Delta \mathbf{m}$ is the resultant magnetic moment in the volume Δv. It is clear that **M** is a vector point function. In general **M** will be a function of the macroscopic **B**-field. Any body for which **M** is not zero everywhere is said to be *magnetized*.

Because of the mathematical equivalence of an infinitesimal current loop and a magnetic dipole, we have two possible ways of representing a magnetized medium; the 'amperian-current' model and the 'equivalent pole' model; the former model uses the magnetic vector potential formulation whilst the latter uses the scalar potential formulation. We shall now give a complete analysis of each of these models in turn and then assess the relative merits of each.

8.2 AMPERIAN-CURRENT MODEL

Consider a finite volume v of magnetized matter. Assume that the magnetization is non-uniform and let its value at any point $Q(x', y', z')$ be $\mathbf{M}$ (Fig. 8.1). The magnetic vector potential at any point $P(x, y, z)$ outside v and at a distance r from Q is given by (equation 7–73),

$$d\mathbf{A}_M = -\frac{\mu_0}{4\pi} \mathbf{M} \times \nabla_P\left(\frac{1}{r}\right) dv$$

or

$$\mathbf{A}_M = \frac{\mu_0}{4\pi} \int_v \mathbf{M} \times \nabla_Q\left(\frac{1}{r}\right) dv \qquad (8\text{--}3)$$

By using the vector identity

$$\nabla \times (\mathbf{M}/r) = (1/r)\nabla \times \mathbf{M} + \nabla(1/r) \times \mathbf{M}$$

equation 8–3 may be written as

$$\mathbf{A}_M = \frac{\mu_0}{4\pi} \int \left[\frac{\nabla_Q \times \mathbf{M}}{r} - \nabla_Q \times \left(\frac{\mathbf{M}}{r}\right)\right] dv \qquad (8\text{--}4)$$

and making use of the relation*

$$-\int_v \nabla_Q \times (\mathbf{M}/r)\, dv = \oint_S (\mathbf{M}/r) \times d\mathbf{S} \qquad (8\text{--}5)$$

---

* In Gauss's theorem

$$\oint_S \mathbf{G} \cdot d\mathbf{S} = \int_v \nabla \cdot \mathbf{G}\, dv$$

let $\mathbf{G} = \mathbf{F} \times \mathbf{C}$ where $\mathbf{C}$ is a constant vector. Hence

$$\oint_S (\mathbf{F} \times \mathbf{C}) \cdot d\mathbf{S} = \int_v \nabla \cdot (\mathbf{F} \times \mathbf{C})\, dv$$

$$= \int_v [\mathbf{C} \cdot (\nabla \times \mathbf{F}) - \mathbf{F} \cdot (\nabla \times \mathbf{C})]\, dv$$

$$= \int_v \mathbf{C} \cdot \nabla \times \mathbf{F}\, dv$$

Since the L.H.S. is a scalar triple product, the vector can be cyclically permuted and we have that

$$\oint_S \mathbf{C} \cdot d\mathbf{S} \times \mathbf{F} = \int_v \mathbf{C} \cdot \nabla \times \mathbf{F}\, dv$$

$$\mathbf{C} \cdot \oint_S d\mathbf{S} \times \mathbf{F} = \mathbf{C} \cdot \int_v \nabla \times \mathbf{F}\, dv$$

and since $\mathbf{C}$ is constant it follows that

$$\oint_S \mathbf{F} \times d\mathbf{S} = -\int_v \nabla \times \mathbf{F}\, dv$$

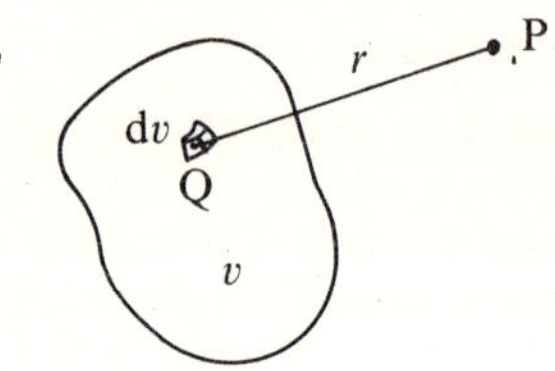

Fig. 8.1 Finite volume v of magnetized matter

we obtain

$$\mathbf{A}_M = \frac{\mu_0}{4\pi} \int_v \frac{\nabla_Q \times \mathbf{M}}{r}\, dv + \frac{\mu_0}{4\pi} \oint_S \left(\frac{\mathbf{M}}{r}\right) \times d\mathbf{S} \qquad (8\text{-}6)$$

where S is the surface bounding the volume v. This equation may now be written as

$$\mathbf{A}_M = \frac{\mu_0}{4\pi} \int_v \frac{\mathbf{j}_M}{r}\, dv + \frac{\mu_0}{4\pi} \oint_S \frac{\mathbf{K}_M}{r}\, d\mathbf{S} \qquad (8\text{-}7)$$

provided that we set

$$\nabla_Q \times \mathbf{M} = \mathbf{j}_M \qquad (8\text{-}8)$$

and

$$\mathbf{M} \times \mathbf{a}_n = \mathbf{K}_M \qquad (8\text{-}9)$$

where $\mathbf{a}_n$ is the unit vector along the outward normal to the surface S. Thus the total magnetic vector potential at the point P is equivalent to the sum of the potentials produced by a volume current density $\mathbf{j}_M$ and a surface current density $\mathbf{K}_M$ (cf. equations 7–53 and 7–57). The current densities $\mathbf{j}_M$ and $\mathbf{K}_M$ are referred to as the amperian current densities to distinguish them from the conduction current densities $\mathbf{j}$ and $\mathbf{K}$. By analogy with the phenomenon of polarization in dielectrics we could equally well refer to $\mathbf{j}_M$ and $\mathbf{K}_M$ as bound current densities since they arise from the orbiting and spinning of electrons bound to the individual atoms or molecules of matter, that is from the motion of bound charges.

Let us now examine in detail the significance of the two terms in equation 8–7. If the surface of the volume of integration is taken just outside the volume v, where $\mathbf{M} = 0$, then the surface integral drops out. In this case only the first integral supplies a finite contribution of the magnetization of the volume v to the magnetic potential. To avoid mathematical difficulties in evaluating this integral, however, it is necessary that the magnetization $\mathbf{M}$ has continuous special derivatives throughout v. Since $\mathbf{M}$ changes abruptly as we pass from one side of the bounding surface to the other, we get around this difficulty by supposing that the change in $\mathbf{M}$ at the boundary takes place continuously throughout a very thin transition layer of thickness δ. In this layer it is then more convenient to replace the volume current density by a surface current density so that there

$$\nabla \times \mathbf{M}\, dv = \nabla \times \mathbf{M}\, dS\, \delta = \mathbf{j}_M\, \delta\, dS = \mathbf{K}_M\, dS \qquad (8\text{--}10)$$

We thus see that the surface density $\mathbf{K}_M$ is not different in any fundamental respect from the volume current density $\mathbf{j}_M$. In fact by introducing the transition layer in which the magnetization changes continuously from some value $\mathbf{M}$ just inside the surface to zero just outside, we do not need to retain the surface integral since the contribution of this layer will then be implicitly contained in the volume integral. To show explicitly that this is so we divide the volume v into two parts: a volume

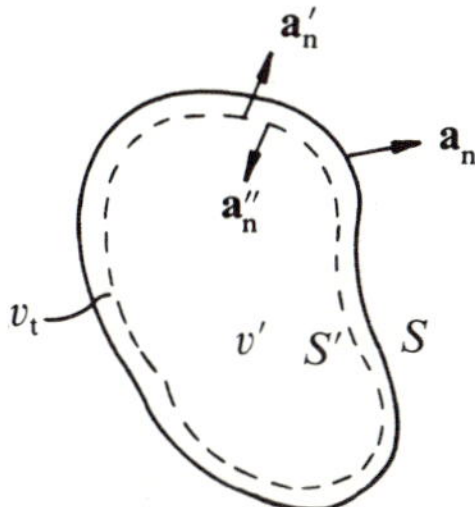

Fig. 8.2 Transition volume near surface of a magnetized body

v' bounded by the surface S' and a transition volume v_t bounded internally by S' and externally by S (Fig. 8.2). Equation 8–6 may then be written as

$$\mathbf{A} = \frac{\mu_0}{4\pi}\left\{ \int_{v'} \frac{\nabla \times \mathbf{M}}{r}\, dv' + \oint_{S'} \frac{\mathbf{M} \times \mathbf{a}'_n}{r}\, dS' + \int_{v_t} \frac{\nabla \times \mathbf{M}}{r}\, dv_t \right.$$
$$\left. + \oint_{S'} \frac{\mathbf{M} \times \mathbf{a}''_n}{r}\, dS' + \oint_{S} \frac{\mathbf{M} \times \mathbf{a}_n}{r}\, dS \right\}$$

and since $\mathbf{a}'_n = -\mathbf{a}''_n$, and $\mathbf{M} = 0$ over S, we have that

$$\mathbf{A} = \frac{\mu_0}{4\pi}\left\{ \int_{v'} \frac{\nabla \times \mathbf{M}}{r}\, dv' + \int_{v_t} \frac{\nabla \times \mathbf{M}}{r}\, dv_t \right\}$$
$$= \frac{\mu_0}{4\pi} \int_{v} \frac{\nabla \times \mathbf{M}}{r}\, dv = \frac{\mu_0}{4\pi} \int_{v} \frac{\mathbf{j}_M}{r}\, dv \qquad (8\text{--}11)$$

where the volume v now includes the transition layer. If the magnetization inside a body is uniform, that is $\mathbf{M}$ is constant, its curl will be zero. However, because $\mathbf{M}$ is always discontinuous at the surface it is always non-uniform throughout the transition layer. In this case the only contribution to $\mathbf{A}_M$ comes from this transition layer and it is then more convenient to express this contribution by means of the surface integral. Thus for uniform $\mathbf{M}$

$$\mathbf{A}_M = \frac{\mu_0}{4\pi} \oint_S \frac{\mathbf{M} \times \mathbf{a}_n}{r} \, dS = \frac{\mu_0}{4\pi} \oint_S \frac{\mathbf{K}_M \, dS}{r} \tag{8-12}$$

It should be pointed out here that the use of amperian currents gives the correct value of $\mathbf{A}_M$ (equation 8–11) at points both outside and inside a material body. This can be easily demonstrated as follows. Each rectangular component of equation 8–11 satisfies Poisson's equation

$$\nabla^2(A_M)_i = \mu_0(j_M)_i; \; (i = x, y, z) \tag{8-13}$$

and if the field point P lies inside the volume, then an analysis similar to that given in section 2–17 for the electrostatic potential shows that equation 8–11 is still valid at all such points.

For the general case in which there exists a conduction current distribution $\mathbf{j}$ inside a magnetized material volume, the magnetic vector potential is given by

$$\mathbf{A} = \mathbf{A}_0 + \mathbf{A}_M = \frac{\mu_0}{4\pi} \int_v \left(\frac{\mathbf{j}}{r}\right) dv + \frac{\mu_0}{4\pi} \int_v \left(\frac{\mathbf{j}_M}{r}\right) dv \tag{8-14}$$

The magnetic flux density is given by

$$\mathbf{B} = \nabla \times \mathbf{A} = \nabla \times \mathbf{A}_0 + \nabla \times \mathbf{A}_M = \nabla \times \left\{ \frac{\mu_0}{4\pi} \int_v \frac{\mathbf{j} + \mathbf{j}_M}{r} \, dv \right\}$$

or

$$\mathbf{B} = \frac{\mu_0}{4\pi} \int_v \frac{(\mathbf{j} + \mathbf{j}_M) \times \mathbf{a}_r}{r^2} \, dv \tag{8-15}$$

If we now compare the expressions for $\mathbf{A}_0$ and $\mathbf{B}_0$ with those for $\mathbf{A}$ and $\mathbf{B}$ respectively, we see that the only difference between them is that in the presence of a material body the conduction current density $\mathbf{j}$ is replaced by the term $(\mathbf{j} + \mathbf{j}_M)$, where $\mathbf{j}_M$ is the equivalent magnetization current density. It follows that the equations

$$\nabla \cdot \mathbf{A} = 0 \tag{8-16}$$

$$\nabla \cdot \mathbf{B} = 0 \tag{8-17}$$

are always valid, whether we have magnetized matter or not, and that equation 7–48 may be written in the more general form*

$$\nabla \times \mathbf{B} = \mu_0(\mathbf{j}+\mathbf{j}_M) \qquad (8\text{–}18)$$

We see from this equation that the magnetic flux density $\mathbf{B}$ has as its source both the conduction currents and the amperian currents. This situation is analogous to the one in electrostatics in which the electric field intensity $\mathbf{E}$ has as its sources both the free charges and the bound charges.

Since the divergence equation $\nabla \cdot \mathbf{B} = 0$ is always valid irrespective of the medium, it follows that its integral form

$$\oint_S \mathbf{B} \cdot d\mathbf{S} = 0 \qquad (8\text{–}19)$$

is also valid for any medium. This simply means that the magnetic flux through any closed surface is always zero whether this surface encloses totally or partially a material medium.

The discontinuity of $\mathbf{M}$ (and hence of $\mathbf{j}_M$) at the surface of a material body may be explicitly taken into consideration by adding the surface integral contribution to both $\mathbf{A}_M$ and $\mathbf{B}_M$. For the general case, therefore, in which there is both a volume conduction density $\mathbf{j}$ and a surface conduction density $\mathbf{K}$ we have

$$\mathbf{A} = \frac{\mu_0}{4\pi} \int_v \frac{\mathbf{j}+\mathbf{j}_M}{r} \, dv + \frac{\mu_0}{4\pi} \oint_S \frac{\mathbf{K}+\mathbf{K}_M}{r} \, dS \qquad (8\text{–}20)$$

and

$$\mathbf{B} = \frac{\mu_0}{4\pi} \int_v \frac{(\mathbf{j}+\mathbf{j}_M) \times \mathbf{a}_r}{r^2} \, dv + \frac{\mu_0}{4\pi} \oint_S \frac{(\mathbf{K}+\mathbf{K}_M) \times \mathbf{a}_r}{r^2} \, dS \qquad (8\text{–}21)$$

The form of equations 8–20 and 8–21 shows that $\mathbf{A}$ or $\mathbf{B}$, at all points inside and outside a body, is exactly the same as that which would be produced in free space by supplementing the conduction current densities $\mathbf{j}$ and $\mathbf{K}$ with amperian currents of volume density $\mathbf{j}_M$ inside the body and surface density $\mathbf{K}_M$ on the surface of the body.

The magnetic flux density $\mathbf{B}$ as given by equation 8–21 may be expressed as the sum of two components

$$\mathbf{B} = \mathbf{B}_0 + \mathbf{B}_M \qquad (8\text{–}22)$$

---

* Equation 8–18 can also be arrived at as follows. Equation 8–12 is a solution of Poisson's equation

$$\nabla^2 \mathbf{A} = -\mu_0(\mathbf{j}+\mathbf{j}_M)$$

and since $\nabla \cdot \mathbf{A} = 0$, then

$$\nabla \times \nabla \times \mathbf{A} = -\nabla^2 \mathbf{A}$$

or

$$\nabla \times \mathbf{B} = -\nabla^2 \mathbf{A} = \mu_0(\mathbf{j}+\mathbf{j}_M)$$

where

$$\mathbf{B}_0 = \frac{\mu_0}{4\pi}\left\{\int_v \frac{\mathbf{j}\times\mathbf{a}_r}{r^2}\,dv + \oint_S \frac{\mathbf{K}\times\mathbf{a}_r}{r^2}\,dS\right\} \qquad (8\text{--}23)$$

and

$$\mathbf{B}_M = \frac{\mu_0}{4\pi}\left\{\int_v \frac{\mathbf{j}_M\times\mathbf{a}_r}{r^2}\,dv + \oint_S \frac{\mathbf{K}_M\times\mathbf{a}_r}{r^2}\,dS\right\} \qquad (8\text{--}24)$$

and equation 8–18 may be rewritten as

$$\nabla\times\mathbf{B} = \nabla\times\mathbf{B}_0 + \nabla\times\mathbf{B}_M = \mu_0\mathbf{j} + \mu_0\mathbf{j}_M$$

so that,

$$\nabla\times\mathbf{B}_0 = \mu_0\mathbf{j} \qquad (8\text{--}25)$$

$$\nabla\cdot\mathbf{B}_0 = 0 \qquad (8\text{--}26)$$

and

$$\nabla\times\mathbf{B}_M = \mu_0\mathbf{j}_M = \mu_0\nabla\times\mathbf{M}$$

$$\nabla\times\left(\frac{\mathbf{B}_M}{\mu_0}-\mathbf{M}\right) = 0 \qquad (8\text{--}27)$$

$$\nabla\cdot\mathbf{B}_M = 0 \qquad (8\text{--}28)$$

Let us now introduce two new auxiliary vectors $\mathbf{H}_0$ and $\mathbf{H}_M$ defined by

$$\mathbf{H}_0 = \mathbf{B}_0/\mu_0 \ [\text{A m}^{-1}] \qquad (8\text{--}29)$$

$$\mathbf{H}_M = \frac{\mathbf{B}_M}{\mu_0}-\mathbf{M} \ [\text{A m}^{-1}] \qquad (8\text{--}30)$$

These two vectors will now *always* obey the relationships

$$\nabla\times\mathbf{H}_0 = \mathbf{j} \qquad (8\text{--}31)$$

$$\nabla\times\mathbf{H}_M = 0 \qquad (8\text{--}32)$$

and it follows that

$$\nabla\times(\mathbf{H}_0+\mathbf{H}_M) = \mathbf{j}$$

which may be written as

$$\nabla\times\mathbf{H} = \mathbf{j} \qquad (8\text{--}33)$$

The vector $\mathbf{H}$ is related to the vector $\mathbf{B}$ by the relation

$$\mathbf{H} = \mathbf{H}_0+\mathbf{H}_M = \frac{\mathbf{B}_0}{\mu_0}+\frac{\mathbf{B}_M}{\mu_0}-\mathbf{M}$$

$$\mathbf{H} = \frac{\mathbf{B}}{\mu_0}-\mathbf{M} \qquad (8\text{--}34)$$

Since at points outside a material body $\mathbf{M} = 0$, the relationship between $\mathbf{H}$ and $\mathbf{B}$ at such points is given by

$$\mathbf{H} = \frac{\mathbf{B}}{\mu_0} \tag{8-35}$$

It follows from the definition of $\mathbf{H}$ that its units must be the same as those of $\mathbf{B}/\mu_0$ or $\mathbf{M}$, that is ampere per metre. It is apparent from equation 8-33 that the new vector $\mathbf{H}$ has as its vortex source only* the true or conduction current density $\mathbf{j}$. The curl of $\mathbf{H}$ describes the spatial distribution and concentration of free currents. In magnetostatics $\mathbf{H}$ is analogous to the vector $\mathbf{D}$ in electrostatics which describes, through its divergence, the spatial distribution and concentration of free charges. Opinions differ widely on the name which should be attached to $\mathbf{H}$. In the vast majority of texts $\mathbf{H}$ is called the magnetic field intensity (or strength) but this name is evidently inappropriate since $\mathbf{H}$ does not represent a force field in the same sense that the electric field $\mathbf{E}$ does. *Magnetic displacement density* or *magnetizing force* are fundamentally more appropriate appellations and we shall adopt the latter consistently through this book.

Ampère's circuital law for the magnetizing force may be readily obtained from equation 8-33 and Stokes's theorem. Thus

$$\int_S \nabla \times \mathbf{H} \cdot d\mathbf{S} = \int_S \mathbf{j} \cdot d\mathbf{S} = I$$

and

$$\oint_C \mathbf{H} \cdot d\mathbf{l} = I \tag{8-36}$$

so that the integral of the tangential component of $\mathbf{H}$ around any closed path C is equal to the total conduction current linked by that path. If the path links a N-turn coil which carries a current I, then the circuital law for $\mathbf{H}$ is given by

$$\oint_C \mathbf{H} \cdot d\mathbf{l} = NI \tag{8-37}$$

and this is why the magnetizing force $\mathbf{H}$ is quite commonly referred to as the ampere-turns per metre (At m^{-1}). The relationship between the direction of integration and the algebraic sign in equations 8-36 and 8-37 is the same as that given in Section 7.6 for equation 7-36.

In the foregoing analysis of the magnetic field, the presence of magnetized bodies has been taken into consideration by replacing the bodies by an equivalent volume and surface distribution of amperian

---

* In the case of time-varying fields, $\mathbf{j}$ must be supplemented by the displacement current density. This will be discussed in Section 9.8.

currents. Thus the conduction current density is supplemented by a volume magnetization current density $\mathbf{j}_M = \nabla \times \mathbf{M}$ throughout the bodies, and a surface magnetization current density $\mathbf{K}_M = \mathbf{M} \times \mathbf{a_n}$ at the surfaces bounding these bodies. The magnetic flux density $\mathbf{B}$ at any point may then be found either by first calculating $\mathbf{A}$ from equation 8–20 or directly from equation 8–21. The magnetizing force may then be obtained from equation 8–34 for points inside a body and equation 8–35 for points outside a body. The following table summarizes the amperian current approach.

Medium	$\mathbf{B} =$	$\mathbf{H} =$
Vacuum	$\mathbf{B}_0$ (any point); Eqn 8–23	$\mathbf{B}_0/\mu_0$ (any point)
Infinitely extended material medium	$\mathbf{B}_0 + \mathbf{B}_M$ (any point); Eqn 8–15	$\dfrac{\mathbf{B}}{\mu_0} - \mathbf{M}$ (any point)
Material medium of finite extent	$\mathbf{B}_0 + \mathbf{B}_M$ (any point); Eqn 8–21	$\mathbf{B}/\mu_0$ (points outside medium) $\dfrac{\mathbf{B}}{\mu_0} - \mathbf{M}$ (points inside medium)

8.3 EQUIVALENT POLE MODEL

In this model it will be shown that a magnetized body can be replaced by an equivalent distribution of magnetic poles having a volume density ρ_m and a surface density σ_m. Although the equivalence is strictly mathematical, since magnetic monopoles do not exist, it is a particularly useful one for solving problems involving permanent magnets or material media placed in an external magnetic field. This is because the methods of solution are completely analogous to those used in solving electrostatic problems.

From equations 8–26 to 8–30 we have that

$$\nabla \cdot \mathbf{H}_M = -\nabla \cdot \mathbf{M} \qquad \text{(always, for } \mathbf{M} \neq 0) \qquad (8–38)$$

$$\nabla \cdot \mathbf{H}_0 = 0 \qquad \text{(always)} \qquad (8–39)$$

$$\nabla \cdot \mathbf{H} = -\nabla \cdot \mathbf{M} \qquad \text{(always, for } \mathbf{M} \neq 0) \qquad (8–40)$$

$$\nabla \times \mathbf{H}_M = 0 \qquad \text{(always)} \qquad (8–41)$$

Since $\mathbf{H}_0$ is solenoidal (equation 8–39), it may be expressed as the curl of a vector function $\mathbf{A}_H$ whilst since $\mathbf{H}_M$ is irrotational (equation 8–41) it may be expressed as the gradient of a scalar function Φ_H. Hence the vector $\mathbf{H}$ may be expressed as

$$\mathbf{H} = \mathbf{H}_M + \mathbf{H}_0 = -\nabla \Phi_H + \nabla \times \mathbf{A}_H \qquad (8–42)$$

where the negative sign before the gradient has been introduced for convenience only.

The splitting of a vector field into a gradient and a curl is by no means a characteristic property of the vector field **H**. Any general vector field may be so decomposed. This is Helmholtz's theorem*; it states that any continuous vector point function **F**, which has continuous first derivatives throughout a region of space, is uniquely defined if its source density $\nabla \cdot \mathbf{F}$ and its circulation density $\nabla \times \mathbf{F}$ are given at all points in this space, provided that the totality of sources, as well as the source and circulation densities, are zero at infinity. In this case we may always express the vector **F** at some point P as

$$\mathbf{F} = -\nabla_P \Phi_F + \nabla_P \times \mathbf{A}_F \tag{8-43}$$

where

$$\Phi_F = \frac{1}{4\pi} \int_v \frac{\nabla_Q \cdot \mathbf{F}}{r} \, dv \tag{8-44}$$

and

$$\mathbf{A}_F = \frac{1}{4\pi} \int_v \frac{\nabla_Q \times \mathbf{F}}{r} \, dv \tag{8-45}$$

where the volume of integration is the whole space in which the vector **F** exists and the subscripts P and Q indicate whether the differentiation operations are to be carried out with respect to the coordinates of the field point P or of the source point Q (Fig. 8.1). If S is any surface of discontinuity within v, for example the interface between two media, then although the vector **F** is continuous inside and outside S, it will vary discontinuously in crossing S. In Helmholtz's theorem, allowance must be made for this fact either implicitly by assuming that the volume integrals include a thin transition volume enclosing S and consisting of very thin layers on opposite sides of S in which **F** and its derivatives vary rapidly but continuously (see Section 8.1), or explicitly by supplementing the volume divergence and volume curl of **F** by the surface divergence and surface curl of **F** respectively. In the latter case the functions Φ_F and $\mathbf{A}_F$ are given by (Fig. 8.3),

$$\Phi_F = \frac{1}{4\pi} \int_v \frac{\nabla_Q \cdot \mathbf{F}}{r} \, dv + \frac{1}{4\pi} \oint_S \frac{(\mathbf{F}_2 - \mathbf{F}_1) \cdot \mathbf{a}_n}{r} \, dS \tag{8-46}$$

and

$$\mathbf{A}_F = \frac{1}{4\pi} \int_v \frac{\nabla_Q \times \mathbf{F}}{r} \, dv + \frac{1}{4\pi} \oint_S \frac{\mathbf{a}_n \times (\mathbf{F}_2 - \mathbf{F}_1)}{r} \, dS \tag{8-47}$$

* A formal proof of this theorem may be found in *Classical Electricity and Magnetism* by W. Panofsky and M. Phillips, pp. 2–5 (Addison-Wesley, 1962).

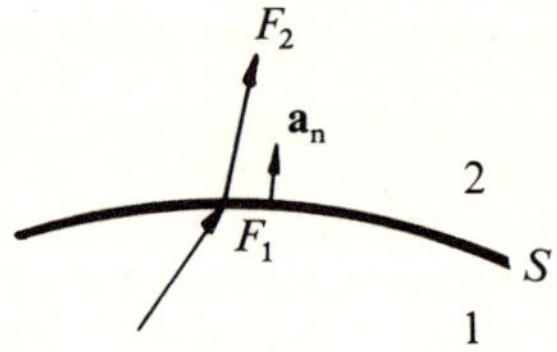

Fig. 8.3 Discontinuity of **F** at a surface

If the vector **F** vanishes outside a volume v bounded by a surface S, then equations 8–46 and 8–47 become

$$\Phi_F = \frac{1}{4\pi}\int_v \frac{\nabla_Q \cdot \mathbf{F}}{r}\,\mathrm{d}v + \frac{1}{4\pi}\oint_S \frac{(-\mathbf{F}\cdot\mathbf{a_n})}{r}\,\mathrm{d}S \qquad (8\text{–}48)$$

and

$$\mathbf{A}_F = \frac{1}{4\pi}\int_v \frac{\nabla_Q \times \mathbf{F}}{r}\,\mathrm{d}v + \frac{1}{4\pi}\oint_S \frac{\mathbf{F}\times\mathbf{a_n}}{r}\,\mathrm{d}S \qquad (8\text{–}49)$$

Consider now a material volume v bounded by a closed surface S surrounded by free space. Let the magnetization at any point inside the volume be **M**; outside v the magnetization is zero. From Helmholtz's theorem we have that

$$\mathbf{M} = -\nabla_P\Phi_M + \nabla_P \times \mathbf{A}_M \qquad (8\text{–}50)$$

where, from equations 8–48 and 8–49,

$$\Phi_M = \frac{1}{4\pi}\int_v \frac{\nabla_Q \cdot \mathbf{M}}{r}\,\mathrm{d}v + \frac{1}{4\pi}\oint_S \frac{-\mathbf{M}\cdot\mathbf{a_n}}{r}\,\mathrm{d}S \qquad (8\text{–}51)$$

and

$$\mathbf{A}_M = \frac{1}{4\pi}\int_v \frac{\nabla_Q \times \mathbf{M}}{r}\,\mathrm{d}v + \frac{1}{4\pi}\oint_S \frac{\mathbf{M}\times\mathbf{a_n}}{r}\,\mathrm{d}S \qquad (8\text{–}52)$$

Introducing equation 8–50 in equation 8–34 and using the relationship $\mathbf{B} = \nabla \times \mathbf{A}$ where **A** is defined by equation 8–20 we obtain the following expression for the magnetizing force,

$$\mathbf{H} = \nabla_P \times \left(\frac{1}{4\pi}\int_v \frac{\mathbf{j}}{r}\,\mathrm{d}v + \frac{1}{4\pi}\oint_S \frac{\mathbf{K}}{r}\,\mathrm{d}S \right)$$

$$+ \nabla_P \left(\frac{1}{4\pi}\int_v \frac{\nabla_Q \cdot \mathbf{M}}{r}\,\mathrm{d}v + \frac{1}{4\pi}\oint_S \frac{-\mathbf{M}\cdot\mathbf{a_n}}{r}\,\mathrm{d}S \right) \qquad (8\text{–}53)$$

$$= \mathbf{H}_0 + \mathbf{H}_M \qquad (8\text{–}53a)$$

Since ∇_P operates on the field coordinates and ∇_Q operates on the source coordinates, whereas the integrations are with respect to the source coordinates, and since $\nabla_P(1/r) = -\mathbf{a}_r/r^2$, equation 8–53 may be expressed in the alternative form

$$\mathbf{H} = \mathbf{H}_0 - \frac{1}{4\pi}\int_v \frac{\nabla_Q \cdot \mathbf{M}}{r^2}\,\mathbf{a}_r\,\mathrm{d}v + \frac{1}{4\pi}\oint_s \frac{\mathbf{M}\cdot\mathbf{a}_n}{r^2}\,\mathbf{a}_r\,\mathrm{d}S \qquad (8\text{–}54)$$

From equations 8–53 and 8–42 we see that we may define a scalar function Φ_H such that

$$\Phi_H = \frac{1}{4\pi}\int_v \frac{-\nabla_Q \cdot \mathbf{M}}{r}\,\mathrm{d}v + \frac{1}{4\pi}\oint_s \frac{\mathbf{M}\cdot\mathbf{a}_n}{r}\,\mathrm{d}S \qquad (8\text{–}55)$$

Equation 8–53 or 8–42 gives the value of the magnetizing force $\mathbf{H}$ at any point either outside or inside the volume v. The magnetic flux density at points *outside* v is then given by $\mathbf{B} = \mu_0\mathbf{H}$, whilst at points *inside* v it is given by $\mathbf{B} = \mu_0(\mathbf{H}+\mathbf{M})$. For a material medium of infinite extent it is evident that the two surface integrals in equation 8–53 are no longer necessary and the magnetizing force is given by

$$\mathbf{H} = \nabla_P \times \left(\frac{1}{4\pi}\int_v \frac{\mathbf{j}}{r}\,\mathrm{d}v\right) + \nabla_P\left(\frac{1}{4\pi}\int_v \frac{\nabla_Q \cdot \mathbf{M}}{r}\,\mathrm{d}v\right) \qquad (8\text{–}56)$$

Let us now consider the case in which we have a permanently magnetized body of finite volume v located in a region of space where the conduction current density is everywhere zero ($\mathbf{H}_0 = 0$). At any point either outside or inside v the magnetizing force $\mathbf{H}$ is given by equations 8–53 and 8–55

$$\mathbf{H} = -\nabla\Phi_H \qquad (8\text{–}57)$$

so that Φ_H may be considered as the magnetic scalar potential of $\mathbf{H}$. At all points *outside* the volume v the magnetic flux density is given by

$$\mathbf{B} = \mu_0\mathbf{H} = -\nabla\mu_0\Phi_H \qquad (8\text{–}58)$$

If we now compare this equation with equation 7–79 we see immediately that the quantity $\mu_0\Phi_H$ is precisely the magnetic scalar potential V_m, that is*

$$V_m = \mu_0\Phi_H \qquad (8\text{–}59)$$

Since at all points outside the magnetized volume $\mathbf{M} = 0$, then

$$\nabla\cdot\mathbf{B} = \nabla\cdot\mathbf{H} = 0$$

and the magnetic scalar potential satisfies Laplace's equation

$$\nabla^2 V_m = \nabla^2\Phi_H = 0 \qquad (8\text{–}60)$$

---

* In most textbooks the function Φ_H, instead of V_m, is taken to define the magnetic scalar potential (see Chapter 11, p. 404).

At all points *inside* the volume v the magnetic flux density is given by

$$\mathbf{B} = \mu_0(\mathbf{H}+\mathbf{M}) = -\nabla V_{\mathrm{m}}+\mu_0\mathbf{M} \qquad (8\text{-}61)$$

and since $\nabla \cdot \mathbf{B} = 0$ always, it follows that V_{m} (and Φ_H) satisfy the Poisson equation

$$-\nabla^2 V_{\mathrm{m}}+\mu_0\nabla \cdot \mathbf{M} = 0$$

or

$$\nabla^2 V_{\mathrm{m}} = \mu_0\nabla \cdot \mathbf{M} \qquad (8\text{-}62)$$

From equations 8–55 and 8–59 we have that*

$$V_{\mathrm{m}} = \frac{\mu_0}{4\pi} \int_v \frac{-\nabla_Q \cdot \mathbf{M}}{r}\, \mathrm{d}v + \frac{\mu_0}{4\pi} \oint_s \frac{\mathbf{M} \cdot \mathbf{a}_{\mathrm{n}}}{r}\, \mathrm{d}S \qquad (8\text{-}63)$$

If it is assumed that magnetic monopoles exist then equation 8–63 would be recognizable as the magnetic scalar potential produced by a distribution of magnetic poles (sometimes called magnetic masses) in free space of volume density

$$\rho_{\mathrm{m}} = -\nabla \cdot \mathbf{M} \qquad (8\text{-}64)$$

and of surface density

$$\sigma_{\mathrm{m}} = \mathbf{M} \cdot \mathbf{a}_{\mathrm{n}} \qquad (8\text{-}65)$$

The replacement of a magnetized body by a volume and surface pole density distribution, defined by equations 8–64 and 8–65 respectively, constitutes the *equivalent pole model*. The use of such a model leads to a magnetic scalar potential. It will be recalled that in the Amperian current model a magnetized body is replaced by an equivalent volume and surface current density distribution; this model led to the use of a magnetic vector potential.

Although in the above analysis we have arrived at the equivalent pole model through use of the Helmholtz theorem—that is, by purely mathematical manipulations—the real reason for the existence of two possible representations of magnetized bodies is the equivalence of a small current loop with a fictitious magnetic dipole (see Section 7.12). Thus if we consider a body in which the magnetization $\mathbf{M}$ is taken to represent the volume density of fictitious magnetic point dipoles instead of elemental current loops, then it follows from equation 7–82 that the magnetic scalar potential at any point P due to an element of volume $\mathrm{d}v$ at point Q is given by

$$\mathrm{d}V_{\mathrm{m}} = \frac{\mu_0}{4\pi}\, \mathbf{M} \cdot \nabla_Q\!\left(\frac{1}{r}\right) \mathrm{d}v$$

---

* The reader should note the analogy which exists between equation 8–63 and equation 4–6 in electrostatics which expresses the potential due to a polarized dielectric volume.

and

$$V_{\mathrm{m}} = \frac{\mu_0}{4\pi} \int_v \mathbf{M} \cdot \nabla_Q\left(\frac{1}{r}\right) \mathrm{d}v \tag{8-66}$$

This equation is analogous to equation 4–5 and may therefore be transformed into the form of equation 4–6 to give equation 8–63. It should be remembered that if the volume integral includes the thin transition layer v_t near the surface S, then it can be shown, by a manner similar to that given in Section 8.2, that we need not retain the surface integral term and that we can express the scalar potential simply as

$$V_{\mathrm{m}} = \frac{\mu_0}{4\pi} \int_v \frac{-\nabla \cdot \mathbf{M}}{r} \mathrm{d}v \tag{8-67}$$

We shall now give a simple example of the application of the above two representations. Given a circular cylindrical magnet throughout which

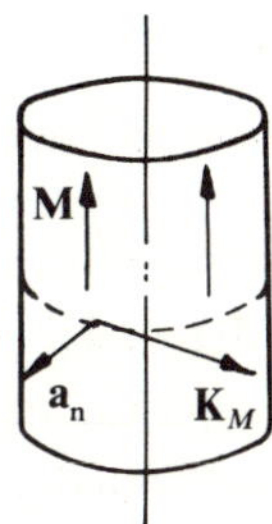

Fig. 8.4 Uniformly magnetized cylindrical magnet

the magnetization is constant and in a direction parallel to the cylinder axis (Fig. 8.4), it is required to find $\mathbf{B}$ and $\mathbf{H}$ at all points on the axis of the magnet.

8.3.1 AMPERIAN CURRENT MODEL

From equations 8–8 and 8–9 we have, bearing in mind that $\mathbf{M}$ is constant, that

$$\mathbf{j}_M = \nabla \times \mathbf{M} = 0$$

$$\mathbf{K}_M = \mathbf{M} \times \mathbf{a}_n$$

The cylinder may therefore be replaced by a cylindrical current sheet of linear density $\mathbf{K}_M$ A m^{-1}. This is equivalent to a closely-wound solenoid of n turns per unit length carrying a current I such that

$$K_M = nI = M$$

The magnetic flux density at any point on the axis may be now obtained

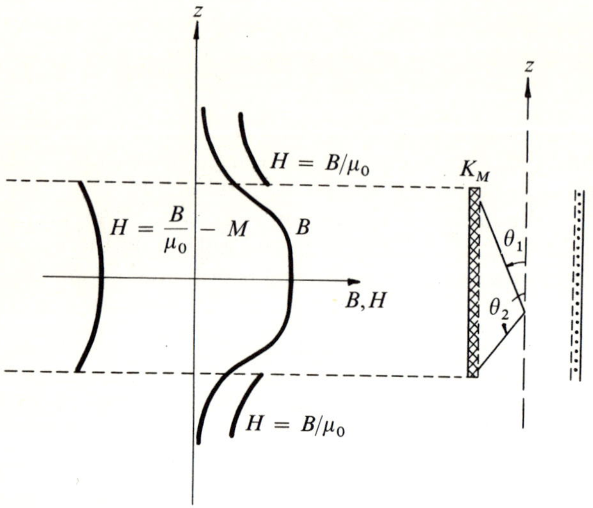

Fig. 8.5 Variations of B and H along the axis of a uniformly magnetized cylinder

directly from equation 7–34a so that

$$\mathbf{B} = \tfrac{1}{2}\mu_0 M(\cos\theta_1 - \cos\theta_2)\mathbf{a}_z \tag{8–68}$$

The magnetizing force is obtained from the relationship $\mathbf{H} = (\mathbf{B}/\mu_0) - \mathbf{M}$ for points on the axis inside the magnet, and the relationship $\mathbf{H} = \mathbf{B}/\mu_0$ for points outside. Figure 8.5 shows the variations of B and H along the cylinder axis.

8.3.2 EQUIVALENT POLE MODEL

From equations 8–64 and 8–65 we have that

$$\rho_{\mathrm{m}} = -\nabla \cdot \mathbf{M} = 0$$

$$\sigma_{\mathrm{m}} = \mathbf{M} \cdot \mathbf{a}_{\mathrm{n}}$$

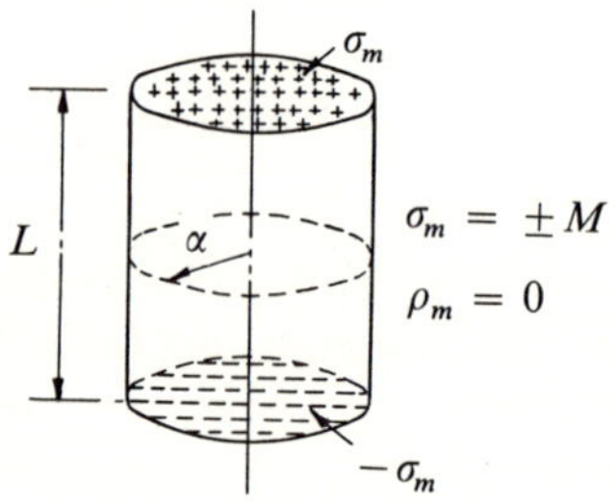

Fig. 8.6 Equivalent pole model of a uniformly magnetized cylinder

and the magnetized cylinder may be replaced by a surface distribution of positive or north poles on the upper circular face and of negative or south poles on the lower circular face (at this face $\mathbf{M}$ and $\mathbf{a_n}$ are oppositely directed). The problem of calculating $\mathbf{H}$ at any point on the axis is equivalent to that of calculating the electric field at any point on the axis of two equally and oppositely charged discs at a distance L apart (Fig. 8.6). The field at any point on the axis of a single charged disc was calculated in Section 2.9. The magnetizing force $\mathbf{H}$ may be obtained directly from equation 2–47 by replacing σ/ϵ_0 by $\sigma_m = M$; for the upper disc

$$\mathbf{H}_M = -\tfrac{1}{2}M\left[\frac{z}{(z^2+a^2)^{1/2}} - \frac{z}{(z^2)^{1/2}}\right]\mathbf{a}_z \qquad (8\text{–}69a)$$

and for the lower disc

$$\mathbf{H}'_M = \tfrac{1}{2}M\left[\frac{z'}{(z'^2+a^2)^{1/2}} - \frac{z'}{(z'^2)^{1/2}}\right]\mathbf{a}_z \qquad (8\text{–}69b)$$

The resultant $\mathbf{H}$ field at any point on the axis is the sum of the above two

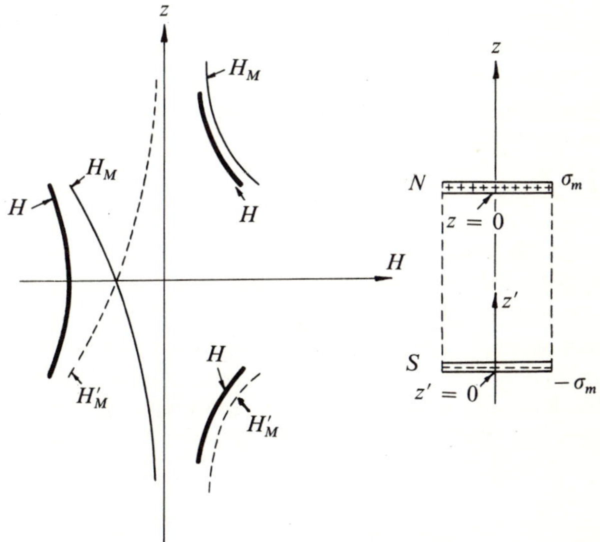

Fig. 8.7 Variation of H along the axis of a uniformly magnetized cylinder

fields (Fig. 8.7). Inside the cylinder $\mathbf{B}$ is equal to $\mu_0(\mathbf{H}+\mathbf{M})$ and outside it is equal to $\mu_0\mathbf{H}$.

Figure 8.8 shows a plot of the B-lines and H-lines both inside and outside the cylindrical magnet. Everywhere outside the magnet the B and

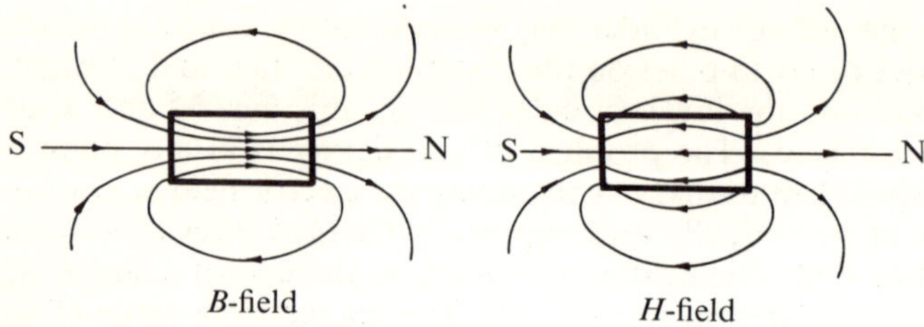

Fig. 8.8 *B*- and *H*-lines inside and outside a cylindrical magnet

H fields are identical since they differ by the constant μ_0 only. Inside the magnet, however, we see that *B* and *H* are oppositely directed. This must be so because the field **B** is solenoidal ($\nabla \cdot \mathbf{B} = 0$) whereas the field **H** is not but has a source density $+\sigma_m$ and a sink density $-\sigma_m$. The *H*-lines therefore begin on the positive poles and end on the negative poles. That **B** and **H** must be oppositely directed inside the magnet may also be concluded from the fact that the line integral of **H** around any closed contour must be zero since there are no conduction currents.

In the above example we have assumed that the magnetization was constant throughout the magnet with the result that the magnetic charges were confined to the end surfaces only. In practice, however, the magnetization in bar magnets is never quite uniform, particularly towards the surfaces, so that there is a volume as well as a surface distribution of magnetic poles in these regions. The assumption of constant magnetization can at best be only an approximation since we are led to a variable **H** field inside the magnet, whereas it is an experimental fact that in all permanently magnetizable materials the magnetization **M** is a function of the magnetizing force **H**.

We have seen from the above example that the field $\mathbf{H}_M$ inside a uniformly magnetized body is oppositely directed to the magnetization **M**. For this reason $\mathbf{H}_M$ is referred to as the *demagnetizing field* (Fig. 8.9). Its value depends on the geometry of the magnetized body and on its magnetization (equation 8–53a). In general it is not possible to cal-

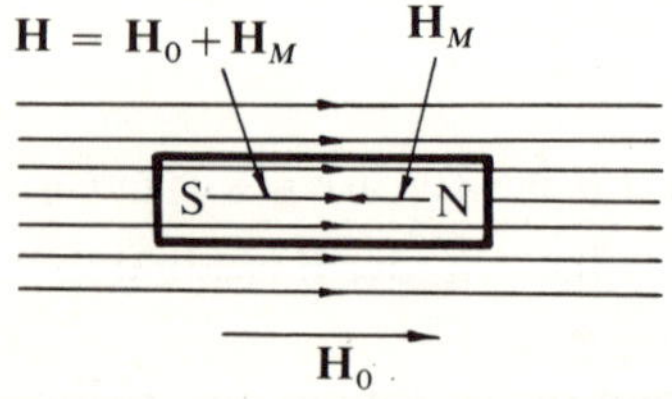

Fig. 8.9 Demagnetizing field inside a magnet

culate $\mathbf{H}_M$ except for certain specific shapes (ellipsoids and their degenerate forms such as spheroids and the sphere*). In such cases the relationship between $\mathbf{H}_M$ and $\mathbf{M}$ is always given by

$$\mathbf{H}_M = -D\mathbf{M} \qquad (8\text{–}70a)$$

and

$$D = (\mathbf{H}_0 - \mathbf{H})/\mathbf{M} \qquad (8\text{–}70b)$$

where D is a positive numerical factor called the *demagnetizing factor*. Its value depends on the shape of the specimen and is identical with the depolarizing factor (Section 4.4).

For a sphere, $D = 1/3$.

For an infinite thin flat plate with its faces perpendicular to the external field, $D = 1$.

For an infinitely long thin rod with its axis parallel to the external field, $D = 0$.

For an infinitely long circular cylinder with its axis perpendicular to the direction of the external field, $D = 1/2$.

8.4 FAR FIELD OF BAR MAGNET AND OF SOLENOID

Let P be any point at a distance r from the centre of a uniformly magnetized bar magnet of radius a and length L (Fig. 8.10). In the above

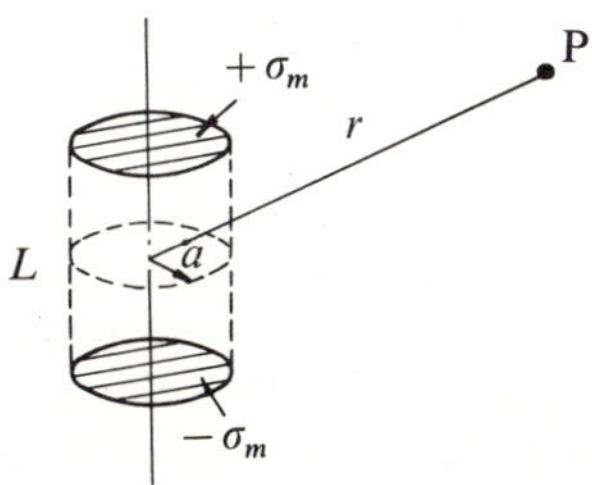

Fig. 8.10 Equivalent pole model of a uniformly magnetized cylinder

example we have seen that the bar may be replaced by equivalent surface pole densities $\sigma_m = \pm M$ at $z = \pm L/2$ respectively. Now if $r \gg a$ we can regard the magnetic poles as concentrated at points at the centre of each face such that

$$Q_m = \pm M(\pi a^2) \qquad (8\text{–}71)$$

The magnetic scalar potential at the point P is then given by (note the

* J. A. Osborn, 'Demagnetizing factors of the general ellipsoid', *Phys. Rev.*, **67**, 351 (1945).

analogy with electrostatics):

$$V_m = \frac{\mu_0}{4\pi} Q_m \left(\frac{1}{r_2} - \frac{1}{r_1} \right) \qquad (8\text{-}72)$$

and the field **B** at P is

$$\mathbf{B} = -\nabla_P V_m$$

$$= -\frac{\mu_0 Q_m}{4\pi} \nabla_P \left(\frac{1}{r_2} - \frac{1}{r_1} \right) \qquad (8\text{-}73)$$

If the distance r of the point P from the origin is large compared with the length L of the bar, the dipole expressions given in Section 2.11 apply. As a first approximation we may therefore write the magnetic scalar potential as

$$V_m = \frac{\mu_0 Q_m L \cos\theta}{4\pi r^2} = \frac{\mu_0 P_m \cos\theta}{4\pi r^2} \qquad (8\text{-}74)$$

and the field **B** as

$$\mathbf{B} = -\frac{\mu_0 P_m}{4\pi} \nabla_P \left(\frac{\cos\theta}{r^2} \right) \qquad (8\text{-}75)$$

where

$$p_m = Q_m L = ML\pi a^2 \qquad (8\text{-}76)$$

is the equivalent magnetic dipole moment of the bar magnet. An exact expression for V_m may be obtained from equation 2–91 by replacing Q/ϵ_0 by $\mu_0 Q_m$.

The magnetic field of a uniformly magnetized cylindrical magnet is the same as that of a closely wound solenoid having the same dimensions as the magnet and a number of ampere-turns per unit length equal to the magnetization, that is, $M = nI$. It follows that the far **B**-field of any solenoid of length L and radius a is given by equation 8–75 in which p_m is now given by

$$p_m = nI\pi a^2 L \qquad (8\text{-}77)$$

which is the equivalent dipole moment of the solenoid.

8.5 BOUNDARY CONDITIONS FOR B AND H

The divergence and curl field equations

$$\nabla \cdot \mathbf{B} = 0$$

$$\nabla \times \mathbf{H} = \mathbf{j}$$

apply to any point in a medium whose properties are continuous. The divergence equation is the point form of Gauss's law for **B**

$$\oint \mathbf{B} \cdot d\mathbf{S} = 0$$

and was derived from it by applying the law to a small volume in the neighbourhood of some point P and then taking the limit of the integral per unit volume as the volume shrinks to zero around the point. This limit has a meaning only if **B** is continuous within the volume element. Suppose now that the point P is chosen on a surface of discontinuity such as the interface between two media, 1 and 2 (Fig. 8.11). Assume

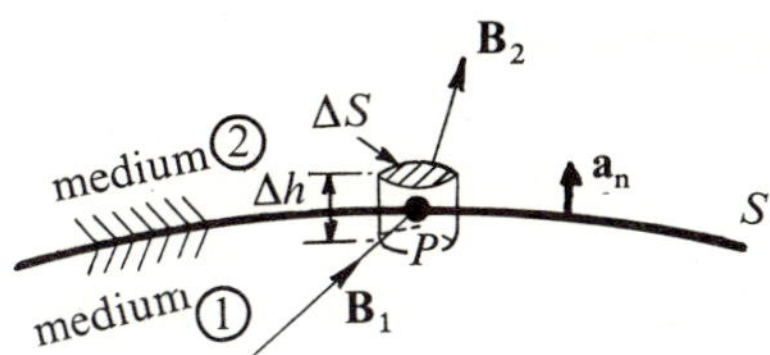

Fig. 8.11 To determine the discontinuity in the normal component of flux density across an interface

that the passage from one medium to the other occurs through a very thin transition layer in which the properties of the media vary rapidly but continuously from their values near S in 1 to their values near S in 2. Within this layer therefore, the vector point functions $\mathbf{B}_1$ and $\mathbf{B}_2$ are continuous. Since the form of the volume element around the point P is entirely arbitrary, we take a small cylinder centred on P, normal to S and reaching through the transition layer on both sides of S. Let Δh be the height of the cylinder, ΔS the area of its bases, and $\mathbf{a}_n$ the unit normal pointing from 1 to 2, that is, medium 2 lies on the positive side of S and medium 1 on the negative side. Integrating over the surface of the cylinder we have,

$$\oint \mathbf{B} \cdot d\mathbf{S} = (\mathbf{B}_2' - \mathbf{B}_1') \cdot \mathbf{a}_n \, \Delta S + \text{contributions of lateral surface} = 0$$

where $\mathbf{B}_1'$ and $\mathbf{B}_2'$ are the average values of **B** over the end faces of the cylinder. Now let the transition layer shrink down to the surface S and at the same time let $\Delta h \to 0$; we obtain

$$\mathbf{a}_n \cdot (\mathbf{B}_2' - \mathbf{B}_1') \, \Delta S = 0$$

If we now let ΔS shrink to zero around the point P then

$$\lim_{\Delta S \to 0} \mathbf{a}_n \cdot (\mathbf{B}_2' - \mathbf{B}_1') \, \Delta S = 0$$

and

$$\mathbf{a}_n \cdot (\mathbf{B}_2 - \mathbf{B}_1) = 0 \tag{8-78}$$

where $\mathbf{B}_2$ and $\mathbf{B}_1$ are the values of $\mathbf{B}$ on the positive and negative side of S at point P.

Equation 8–78 tells us that *the normal component of* $\mathbf{B}$ *is continuous across any interface*. Also, at any point on the interface the vector $(\mathbf{B}_2 - \mathbf{B}_1)$ is perpendicular to $\mathbf{a}_n$ so that $(\mathbf{B}_2 - \mathbf{B}_1)$ lies in the plane tangent to the interface at that point.

It is worth noting that the result given in equation 8–78 could have been written down immediately by applying the definition of surface divergence, given in Section 2.15, equation 2–111, to the vector $\mathbf{B}$ at any point on the interface.

The curl equation $\nabla \times \mathbf{H} = \mathbf{j}$ is the point form of Ampère's circuital law for $\mathbf{H}$,

$$\oint_C \mathbf{H} \cdot d\mathbf{l} = I$$

and is derived from it by applying the law to a small contour C around a point P and then taking the limit of the integral per unit area as the area bounded by C shrinks to zero. Consider again the interface S and the thin transition layer defined in the preceding paragraphs. Let P be a point on S and abcda a small rectangular loop of length ΔL and height Δh; the plane of the loop is normal to the surface S and contains P

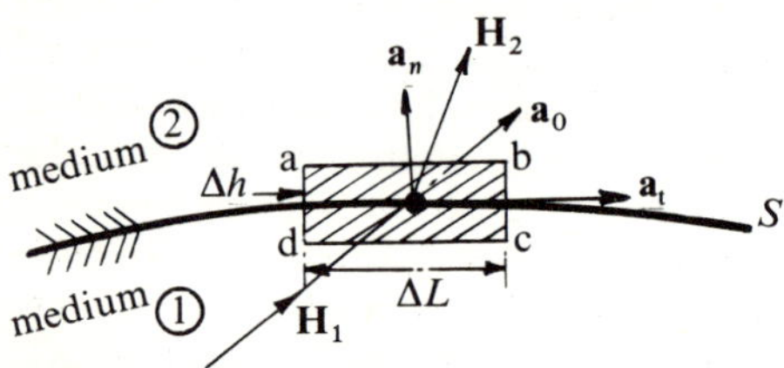

Fig. 8.12 To determine the discontinuity in the tangential component of the magnetizing force across an interface

(Fig. 8.12). Let $\mathbf{a}_0$ be the unit positive normal to the loop and $\mathbf{a}_n$ the unit positive normal to S at P. The unit tangent vector $\mathbf{a}_t$ is defined by

$$\mathbf{a}_0 \times \mathbf{a}_n = \mathbf{a}_t \tag{8–79}$$

From Ampère's law we have that

$$\int_a^d \oint_b^c \mathbf{H} \cdot d\mathbf{l} = \int_{\Delta S} \mathbf{j} \cdot dS \mathbf{a}_0$$

or

$$(\mathbf{H}_2' - \mathbf{H}_1') \cdot \mathbf{a}_t \, \Delta L + \text{contributions from the ends bc and da}$$
$$= (\mathbf{j}_1' + \mathbf{j}_2') \cdot \mathbf{a}_0 \, \Delta L \tfrac{1}{2} \Delta h$$

where $\mathbf{H}_2'$ and $\mathbf{H}_1'$ are the average values of $\mathbf{H}$ over ab and cd respectively, and $\mathbf{j}_2'$ and $\mathbf{j}_1'$ are the average current densities within the upper and lower half of the loop area respectively. Letting the transition layer shrink to the surface S while $\Delta h \to 0$ we have

$$(\mathbf{H}_2' - \mathbf{H}_1') \cdot \mathbf{a}_t \, \Delta L = \lim_{\Delta h \to 0} (\mathbf{j}_1' + \mathbf{j}_2') \cdot \mathbf{a}_0 \tfrac{1}{2}\Delta h \, \Delta L$$

If the conductivities of the two media are finite then the current densities are also finite and the right-hand side of the above equation vanishes and we have that at P

$$\lim_{\Delta L \to 0} (\mathbf{H}_2' - \mathbf{H}_1') \cdot \mathbf{a}_t \, \Delta L = 0$$

or

$$(\mathbf{H}_2 - \mathbf{H}_1) \cdot \mathbf{a}_t = 0 \qquad\qquad (8\text{--}80)$$

where $\mathbf{H}_2$ and $\mathbf{H}_1$ are the values of $\mathbf{H}$ at the positive and negative sides of S at P.

In many important practical problems—e.g., a closely wound solenoid or a good conductor carrying a high frequency current—it is convenient to assume that one of the media is infinitely conducting (perfect conductor), in which case the limit of $\mathbf{j}\,\Delta h$ as $\Delta h \to 0$ defines a surface current density $\mathbf{K}$ (equation 6–5). The boundary condition on $\mathbf{H}$ is then given by

$$(\mathbf{H}_2 - \mathbf{H}_1) \cdot \mathbf{a}_t = \mathbf{K} \cdot \mathbf{a}_0 \qquad\qquad (8\text{--}81)$$

We thus see that in the absence of any surface conduction currents the tangential components of $\mathbf{H}$ are continuous across any interface. In the presence of a surface current density, however, the tangential component of $\mathbf{H}$ is discontinuous, the amount of the discontinuity being equal to the component of the surface current density at right angles to it.

A more convenient expression for the discontinuity in $\mathbf{H}$ can be obtained from equation 8–81 by substituting for $\mathbf{a}_t$ from equation 8–79; noting that $\mathbf{H} \cdot \mathbf{a}_0 \times \mathbf{a}_n = \mathbf{a}_n \times \mathbf{H} \cdot \mathbf{a}_0$, we have that

$$\mathbf{a}_n \times (\mathbf{H}_2 - \mathbf{H}_1) \cdot \mathbf{a}_0 = \mathbf{K} \cdot \mathbf{a}_0$$

or

$$\mathbf{a}_n \times (\mathbf{H}_2 - \mathbf{H}_1) = \mathbf{K} \qquad\qquad (8\text{--}82)$$

The reader should note that the left-hand side of this equation is the surface curl of $\mathbf{H}$ (see equation 7–49). If we now form the scalar product of the two sides of the above equation with the vector $\mathbf{H}_2 - \mathbf{H}_1$ we obtain

$$\mathbf{K} \cdot (\mathbf{H}_2 - \mathbf{H}_1) = 0$$

This result shows that the tangential component of $\mathbf{H}$ parallel to $\mathbf{K}$ is continuous and that it is the tangential component of $\mathbf{H}$ perpendicular to $\mathbf{K}$ which therefore suffers a discontinuity K—cf., Chapter 7, p. 217.

The law of refraction for lines of magnetic flux is given in Section 8.14.

8.6 MAGNETIC SUSCEPTIBILITY AND MAGNETIC PERMEABILITY

It has been shown that in the presence of any material substance the relationship between the magnetic flux density $\mathbf{B}$ at some point and the magnetizing force $\mathbf{H}$ at this same point is given by

$$\mathbf{B} = \mu_0(\mathbf{H}+\mathbf{M})$$

which may be rewritten as

$$\mathbf{B} = \mu_0\mathbf{H}\left(1+\frac{\mathbf{M}}{\mathbf{H}}\right)$$

If the relationship between $\mathbf{M}$ and $\mathbf{H}$ for a given material is linear, then their ratio will be a dimensionless constant. This ratio is called the magnetic susceptibility of the material and is defined by

$$\chi_m = \mathbf{M}/\mathbf{H} \tag{8-83}$$

In this case the relationship between $\mathbf{B}$ and $\mathbf{H}$ will also be linear and given by

$$\mathbf{B} = \mu_0(1+\chi_m)\mathbf{H}$$
$$= \mu_0\mu_r\mathbf{H} \tag{8-84}$$

where

$$\mu_r = 1+\chi_m \tag{8-85}$$

μ_r is called the relative permeability of the material and is, like χ_m, a characteristic property of the material substance; μ_r is a dimensionless number. Equation 8–84 is simple and convenient since it eliminates the necessity of working with $\mathbf{M}$. The permeability μ of a medium is defined as

$$\mu = \mu_0\mu_r \left[\mathrm{H\,m^{-1}}\right] \tag{8-86}$$

and has the same dimensions as μ_0. The relationship between $\mathbf{B}$ and $\mathbf{H}$ may thus be expressed concisely as

$$\mathbf{B} = \mu\mathbf{H} \tag{8-87}$$

For anisotropic single crystals equations 2–83 and 2–87 must be replaced by the more general relationships.

$$M_i = \chi_{mij}H_j \qquad (i,j = 1, 2, 3) \tag{8-88}$$
$$B_i = \mu_{ij}H_j \qquad (i,j = 1, 2, 3) \tag{8-89}$$

where the coefficients χ_{mij} and μ_{ij} represent the magnetic susceptibility tensor and the permeability tensor respectively. The matrix form representation of these tensors is analogous to that of the electric susceptibility and permittivity tensors given in Section 4.3. In the subsequent analysis all materials are assumed to be isotropic, for which χ_m and μ are scalar quantities.

There are two classes of linear magnetic materials: diamagnetics and paramagnetics.

Diamagnetic materials are those whose atoms or molecules have no intrinsic magnetic moments. They are characterized by a negative susceptibility which is very small compared with unity (of the order of -10^{-5}) so that the relative permeability is only very slightly less than unity. Examples of diamagnetic materials are copper, gold, silver, mercury, water and hydrogen gas.

Paramagnetic materials are those whose atoms or molecules have a non-zero total magnetic moment. They are characterized by a positive

Table 8.1 Magnetic susceptibility of some diamagnetic and paramagnetic materials. (The measured value of χ_m depends greatly upon the purity of the substance, especially its freedom from traces of ferromagnetic impurities, so that the values quoted may vary somewhat.)

Material	Susceptibility $10^5 \cdot \chi_m$
Diamagnetic	
Bismuth	$-16 \cdot 7$
Mercury	$-3 \cdot 2$
Silver	$-2 \cdot 64$
Lead	$-1 \cdot 7$
Copper	$-0 \cdot 9$
Water	$-0 \cdot 91$
Argon (s.t.p.)	$-0 \cdot 000945$
Hydrogen (s.t.p.)	$-0 \cdot 00021$
Paramagnetic	
Oxygen (liquid)	$+307 \cdot 0$
Oxygen (s.t.p.)	$+0 \cdot 2$
Air (s.t.p.)	$+0 \cdot 037$
Manganese	$+100 \cdot 0$
Platinum	$+30 \cdot 0$
Tungsten	$+8 \cdot 0$
Aluminium	$+2 \cdot 1$

susceptibility which is small compared with unity, but large if compared with that of diamagnetic materials, so that the relative permeability is slightly greater than unity. The susceptibility is inversely proportional to the absolute temperature and its value at room temperature is in the region of 10^{-3}. Examples of paramagnetic materials are aluminium, tungsten, platinum, manganese and air.

Table 8.1 shows the susceptibilities of a number of dia- and paramagnetic substances. It is apparent that for most practical purposes we may safely assume that the relative permeability of such substances is equal to unity.

In order to understand why some materials may exhibit a diamagnetic or a paramagnetic effect we shall give a simplified (classical) analysis of the basic processes responsible for these two effects. To do this we must examine the interaction between a material and an externally applied magnetic field from the microscopic or atomic point of view.

8.7 THE ORIGIN OF MAGNETIC DIPOLES

8.7.1 ORBITAL DIPOLE MOMENT

An atom of any element consists of a central positively charged heavy nucleus surrounded by electrons whose number is equal to the atomic number of the element. In the classical picture of an atom, the electrons are visualized as moving around the nucleus in fixed orbits. Consider

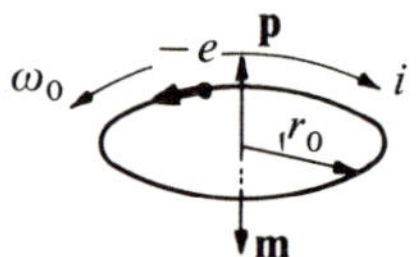

Fig. 8.13 Electron circulating with angular velocity ω_0 in an orbit of radius r_0

an electron circulating with an angular velocity ω_0 in an orbit of radius r_0 (Fig. 8.13). Since the electron has a mass m it possesses a mechanical angular momentum

$$\mathbf{p} = mv_0 r_0 \mathbf{a}_z = m\omega_0 r_0^2 \mathbf{a}_z \tag{8-90}$$

which is represented by a vector normal to the plane of the orbit in the direction of a right-handed screw rotated in the same sense as the electron.

The circulation of the electron in its orbit constitutes a small current loop whose magnetic moment is equal to the loop area multiplied by the equivalent loop current. The latter is equal to the charge passing any given point on the orbit per second, that is,

$$i = \frac{q\omega_0}{2\pi} = -\frac{|e|\omega_0}{2\pi}$$

According to the usual convention for the direction of current flow i circulates in the opposite direction to the negative charge e. The magnetic moment of the orbit is thus given by

$$\mathbf{m} = -\frac{|e|\omega_0}{2\pi}\,\pi r_0^2 \mathbf{a}_z = -\tfrac{1}{2}|e|\omega_0 r_0^2 \mathbf{a}_z \qquad (8\text{--}91)$$

in the direction shown in Fig. 8.13.

For a given direction of rotation it is evident that the two vectors $\mathbf{m}$ and $\mathbf{p}$ point in opposite directions; if the direction of rotation of the electron is reversed the direction of the vectors $\mathbf{m}$ and $\mathbf{p}$ is also reversed. Hence for any orbiting electron we may always write

$$\mathbf{m} = -\frac{|e|}{2m}\,\mathbf{p} \qquad (8\text{--}92)$$

The ratio

$$|\mathbf{m}|/|\mathbf{p}| = \gamma = \frac{|e|}{2m} \qquad (8\text{--}93)$$

is known as the gyromagnetic ratio.

In the simple Bohr theory of the atom, the angular momentum of an electron must be an integral multiple of $h/2\pi$, where h is Planck's constant, that is

$$p = \frac{nh}{2\pi}$$

For the innermost orbit of a hydrogen atom $n = 1$, so that

$$p = h/2\pi$$

and the magnitude of the corresponding magnetic moment

$$\beta = \frac{|e|}{2m}\frac{h}{2\pi} = 9 \cdot 27 \times 10^{-24} \text{ A m}^2 \qquad (8\text{--}94)$$

is called the *Bohr magneton* and is the fundamental unit for the magnitude of any magnetic dipole moment.

8.7.2 SPIN DIPOLE MOMENT

In addition to its orbital motion, an electron with an atom possesses an inherent angular momentum, referred to as spin, which corresponds (in the classical picture) to a rotation of the electron around its own axis. Because of this rotation the electron itself will possess a magnetic moment. If the spin magnetic moment and the spin angular momentum are computed by assuming that the electronic charge is homogeneously distributed throughout a sphere rotating around its own axis it will be found that the spin gyromagnetic ratio is the same as the orbital gyro-magnetic ratio (equation 8.93). There is, however, abundant experimental proof (Zeeman effect, Stern–Gerlach experiment and the Richardson–Einstein–de Haas effect) that this ratio is actually twice that found from classical theory, that is

$$\gamma \,(\text{spin}) = \frac{|e|}{m} \qquad\qquad (8\text{–}95)$$

This result is known as the magneto-mechanical anomaly of the spinning electron and was theoretically predicted by Dirac's relativistic wave-mechanical theory of the electron.

8.7.3 NUCLEAR DIPOLE MOMENT

The nuclei of atoms have angular momenta of the same order of magnitude as those of the electrons. However since mass of the nucleus is 10^3 to 10^5 times greater than that of the electron, the nuclear magnetic moments are so small compared to those associated with the electrons, that their contribution to the magnetic properties of materials may be neglected.

In the next section we shall show that the orbital and spin motions of the electrons in atoms give rise to diamagnetism, that is, a negative magnetic susceptibility.

8.8 DIAMAGNETISM

Consider an electron orbit placed in a magnetic field **B′** which makes an angle α with the normal to the plane of the orbit (Fig. 8.14). When a

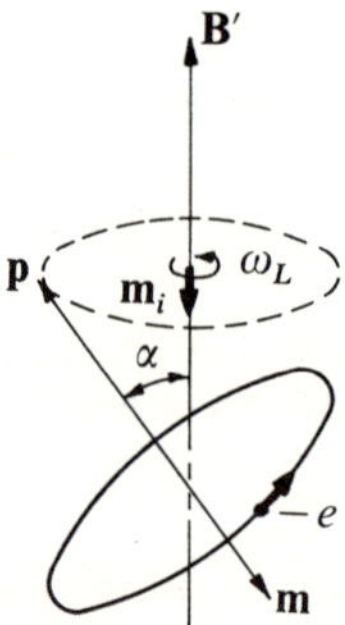

Fig. 8.14 Precession of an electron orbit

magnetic dipole of moment **m** is placed in a magnetic field **B′**, the torque exerted on the dipole is given by (equation 7.90)

$$\mathbf{T} = \mathbf{m} \times \mathbf{B}' = -\frac{|e|}{2m}\mathbf{p} \times \mathbf{B}' = \gamma \mathbf{p} \times \mathbf{B}'$$

and since torque is equal to the rate of change of angular momentum, we have that

$$\frac{d\mathbf{p}}{dt} = -\gamma \mathbf{p} \times \mathbf{B}' \qquad (8\text{-}96)$$

If we take the direction of **B′** as that of the z-axis, the torque components are

$$dp_x/dt = -\gamma p_y B_z' \qquad \text{(i)}$$
$$dp_y/dt = \gamma p_x B_z' \qquad \text{(ii)}$$
$$dp_z/dt = 0 \qquad \text{(iii)}$$

From (iii) it follows that

$$p_z = p\cos\alpha = \text{constant}$$

so that

$$\alpha = \text{constant}$$

From (i) and (ii) we have that

$$\frac{d^2 p_x}{dt^2} = -\gamma B_z' \frac{dp_y}{dt} = -(\gamma B_z')^2 p_x$$

and the solution of this equation is given by

$$p_x = A\cos(\gamma B_z' t + \epsilon)$$

where A and ϵ are constants. From (i) we have that

$$p_y = -\frac{1}{\gamma B_z'}\frac{dp_x}{dt} = A\sin(\gamma B_z' t + \epsilon)$$

It follows that,

$$(p_x^2 + p_y^2)^{1/2} = p\sin\alpha = A$$

Thus the projection of **p** on the xy-plane has a constant magnitude but rotates with an angular velocity

$$\omega_L = \gamma B_z' = \frac{|e|}{2m}B_z' \qquad (8\text{-}97)$$

The direction of rotation is always given by that of a right-handed screw progressing along the field direction. ω_L is the *Larmor angular velocity* with which the angular momentum $\mathbf{p}$, and hence the plane of the orbit, precesses about the B'_z-axis. ω_L is independent of α, of the radius of the orbit and of the direction of rotation of the electron in its orbit.

The precessional motion of an orbit in a magnetic field has the effect of producing an additional magnetic moment given by

$$\mathbf{m}_{\text{ind}} = -\tfrac{1}{2}|e|\omega_L\langle r^2\rangle\mathbf{a}_z$$

$$= -\frac{|e^2|}{4m}B'_z\langle r^2\rangle\mathbf{a}_z \tag{8-98}$$

where $\langle r^2\rangle$ is now the mean square value of the projection of the orbital radius on a plane perpendicular to $\mathbf{B}'$. If the radius of the circular orbit is r_0, then

$$\langle r_0^2\rangle = \langle x^2\rangle + \langle y^2\rangle + \langle z^2\rangle$$
$$\langle r^2\rangle = \langle x^2\rangle + \langle y^2\rangle$$

The direction of the induced moment $\mathbf{m}_{\text{ind}}$ is *always* opposite to that of the external magnetic field (Fig. 8.14). This is the origin of diamagnetism and accounts for the negative susceptibility of diamagnetic materials.

If an atom contains Z electrons we may write

$$\mathbf{m}_{\text{ind}} = -\tfrac{1}{2}|e|\omega_L\sum_i^Z\langle r_i^2\rangle\mathbf{a}_z$$

For a distribution of charge which is spherically symmetrical we have

$$\langle x_i^2\rangle = \langle y_i^2\rangle = \langle z_i^2\rangle = \tfrac{1}{3}\langle r_{oi}^2\rangle$$

and

$$\langle r_i^2\rangle = \tfrac{2}{3}\langle r_{oi}^2\rangle$$

so that

$$\mathbf{m}_{\text{ind}} = -\tfrac{1}{3}|e|\omega_L\sum\langle r_{oi}^2\rangle\mathbf{a}_z$$

If there are N atoms per unit volume, the magnetization $\mathbf{M}$ is given by

$$\mathbf{M} = N\mathbf{m}_{\text{ind}} = -\frac{Ne^2}{6m}\sum\langle r_{oi}^2\rangle\mathbf{B}'$$

Since $\mathbf{M}$ is very small we may write $\mathbf{B}' = \mu_0\mathbf{H} + \mu_0\mathbf{M} \simeq \mu_0\mathbf{H}$ and the magnetic susceptibility is thus,

$$\chi_m = \frac{\mathbf{M}}{\mathbf{H}} = -\frac{\mu_0 Ne^2}{6m}\sum\langle r_{oi}^2\rangle \tag{8-99}$$

A rough estimate of the magnitude of χ_m may be obtained by assuming that r_0 is about 10^{-10} m and N about 5×10^{28} m^{-3}. This gives $\chi_m \simeq 10^{-6} Z$; this order of magnitude agrees with the experimentally determined values (Table 8.1), and shows that the classical interpretation of diamagnetism as given above is phenomenologically correct.

8.9 TYPES OF MAGNETISM

In a multi-electron atom the individual vector orbital momenta and the individual spin momenta are added together according to certain rules (Hund's rules) to give a resultant orbital momentum **L** and a resultant spin momentum **S**. These two vectors are then coupled together to give the total angular momentum **J** of the atom. Because of the coupling between the spin and orbital motions (the electron spins in the magnetic field produced by the orbital motion whilst the electron orbits in the field of the spinning electron), it can be shown that both **L** and **S** precess rapidly about their resultant **J**. As a result of the magneto-mechanical anomaly of the spin (equation 8–95), the total magnetic moment of the atom, which is vectorially composed of $\mathbf{M}_L$ (associated with **L**) and $\mathbf{M}_S$ (associated with **S**), does not coincide with the direction of **J**. This total moment precesses around **J** so that only its component along **J** appears as the effective magnetic moment of the atom. It has been found that the ratio between the magnitude of this effective moment and the magnitude of the resultant angular momentum is given by

$$\gamma = g \frac{|e|}{2m} \qquad (8\text{--}100)$$

where g is a factor known as the Landé factor. For pure orbital motion with no spin $g = 1$ and for spin only $g = 2$. With both orbital and spin motions the factor lies between 1 and 2. The value of g is an indication of the relative contributions of the orbital and spin motions to the magnetic dipole moment.

Whether an atom does or does not have a resultant angular momentum, and hence also a resultant dipole moment, depends on the electronic configuration of the atom. For example, atoms with closed electron shells such as helium, neon or argon, do not have a resultant magnetic moment. When individual atoms whose resultant angular momentum differs from zero combine to form molecules or crystals, their momenta may compensate each other in such a way that their resultant is always zero. In crystals in particular, it is found that the orbital momenta of the electrons are totally or almost totally compensated and that the resulting magnetic moment is essentially due to the uncompensated spin momenta.

All substances whose atoms or molecules have a zero resultant angular momentum, and hence zero magnetic moment, exhibit a net diamagnetism when placed in an external magnetic field. This is because for every electron orbit and every electron spin there is an induced component of magnetic moment which is in opposition to the applied field irrespective of the actual direction of the angular momenta. Thus diamagnetism is a universal property of all materials. However, since diamagnetic effects are very small they can only be observed in the absence of other stronger magnetic effects such as paramagnetism and ferromagnetism which are associated with ·non-zero atomic or molecular resultant dipole moments.

Substances whose atoms or molecules have a non-zero total magnetic moment may be *paramagnetic, ferromagnetic, antiferromagnetic* or *ferrimagnetic*, depending on the degree of interaction between neighbouring dipoles. This interaction is not a classical dipole-dipole interaction but is due to the existence of quantum-mechanical forces which are electrostatic in character and are due to the wave nature of the electron. These forces are called 'exchange forces' because they result from an exchange of electrons between neighbouring atoms. If the exchange forces are zero or very weak, the dipole moments of a substance will be randomly oriented as a result of thermal agitation and the material

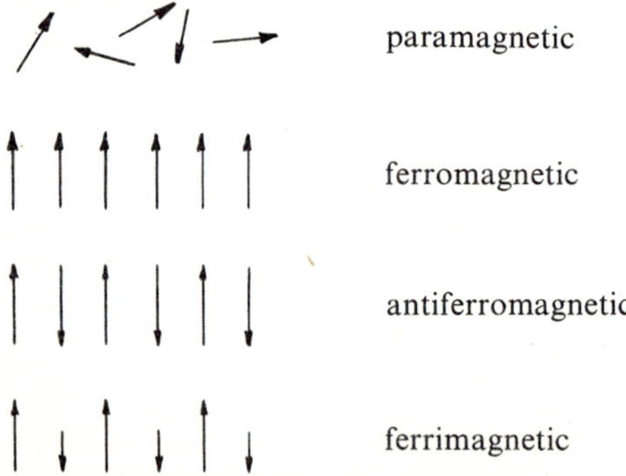

Fig. 8.15 Orientation of magnetic dipoles in the different types of magnetizable media

will be paramagnetic (Fig. 8.15). Under the action of strong exchange forces, neighbouring dipoles may tend to line up in parallel or in an antiparallel arrangement. Materials with a parallel alignment of neighbouring dipoles are ferromagnetic. In the case of an antiparallel alignment, if the adjacent dipoles are of equal magnitude the material is antiferromagnetic and if the magnitudes are unequal the material is ferrimagnetic.

8.10 PARAMAGNETISM

When there is little or no interaction between the individual dipoles, these will be randomly oriented so that the resultant macroscopic dipole moment per unit volume will be zero. When an external magnetic field is applied, there will be a tendency for the magnetic dipoles to align themselves parallel to the field and so produce a net magnetic moment in the body as a whole. The classical theory of paramagnetism is analogous to that of the orientational polarization in a dielectric (Section 4.6.3). If each atom or molecule has a permanent magnetic moment **m** and there are N atoms or molecules per unit volume, the magnetization is given by the Langevin equation,

$$M = Nm\left[\coth a - \frac{1}{a}\right] = Nm\,L(a) \qquad (8\text{--}101)$$

where

$$a = \frac{mB}{kT}$$

$$B = \text{applied field.}$$

At room temperatures, and for fields attainable in the laboratory the value of $a \ll 1$ so that $L(a) \simeq a/3$ and the magnetization is given by

$$M \simeq \frac{Nm^2B}{3kT} \qquad (8\text{--}102)$$

The magnetic susceptibility is therefore

$$\chi_m = \frac{\mu_0 Nm^2}{3kT} = \frac{C}{T} \qquad (8\text{--}103)$$

where the Curie constant $C = \mu_0 Nm^2/3k$. Equation 8–103 is the Curie law of paramagnetism. An estimate of the value of χ_m may be obtained by taking $N \simeq 10^{29}\,\text{m}^{-3}$, $m \simeq 10^{-23}$, $k \simeq 10^{-23}$ so that $\chi_m \simeq 0{\cdot}4/T$. At room temperature this is of the order of 10^{-3}. Table 8.1 shows that the experimentally determined values are of this order of magnitude. It should be mentioned that the measured susceptibility of paramagnetic substances includes a diamagnetic contribution, which, as mentioned in the previous section, is present in all substances. However, since this contribution is of the order of 10^{-5}, it is completely outweighted by the paramagnetic effect.

As the argument a of the Langevin function tends to infinity the function tends to unity and $M \to Nm$. Physically this corresponds to the situation in which there is complete alignment of all the dipoles along the field direction. Under these circumstances the magnetization is said to have attained *saturation*. At room temperatures saturation in paramagnetic materials can only be attained with infinite fields. To reach 60% of saturation the value of a is 2·4 and the corresponding flux density is

$$B = \frac{akT}{m} \simeq 900 \text{ Wb m}^{-2}$$

Such high values of B are unattainable in the laboratory, but magnetic saturation in paramagnetic materials can be attained at temperatures within a few degrees of absolute zero at much lower flux densities ($<5 \text{ Wb m}^{-2}$).

The Curie law given by equation 8–103 was derived with the assumption that the field acting on the magnetic dipoles is the same as the external applied field. This means that any interaction between dipoles has been completely neglected. This is strictly possible only for gases where the molecules are sufficiently far apart for their mutual interaction to be negligible. In liquids and solids, particularly ionic crystals, such interactions may be large and the susceptibility is found to obey the Curie–Weiss law,

$$\chi_m = \frac{C}{T-\theta_p} \text{ for } T > |\theta_p| \qquad (8\text{–}104)$$

where θ_p is the paramagnetic Curie temperature and can be either positive or negative. The law is only valid for temperatures $T > |\theta_p|$ because for $T = \theta_p$ the susceptibility would become infinite.

The Curie–Weiss law may be obtained by assuming that the field acting on the dipoles is equal to the external field plus an 'internal' or 'molecular' field which is proportional to the intensity of magnetization. Weiss was the first to postulate the existence of such a field in order to account for the occurrence of spontaneous magnetization in ferromagnetic substances. The local field may be expressed as

$$\mathbf{B'} = \mathbf{B} + \mathbf{B}_{int}$$

$$= \mu_0(\mathbf{H} + \lambda\mathbf{M}) \qquad (8\text{–}105)$$

where λ is the Weiss molecular field constant. If we assume that λ is small so that the magnetization is still proportional to the effective field we may replace $\mathbf{B}$ in equation 8–102 by $\mathbf{B'}$ so that,

$$M = \frac{Nm^2\mu_0}{3kT}(H+\lambda M)$$

from which

$$M = \frac{Nm^2\mu_0 H}{3kT[1-(Nm^2\mu_0\lambda)/3kT]}$$

and

$$\chi_{\mathrm{m}} = \frac{M}{H} = \frac{\mu_0 N m^2}{3k[T-(\mu_0 N m^2\lambda)/3k]} = \frac{C}{T-\theta_{\mathrm{p}}}$$

which is the Curie–Weiss law. It will be noted that $\theta_{\mathrm{p}} = \lambda C$.

8.11 FERROMAGNETISM

The most important magnetic materials for practical purposes are the ferromagnetic materials, so called because the best-known member is iron. Iron, cobalt, nickel and the rare element gadolinium are the only elements which are ferromagnetic. However many related alloys as well as some alloys composed wholly of non-ferromagnetic elements (for example the Heuster alloys formed from the non-ferrous metals copper, manganese and aluminium) are ferromagnetic. One great difference between paramagnetic and ferromagnetic materials is that the latter are capable of being magnetized very strongly by a magnetic field which is only a small fraction of that needed to produce a comparable magnetization in a paramagnetic material. For example, at room temperature the saturation flux density for a simple of polycrystalline iron is about $2\,\mathrm{Wb\,m^{-2}}$. The outstanding fact about ferromagnetic materials is that they are capable of retaining a considerable fraction of their magnetization when removed from the field. Furthermore the relationship between **B** and **H** is not linear for any one material but depends on the previous mechanical, thermal and magnetic history of the sample being tested. A property common to all ferromagnetic materials is that each material has a characteristic temperature (the ferromagnetic Curie temperature) above which it loses its ferromagnetic properties and behaves as a normal paramagnetic.

The relationship between **B**, **H** and **M** as given by

$$\mathbf{B} = \mu_0(\mathbf{H}+\mathbf{M})$$

is exact for all substances, including ferromagnetics. However, for ferromagnetics the relationship

$$\mathbf{B} = \mu_0\mu_{\mathrm{r}}\mathbf{H}$$

is *not linear* and the relative permeability μ_{r} cannot be represented by a single value. Instead, the relationship between **B** and **H** for a given ferromagnetic material is represented by a *magnetization curve* which is obtained experimentally from a test specimen which is initially demagnetized.* Figure 8.16 shows such a typical curve which is also known as the *initial* or *virgin magnetization curve*. For small values of

* See for example *Electric and Magnetic Measurements* by P. Vigoureux and C. E. Webb (Blackie, 1946).

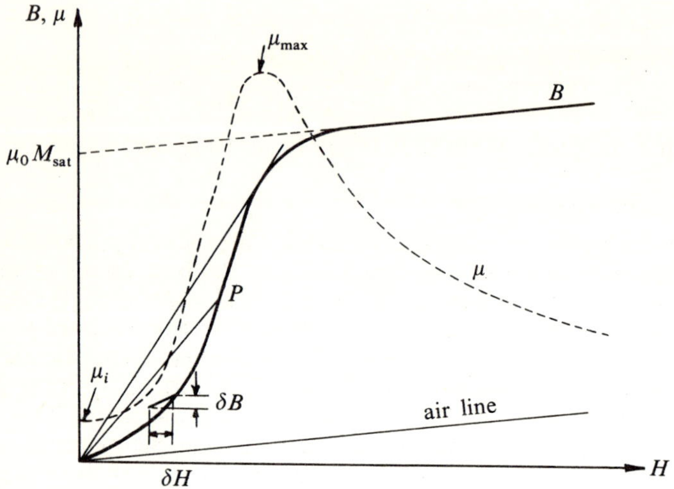

Fig. 8.16 Typical magnetization curve $(B-H)$ and the corresponding $\mu-H$ curve

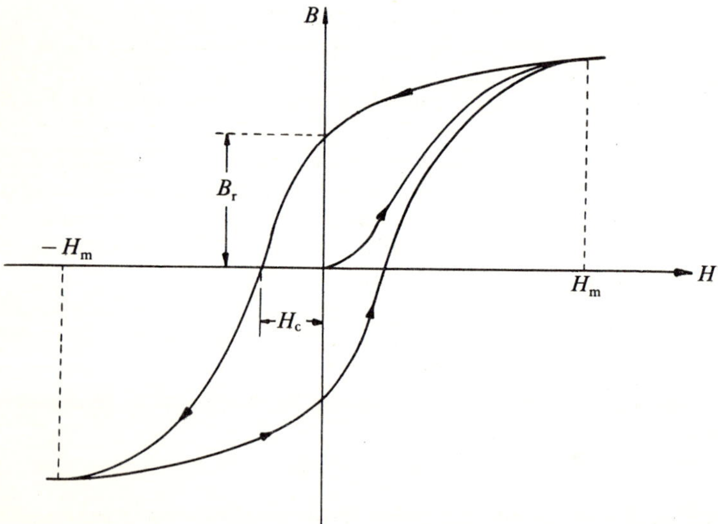

Fig. 8.17 Major hysteresis loop

H, B varies linearly with H. As H increases B increases rapidly at first but then tends asymptotically to the value $B = (H + M_{sat})$, where M_{sat} is the saturation magnetization of the material. It is also possible to draw the magnetization curve by plotting **M** versus **H**; thus the difference between the B ordinate on the $B-H$ curve at a given value of H and the ordinate on the $B = \mu_0 H$ line (vacuum or airline) at this same H value is equal to the magnetization times $\mu_0 \cdot \mu_0 M$ reaches a definite saturation value $\mu_0 M_{\text{sat}}$ whereas when saturation has been reached B increases in accordance with the air line $B = \mu_0 H$. For most ferromagnetic materials the difference between the two curves is negligibly small.

If after reaching saturation on the virgin curve the magnetizing force H is steadily reduced to zero, it is found that the flux density does not follow the original curve but lags behind H (Fig. 8.17). This phenomenon is called *hysteresis* and is a characteristic of all ferromagnetic substances. When the magnetizing force is reduced to zero, a finite flux density B_r remains. This flux density is known as the *remanence* and the original sample is now permanently magnetized. In order to reduce the remnant flux density to zero, a negative magnetizing force of magnitude H_c must be applied to the sample. This is known as the *coercivity*. The application of a steadily increasing magnetizing force of opposite polarity will eventually saturate the specimen with the magnetization reversed. By now restoring the magnetizing force to its original value for positive saturation, a closed loop will have been traced as shown in Fig. 8.17. This loop is called the *saturation* or *major hysteresis loop*. It is the largest hysteresis loop obtainable for a given material. The shapes of hysteresis loops differ widely from one substance to another. Substances for which H_c is small are magnetically 'soft' and have a narrow hysteresis loop and those with large H_c are magnetically 'hard' and have a wide hysteresis loop (Fig. 8.18). Typical values for the coercivity H_c are as follows (in A/m),

Very soft	Pure iron	2·0
	Permalloy	2·4
	Mumetal	2·4
Soft	Silicon-iron alloys	16
	Armco iron	80
Hard	Steel	1700
	Tungsten steel	5600
	Alnico alloys	30 000–120 000
Very hard	Platinum-cobalt alloys	340 000

Figure 8.19 shows a series of hysteresis loops of different sizes obtained for one sample but with different values of H_m. For the separate hysteresis loop to be repeatable—i.e., have a single contour—they should be cycled slowly several times between $\pm H_m$. The curve obtained by joining the tips of these minor loops is called the *normal magnetization* curve of the material. It is not identical with the initial

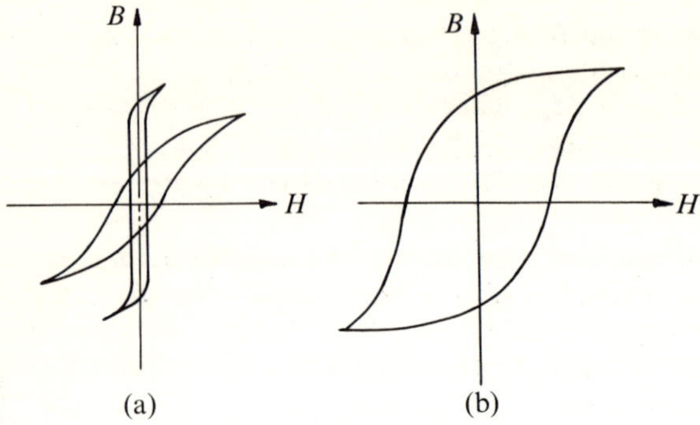

Fig. 8.18 Typical hysteresis loop for (a) magnetically soft materials, (b) magnetically hard materials

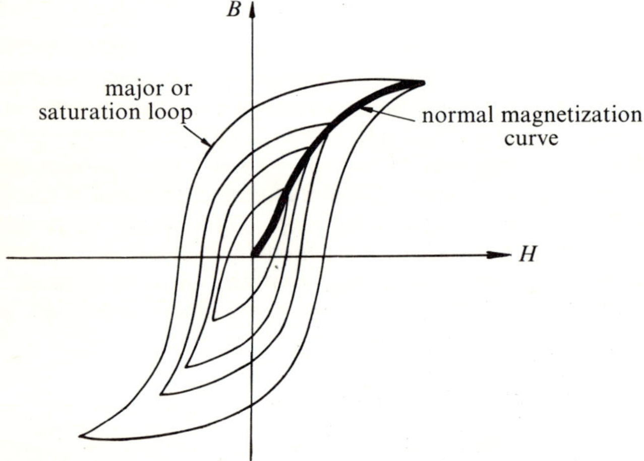

Fig. 8.19 Dependence of size and shape of hysteresis loop on H_m

magnetization curve but differs only very slightly from it, particularly near the origin.

For hysteresis loops smaller than the saturation loop, the value of $|B|$ for $H = 0$ is called the *remanent magnetism* whilst the value of $|H|$ for $B = 0$ is called the *coercive force*.

In order to demagnetize a ferromagnetic specimen—i.e., restore it back to the condition $B = 0$ at $H = 0$—it is made to follow a $B-H$ curve such as the one shown in Fig. 8.20. H_m must be reduced very slowly for each pseudo-cycle otherwise the origin will not be attained. The process can be easily and quickly carried out by using a 50 Hz alternating current.

It will be shown in Chapter 10 that the area of a hysteresis loop represents the energy, per unit volume of specimen, which is expended in magnetizing and demagnetizing the specimen. This energy appears as heat (perceived as an increase in the temperature of the specimen) and thus represents a loss of energy. For this reason it is highly desirable that all electrical machinery, such as generators, motors and transformers, in which the ferromagnetic armature or core is subjected to an alternating magnetic field, have materials with low hysteresis loss. Iron alloys containing from 0.8% up to 5% silicon have relatively less hysteresis losses and are extensively used in most electrical machines. For special applications where expense is not a limiting factor and losses must be very low, special low-loss alloys such as Permalloy (78.5% Ni, 21.5% Fe) are used.

8.11.1 PERMEABILITY OF FERROMAGNETICS

Although the permeability is defined as the ratio B/H, it is evident that in the case of ferromagnetic materials little significance can be attached to any constant value of μ. For a given material various specific permeabilities have been defined using the initial magnetization curve as reference. By definition the permeability is the slope of the line OP between the origin and any point P on the curve (Fig. 8.16). It is seen that the slope starts from a definite value, rises rapidly to a maximum at the knee of the magnetization curve and then decreases tending to the value μ_0 as saturation is approached.

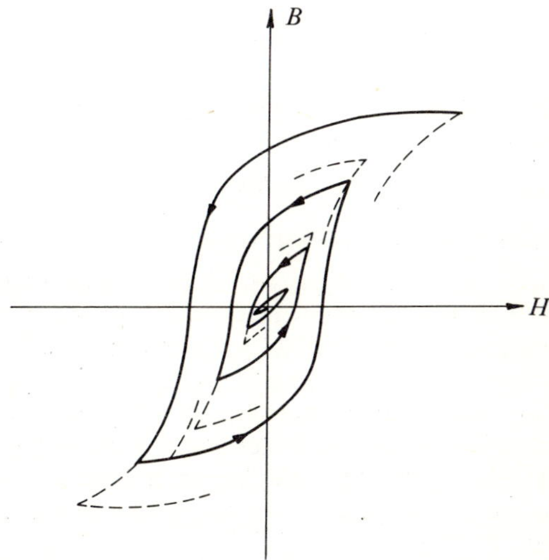

Fig. 8.20 Demagnetization of a magnetized ferroelectric specimen

The slope at the origin defines the *initial permeability* μ_i. The slope at the knee of the curve defines the *maximum permeability* μ_{max}. The slope at the tangent at any point on the curve is the *differential permeability*

$$\mu_{diff} = dB/dH \tag{8-106}$$

If at some point on the magnetization curve H is varied reversibly by δH and the flux density varies reversibly by δB, the *reversible permeability* is defined by

$$\mu_{rev} = \delta B/\delta H \tag{8-107}$$

If a large constant magnetizing force (H_{DC}) and a small alternating one (H_{AC}) are superimposed the resulting hysteresis loop is known as a

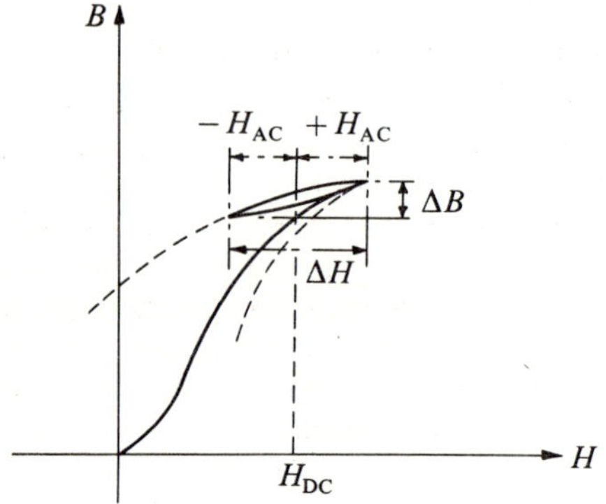

Fig. 8.21 Minor hysteresis loop

minor hysteresis loop (Fig. 8.21). The slope of the line connecting the tips of this minor loop is the *incremental permeability*,

$$\mu_{inc} = \Delta B/\Delta H \tag{8-108}$$

8.11.2 PHENOMENOLOGICAL THEORY OF FERROMAGNETISM

As already mentioned, an important characteristic of ferromagnetic substances is that the magnetization will approach saturation at ordinary temperatures with applied magnetic fields which are only a very small fraction of those required to produce a comparable degree of magnetization in paramagnetic substances. This means that the Weiss molecular field in ferromagnetic materials must be very large. Thus the assumption that the argument of the Langevin function is small so that the magnetization is proportional to the field (equation 8–102) is no longer valid and we must write

$$M = Nm\,L(a') \tag{8-109}$$

where

$$a' = \frac{\mu_0 m}{kT}(H + \lambda M) \tag{8-110}$$

and since at saturation $M_{\text{sat}} = Nm$,

$$M/M_{\text{sat}} = L(a') \tag{8-111}$$

Equation 8–110 may be rewritten as

$$\frac{M}{M_{\text{sat}}} = \frac{kT}{\mu_0 m^2 N\lambda}a' - \frac{H}{mN\lambda} \tag{8-112}$$

The value of M/M_{sat} can now be obtained graphically from the intersection of the Langevin function given by equation 8–111 and the straight line given by equation 8–112, as shown in Fig. 8.22.

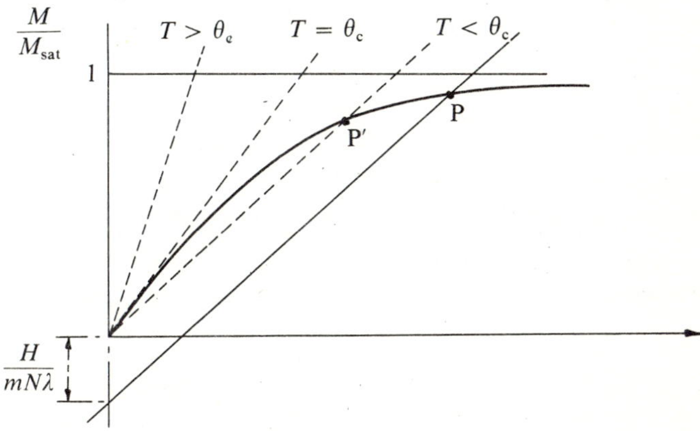

Fig. 8.22 Graphical solution of equation 8–112

The slope of the line is determined by the temperature and its intercept by H. When $H = 0$ the line passes through the origin. For temperatures such that the slope of the line is smaller than the slope of the tangent to the Langevin function at the origin, a point of intersection will exist. This means that even in the absence of an external field the material can be magnetized spontaneously to some value M which is a function of the temperature T, so long as T is below a certain critical temperature θ_F. Since the slope of the Langevin function at the origin is 1/3 we have that

$$\frac{k\theta_F}{\mu_0 m^2 N\lambda} = \frac{1}{3}$$

or

$$\theta_F = \frac{\mu_0 N m^2 \lambda}{3k}$$

which is the ferromagnetic Curie temperature. If the temperature is greater than θ_F there can be no magnetization with zero external field, and this corresponds to paramagnetic behaviour. It will be noted that on the basis of the simple Weiss theory the temperature θ_F above which a substance ceases to be ferromagnetic is the same as the temperature θ_p above which the Curie–Weiss law for paramagnetism (equation 8–104) is valid. Experimental values for the Curie temperatures show

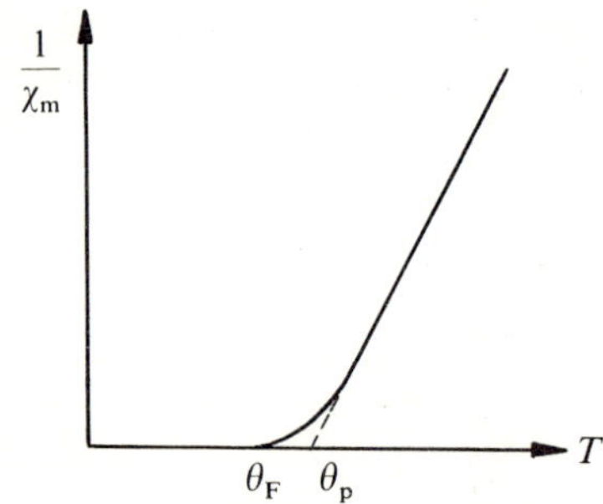

Fig. 8.23 Variation of $1/\chi_m$ with temperature

that θ_p is usually somewhat higher (10–20°C) than θ_F (Fig. 8.23). The Curie temperatures for the three most important ferromagnetic elements are

	Fe	Co	Ni
θ_p	1093	1428	650
θ_F	1043	1393	631

For a given temperature the value of M/M_{sat} where M is the spontaneous magnetization at this temperature (and is equal to M_{sat} at 0°K), is determined by the intersection of the Langevin function with the line corresponding to equation 8–112 for $H = 0$. When this procedure is repeated for different temperatures one can plot M/M_{sat} as a function of T/θ_F. Figure 8.24 shows the general shape of the curve. The dotted curve is the one obtained when quantum mechanical corrections are introduced* and is in good agreement with experiment.

* The Langevin function of the classical theory is a result of the assumption that the magnetic dipoles are free to rotate and may thus take any angular position with respect to the direction of the applied field. In the quantum theory the dipoles are not freely rotating but are restricted to a finite set of orientations relative to the applied field. In both theories the Weiss internal-field postulate leads to the same phenomenological conclusions, namely the existence of a spontaneous polarization and of a Curie temperature.

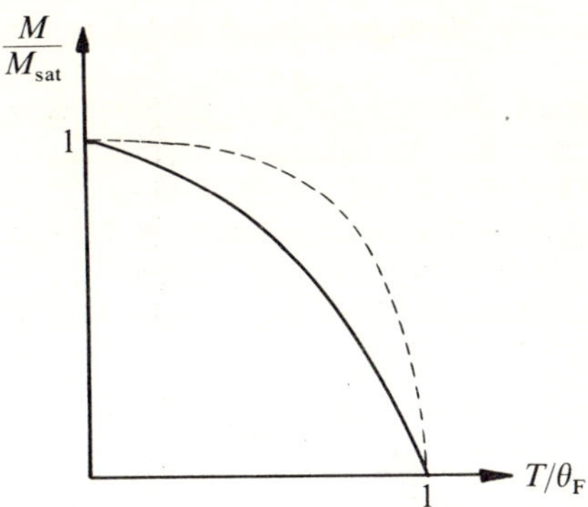

Fig. 8.24 Plot of M/M_{sat} as a function of T/θ_F. Solid curve from equation 8–112; broken curve from quantum mechanical consideration and from experiment

The Curie constant and the Curie temperature can be determined from measurements of the susceptibility as a function of temperature and the Weiss molecular field constant can be calculated from the relationship $\theta = \lambda C$. For ferromagnetic elements λ is found to be of the order of 10^3. This is 10^3 times more than one would expect if one assumed that the internal field is due to the classical magnetic interaction of atomic dipoles. It is not therefore possible to attribute the existence of large internal fields to dipole-dipole interaction. The first theoretical explanation of the existence of such fields in ferromagnetic materials was given by Heisenberg in 1928. He showed that the internal field is due to exchange interactions between spinning electrons; these interactions are basically electrostatic in character and arise from the wave-nature of the electrons. Discussion of the Heisenberg theory and other quantum-mechanical theories of ferromagnetism is outside the scope of this book and the interested reader is referred to specialized books on magnetism*.

8.11.3 FERROMAGNETIC DOMAINS

The task of any theory of ferromagnetism is to explain the existence of spontaneous magnetization, the existence of a Curie temperature, the possibility of a ferromagnetic material to exist in an unmagnetized (virgin) or a demagnetized condition and finally the existence of an hysteresis loop. We have seen from the above discussion that Weiss's internal-field postulate was successful in explaining the first two phenomena. To account for the remaining experimentally observed phenomena, Weiss introduced a second postulate: the existence of *domains*. According to this postulate, any ferromagnetic material is

* See *Physical Principles of Magnetism*, by F. Brailsford (Van Nostrand, London, 1966).

composed of a number of regions, each spontaneously magnetized to saturation corresponding to a particular temperature (in accordance with the results given above) and in some particular 'easy' direction; these regions are known as domains and each domain may have several easy directions. This is because single crystals of ferromagnetic substances exhibit *magnetic anisotropy*, that is, it is easier to magnetize the crystal along certain directions than along others. For example, in a single iron crystal, which is of the body-centred cubic type (Fig. 8.25),

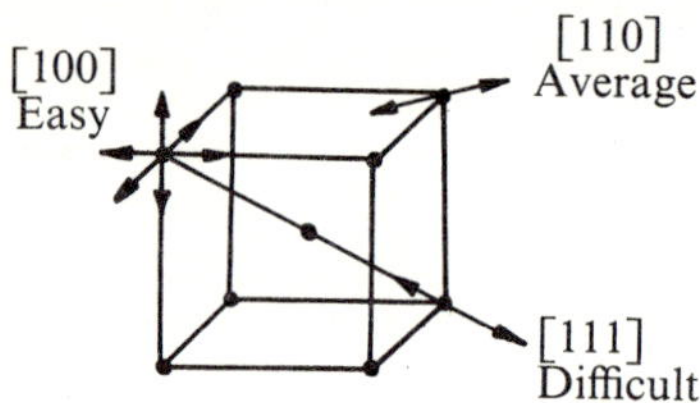

Fig. 8.25 Easy and hard directions of magnetization for a body-centred iron crystal

the six cube edge directions [100] are the directions of easy magnetization whereas the most difficult or hard direction is along the cube diagonal [111]. A single crystal will itself consist of a number of domains, with each domain spontaneously magnetized to saturation along an easy direction.

In a demagnetized polycrystalline specimen, which is composed of a large number of single-crystal fragments called crystallites, each crystallite consists of a number of domains spontaneously magnetized along some easy direction. The directions of domain magnetization are distributed at random among the various possible easy directions so that the net magnetization of the specimen is zero. The demarcation between two domains is called a *domain wall* and a wall is usually thought of as a very thin transition region whose volume is much smaller than that of the domains and throughout which the direction of magnetization changes gradually as we move from one domain to another (Fig. 8.26). The thickness of a domain wall may vary from material to material but a typical wall between two domains magnetized

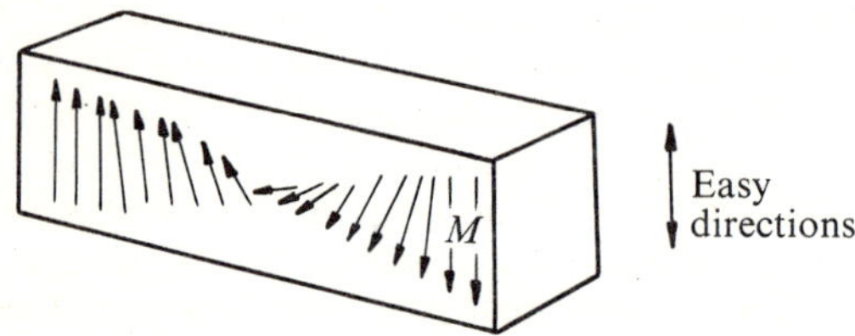

Fig. 8.26 Gradual change in direction of magnetization throughout a domain wall

at 180° to each other extends over about 100 atoms. Domain sizes may vary from 10^{-6} cm to 1 cm in linear dimensions.

When a demagnetized specimen is subjected to an external magnetic field, at first those domains whose magnetizations are in the same sense as the external field grow by boundary displacement at the expense of the unfavourably oriented domains (Fig. 8.27). If the field is weak the

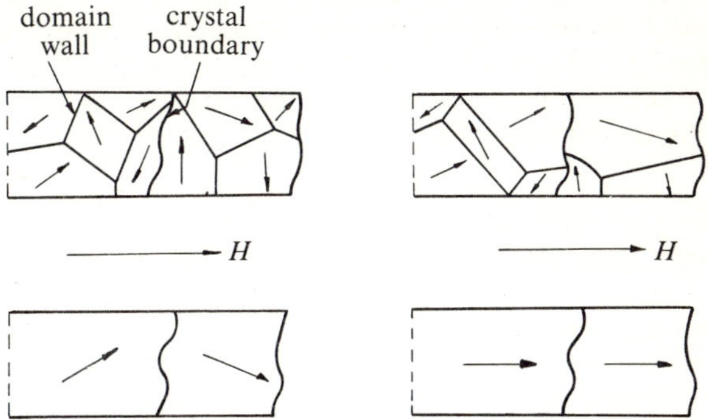

Fig. 8.27 Domain growth by boundary displacement

displacements are reversible (this corresponds to the initial part of the magnetization curve shown in Fig. 8.28); as the field increases the boundary displacements increase discontinuously by jumps and are irreversible (this corresponds to the steep part of the magnetization curve). At high fields the domain magnetization vectors rotate towards the direction of the field if the latter does not coincide with an easy direction of magnetization; rotation of the magnetization away from an easy direction is a difficult process and a large increase in H is neces-

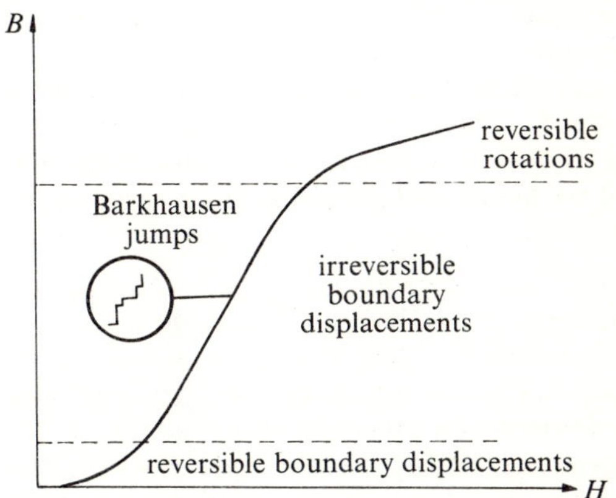

Fig. 8.28 Relationship between boundary displacements and magnetization curve

sary to produce a small increase in M. Magnetic saturation is reached when the domain growth and rotation processes are complete. If the field is now reduced, the magnetization vectors rotate back to the easy directions nearest the field and when the external field is zero there is a remnant magnetization (remanence) due essentially to the irreversible domain growth which has taken place during the initial magnetization process. A reverse field is now necessary to reduce the magnetization and the field required to produce zero magnetization is the coercive field (coercivity). We thus see that by means of the Weiss domain postulate it is possible to account for the existence of ferromagnetic materials in an unmagnetized state and for the phenomenon of hysteresis.

The existence of domains can be demonstrated experimentally. If a ferromagnetic specimen is surrounded by an induction coil connected to headphones or to a loudspeaker via an amplifier, on increasing the applied field a series of clicks are produced (Barkhausen effect). These clicks are especially pronounced in the steep portion of the magnetization curve and are caused by emfs of very short duration induced in the coil by sudden changes in the magnetization of the specimen. These changes are caused by discontinuous jumps in the movement of the domain walls; the jumps occur at structural imperfections such as impurities or regions of local strain. Such imperfections may be assumed to constitute a barrier to domain wall movement and a certain minimum field would be required for the domain wall to overcome it. Direct evidence for the existence of domains can be obtained by using a technique originally developed by Bitter in 1931. It consists of sprinkling a colloidal suspension of fine iron oxide of grain size of the order of 1 μm or smaller onto a smooth electrolytically polished surface. At the domain boundaries there is a little leakage flux out of the surface and patterns are formed due to the aggregation of particles at these boundaries. These patterns, called Bitter patterns, can be observed under a microscope. There are now many other techniques for revealing domain walls; these include the use of the electron transmission microscope (thin films), polarized light (Kerr and Faraday effects) and photo-electronic emission.

The ease with which a ferromagnetic material can be magnetized or demagnetized depends to a large extent on the ease with which domain walls can move reversibly. The movement of domain walls is impeded by local strains or small impurity particles and neither reversible boundary movements nor reversals of magnetization can take place as readily as in a perfect material. Materials in which domain walls can move easily and reversibly are the magnetically soft materials which are characterized by a high permeability (the magnetization changes by large amounts for small changes in the magnetizing force) and a low coercive force. In magnetically hard materials domain wall movement is difficult so that the permeability is relatively low and the coercive force is high.

8.12 MAGNETOSTRICTION

When ferromagnetic monocrystals are magnetized a change occurs in their linear dimensions; this phenomenon is called magnetostriction and is an elastic effect. The magnitude of the dimensional changes depends on the direction of magnetization with respect to the crystal axes. An overall magnetostrictive effect is also exhibited by poly-crystalline materials. Figure 8.29 shows the relationship between

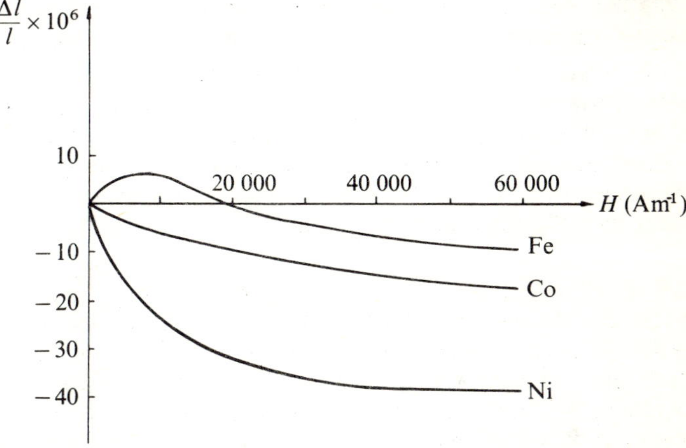

Fig. 8.29 Relationship between longitudinal strain in the direction of magnetization and magnetizing force for polycrystalline bars of iron, cobalt and nickel

longitudinal strain in the direction of magnetization and magnetizing force for polycrystalline bars of iron, cobalt and nickel; of these three basic ferromagnetic elements nickel exhibits the largest magnetostriction. It will be noted that for all three elements large values of H are required to reach magnetostriction saturation.

If tension is applied to a ferromagnetic material such as nickel, which normally contracts when magnetized, it is found that a much larger magnetizing force is required to produce a given flux density—i.e., the permeability decreases—and that there is a large decrease in the remanent magnetism. Both these effects are used in certain magneto-strictive transducers used in force-measuring instruments and in ultrasonic generators.

When a ferromagnetic material is subjected to an alternating magnetizing force, the same dimensional changes occur during each half-cycle. These small but rapid changes in dimensions produce a noise whose frequency is twice that of the electrical supply; this is the principal source of the familiar hum associated with power transformers.

8.13 FERRITES

When an alternating current flows through a coil wound on a magnetic core, a time-varying flux is set up in the core material and this flux induces an electric field. Since ferromagnetic core materials are normally good conductors of electricity, this electric field produces circulating parasitic currents known as eddy currents (see Section 10.8). The flow of these currents results in a loss of energy known as eddy-current loss. This loss increases as the square of the frequency and at radio and microwave frequencies it becomes so overwhelmingly large that the use of ferromagnetic materials becomes impractical.

The search for materials with high resistivities (and hence low eddy losses) but still possessing ferromagnetic properties led to the development of non-metallic ferromagnetic materials, known as ferrites, having resistivities of from 10^{-1} to 10^4 ohm metre, an improvement over conventional metallic ferromagnetic materials of between 10^6 and 10^{12} times.

Ferrites are chemical compounds of the general form MFe_2O_4, where M represents a divalent metal. Magnetite (Fe_3O_4) is the earliest known ferrite but the divalent metal can be either Cu, Mg, Mn, Ni, Co, Zn or Cd. Zinc and cadmium ferrites are the only simple ferrites which are non-magnetic; all other ferrites are ferromagnetic. Mixed ferrite cores are produced by sintering one or more single ferrites into a hard dense substance by processes similar to those used in the production of common insulating ceramics. By varying the composition and the manufacturing procedures it is possible to vary the properties (resistivity, permeability, shape of hysteresis loop) of the final product. The most important of the mixed ferrites are the Mn–Zn ferrites (trade name Ferroxcube 3), the Ni–Zn ferrites (Ferroxcube 4) and the Mg–Mn Ferrites (Ferroxcube 6). The first two types are used for all high frequency engineering applications as cores for inductance coils, pulse transformers, television line output transformers, deflection coils in television receivers, as rods for aerials and as unidirectional isolators in wave-guides. The third type of ferrites have an almost rectangular hysteresis loop and are used as switches or as storage (memory) elements in digital computers. Ferrites in general are not considered to be substitutes for iron used in machines and transformers because their saturation flux density ($<0{\cdot}4\,\mathrm{Wb\,m^{-2}}$) is appreciably lower than that of iron ($\sim 2\,\mathrm{Wb\,m^{-2}}$).

8.13.1 FERRITE CORE MEMORY

Because Mn–Zn ferrites possess an almost rectangular hysteresis loop, they are particularly suitable for use as magnetic information storage devices and as such find wide application in high-speed digital computers. Consider the ferrite core shown in Fig. 8.30a; the core material has the hysteresis loop shown in Fig. 8.30b. The loop intersects the

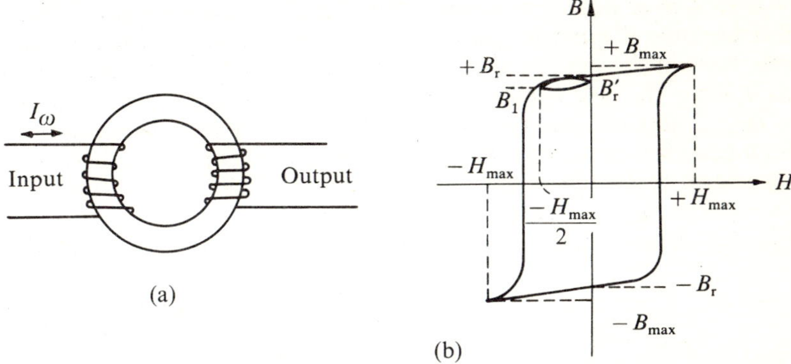

(a)

(b)

Fig. 8.30 (a) Ferrite core with input and output windings; (b) hysteresis loop of core material

B-axis at the points $+B_r$ and $-B_r$. At these points $H = 0$, that is $I_\omega = 0$ and the core acts like a small permanent magnet with either positive or negative remanence. The two states of the core $(+B_r, 0)$ and $(-B_r, 0)$ may be used to represent the binary digits one and zero respectively. No change in the state of the core can occur until the magnetizing force (current) reaches a value close to $\pm H_c$; the magnetization then switches abruptly to the opposite direction. This abrupt reversal of magnetization, and hence of flux density and flux, produces an induced emf across the output winding.

We thus see that by applying a positive or negative current pulse to the input winding, a binary digit may be 'written' in the core; this number will then be stored in the core for an indefinite period. When needed the digit can be 'read' by observing the emf produced in the output winding when a positive or negative current pulse is applied to the input winding. If the core was initially in the state $(+B_r, 0)$ and is now subjected to a magnetizing pulse $-H_{max}$, the flux change produced will be proportional to $(B_{max} + B_r)$. If, however, the core was initially in the negative remanent state $(-B_r, 0)$ the flux change will be proportional to $(B_{max} - B_r)$. In the first case the emf induced in the output winding will be large and in the second case it will be negligibly small. It is important to note that energy is only required to change the state of the core but that no energy is needed to maintain the core in either of its two remanent states.

The storage and read properties of ferrite cores are utilized to form a so-called Magnetic Matrix Store which may contain millions of separate cores each representing one binary digit of information. The successful operation of such a matrix depends on the fact that a magnetizing force of magnitude H_{max} can switch the state of a core from $-B_r$ to $+B_r$ or vice versa, whereas half that force has a negligible effect. For example if a magnetizing force of $-H_{max}/2$ is applied to a core originally in the state $(+B_r, 0)$, the flux density falls to B_1 and returns to B'_r when the field is removed (Fig. 8.30b). Repeated applications of $-H_{max}/2$ take the core round a closed minor hysteresis loop and since for a good ferrite material B'_r is only slightly smaller than B_r, there is no significant reduction in the remanent flux density. As an example of the operation of a storage matrix consider the 4×4 matrix shown in Fig. 8.31. Each core is threaded by one horizontal and one vertical wire

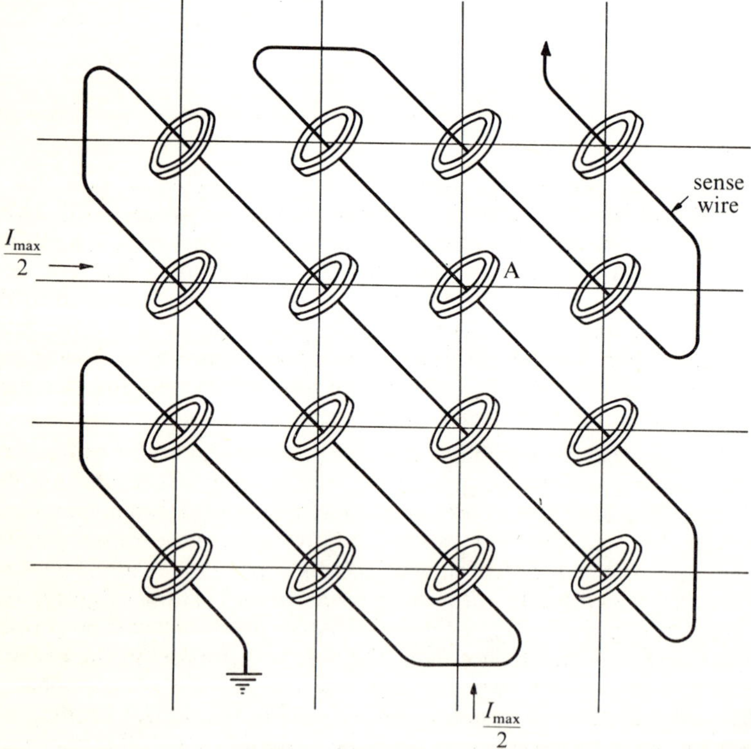

Fig. 8.31 A 4×4 storage matrix

while a single 'sense' wire, threaded diagonally, links all the cores.

Assume that all the cores are initially in the negative state $(-B_r, 0)$, that is storing zeros. To write 1 in core A say, current pulses $+I_{max}/2$ corresponding to a magnetizing force of $+H_{max}/2$ are applied to the horizontal and vertical wires which pass through core A. This core will be subjected to the total field $+H_{max}$ and will switch from the state '0' to the state '1'. The other cores threaded by the same wires will experience a magnetizing force of only $+H_{max}/2$ and their states will not be affected. To read the information stored in a core, current pulses of amplitude $-I_{max}/2$ are applied to the two wires linking that core. If a '1' has been stored in the core the flux density will change from $+B_r'$ to $-B_{max}$ and an emf will be induced in the output winding. If a '0' has been stored the flux density changes from $-B_r$ to $-B_{max}$ and the induced emf is negligibly small. Since only one core is switched at any time the output winding takes the form of a single wire linking all cores.

8.14 LAW OF REFRACTION FOR LINES OF MAGNETIC FLUX

When the boundary between two media carries no surface current density, the boundary conditions given by equations 8–77 and 8–81 may be written as

$$B_{n1} = B_{n2} \tag{8-113}$$

$$H_{t1} = H_{t2} \tag{8-114}$$

In terms of Φ_H (see equation 8–57), equations 8–113 and 8–114 may be written as

$$\mu_1 \frac{\partial \Phi_{H1}}{\partial n} = \mu_2 \frac{\partial \Phi_{H2}}{\partial n} \tag{8-113a}$$

$$\Phi_{H1} = \Phi_{H2} \tag{8-114a}$$

With reference to Fig. 8.32 it can be seen that

$$B_{n1} = B_1 \cos \theta_1, \; B_{n2} = B_2 \cos \theta_2$$

$$B_{t1} = B_1 \sin \theta_1, \; B_{t2} = B_2 \sin \theta_2$$

Introducing the permeabilities of the two media using the constitutive relation $\mathbf{B} = \mu \mathbf{H}$, the boundary relations may be written as

$$B_1 \cos \theta_1 = B_2 \cos \theta_2 \tag{8-115}$$

$$\frac{1}{\mu_1} B_1 \sin \theta_1 = \frac{1}{\mu_2} B_2 \sin \theta_2 \tag{8-116}$$

and it follows that

$$\mu_1 \cot \theta_1 = \mu_2 \cot \theta_2 \tag{8-117}$$

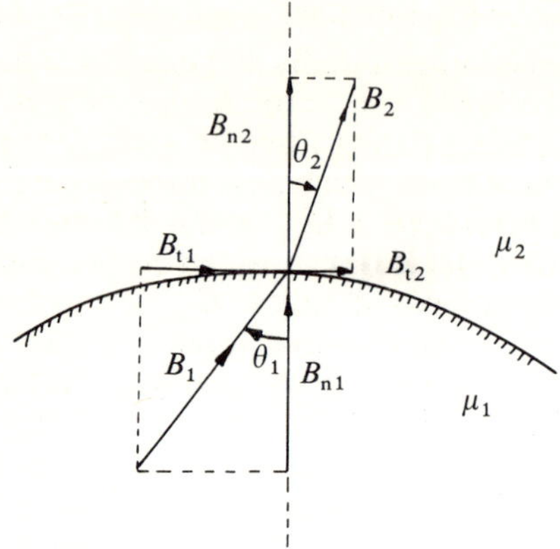

Fig. 8.32 Refraction of **B**-lines at a boundary

which is the refraction law for magnetic flux lines. The flux lines are bent away from the normal in the medium of higher permeability. When $\mu_1 \gg \mu_2$ then θ_2 will be very small and the direction of the flux lines in medium 2 may be taken as normal to the boundary surface, provided that θ_1 does not equal exactly 90°. Such a case is encountered in practice when flux lines traverse the boundary between a ferromagnetic material and air (or a non-magnetic medium); for example, if for a given flux density the relative permeability of medium 1 is 8000 and that of medium 2 is unity, then for $\theta_1 = 87°$ we find that $\theta_2 = 0.14°$.

Let us now examine the boundary conditions for the limiting case in which the permeability of one of the media becomes infinite. Thus if $\mu_1 \to \infty$ with μ_2 finite, the boundary condition (equation 8–117) will be satisfied in the two following cases,

(a) $\theta_2 \to 0$, θ_1 can have any value. In this case both B_1 and B_2 have finite values and $H_1 = B_1/\mu_1 = 0$.

(b) $\theta_1 = \pi/2$, θ_2 can have any value including $\pi/2$. For a finite value of B_2 it follows from equations 8–115 and 8–116 that $B_1/\mu_1 = H_1$ is finite and that $B_1 \to \infty$.

Examples in which boundary conditions corresponding to the above two cases will be found in Section 11.5.

8.15 MAGNETIC CIRCUITS

The functioning and the performance characteristics of electromagnetic devices such as transformers, motors, generators, reactors, relays, measuring instruments and loudspeakers, depend on the production of intense magnetic fields by current-carrying coils wound on ferromagnetic cores and on the distribution of these fields in their magnetic structure. In order to study the performance characteristics of such devices, it is necessary to determine the distribution of the magnetic flux density which corresponds to the geometric configuration of the device. This is essentially a three-dimensional boundary-value field problem whose solution is in general so complex that very few practical cases can be solved analytically. In many cases, however, it is possible, by introducing certain simplifying assumptions, to obtain solutions which are sufficiently accurate for most practical purposes. These assumptions, which are valid for constant or slowly varying magnetic flux, are:

(*a*) The magnetic flux tends to confine itself almost entirely to the high permeability paths of the ferromagnetic structure.

(*b*) The dimensions of the magnetic structure are such that the magnetic flux is uniform over any cross-section of the structure. This assumption is justified as long as the linear dimensions of the cross-section are much smaller than the length of the mean flux line.

(*c*) For soft ferromagnetic materials hysteresis is completely ignored so that the relation between B and H is determined by the magnetization curve of the material. Overall hysteresis effects, if present, are taken into consideration separately. It should be noted that the first two assumptions are analogous to the ones implicitly made in calculating an electric current flowing in a cylindrical conductor, namely that the current flow is confined to follow the conductor path and that the current density is uniform over the conductor cross-section. These two approximations form the very basis of the electric circuit concept and we shall see that in the magnetic case the assumptions made lead to the magnetic circuit concept.

Consider as a simple example a toroid having N uniformly wound turns through which a current I is flowing (see Fig. 7.23). As discussed in Section 7.7, the magnetic flux density is constant in magnitude at all points on a circular path of radius r and its direction is tangential to that path. From equation 7–42 we have that

$$B_0 = \frac{\mu_0 NI}{2\pi r}$$

where the subscript indicates that the core of the toroid is air. The magnetizing force is

$$H = \frac{NI}{2\pi r}$$

Now if instead of air the core of the toroid is a ferromagnetic material of relative permeability μ_r, it is evident that the magnetizing force H

will not change but that the flux density $B = \mu_0\mu_r H$ will not be μ_r times B_0, that is,

$$B = \frac{\mu_0\mu_r NI}{2\pi r}$$

If we consider a point just inside the core surface and a point just outside it then the boundary condition given by equation 8–114 tells us that H at these two points is the same and hence the magnetic flux density at the outer point will be μ_r times smaller than its value at the inner point. Since the magnitude of μ_r for ferromagnetic materials (in the usual operation range of flux densities) is of the order of 10^3–10^4, it is a sensible approximation to assume that practically the whole of the magnetic flux will be confined to the high permeability section of the core (assumption a).

If R is the radius of the toroid axis, $L = 2\pi R$ the mean length of the toroid and S its cross-sectional area, the magnetic flux in the core is given by (assumption b),

$$\Phi = BS = \mu_0\mu_r NIS/L$$

which may be conveniently rewritten as,

$$\Phi = \frac{NI}{L/\mu_0\mu_r S}$$

If this equation is compared with Ohm's law for electric circuits,

$$\text{current} = \frac{\text{emf}}{\text{resistance}}$$

a formal mathematical analogy may be established between a *magnetic circuit* and an electric circuit if we compare the flux Φ with electric current, the term NI with electromotive force and the term $L/\mu_0\mu_r S$ with resistance. We define *magnetomotive force* (mmf) as

$$\text{mmf} = NI \qquad \text{ampere-turn} \qquad (8\text{--}118)$$

$$= \oint \mathbf{H} \cdot \mathbf{dl} \qquad \text{(from equation 8--37)}$$

and *reluctance* as

$$\mathcal{R} = \frac{L}{\mu_0\mu_r S} \qquad \text{ampere-turn metre}^{-1} \qquad (8\text{--}119)$$

The reciprocal of $\mathcal{R}$ is called the *permeance*. 'Ohm's law' for a magnetic circuit may thus be written as,

$$\text{flux} = \frac{\text{mmf}}{\text{reluctance}} = \text{mmf} \times \text{permeance} \qquad (8\text{--}120)$$

The rules for the addition of emfs and for the combination of series and parallel resistances, which are derived from Ohm's law, also hold for magnetomotive forces and reluctances. It is important to realize, however, that reluctance is only a qualitative concept and no significance can be attached to its numerical value. This is because the permeability is not constant for a given ferromagnetic material but its value depends on the point of operation on the magnetization curve or on the hysteresis curve of the material. Thus the magnetic circuit is a non-linear circuit, that is the flux is not proportional to the mmf. Reluctance can only be exactly defined as the ratio between the mmf and the flux across the uniform area of a finite length of magnetic circuit, for a particular value of flux.

8.15.1 MAGNETIC CIRCUIT WITH AIR GAP (SERIES CIRCUIT)

Many electromagnetic and electromechanical devices require an air gap in their magnetic circuit. Examples are electric motors and generators, relays, measuring instruments and iron-core inductances of large inductance; in the latter case the air gap serves to make the inductance less sensitive to changes in the current flowing in the exciting coil. Con-

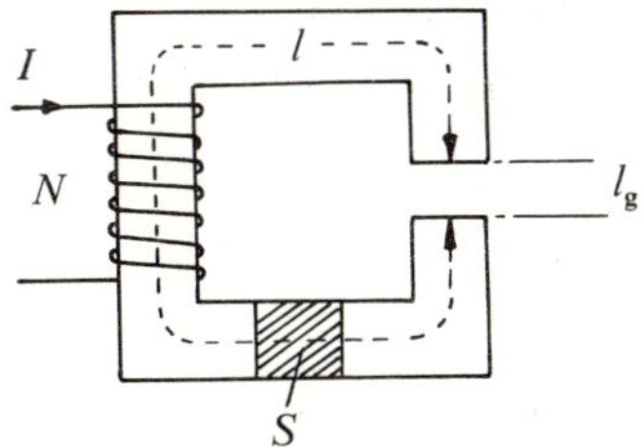

Fig. 8.33 Magnetic circuit with air gap

sider the circuit shown in Fig. 8.33. We shall assume for the moment that in the air gap the magnetic flux is confined to the gap volume. From equation 8–118 we have that

$$NI = \oint \mathbf{H} \cdot d\mathbf{l} = Hl + H_g l_g$$

where H and H_g are the values of the magnetizing force in the core and in the gap respectively. Since

$$H = B/\mu, \quad H_g = B_g/\mu_0$$

and

$$B = \Phi/S, \quad B_g = \Phi/S_g$$

then

$$NI = \Phi\left(\frac{l}{\mu S} + \frac{l_g}{\mu_0 S_g}\right)$$

$$= \Phi(\mathscr{R} + \mathscr{R}_g)$$

Since the reluctances of the two parts of the magnetic circuit are added together, the core and the air gap are considered to be in series.

Example. In the circuit shown in Fig. 8.33 the core is made of 100 quality steel and has a uniform cross-sectional area of 30 cm^2 and a mean length of 1·0 m. If the length of the air gap is 0·5 cm, find the mmf required to produce a flux of 0·003 weber in the air gap.

$$\text{mmf} = NI = Hl + H_g l_g$$

$$\Phi = BS = B_g S_g$$

Since $\Phi = 0·003$ Wb and $S = S_g = 30 \times 10^{-4}\,\text{m}^2$

$$B = B_g = \Phi/S = 1\,\text{Wb m}^{-2}$$

The corresponding value of H for the core is obtained from the B–H curve for 100 quality steel (Appendix V)

$$H = 140\,\text{At m}^{-1}$$

For the air gap,

$$H_g = B/\mu_0 = 79·6 \times 10^4$$

Hence,

$$\text{mmf} = 140 \times 1 + 79·6 \times 10^4 \times 0·5 \times 10^{-2}$$

$$= 140 + 3980$$

$$= 4120\,\text{At}$$

It will be noted that although the air-gap length is 200 times smaller than the length of the path in the steel core, the mmf required to drive the flux across the air gap is approximately 22 times that required to drive the same flux in the ferromagnetic core. Unless the gap length is very small, the core term Hl may be neglected provided the flux density in the core is well below its saturation value; in such cases the flux in the air gap can be approximated by

$$[\Phi_g]_{\text{approx}} = NI\frac{\mu_0 S_g}{l_g} \tag{8–121}$$

Fringing. Unless the length of the air gap is small, the assumption that the flux will be confined to the gap volume is not justified. This is

Fig. 8.34 Fringing

because once the lines of flux are in air they are not forced to follow any specific path but are free to spread out. This spreading out of flux lines (Fig. 8.34) is called fringing. The effect of fringing is to make the flux density in the air gap less than that in the ferromagnetic core, that is, the effective cross-section of the gap is actually greater than that of the core at the air gap. In general it is difficult to calculate the area of an air gap with any degree of accuracy but the following empirical formulas give satisfactory results if the length of the gap is less than about 20% of its cross-sectional dimensions. If a and b are the cross-sectional dimensions of the parallel faces of an air gap, the effective area is given by

$$S_g = (a + l_g)(b + l_g) \tag{8-122}$$

For circular cross-sections the diameter is increased by the length of the air gap so that,

$$S_g = \tfrac{1}{4}\pi(d + l_g)^2 \tag{8-123}$$

8.15.2 MAGNETIC CIRCUITS WITH PARALLEL BRANCHES

A typical magnetic circuit with parallel branches and single excitation is shown in Fig. 8.35. Since $\oint \mathbf{B} \cdot d\mathbf{S} = 0$ and $\nabla \cdot \mathbf{B} = 0$ it follows that

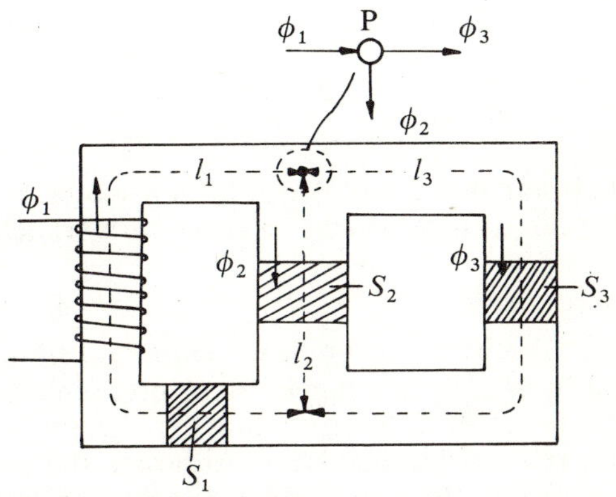

Fig. 8.35 Magnetic circuit with parallel branches

at a junction of several branches, such as P, the sum of the fluxes directed towards P is equal to the sum of the fluxes directed away from P, that is,

$$\Phi_1 = \Phi_2 + \Phi_3 \tag{8-124}$$

Applying Ampère's law to branches 2 and 3, going round in a clockwise direction, we obtain

$$H_3 l_3 - H_2 l_2 = 0 \tag{8-125}$$

since the path does not enclose any current. For branches 1 and 2 we have

$$NI = H_1 l_1 + H_2 l_2 \tag{8-126}$$

Using the relationships

$$H_1 = B_1/\mu_1 = \Phi_1/\mu_1 S_1$$
$$H_2 = B_2/\mu_2 = \Phi_2/\mu_2 S_2$$
$$H_3 = B_3/\mu_3 = \Phi_3/\mu_3 S_3$$

and substituting in equations 8–125 and 8–126 we obtain

$$\Phi_3 \mathscr{R}_3 = \Phi_2 \mathscr{R}_2$$
$$NI = \Phi_1 \mathscr{R}_1 + \Phi_2 \mathscr{R}_2$$

so that

$$\Phi_1 = \Phi_2(\mathscr{R}_2 + \mathscr{R}_3)/\mathscr{R}_3$$

and

$$NI = \Phi_1 \mathscr{R}_1 + \Phi_1 \mathscr{R}_2 \mathscr{R}_3(\mathscr{R}_2 + \mathscr{R}_3)$$

or

$$\Phi_1 = \frac{NI}{\mathscr{R}_1 + \mathscr{R}_2 \mathscr{R}_3/(\mathscr{R}_2 + \mathscr{R}_3)} \tag{8-127}$$

We thus see that branches 2 and 3 may be considered as connected in parallel and these in turn connected in series with branch 1.

8.15.3 LEAKAGE FLUX

The assumption that the magnetic flux is entirely confined to the high permeability paths of a ferromagnetic structure, implies that the flux density at any point in the air surrounding the structure is zero, that is, that the air is a perfect flux insulator. Although the permeability of most ferromagnetics is 10^3 to 10^5 times the permeability of air, this ratio is not large enough to prevent some flux lines from following

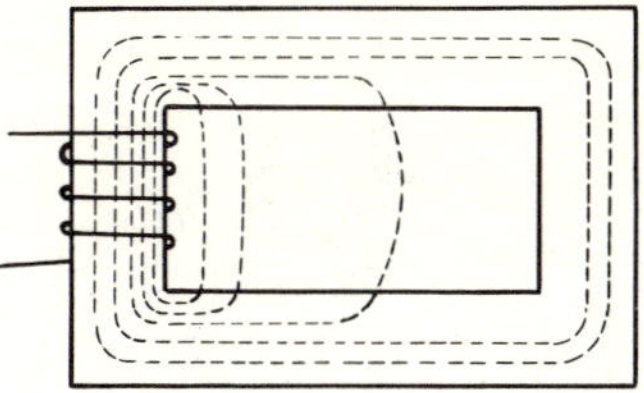

Fig. 8.36 Magnetic circuit with leakage flux

paths not wholly inside the material (Fig. 8.36). This flux is called the leakage flux whereas the flux that produces the desired magnetic effect is called the useful flux.

An analogous situation is present in an electrical circuit except that in this case the ratio of the conductivity of the conductor to that of the surrounding insulating medium is 10^{15} or greater. Nevertheless, there are some cases in which the flow of current outside the path prescribed by the conductor must be taken into consideration, as for example in an insulated cable of considerable length; as a result of the leakage current flowing through the insulating material the current at the load terminals—i.e., the useful current—will be less than that at the source terminals.

In a magnetic circuit it is difficult to express leakage flux in precise mathematical terms because the paths taken by the leakage flux are usually very irregular and depend on the geometry and magnetic state of the core. Leakage may be considerable when the core is operating at flux densities near saturation or when the core has several opposing windings. When leakage flux must be taken into consideration one has to rely on empirical formulas or on experience. In order to reduce leakage flux to a minimum the exciting winding—i.e., the mmf—must be distributed along the magnetic circuit in the same manner as the reluctance drop in the circuit is distributed.

8.16 PERMANENT MAGNETS

Consider a ring-shaped permanent magnet (Fig. 8.37) originally magnetized in a clockwise direction. Let l_{m} be the average length of the magnetic core and l_{g} the length of the air gap between its two poles (see Section 8.3). Neglecting any leakage flux we may write that

$$\Phi_{\mathrm{m}} = \Phi_{\mathrm{g}}$$

$$B_{\mathrm{m}}S_{\mathrm{m}} = B_{\mathrm{g}}S_{\mathrm{g}} \tag{8-128}$$

and since the flux does not link any current

$$H_{\mathrm{m}}l_{\mathrm{m}} + H_{\mathrm{g}}l_{\mathrm{g}} = 0$$

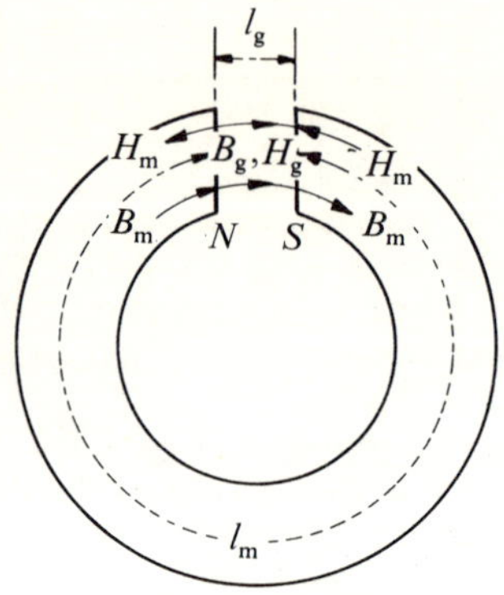

Fig. 8.37 Ring-shaped permanent magnet

or

$$H_m l_m = -H_g l_g \qquad (8-129)$$

Multiplying equation 8–128 by equation 8–129 we obtain

$$H_m B_m v_m = -H_g B_g v_g$$

or

$$\frac{v_m}{v_g} = -\frac{H_g B_g}{H_m B_m} \qquad (8-130)$$

where $v_m = S_m l_m$ and $v_g = S_g l_g$ are the volumes of the magnetic material and of the air gap respectively. Equation 8–130 shows that for a given air-gap volume and air-gap flux density, the greater the product $B_m H_m$ the smaller the volume of the magnetic material.

It is evident from equation 8–129 that the direction of H_m and H_g are opposite to each other, a result already demonstrated in Section 8.3. $-H_m$ thus represents the self-demagnetizing field of the magnet. The

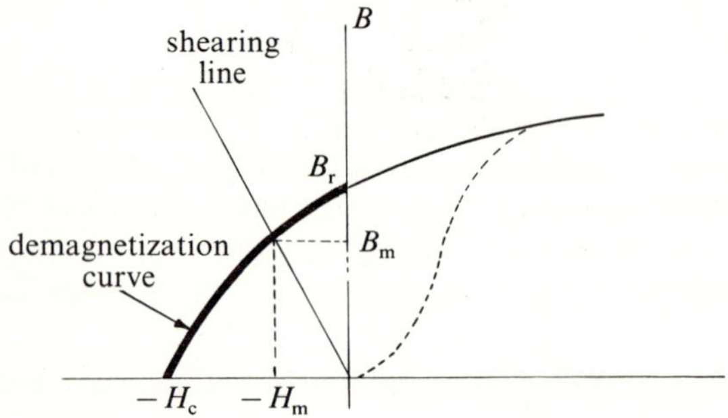

Fig. 8.38 Demagnetization curve and shearing line

equilibrium magnetic state of the magnet is therefore not represented by the point $(0, B_r)$ on the hysteresis loop but by a point $(B_m, -H_m)$ on the portion of the hysteresis curve which lies in the second quadrant (Fig. 8.38); this portion is known as the *demagnetization curve*. Since the BH product is zero at the two points $(B_r, 0)$ and $(0, -H_c)$ it is evident that at some intermediate point on the demagnetization curve the product will be a maximum. Thus $(BH)_{max}$ may be considered as a quality factor of the magnetic material used, and in general magnets

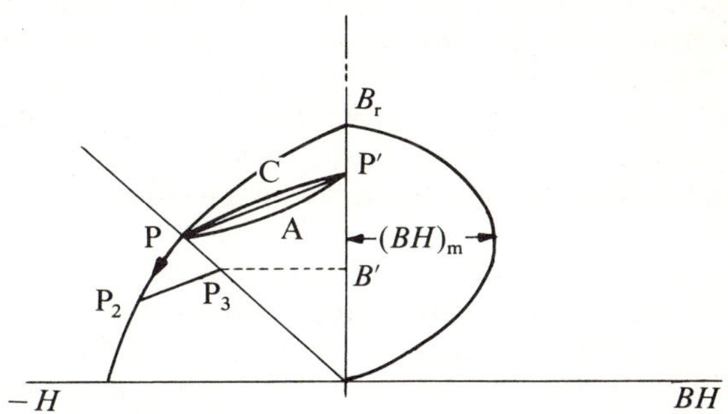

Fig. 8.39 Energy product curve and recoil loop

are designed to operate as close to the $(BH)_{max}$ point as possible (Fig. 8.39). The product $(BH)_{max}$ is often called the maximum energy product.

From equation 8–129 we have that

$$H_m l_m = -B_g l_g / \mu_0$$

or

$$B_g / H_m = -\mu_0 l_m / l_g$$

Substituting for B_g from equation 8–128 we obtain

$$\frac{B_m}{H_m} = -\frac{\mu_0 l_m S_g}{l_g S_m} \tag{8–131}$$

which is the equation of a straight line passing through the origin and having a slope of $-\mu_0 l_m S_g / l_g S_m$. It is called the *load* or *shearing line* and the point at which it intersects the demagnetization curve represents the equilibrium state of the magnet.

Suppose now that the self-demagnetizing field H_m is reduced to zero by inserting between the poles of the magnet a well-fitting soft-iron keeper (this is the function of permanent magnet keepers). The flux density will not follow back along the major hysteresis loop but will trace a new path PAP′ (Fig. 8.39) inside the major loop. If the original gap is now restored by removing the keeper, the flux density will return to the point P via the path P′CP forming a minor hysteresis loop PAP′CP known as the *recoil loop*. In general the two paths are very close together so that the loop can be replaced by the straight line PP′ known as the *recoil line*.

In many practical applications permanent magnets are required to operate in the vicinity of external magnetic fields such as those set up by current-carrying conductors, arcs or coils. The presence of such stray fields will affect the air-gap flux supplied by the magnet and this is un-acceptable where final magnet performances within narrow limits are required. In order to *stabilize* a magnet, it is first magnetized to give a higher flux density than is needed and is then deliberately demagnetized to the required value. When a moderate demagnetizing field is applied to a magnet whose effective load line intersects the demagnetization curve at point P (Fig. 8.39), the operating point will shift along this curve to a point P_2. When the demagnetizing field is removed, the operating point recoils to the point P_3 on the load line, thus reducing the flux density to the required value B'. When fields less than the demagnetizing field are applied to the magnet and then removed the operating point will recoil back to the point P_3. The degree of deliberate demagnetization is normally from 1% to 5% of the initial magnetized condition.

The materials suitable for permanent magnets are the magnetically hard materials which are essentially characterized by a large coercivity and a large $(BH)_{\max}$ value. The final hard magnetic properties of such materials are not determined by their chemical composition alone but also by the method by which they are produced. The simplest and most readily available materials are the carbon, chromium, tungsten and cobalt martensitic steel alloys. However, except for limited applications these steels have been superseded by a number of other alloys with greatly improved magnetic properties. The alloys of copper, nickel and iron (Cunife) and of cobalt, vanadium and iron (Vicalloy) are examples; they have the advantage of being mechanically soft and may be cold rolled into thin strips or drawn into thin wires.

A large group of permanent magnet materials which have excellent magnetic properties are the so-called Alnico (Ticonal) alloys. However, since such alloys are difficult to machine as they are extremely hard and brittle, Alnico magnets are manufactured either by casting or by using ceramic techniques. The latter method consists of pressing the finely ground alloy powder into the desired shape and then sintering it at high temperatures in a manner similar to that used in manufacturing ceramics. After casting or sintering, the alloy is cooled in a strong magnetic field; this produces a directional or anisotropic magnet having the best properties in the direction of the field applied during cooling. The maximum heating temperature, the holding time at this temperature and the rate of cooling differ according to the alloy composition and these must be carefully controlled to obtain the best magnetic characteristics.

Oxide-type (ferrite) permanent magnets are manufactured by sintering oxides of ferromagnetic elements. The most widely used type is barium ferrite, $BaO.6Fe_2O_3$ also called Ferroxdure or Ferroba. They have a high coercive force, excellent recoil properties and low density but poor mechanical properties.

By far the best material with isotropic properties is the 77–23 platinum–cobalt alloy, but as it is very expensive its use is restricted to very special applications requiring high resistance to demagnetization.

Quite recently Philips in Europe and General Electric in the USA have

Table 8.2 Chemical composition and magnetic properties of some permanent magnet steels and alloys

Material	Composition % (remainder Fe)	B_r tesla	H_c A m^{-1}	$(BH)_{max}$ J m^{-3}
Carbon steel	1·0 C	0·9	3980	1600
3% Chromium steel	0·9 C, 3·0 Cr, 0·3 Mn	0·94	5000	2300
6% Tungsten steel	0·7 C, 0·3 Cr, 6·0 W	1·0	5175	2380
2% Cobalt, 4% Chromium steel	1·0 C, 4·0 Cr, 2·0 Co, 0·5 W, 0·4 Mn	0·98	6400	2740
35% Cobalt steel	0·9 C, 6·0 Cr, 35 Co, 4·0 W	0·9	19900	7560
Cunife*	60 Cu, 20 Ni	0·54–0·6	47000–28000	8000–15000
Vicalloy*	52 Co, 13 V	0·95–1·05	29450–40600	16000–28000
Silmanal	86·8 Ag, 8·8 Mn, 4·4 Al	0·6	43000	6000
Alnico 5, Ticonal G	8·0 Al, 14 Ni, 24 Co, 3 Cu	1·26	47000	39700
Alnico 8, Ticonal X	7·0 Al, 14·5 Ni, 35 Co, 5 Cu, 5 Ti	0·8	110000	32000
Anisotropic barium ferrite (Ferroxdure)	$BaO.6Fe_2O_3$	0·4	140000	24000
Platinum–cobalt alloy	77 Pt, 23 Co	0·64	340000	75500

* magnetic properties vary with amount and type of cold working.

developed powerful permanent magnets made from lanthanide/transition-metal compounds of composition RT_5, where R is a lanthanide (rare earth) and T is a transition metal. Examples are $GdCo_5$ (gandolinium–cobalt) and $SmCo_5$ (samarium–cobalt). These materials have a maximum energy product of over $160\,\mathrm{kJ\,m^{-3}}$ and their extremely high coercivities ($>500{,}000\,\mathrm{A\,m^{-1}}$) make them particularly suitable for dynamic applications where strong demagnetizing fields are present. The lanthanide/transition-metal compounds are, at present, very expensive, brittle, and difficult to machine.

The composition and magnetic properties of some common permanent magnet materials are given in Table 8.2.

PROBLEMS

Chapter Eight

8.1   The magnetization inside a permeable sphere of radius a is uniform and given by $\mathbf{M} = M_0 \mathbf{a}_z$. Find (a) the equivalent amperian current densities and the equivalent magnetic pole densities, (b) the magnetic vector and a scalar potentials at a point $(0, 0, z)$ for $z > a$ and $z < a$, (c) the magnetic flux density at $(0, 0, z)$ using the amperian current and the equivalent pole models.

8.2   Find the demagnetizing factor for
   (a) an infinitely thin flat plate with its faces perpendicular to the external field,
   (b) an infinitely long thin rod with its axis parallel to the external field,
   (c) an infinitely long circular cylinder with its axis perpendicular to the direction of the external field.

8.3   The magnetization $\mathbf{M}$ inside a permeable spherical shell of internal and external radii a and b respectively is uniform. Find the magnetic flux density at points inside the cavity and points outside the shell.

8.4   Show that the tangential component of $\mathbf{A}$ is continuous across the interface between two media.

8.5   For the configuration shown in Fig. P8.5 show that the flux linking the loop is given by

$$\Phi = \frac{Ib(\mu_0 + \mu)}{4\pi} \log \frac{d+a}{d}$$

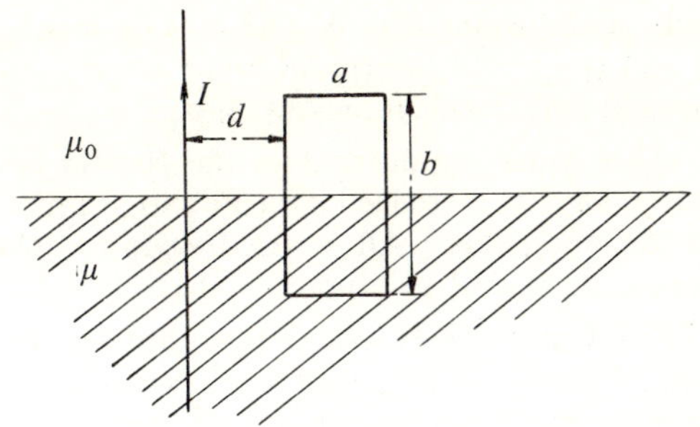

Fig. P. 8.5

8.6   A cast steel toroid has a uniform cross-sectional area of $10\,\text{cm}^2$ and a mean circumference of $0\cdot7$ m. A coil of 300 turns is wound uniformly around the toroid; if the coil carries a direct current of $2\,\text{A}$, find the flux in the toroid. Find the current in the coil to produce a flux of $8 \times 10^{-4}\,\text{Wb}$ in the toroid.

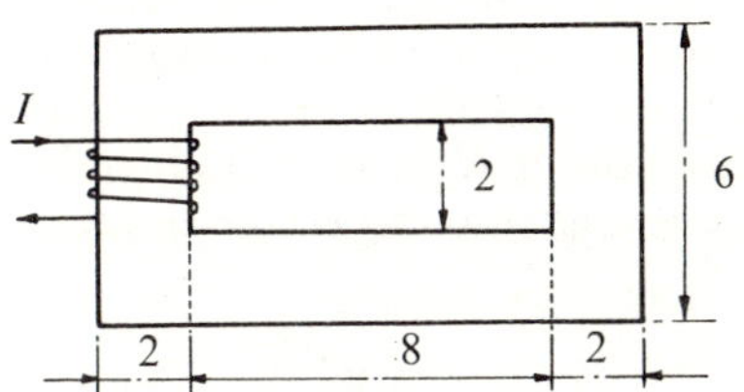

Fig. P. 8.7

8.7   The core shown in Fig. P8.7 is made of 100 quality steel. Find the direct current in the 150 turn winding to produce a flux of $10^{-3}\,\text{Wb}$ in the core which has a uniform cross section of $2 \times 5\,\text{cm}^2$. What would this current be if the core were made of cast steel? Neglect leakage flux. (Dimensions are in cm).

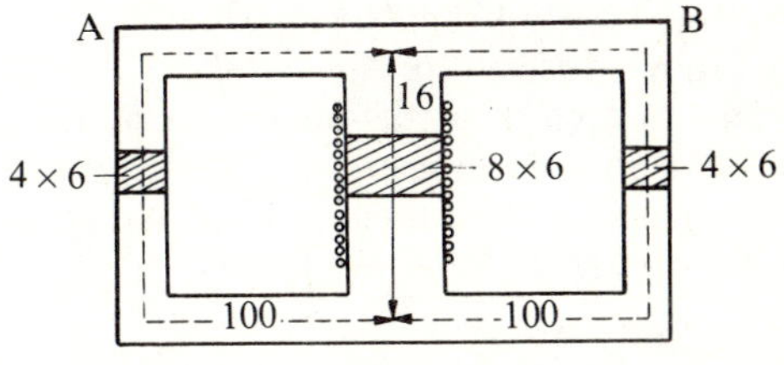

Fig. P. 8.8

8.8   The core shown in Fig. P8.8 is made of 100 quality steel. Find the current in the central 100 turn coil to produce a flux density of

1 T in each of the outer legs A and B. The mean length of the branches and their cross-sectional areas are shown in the figure (dimensions in cm). Neglect leakage flux.

8.9   To what value must the current in the central coil of the core shown in Fig. P8.8 be increased if an air gap of 1 mm is introduced in one of the outer legs? Allow for fringing at the air gap but neglect leakage flux and assume $B_g = 1\mathrm{T}$.

8.10 A long solenoid of N closely wound turns and length l has a closely fitting core of permeability μ; the core is made up of two halves separated by an air gap of length d at the centre of the coil. If the coil carries a current I and all end effects are neglected, show that the magnetic flux density in the air gap is $\mu NI/[l+d(\mu_r-1)]$.

8.11 A long solenoid has a cross sectional area S. If a cylindrical magnetic core of permeability μ and cross-section $S'(S' < S)$ is introduced coaxially into the solenoid, show that the inductance of the solenoid is increased by $1+(\mu_r-1)(S'/S)$.

8.12 An iron ring with a rectangular cross-section of width d internal and external diameters a and b respectively. A radial air gap which subtends an angle θ radians at the centre has been cut out of the ring. If a coil of N turns is uniformly wound on the ring show that, neglecting end effects, the inductance of the system is given by

$$L = \frac{\mu N^2 d}{[2\pi+(\mu_r-1)\theta]}\log\frac{b}{a}$$

8.13 The demagnetization curve of a permanent magnet material is given by $B = (B_0/H_0)H+B_0$. A permanent magnet made from this material is to have a fixed volume; it is to be fitted with pole shoes of negligible reluctance to produce flux in an air gap of length l_g and area S_g. Show that the flux density in the gap will be a maximum if the ratio of the length l of the magnet to its area S is $B_0 l_g/\mu_0 S_g H_0$. Neglect leakage and fringing.

8.14 It is required to produce a flux density of $0\cdot5\,\mathrm{T}$ in an air gap 1 mm long and $1\,\mathrm{cm}^2$ in area. For this purpose a permanent magnet is fitted with soft iron pole shoes of negligible reluctance. Find the optimum dimensions of the magnet if its demagnetization curve is given by $B = 0\cdot9[(H/19900)+1]$.

CHAPTER NINE

ELECTROMAGNETIC INDUCTION

In Chapter 7 we discussed the laws which govern the relationship between a magnetostatic field and the stationary current source producing this field. These laws, namely the Biot–Savart law and Ampère's law, are the mathematical formulations of Oersted's experimental discovery that a steady electric current produces magnetism. Following this discovery, the argument that if electricity produces magnetism then magnetism should produce electricity seemed a logical one and many workers set themselves the task of finding whether a steady current could be produced by stationary magnetism. All efforts in this direction, however, remained unsuccessful (today of course we know why) and it was only in 1831, eleven years after Oersted's discovery, that Faraday finally succeeded in showing that an electric current would, under certain conditions, be generated by magnetism. By means of a simple experiment, yet surely one of the great ones in the realm of science, Faraday discovered the phenomenon of electromagnetic induction. Essentially the apparatus consisted of two separate coils wound on a soft-iron toroid. In one circuit was placed a switch in series with a battery and in the other a galvanometer at a sufficient distance from the toroid so as to be unaffected by the magnetic field of the latter. Faraday noticed that when the current was switched on the galvanometer was momentarily deflected in one direction whilst when the battery was switched off the galvanometer was momentarily deflected in the opposite direction. A steady flow of current produced no effect. It was the *changing* current, and hence the changing magnetic field produced by it, that was responsible for the effect. Faraday carried out a number of other fundamental experiments the results of which showed that a current could be 'induced' in a coil by moving a magnet relative to the coil or by moving the coil relative to a magnet or simply by changing the current in a neighbouring coil, both coils remaining stationary. Since any agency that tends to cause charge to circulate around a closed path is said to possess an electromotive force, the induced current is interpreted as being caused by an emf induced in the coil. The source of this emf is the relative motion which exists between the coil and a magnet or the changing magnetic field generated by a changing current in a neighbouring stationary circuit.

The qualitative results of Faraday's experimental findings were put in mathematical form by Neumann, a contemporary of Faraday. In this form *all* induction phenomena may be quantitatively accounted for by the single equation

$$\mathscr{V} = -\frac{d\Phi}{dt} \tag{9-1}$$

In words this equation states that the magnitude of the emf $\mathscr{V}$ (measured in volts) induced in any closed circuit is equal to the time rate of change of the total magnetic flux Φ (measured in webers) enclosed by the circuit. The total flux is the flux produced by currents or magnets external to the circuit plus the flux produced by any currents induced in the circuit itself. The negative sign gives the direction of the induced emf. This direction is determined by Lenz's law which states that

'induction effects always tend to oppose the actions which produce them'.

In spite of the apparent simplicity of equation 9–1, both its interpretation and application as such have been a source of constant controversy ever since its formulation by Neumann in 1845. For this reason we shall not use it as our starting point in the discussion of induction phenomena. Instead we shall discuss separately the only two independent physical actions which, relative to any particular suitable frame of reference, are known to produce electromagnetic induction. These are the *variational* or *transformer action* and the *motional* or *flux-cutting action*.

9.1 VARIATIONAL ACTION

Suppose that in a region of space there is established a time-varying magnetic field whose source—e.g., a solenoid carrying a time-varying current—is stationary with respect to a fixed system of coordinate axes. If we now place in this field a stationary conducting loop closed through an ammeter, it is an experimental fact that a current flow around the loop will be indicated by the meter. This means that the variation of the magnetic field with time has the same effect on the loop as if an external current-driving force or emf had been placed in series with it. We therefore say that the time-varying flux induces a variational emf in the circuit. Now since a current can only flow in a stationary conductor if there exists in it an electric field, it follows that in the present case there must exist an induced field in the conductor such that its line integral taken around the *closed* contour of the loop is equal to the induced emf. That is,

$$\mathscr{V} = \oint \mathbf{E} \cdot d\mathbf{l}$$

The integral represents the work per coulomb done in encircling the path once, so that the unit of $\mathscr{V}$ is obviously the volt, that is, the same as for potential difference. To distinguish emf from the scalar potential V, the symbol $\mathscr{V}$ will be used to denote emf.

Since the induced emf has as its origin the time-varying magnetic field, there must exist a quantitative relationship between an induced emf and the magnetic field producing it. This relationship is given by *Faraday's law for variational induction*. According to this law the magnitude of the emf induced in a closed conducting loop is proportional to the time rate of change of magnetic flux which crosses any surface bounded by the loop. In SI units the constant of proportionality is unity. The direction of the induced emf is given by Lenz's law. For the variational action which we are considering, this law tells us that the direction of the induced emf must be such that the magnetic flux of the current it tends to produce always tends to oppose the change in the external flux.

The above two laws are combined in the equation

$$\mathscr{V} = \oint_C \mathbf{E} \cdot \mathbf{dl} = -\frac{\partial \Phi}{\partial t} \quad [\text{Wb s}^{-1} \equiv \text{V}] \qquad (9\text{--}2)$$

where Φ is the flux enclosed by the loop C. The negative sign accounts for the direction of the emf in accordance with Lenz's law. The sign of the derivative itself will be positive for increasing flux and negative for decreasing flux. E has the same direction as the induced current. The direction of integration around the closed loop is that direction in which a right-handed screw advances in the positive direction of the flux Φ. This automatically gives the correct sign of the induced voltage.

If the conducting loop consists of N turns in series then if the same flux Φ links each of the turns, Faraday's law may be expressed as

$$\mathscr{V} = -N\frac{\partial \Phi}{\partial t} \qquad (9\text{--}3)$$

On the other hand if every turn is not linked by the same value of flux, we have that

$$\mathscr{V} = -\sum_{i=1}^{N} \frac{\partial \Phi_i}{\partial t} \qquad (9\text{--}4)$$

where Φ_i is the magnetic flux linking the ith turn.

As a simple example illustrating the application of the foregoing rules and regulations so to speak, consider a stationary wire loop encircling a changing magnetic flux which is normal to the plane of the loop. The four possible cases giving rise to an induced emf by variational action are shown in Fig. 9.1. We have indicated on each diagram the directions of the induced field and current as well as the direction of integration and the sign of the induced emf.

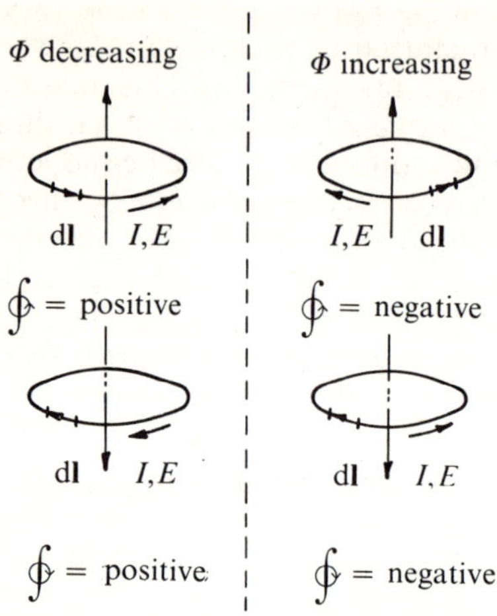

Fig. 9.1 Four possible cases giving rise to an induced emf by variational action

A direct consequence of the variational action is that a time-varying magnetic field produces an electric field. It is the field which moves the free conduction electrons around the stationary wire loop giving rise to a current. The conducting loop detects the presence of the field but is in no way necessary for its existence. That is, the field E exists whether the loop is present or not. Because of this it is possible for us to restate Faraday's law for variational induction in a more generalized form as follows.

Any change with time of a magnetic flux gives rise to a non-conservative electric field having the property that its integral around any closed contour in space, whether in conducting or non-conducting media or in free space, is equal to the negative time derivative of the magnetic flux enclosed by that contour.

It is important to note that, in contrast to the electrostatic field, the induced field cannot be expressed as the negative gradient of a scalar potential. One proof of the existence of an induced field in the absence of any material media is the electron accelerator known as the betatron, in which electrons are accelerated in a vacuum to high velocities by means of a circulating electric field generated by a rapidly changing magnetic flux.

The magnetic flux enclosed by any closed contour C is given by

$$\Phi = \int_S \mathbf{B} \cdot d\mathbf{S}$$

where S is any surface bounded by the contour C. Equation 9–2 may therefore be rewritten as

$$\oint_C \mathbf{E} \cdot d\mathbf{l} = -\int_S \frac{\partial \mathbf{B}}{\partial t} \cdot d\mathbf{S} \tag{9–5}$$

where the derivative has been taken inside the integral sign since both C and S are fixed in space and the magnetic flux is the only time-varying quantity. Equation 9–5 is usually referred to as the integral form of Faraday's variational law. Since the flux Φ can also be expressed in terms of the magnetic vector potential $\mathbf{A}$ (equation 7–58), the above equation may be expressed as

$$\oint_C \mathbf{E} \cdot d\mathbf{l} = -\oint_C \frac{\partial \mathbf{A}}{\partial t} \cdot d\mathbf{l} \tag{9–6}$$

By applying Stokes's theorem to the left-hand side of equation 9–5 we have that

$$\int_S \nabla \times \mathbf{E} \cdot d\mathbf{S} = -\int_S \frac{\partial \mathbf{B}}{\partial t} \cdot d\mathbf{S}$$

and since this equation is valid for any surface S bounded by C, it follows that

$$\nabla \times \mathbf{E} = -\frac{\partial \mathbf{B}}{\partial t} \tag{9–7}$$

This is the differential or point form of Faraday's law. It replaces the relation $\nabla \times \mathbf{E} = 0$ which is only valid for purely electrostatic fields. By introducing the relation $\mathbf{B} = \nabla \times \mathbf{A}$ into equation 9–7 we obtain

$$\nabla \times \mathbf{E} = -\nabla \times \frac{\partial \mathbf{A}}{\partial t}$$

or

$$\nabla \times \left(\mathbf{E} + \frac{\partial \mathbf{A}}{\partial t} \right) = 0 \tag{9–8}$$

and since the curl of any gradient is identically zero it follows that

$$E + \frac{\partial \mathbf{A}}{\partial t} = -\nabla U$$

and

$$E = -\frac{\partial \mathbf{A}}{\partial t} - \nabla U \qquad (9\text{–}9)$$

where U is a scalar potential function which will in general be a function of time. If the time variations are slow enough the function U will represent the electrostatic potential V, so that $-\nabla V$ will be simply the component of the electric field which is due to static charges; $-\partial \mathbf{A}/\partial t$ is that part of the electric field strength which is of magnetic origin. Slowly varying fields, for which we may write

$$\mathbf{E} = -\frac{\partial \mathbf{A}}{\partial t} - \nabla V \qquad (9\text{–}10)$$

where V is the electrostatic potential, are usually referred to as *quasi-stationary* fields; a more formal definition of such fields will be given in Section 9.7.

9.2 POTENTIAL DIFFERENCE AND INDUCED EMF

Figure 9.2 shows a magnetic flux which is increasing out of the paper at a constant rate of $e\,\mathrm{Wb\,s^{-1}}$. The induced electric field encircles the

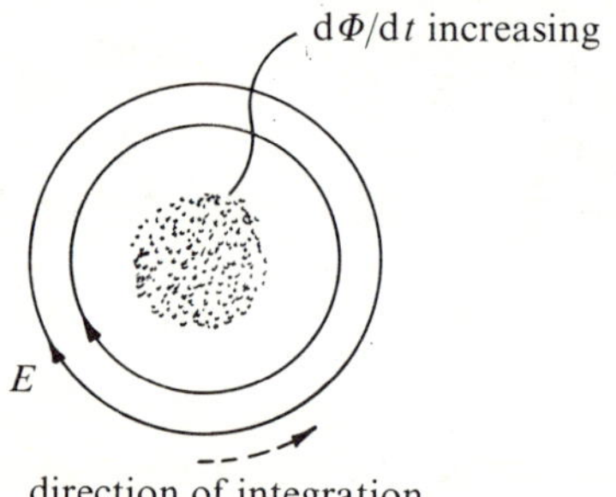

Fig. 9.2 Induced field encircling a changing magnetic flux

changing magnetic flux as indicated on the figure. For any closed path encircling the flux we must have that

$$\oint_C \mathbf{E} \cdot \mathbf{dl} = -e \qquad (9\text{–}11)$$

the direction of integration being taken in this case counterclockwise. Suppose now that we encircle the flux with a circular conducting loop which has two open terminals 1 and 2 as shown in Fig. 9.3. The electric field will immediately redistribute the free charges in the conductor so as to establish an equilibrium condition in which the resultant field is zero everywhere inside the conductor. Thus terminal 2

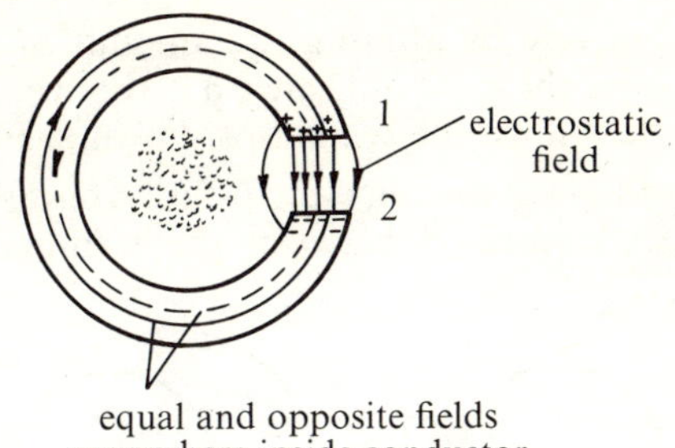

Fig. 9.3 Two-terminal conducting loop encircling a changing magnetic flux

will be negatively charged (accumulation of electrons) whilst terminal 1 will be positively charged (dearth of electrons). These charges now give rise to an electrostatic field which bridges the gap between the two terminals, extending from terminal 1 to terminal 2. As the electric field produced by these charges alone is conservative ($\oint \mathbf{E} \cdot \mathbf{dl} = 0$), the presence of the charges cannot alter the value of the emf generated by the changing magnetic flux. Since the field inside the conductor is zero the line integral in equation 9–11 reduces to

$$\int_2^1 \mathbf{E} \cdot \mathbf{dl} = -e$$

or

$$-\int_2^1 \mathbf{E} \cdot \mathbf{dl} = e = V_{12}$$

so that the potential difference between the terminal is equal to the emf induced *in the circuit*. It must be emphasized here that although the emf appears as a potential difference between the terminals, it is physically located in the circuit itself and not in the air gap between the terminals. To see this it is only necessary to realize that the emf between the same two fixed points 1 and 2 in space may be changed in

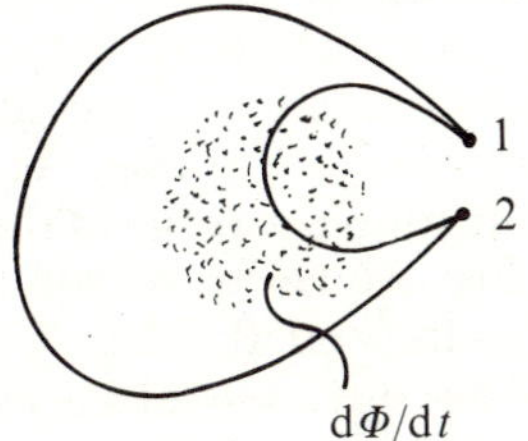

Fig. 9.4 The emf appearing between points 1 and 2 depends on the amount of flux encircled by a loop

an arbitrary way merely by altering the amount of flux enclosed by the loop. For example the emf (and hence p.d.) between the same points 1 and 2 will be different for the two loops shown in Fig. 9.4.

The reader will readily see that an electrostatic voltmeter (which draws negligible current) connected between the terminals 1 and 2 will

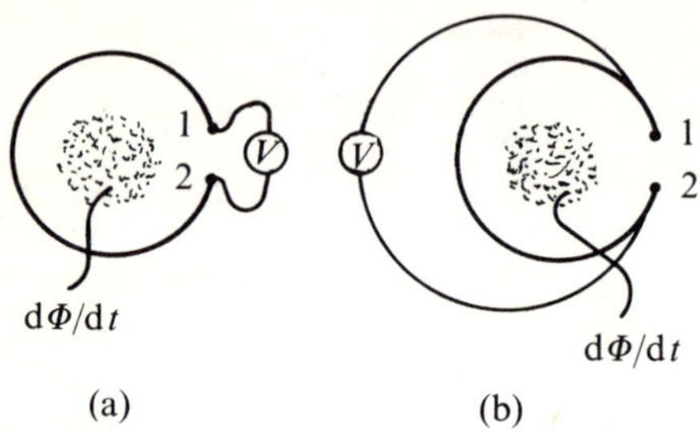

Fig. 9.5 Two different ways of connecting a voltmeter across terminals 1 and 2

measure the induced emf if connected as in Fig. 9.5a but will read zero if connected as in Fig. 9.5b.

9.3 MOTIONAL ACTION

Suppose that in a region of space there exists a magnetic field of density **B** which is constant in time and whose source is stationary with respect to a fixed frame of reference (coordinate axes). We know (Section 7.2) that a charge q moving with a velocity **v** relative to the reference frame will experience a magnetic force:

$$F = q\mathbf{v} \times \mathbf{B}$$

Suppose now that instead of a single charge of q we have a conducting segment AB which is moving with uniform velocity **v** relative to the fixed frame of reference (Fig. 9.6). As far as the freely moving or conduction electrons are concerned, their average velocity, at any given instant, with respect to the element itself is zero. Their average velocity with respect to the stationary field **B** is therefore **v**. Each electron will thus experience an average magnetic force $-q\mathbf{v} \times \mathbf{B}$ relative to the stationary reference frame. As a result there will be a net displacement of conduction electrons in the direction of this force. For the case shown in Fig. 9.6 electrons will be transferred from the end A of the segment to the end B. That is, there will be a flow of current from B to A. This direction of current flow, in accordance with Lenz's law, produces a force (electromagnetic reaction) which tends to slow down the segment. There will thus be a gradual build-up of a net positive charge at A and a net negative charge at B. The process

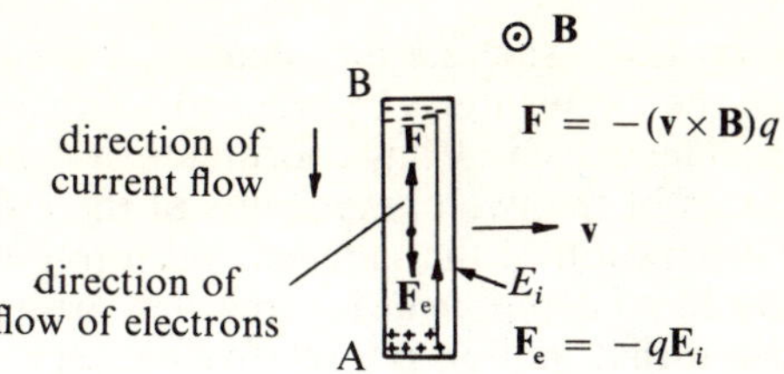

Fig. 9.6 Conducting segment moving with uniform velocity in a magnetic field

of charge separation will continue until the redistribution of charge is such that the resultant *force per unit charge* everywhere inside the conductor is zero (steady state conditions). The charge redistribution will give rise to an electric field which in itself is a conservative field. Everywhere inside the conductor this field is exactly equal and opposite to the force per unit charge generated by the motion of the conductor. The work done in taking a unit charge around any closed path such that the conductor segment itself is a part of that path, is zero as far as the electric field is concerned but is equal to

$$\mathscr{V} = \int_{B}^{A} \mathbf{v} \times \mathbf{B} \cdot d\mathbf{l} \tag{9-12}$$

as far as the motional force per unit charge inside the conductor is concerned. Equation 9-12 represents the motional or flux-cutting emf induced in the conductor. If the conductor forms a closed loop then the total induced emf around the loop is given by

$$\mathscr{V} = \oint \mathbf{v} \times \mathbf{B} \cdot d\mathbf{l} \tag{9-13}$$

In this case a current will flow around the loop; the direction of current flow will be determined by the direction of the force per unit charge $\mathbf{v} \times \mathbf{B}$. The direction of integration is that direction in which a right-handed screw advances in the positive direction of the magnetic flux.

9.4 GENERAL CASE OF INDUCTION

Whenever we have a variational action and a motional action occurring simultaneously—that is, we have a circuit moving through a magnetic flux which is changing with time—the resultant induced emf is the algebraic sum of the variational emf as given by equation 9–5 and the motional emf as given by equation 9–13. That is

$$\mathscr{V} = -\int_{s} \frac{\partial \mathbf{B}}{\partial t} \cdot d\mathbf{S} + \oint \mathbf{v} \times \mathbf{B} \cdot d\mathbf{l} \tag{9-14}$$

In applying the above equation it is of fundamental importance to realize that all variables such as flux density, velocity, position and configuration of the circuit must be referred to a specified system of coordinate axes. The choice of this coordinate system determines the relative magnitudes of the two components of the induced voltage so that the actual separation of the induced emf into a variational component and a motional component is somewhat arbitrary. The reason for this of course is that motion is a relative concept and any velocity must necessarily be referred to a specified frame of reference. In the next section we shall show explicitly how the phenomenon of induction may be interpreted either in terms of both an electric and magnetic property of space or in terms of a purely electric property of space, depending entirely on the frame of reference in which an observer finds himself.

9.5 THE INDUCED EMF VIEWED FROM STATIONARY AND MOVING FRAMES OF REFERENCE

It should be stated at the outset that the displacement velocity of all moving charges or circuits relative to any observer will be assumed small as compared with the velocity of light. This means that deviations from Newtonian mechanics, due to relativistic effects, are negligibly small. Under this assumption Newton's laws will be the same to an observer in any inertial frame of reference.*

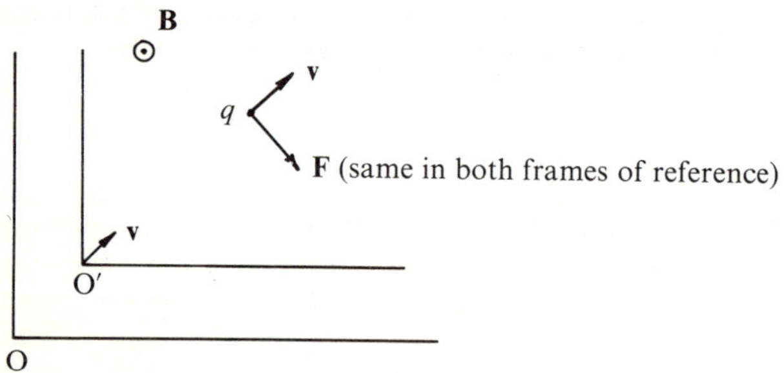

Fig. 9.7 Point charge moving with velocity v with respect to stationary reference frame O; the sources of B are stationary in O

Let O be a laboratory frame of reference† and suppose that at all points in space (defined with respect to O) there exists a magnetic field

---

* An inertial frame of reference is one which travels with a constant speed in a fixed direction.

† Since effects associated with the rotation of the earth can be usually neglected, the earth—and hence the laboratory—may be considered as a satisfactory inertial frame of reference.

of flux density $\mathbf{B}$, whose sources are stationary in O (Fig. 9.7). A point charge q moving with velocity $\mathbf{v}$ relative to the frame O will experience a force, of magnetic origin, given by $\mathbf{F} = q\mathbf{v} \times \mathbf{B}$. Suppose now that we have a second inertial frame O′ moving with a velocity $\mathbf{v}$ relative to O. The test charge q is now stationary relative to O′. Because, however, the force acting on q must be the same in both frames of reference it follows that

$$\mathbf{F}' = q\mathbf{E}' = \mathbf{F} = q\mathbf{v} \times \mathbf{B}$$

Although these two forces are equal the interpretation of their origin will differ according to whether the charge is viewed by a stationary observer in O or by a moving observer in O′. To the stationary observer the force will be interpreted as a magnetic force whose very existence depends on the presence of the *moving* charge in the stationary magnetic field. To the observer moving with O′ the charge appears stationary and since the only force which can act on a stationary charge is that due to an electric field, it follows that the observer will interpret the force as due to the existence, relative to O′, of an external electric field E'. To him this field exists in space even in the absence of any charge; the origin of the field can only be interpreted as due to the relative motion which exists between himself and the source of the magnetic field. He will correctly conclude, therefore, that a moving source of magnetic field produces an electric field. We thus see that in the frame O the force on q is of a purely magnetic nature whereas in the frame O′ this same force is of a purely electric nature.

Let us now substitute the point charge q by a segment dl of conducting wire which moves with a uniform velocity $\mathbf{v}$ relative to O. To an observer in O a magnetic force per unit charge $\mathbf{v} \times \mathbf{B}$ will act on the conduction electrons leading to a displacement of charge which, as discussed in Section 9.3, gives rise to an electric field outside the moving element as well as to an electric field $-\mathbf{v} \times \mathbf{B}$ at all points inside the element. An observer in O′ will simply see a stationary conducting element placed in an external field $\mathbf{E}'$. This field gives rise to a redistribution of charge over the surface of the element such that the field everywhere inside it is zero. The electric field seen by this observer will be the superposition of the external field $\mathbf{E}'$ and the field produced by the surface charge distribution.

Suppose now that we keep adding elements dl so as to form a rigid closed loop C of arbitrary shape which moves with the velocity $\mathbf{v}$ relative to the frame O. Let us examine this system from the stationary and moving reference frames in turn, for the two cases when the magnetic field does not vary with time and when the magnetic field varies with time.

9.5.1 STEADY MAGNETIC FIELD

Relative to frame O the force per unit charge acting at any point *on the loop* is $\mathbf{f} = \mathbf{v} \times \mathbf{B}$ (Fig. 9.8a). It should be stressed that this force exists because the loop exists *and is moving*. The induced electromotive force around the complete circuit is the line integral of $\mathbf{f}$ around the circuit, that is

$$\mathscr{V} = \oint \mathbf{f} \cdot \mathrm{d}\mathbf{l} = \oint \mathbf{v} \times \mathbf{B} \cdot \mathrm{d}\mathbf{l} \tag{9-15}$$

If the field $\mathbf{B}$ is uniform then the induced emf is zero since

$$(\mathbf{v} \times \mathbf{B}) \cdot \oint \mathrm{d}\mathbf{l} = 0$$

Relative to frame O' the circuit is at rest in an external electric field $\mathbf{E}'$ whose value at any point *in space* is $\mathbf{v} \times \mathbf{B}$. This field exists irrespective of whether the loop exists or not (Fig. 9.8b). Except for the case in

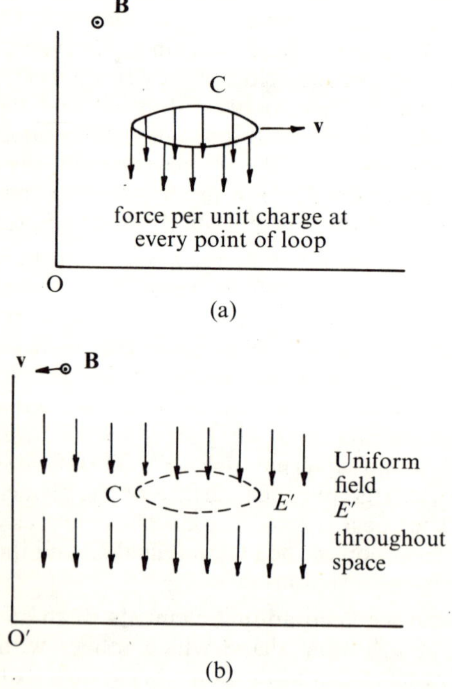

Fig. 9.8 (a) Rigid loop moving with velocity $\mathbf{v}$ relative to frame O (B sources stationary relative to O). (b) Rigid loop stationary relative to frame O' (B sources moving with velocity $-\mathbf{v}$ relative to O')

which $\mathbf{E}'$ is uniform—i.e., when $\mathbf{B}$ is uniform—the integral of $\mathbf{E}'$ around any closed path is not zero. When the closed path coincides with that of the circuit C the integral gives the emf induced in the circuit. That is

$$\mathscr{V} = \oint \mathbf{E}' \cdot d\mathbf{l} = \oint \mathbf{v} \times \mathbf{B} \cdot d\mathbf{l} \qquad (9\text{--}16)$$

Although equations 9–15 and 9–16 are identical their physical interpretation as we have seen is basically different.

9.5.2 TIME-VARYING MAGNETIC FIELD

Relative to frame O the induced emf in the circuit, at any particular instant of time, is given by

$$\mathscr{V} = -\int_S \frac{\partial \mathbf{B}}{\partial t} \cdot d\mathbf{S} + \oint_C \mathbf{v} \times \mathbf{B} \cdot d\mathbf{l} \qquad (9\text{--}17)$$

where S is any surface bounded by the contour C.

Relative to O′ the circuit is stationary and the induced emf must be attributed to the existence in O′ of an electric field $\mathbf{E}'$ whose value must be such that

$$\mathscr{V} = \oint \mathbf{E}' \cdot d\mathbf{l} \qquad (9\text{--}18)$$

$\mathbf{E}'$ is a 'true' electric field in the sense that it is the force per unit charge which acts on stationary charges. From equations 9–17 and 9–18 we have that

$$\oint_C \mathbf{E}' \cdot d\mathbf{l} = -\int_S \frac{\partial \mathbf{B}}{\partial t} \cdot d\mathbf{S} + \oint_C \mathbf{v} \times \mathbf{B} \cdot d\mathbf{l}$$

By applying Stokes's theorem to the two line integral terms we have that

$$\int_S \nabla \times \mathbf{E}' \cdot d\mathbf{S} = -\int_S \frac{\partial \mathbf{B}}{\partial t} \cdot d\mathbf{S} + \int_S \nabla \times (\mathbf{v} \times \mathbf{B}) \cdot d\mathbf{S}$$

and it follows that

$$\nabla \times \mathbf{E}' = -\frac{\partial \mathbf{B}}{\partial t} + \nabla \times (\mathbf{v} \times \mathbf{B})$$

and

$$\nabla \times (\mathbf{E}' - \mathbf{v} \times \mathbf{B}) = -\frac{\partial \mathbf{B}}{\partial t} \qquad (9\text{--}19)$$

Now the total force per unit charge in the stationary frame of reference is the sum of any electric field $\mathbf{E}$ present and the force per unit charge due to motion, that is:

$$\mathbf{f} = \mathbf{E} + \mathbf{v} \times \mathbf{B}$$

whilst the total force per unit charge in the moving frame of reference is simply equal to the electric field $\mathbf{E}'$ present there

$$\mathbf{f}' = \mathbf{E}'$$

Since these two forces must be identical in all inertial reference frames it follows that

$$\mathbf{E}' = \mathbf{E} + \mathbf{v} \times \mathbf{B} \tag{9-20}$$

Combining this equation with equation 9-19 we obtain

$$\nabla \times \mathbf{E} = -\frac{\partial \mathbf{B}}{\partial t}$$

Thus as far as a stationary observer in O is concerned Faraday's law applied to any point in his frame of reference is the same whether the point is stationary or moving with respect to that frame. This is because the field $\mathbf{E}$ at any point in O is a true field whose presence is due solely to the time variation of $\mathbf{B}$ and to any static charge distribution* which may be present, irrespective of whether there is moving or stationary matter at that point.

Now it can be formally shown that†

$$-\int_s \frac{\partial \mathbf{B}}{\partial t} \cdot \mathrm{d}\mathbf{S} + \oint_c \mathbf{v} \times \mathbf{B} \cdot \mathrm{d}\mathbf{l} = -\frac{\mathrm{d}}{\mathrm{d}t} \int_s \mathbf{B} \cdot \mathrm{d}\mathbf{S} = -\frac{\mathrm{d}\Phi}{\mathrm{d}t} \tag{9-21}$$

so that in the moving frame of reference

$$\oint_c \mathbf{E}' \cdot \mathrm{d}\mathbf{l} = -\frac{\mathrm{d}}{\mathrm{d}t} \int_s \mathbf{B} \cdot \mathrm{d}\mathbf{S} \tag{9-22}$$

Although throughout the above analysis we have assumed that the loop C is rigid, that is the velocity $\mathbf{v}$ is the same for all points on the loop, equations 9-17 and 9-21 are in fact valid for any closed loop the points of which move with an arbitrary velocity $\mathbf{v}$ so that the loop may be translating, rotating and deforming at the same time. Equation 9-20 expresses the force per unit charge at a point in a moving reference frame with respect to which the point is at rest and at this same point

---

* The presence of such charges in no way affects the induction equation because the electric field of these charges is conservative, that is:

$$\oint \mathbf{E}_0 \cdot \mathrm{d}\mathbf{l} = 0 \qquad \text{and} \qquad \nabla \times \mathbf{E}_0 = 0$$

† See H. H. Woodson and J. R. Melcher, *Electromechanical Dynamics*, Part 1, Appendix B (Wiley, New York, 1968).

in a stationary reference frame with respect to which the point is moving with an instantaneous velocity $\mathbf{v}$. The force is independent of the choice of the inertial reference frame so that at any one instant the expression $\mathbf{E}' = \mathbf{E} + \mathbf{v} \times \mathbf{B}$ is valid for any point on the loop. Now, however, there is not a common moving frame of reference for the entire loop, but a different instantaneous inertial frame for each point on the loop. In equation 9–22 $\mathbf{E}'$ is now the electric field measured with respect to a frame of reference which at any one instant is moving with the velocity of $\mathbf{dl}$ at the point of integration. Equation 9–22 can be considered as the most general law of induction—that is, as the most general quantitative formulation of Faraday's experimental findings. Since it is more convenient to express all variables with respect to a common reference frame, we use equation 9–20 to transform equation 9–22 to our fixed laboratory frame. Thus the law of induction becomes

$$\oint_C (\mathbf{E} + \mathbf{v} \times \mathbf{B}) \cdot \mathbf{dl} = -\frac{d}{dt} \int_S \mathbf{B} \cdot \mathbf{dS} = -\frac{d\Phi}{dt} \qquad (9\text{–}23)$$

In words this law may be stated as follows: In a region of space, defined with respect to a laboratory frame of reference, where there exists a true electric field $\mathbf{E}$ and a magnetic field $\mathbf{B}$, the integral of $\mathbf{E} + \mathbf{v} \times \mathbf{B}$ around any closed loop the points of which move with an arbitrary velocity $\mathbf{v}$, is equal to the rate of change of the magnetic flux linked by the moving loop. Either quantity represents the emf induced in the loop.

9.6 SOME EXAMPLES ON THE CALCULATION OF INDUCED EMFS

9.6.1 LOOP FORMED BY SLIDING CONDUCTOR

Consider the configuration shown in Fig. 9.9. The sliding conductor is moving to the right with a uniform velocity $\mathbf{v} = v\mathbf{a}_y$ relative to the fixed

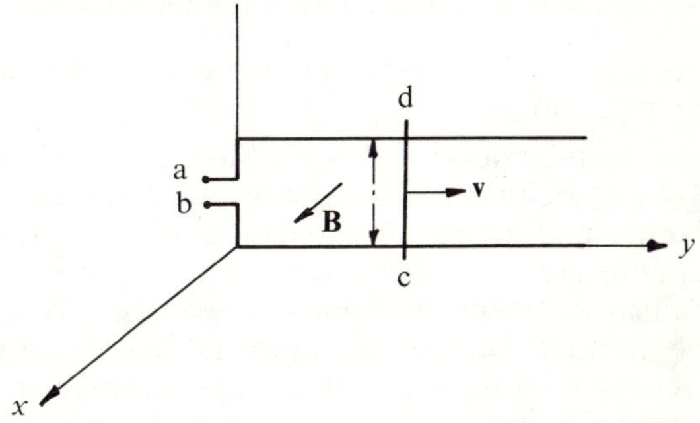

Fig. 9.9 Sliding conductor moving with uniform velocity on two parallel conducting rails

frame x, y, z. The flux density $\mathbf{B}$ is assumed uniform everywhere and normal to the plane of the circuit $(x = 0)$. Assume that $\mathbf{B}$ varies harmonically with time so that

$$\mathbf{B} = B_0 \cos \omega t \mathbf{a}_x$$

At any instant t the emf induced around the path abcd is given by

$$\mathscr{V} = -\int_s \frac{\partial \mathbf{B}}{\partial t} \cdot d\mathbf{S} + \oint_c \mathbf{v} \times \mathbf{B} \cdot d\mathbf{l}$$

The variational component of the induced emf is given by

$$-\int_s \frac{\partial \mathbf{B}}{\partial t} \cdot d\mathbf{S} = \int_0^l \int_0^y B_0 \omega \sin \omega t \mathbf{a}_x \cdot dy\, dz \mathbf{a}_x$$

$$= \omega B_0 yl \sin \omega t$$

The direction of this emf component is anticlockwise (terminal a positive, terminal b negative) whenever $\sin \omega t$ is positive and clockwise whenever $\sin \omega t$ is negative. The motional component of the induced emf is given by

$$\overset{d}{\underset{a}{\oint}} \overset{c}{\underset{b}{}} \mathbf{v} \times \mathbf{B} \cdot d\mathbf{l} = \int_c^d (v\mathbf{a}_y \times \mathbf{a}_x B_0 \cos \omega t) \cdot dz \mathbf{a}_z$$

$$= -vB_0 \cos \omega t \int_0^l dz = -vlB_0 \cos \omega t$$

The negative sign means that the direction of this emf component is clockwise (terminal a negative, terminal b positive) for $\cos \omega t$ positive and anticlockwise for $\cos \omega t$ negative. The total emf induced in the circuit abcd at any given instant is thus

$$\mathscr{V} = \omega yl B_0 \sin \omega t - vl B_0 \cos \omega t$$

This is the voltage which appears across the terminals a and b.

9.6.2 RECTANGULAR COIL ROTATING IN A UNIFORM MAGNETIC FIELD

Let a rigid rectangular loop of length l and width b rotate with a uniform angular velocity ω radians per second about an axis which is perpendicular to a uniform steady magnetic field $\mathbf{B}$. The loop terminals are brought out to slip rings on the rotation axis as shown in Fig. 9.10. This arrangement represents a simple a.c. generator. At any instant t let $\theta = \omega t$ be the angle between $\mathbf{B}$ and the normal to the plane of the loop. With reference to the set of stationary rectangular axes shown in the figure, we have that

$$\mathbf{B} = B\mathbf{a}_y; \quad \omega = \omega \mathbf{a}_z$$

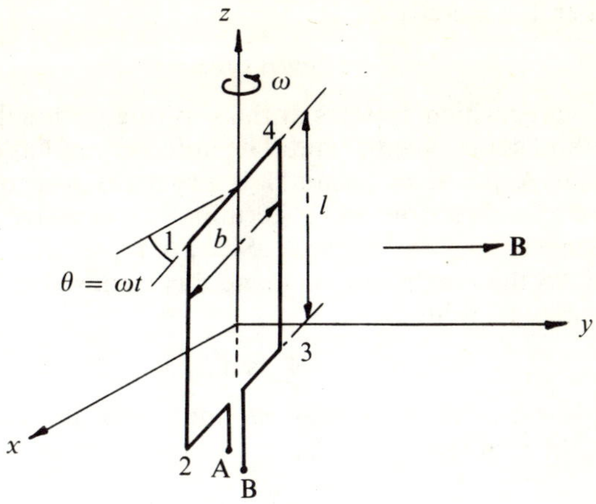

Fig. 9.10 Rectangular coil rotating in a uniform magnetic field

The velocity at any point on the loop whose position vector is **r** is

$$\mathbf{v} = \boldsymbol{\omega} \times \mathbf{r} = \omega\mathbf{a}_z \times (x\mathbf{a}_x + y\mathbf{a}_y + z\mathbf{a}_z)$$

$$= \omega(x\mathbf{a}_y - y\mathbf{a}_x)$$

so that

$$\mathbf{v} \times \mathbf{B} = \omega(x\mathbf{a}_y - y\mathbf{a}_x) \times B\mathbf{a}_y$$

$$= -\omega y B\mathbf{a}_z$$

Along the top and bottom sides of the loop $\mathbf{v} \times \mathbf{B} \cdot d\mathbf{l} = 0$ since both are parallel to the xy plane. Taking the direction 1234 as positive we have that

$$\oint \mathbf{v} \times \mathbf{B} \cdot d\mathbf{l} = \int_1^2 \mathbf{v} \times \mathbf{B} \cdot d\mathbf{l} + \int_3^4 \mathbf{v} \times \mathbf{B} \cdot d\mathbf{l}$$

along 12, $y = \tfrac{1}{2}b \sin \omega t$ and

$$\int_1^2 \mathbf{v} \times \mathbf{B} \cdot d\mathbf{l} = -\tfrac{1}{2}\int_{l/2}^{-l/2} \omega b B \sin \omega t\,\mathbf{a}_z \cdot dz\mathbf{a}_z$$

$$= \tfrac{1}{2}\,\omega b l B \sin \omega t$$

along 34, $y = -\tfrac{1}{2}b \sin \omega t$ and

$$\int_3^4 \mathbf{v} \times \mathbf{B} \cdot d\mathbf{l} = \tfrac{1}{2}\int_{-l/2}^{l/2} \omega b B \sin \omega t\,\mathbf{a}_z \cdot dz\mathbf{a}_z$$

$$= \tfrac{1}{2}\,\omega b l B \sin \omega t$$

The induced emf is therefore

$$\mathscr{V} = \omega b l B \sin \omega t$$

This is the voltage which appears at the slip-ring terminals and we see that it varies sinusoidally with time. The induced emf has the direction 1234 (terminal A positive, terminal B negative) whenever $\sin \omega t$ is positive, and the direction 1432 (terminal A negative, terminal B positive) whenever $\sin \omega t$ is negative. Since bl represents the area S of the loop and BS the maximum magnetic flux enclosed by the loop the induced emf may be written as,

$$\mathscr{V} = \omega \Phi_{\mathrm{m}} \sin \omega t$$

If the loop consists of N closely wound turns in series, the emf appearing at the terminal would be

$$\mathscr{V} = N\omega \Phi_{\mathrm{m}} \sin \omega t$$

9.6.3 THE HOMOPOLAR GENERATOR (FARADAY DISC)

This device, shown in Fig. 9.11, consists essentially of a circular conducting disc which is rotated about its axis with a constant angular

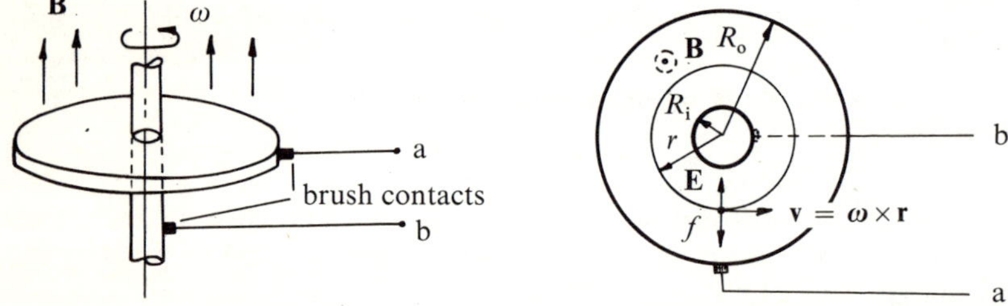

Fig. 9.11 A homopolar generator (Faraday disc)

velocity ω in a uniform axial stationary magnetic field of flux density $\mathbf{B}$. It is required to find the emf which appears between the two terminals a and b. We have that

$$\boldsymbol{\omega} = \omega \mathbf{a}_z; \quad \mathbf{B} = B\mathbf{a}_z$$

and the velocity at any point at a distance r from the centre of the disc is

$$\mathbf{v} = \boldsymbol{\omega} \times \mathbf{r} = \omega r \mathbf{a}_z \times \mathbf{a}_r = \omega r \mathbf{a}_\theta$$

The force per unit charge at any point in the disc is radial and given by

$$\mathbf{f} = \mathbf{v} \times \mathbf{B} = \omega r B \mathbf{a}_r$$

This force will drive the free electrons to the centre of the disc, building up a negative charge distribution there and leaving a positive charge distribution on the outer surface of the disc. Under equilibrium conditions this charge distribution will be static and will give rise to an electric field inside the disc such that at any point it is exactly equal and opposite to the force $\mathbf{f}$. Thus

$$\mathbf{E} = -\mathbf{f} = -\omega r B \mathbf{a}_r$$

and the potential difference between the terminals a and b is

$$V_{ab} = -\int_b^a \mathbf{E} \cdot \mathbf{dr} = \int_{R_i}^{R_o} \omega r B \, dr$$

$$= \tfrac{1}{2}\omega B(R_o^2 - R_i^2)$$

This is the open circuit voltage of the generator (see also Problem 9.2). If $R_i = 0$ the terminal voltage is

$$V_{ab} = \tfrac{1}{2}\omega B R_o^2$$

9.7 THE QUASI-STATIONARY STATE

The fundamental difference between static and dynamic or time-varying fields is that electrostatic and magnetostatic fields can exist independently whereas in a dynamic field the electric and magnetic fields are interrelated. Given a static charge distribution or a distribution of steady currents, the electric scalar and magnetic vector potentials are given by the respective expressions

$$V = \frac{1}{4\pi\epsilon} \int_v \left(\frac{\rho}{r}\right) dv \tag{9-24}$$

and

$$\mathbf{A} = \frac{\mu}{4\pi} \int_v \left(\frac{\mathbf{j}}{r}\right) dv \tag{9-25}$$

and the corresponding electric field intensity and magnetic flux density may be obtained from the equations

$$\mathbf{E} = -\nabla V$$

and

$$\mathbf{B} = \nabla \times \mathbf{A}$$

In static cases, therefore, electric and magnetic fields are decoupled; that is, they are completely independent of one another. In the present chapter we have seen that the physical phenomenon of electromagnetic induction is associated with a time-dependent magnetic field. This field produces an electric field and the two fields are related by the point form of Faraday's law

$$\nabla \times \mathbf{E} = -\frac{\partial \mathbf{B}}{\partial t}$$

from which we concluded (Section 9.1) that

$$\mathbf{E} = -\frac{\partial \mathbf{A}}{\partial t} - \nabla U \qquad (9\text{-}26)$$

Now since $\nabla \cdot \mathbf{D} = \rho$, we have that for a homogeneous medium

$$\nabla \cdot \mathbf{E} = -\frac{\partial}{\partial t}(\nabla \cdot \mathbf{A}) - \nabla^2 U = \frac{\rho}{\epsilon}$$

so that

$$\nabla^2 U + \frac{\partial}{\partial t}(\nabla \cdot \mathbf{A}) = -\frac{\rho}{\epsilon} \qquad (9\text{-}27)$$

If $\nabla \cdot \mathbf{A} = 0$, this equation reduces to the Poisson equation for electrostatics in which case the function U will be identified as the electrostatic potential V. The solution of this equation is given by equation 9–24. Now it has been shown in Section 7.10 that the condition $\nabla \cdot \mathbf{A} = 0$ is satisfied when the magnetic vector potential is given by equation 9–25 and $\nabla \cdot \mathbf{j} = 0$; it was also shown, alternatively, that if the condition $\nabla \cdot \mathbf{A} = 0$ is imposed, then $\mathbf{A}$ is given by equation 9–25 and $\nabla \cdot \mathbf{j} = 0$. Thus if the differential equations that relate the field quantities to the sources are given by

$$\left.\begin{aligned} \nabla \cdot \mathbf{D} &= \rho \\ \nabla \cdot \mathbf{B} &= 0 \\ \nabla \times \mathbf{E} &= -\frac{\partial \mathbf{B}}{\partial t} \\ \nabla \times \mathbf{B} &= \mu \mathbf{j} \end{aligned}\right\} \qquad (9\text{-}28)$$

then it is implied that $\mathbf{B}$ is given by the Biot–Savart law, so that $\nabla \cdot \mathbf{A} = 0$, $\nabla \cdot \mathbf{j} = 0$ and $\mathbf{A}$ and V are, at any instant, given by equations 9–24 and 9–25 respectively, that is by the same equations which apply in the static case. The field $\mathbf{E}$, however, is no longer conservative but is now given by

$$\mathbf{E} = -\frac{\partial \mathbf{A}}{\partial t} - \nabla V \qquad (9\text{-}29)$$

Under these conditions the set of equations defined by 9–28 are said to define a quasi-stationary magnetic field system. We shall discuss in the next section the conditions under which such a set of equations are applicable.

9.8 MAXWELL'S EQUATIONS

Let us examine the equation $\nabla \times \mathbf{B} = \mu \mathbf{j}$ more critically. If we take the divergence of both sides we have that for a homogeneous medium

$$\nabla \cdot \nabla \times \mathbf{B} = \mu \nabla \cdot \mathbf{j}$$

The left-hand side is identically zero whilst the divergence of $\mathbf{j}$ is zero only for steady currents (equation 6–19). In a system where the charge density varies with time the equation of continuity of current (equation 6–20) requires that

$$\nabla \cdot \mathbf{j} = -\partial \rho / \partial t$$

As shown in Section 6.7 this equation may be written as

$$\nabla \cdot \left(\mathbf{j} + \frac{\partial \mathbf{D}}{\partial t} \right) = 0$$

where $\partial \mathbf{D}/\partial t$ represents the maxwellian displacement current density. We thus see that whenever we have a time variation of charge, Ampère's law as given by $\nabla \times \mathbf{B} = \mu \mathbf{j}$ is inconsistent with the equation of continuity. To make Ampère's law consistent we must add to the normal conduction current density the displacement current density and write the modified law as

$$\nabla \times \mathbf{B} = \mu \left(\mathbf{j} + \frac{\partial \mathbf{D}}{\partial t} \right) \tag{9–30}$$

This equation shows that even in regions of space where there is no material circuit ($\mathbf{j} = 0$), a time-varying electric field gives rise to an induced magnetic field. (Recall that in the case of variational induction a time-varying magnetic field gives rise to an induced electric field even in the absence of a material circuit.)

Although the above reasoning leading to equation 9–30 may appear quite obvious to us, it required no lesser a genius than Maxwell to see the inconsistency of the point form of Ampère's law in the first place and then to introduce the vital concept of displacement current in order to reconcile Ampère's law with the equation of continuity. In his *Treatise on Electricity and Magnetism* Maxwell expressed his views on this problem as follows:

'We have very little experimental evidence relating to direct electro-magnetic action of current due to the variation of the electric displacement in dielectrics, but the extreme difficulty of reconciling the laws of electromagnetism with the existence of currents which are not closed is one reason among many why we must admit the existence of transient (time-varying) currents due to the variation of the displacement.' Further on he states that

'One of the chief peculiarities of this treatise is the doctrine which asserts that the true electric current $\mathbf{J}$, that on which the electromagnetic phenomena depend, is not the same thing as $\mathbf{j}$, the current of conduction, but that the time variation of $\mathbf{D}$, the electric displacement, must be taken into account in estimating the total movement of electricity so that we must write

$$\mathbf{J} = \mathbf{j} + \frac{\partial \mathbf{D}}{\partial t} \quad .'$$

Maxwell called the time rate of change of electric displacement a displacement 'current' because at that time space or the aether was assumed to be a dielectric medium which contained displaceable bound charges. To Maxwell, therefore, the term $\partial\mathbf{D}/\partial t$ represented a current in the literal and not in the metaphorical sense of the word. As shown in Section 6.7 the term $\partial\mathbf{D}/\partial t$ actually consists of two parts

$$\frac{\partial \mathbf{D}}{\partial t} = \frac{\partial \mathbf{P}}{\partial t} + \epsilon_o \frac{\partial \mathbf{E}}{\partial t}$$

The first part is a real current; that is, it arises from a movement of electric charge (bound) and exists only in a material medium. The second part alone can exist in free space and to call it a current when there is no physical flow or displacement of charge is somewhat misleading. To us today, therefore, it represents a current only in the metaphorical sense. In the context of Maxwell's equation 9–30, however, we may always define 'current' as that which causes a magnetic field. That a magnetic field may be produced either by the motion of electric charge or by a time-varying electric field may be regarded as a manifestation of the dual nature of electric current.

It is of course always possible not to try to attach any physical significance to the concept of displacement current and consider its existence merely as a hypothesis which Maxwell was perfectly entitled to make, provided that the consequences of this hypothesis lead to a result which can be verified experimentally. Maxwell's hypothesis implies that a time-varying electric field generates a magnetic field so that time-varying electric and magnetic fields are interdependent. Maxwell showed that this interdependence led to the mathematical conclusion that energy could be transmitted as electromagnetic waves —that is, electromagnetic radiation could take place. Every time we switch a television or a wireless set on we have an experimental confirmation of the propagation of energy and hence of the physical existence of electromagnetic waves. What better justification is there for accepting the validity of Maxwell's hypothesis? The fact that we do not accept Maxwell's physical explanation of the displacement current in no way invalidates his hypothesis.

The following group of equations

$$\left.\begin{aligned}
\nabla \cdot \mathbf{D} &= \rho \\
\nabla \cdot \mathbf{B} &= 0 \\
\nabla \times \mathbf{E} &= -\frac{\partial \mathbf{B}}{\partial t} \\
\nabla \times \mathbf{B} &= \mu\left(\mathbf{j} + \frac{\partial \mathbf{D}}{\partial t}\right)
\end{aligned}\right\} \qquad (9\text{–}31)$$

are collectively known as Maxwell's electromagnetic field equations and we may look upon them as being complete generalizations of the laws of Coulomb, Gauss, Faraday and Ampère. These equations together with the constitutive relations

$$\mathbf{D} = f_1(\mathbf{E})$$
$$\mathbf{B} = f_2(\mathbf{H})$$
$$\mathbf{j} = f_3(\mathbf{E})$$

and the Lorentz force equation $\mathbf{F} = q(\mathbf{E} + \mathbf{v} \times \mathbf{B})$ may be regarded as the distillate of our knowledge concerning macroscopic electricity and magnetism. For linear isotropic media the constitutive relations are

$$\left.\begin{aligned}
\mathbf{D} &= \epsilon\mathbf{E} \\
\mathbf{B} &= \gamma\mathbf{H} \\
\mathbf{j} &= \gamma\mathbf{E}
\end{aligned}\right\} \qquad (9\text{–}32)$$

It is important to realize that all the variables which appear in Maxwell's equations are those which exist at one macroscopic point in space at any particular instant of time. Maxwell's equations then give the relationships which exist between the rates of changes of the electric and magnetic vector field quantities at a given time-point with position and time and the charge and current densities at this same time-point. Maxwell's equations are therefore valid irrespective of the position of the sources which give rise to the $\mathbf{E}$ and/or $\mathbf{B}$ fields.

It is apparent that for those cases in which the displacement current can be neglected the equations in set 9–31 reduce to the ones given in set 9–28 which define a quasi-stationary magnetic field system. The displacement current may be neglected whenever the time rate of change represented by the term $\partial\rho/\partial t$, and hence by the term $\partial\mathbf{D}/\partial t$, is negligibly small compared with $\nabla \cdot \mathbf{j}$. This simply means that electromagnetic radiation is ignored and that conduction currents alone are the only sources of the magnetic field.

In the majority of physical and engineering problems the temporal variations of any of the variables which appear in Maxwell's equations are harmonic, that is they vary sinusoidally with time. (If not, it is usually possible to analyse a time-dependent function into a Fourier series of harmonic components.) We shall henceforth assume, therefore, that all electric and magnetic field quantities vary harmonically with time. Thus with

$$\mathbf{D} = \mathbf{D}_0 e^{i\omega t}; \quad \mathbf{B} = \mathbf{B}_0 e^{i\omega t}$$

we have that

$$\frac{\partial \mathbf{D}}{\partial t} = i\omega \mathbf{D}_0 e^{i\omega t} = i\omega \mathbf{D}; \quad \frac{\partial \mathbf{B}}{\partial t} = i\omega \mathbf{B}_0 e^{i\omega t} = i\omega \mathbf{B}$$

and Maxwell's equations become (we have replaced the time functions by the corresponding complex quantities and the operator $\partial/\partial t$ by the imaginary quantity $i\omega$)

$$\nabla \cdot \mathbf{D} = \rho$$
$$\nabla \cdot \mathbf{B} = 0$$
$$\nabla \times \mathbf{E} = -i\omega \mathbf{B} \tag{9-33}$$
$$\nabla \times \mathbf{B} = \mu(\mathbf{j} + i\omega \mathbf{D})$$

By using the constituent relations 9–32 the last two of the above equations may be written as

$$\nabla \times \mathbf{E} = -i\omega \mu \mathbf{H} \tag{9-34}$$

$$\nabla \times \mathbf{H} = (\gamma + i\omega \epsilon)\mathbf{E} = \left(1 + i\frac{\omega \epsilon}{\gamma}\right)\gamma \mathbf{E} \tag{9-35}$$

We have already shown in Section 6.8 that for good conductors the ratio $\omega \epsilon / \gamma$ becomes comparable with unity only for frequencies of the order of 10^{19} Hz, so that for fields inside metals it is entirely justifiable to neglect the displacement current for all but the highest frequencies. In general, electromagnetic wave propagation (and hence the displacement current) can be ignored in any system whose largest dimension is small compared to the wavelength (velocity of propagation divided by frequency of operation) of the harmonic time dependence.

A summary of Maxwell's equations in both their integral and point forms is given in Table 9.1.

	From Gauss's Law		From Gauss's Law	
	Integral	Point	Integral	Point
General	$\oint \mathbf{D} \cdot d\mathbf{S} = Q$	$\nabla \cdot \mathbf{D} = \rho$	$\oint \mathbf{B} \cdot d\mathbf{S} = 0$	$\nabla \cdot \mathbf{B} = 0$
Free space	$\oint \mathbf{D} \cdot d\mathbf{S} = 0$	$\nabla \cdot \mathbf{D} = 0$	$\oint \mathbf{B} \cdot d\mathbf{S} = 0$	$\nabla \cdot \mathbf{B} = 0$
Harmonic variation	$\oint \mathbf{D} \cdot d\mathbf{S} = Q$	$\nabla \cdot \mathbf{D} = \rho$	$\oint \mathbf{B} \cdot d\mathbf{S} = 0$	$\nabla \cdot \mathbf{B} = 0$
Steady	$\oint \mathbf{D} \cdot d\mathbf{S} = Q$	$\nabla \cdot \mathbf{D} = \rho$	$\oint \mathbf{B} \cdot d\mathbf{S} = 0$	$\nabla \cdot \mathbf{B} = 0$
Static	$\oint \mathbf{D} \cdot d\mathbf{S} = Q$	$\nabla \cdot \mathbf{D} = \rho$	$\oint \mathbf{B} \cdot d\mathbf{S} = 0$	$\nabla \cdot \mathbf{B} = 0$

	From Ampère's Law		From Faraday's Law	
	Integral	Point	Integral	Point
General	$\oint \mathbf{B} \cdot d\mathbf{l} = \mu I_t$	$\nabla \times \mathbf{B} = \mu\left(\mathbf{j} + \dfrac{\partial \mathbf{D}}{\partial t}\right)$	$\oint \mathbf{E} \cdot d\mathbf{l} = -\int \dfrac{\partial \mathbf{B}}{\partial t} \cdot d\mathbf{S}$	$\nabla \times \mathbf{E} = -\dfrac{\partial \mathbf{B}}{\partial t}$
Free space	$\oint \mathbf{B} \cdot d\mathbf{l} = \mu I_d$	$\nabla \times \mathbf{B} = \mu \dfrac{\partial \mathbf{D}}{\partial t}$	$\oint \mathbf{E} \cdot d\mathbf{l} = -\int \dfrac{\partial \mathbf{B}}{\partial t} \cdot d\mathbf{S}$	$\nabla \times \mathbf{E} = -\dfrac{\partial \mathbf{B}}{\partial t}$
Harmonic variation	$\oint \mathbf{B} \cdot d\mathbf{l} = \mu I_t$	$\nabla \times \mathbf{B} = \mu(\gamma + i\omega\epsilon)\mathbf{E}$	$\oint \mathbf{E} \cdot d\mathbf{l} = -i\omega \int \mathbf{B} \cdot d\mathbf{S}$	$\nabla \times \mathbf{E} = -i\omega\mathbf{B}$
Steady	$\oint \mathbf{B} \cdot d\mathbf{l} = \mu I_c$	$\nabla \times \mathbf{B} = \mu\mathbf{j}$	$\oint \mathbf{E} \cdot d\mathbf{l} = 0$	$\nabla \times \mathbf{E} = 0$
Static	$\oint \mathbf{B} \cdot d\mathbf{l} = 0$	$\nabla \times \mathbf{B} = 0$	$\oint \mathbf{E} \cdot d\mathbf{l} = 0$	$\nabla \times \mathbf{E} = 0$

PROBLEMS

Chapter Nine

9.1 A closed uniform wire loop of resistance R encircles a sinusoidal time-varying flux. An electrostatic voltmeter is connected between two diametrically opposite points A and B once, as shown in Fig. P9.1a, and once as in Fig. P9.1b. What will the voltmeter

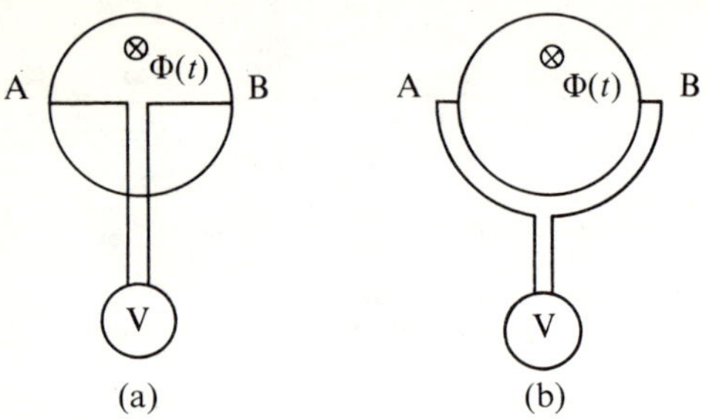

Fig. P. 9.1

read in each case? See the *Students' Quarterly Journal* (The Institution of Electrical Engineers) June 1966, pp. 227–32. 'More trouble with flux.'

9.2 A load resistance R is connected across the terminals of a Faraday disc generator (Fig. P9.2). If the disc rotates with an

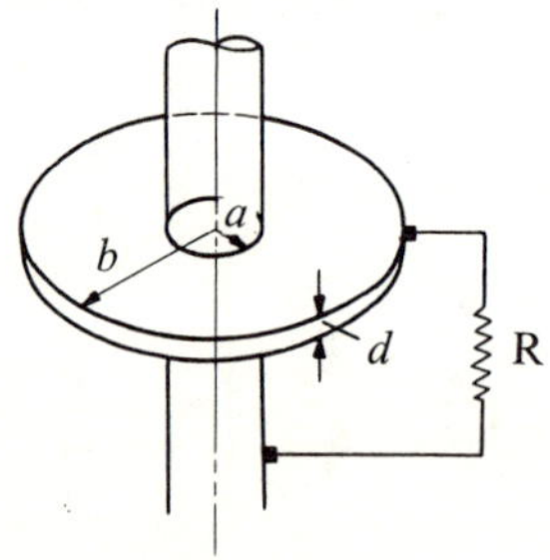

Fig. P. 9.2

angular velocity ω in a uniform magnetic field of flux density B parallel to the disc axis, show that the terminal voltage is given by

$$V = -\frac{I}{2\pi\gamma d}\log(b/a) + \tfrac{1}{2}\omega B(b^2 - a^2)$$

where γ is the conductivity of the disc material. Calculate the

value of V for a copper disc having the following parameters:
$\gamma = 5 \cdot 9 \times 10^7 \, \text{ohm}^{-1} \, \text{m}^{-1}$, $a = 0 \cdot 02 \, \text{m}$. $b = 0 \cdot 3 \, \text{m}$, $d = 0 \cdot 006 \, \text{m}$, $B = 1 \, \text{T}$, $\omega = 500 \, \text{rad s}^{-1}$, $R = 1 \, \text{ohm}$.

9.3   A Faraday disc rotates with an angular velocity ω in the presence of a uniform magnetic field B parallel to the disc axis. Show that the charge density at any point at a distance r from the centre of the disc is $\rho = -2\epsilon_0 \omega B$. This charge density rotates with the disc and thus produces a convection current; show that the magnetic field of this current is negligibly small compared with B.

9.4   A stationary loop of radius a carries a steady current. A second loop whose radius is much smaller than a travels with uniform velocity along the axis of the first loop. Show that the maximum voltage induced in the moving loop occurs when the latter is at a distance $a/2$ from the centre of the stationary loop.

9.5   A coil of 1000 turns and cross-sectional area $1 \, \text{cm}^2$ is placed with its axis parallel to a uniform magnetic field of flux density $1 \cdot 2 \, \text{T}$. If the coil is now rotated with a constant speed of 100 rev/min about an axis passing through the coil centre and perpendicular to the coil axis, find the voltage between the coil terminals.

9.6   A small magnet of magnetic moment $\mathbf{m}$ is placed at the origin of a system of rectangular coordinates. A circular coil of radius a and with Oy as axis is placed at a distance d from O. If the magnet is now rotated with an angular velocity ω about the Ox axis, show that the emf induced in the coil is given by

$$\frac{\mu_0 \omega m}{2d} \left[\frac{4}{15} + \frac{d^3}{3(d^2 + a^2)^{3/2}} - \frac{3d^5}{5(d^2 + a^2)^{5/2}} \right]$$

9.7   Discuss briefly the derivation of Maxwell's equations and show that in a uniform isotropic medium both $\mathbf{H}$ and $\mathbf{E}$ satisfy an equation of the form

$$\left(\nabla^2 - \mu\epsilon \frac{\partial^2}{\partial t^2} - \mu\gamma \frac{\partial}{\partial t} \right) \binom{\mathbf{H}}{\mathbf{E}} = 0$$

9.8   Find the maximum amplitude of the displacement current density associated with each of the following (a) a 10 MHz radio wave of amplitude $3 \, \text{mv m}^{-1}$ in air (assume that for air $\gamma = 0$), (b) a copper conductor carrying a conduction current of density $10 \, \text{A m}^{-2}$ at 500 Hz (assume $\epsilon = \epsilon_0$ for copper), (c) the air gap of an electric machine in which the flux density is given by $\mathbf{B} = 10^{-6} \sin[337t - (x/c)]\mathbf{a}_z$, where c is the velocity of light.

9.9   A flat cylindrical disc of internal radius a, external radius b, thickness d and conductivity γ is placed with its axis parallel to a uniform alternating field $B = B_0 \sin \omega t$. Find the magnitude and direction of the current density at any point on the disc when (a) the disc is stationary and (b) the disc rotates about its axis with a uniform angular velocity ω_0.

9.10 Show that for quasi-stationary conditions the current density in a conducting medium of uniform permeability μ and conductivity γ satisfies the equation

$$\nabla \cdot (\nabla \mathbf{j}) = \gamma \mu \frac{\partial \mathbf{j}}{\partial t}$$

9.11 A small magnet of constant moment $\mathbf{m}$ rotates about its centre with an angular velocity $\boldsymbol{\omega}$. Show that the electric field intensity at a distance $\mathbf{r}$ from $\mathbf{m}$ is given by

$$\mathbf{E} = \frac{\mu_0}{4\pi} \frac{\boldsymbol{\omega}(\mathbf{m} \cdot \mathbf{r}) - \mathbf{m}(\boldsymbol{\omega} \cdot \mathbf{r})}{r^3}$$

9.12 A sinusoidal voltage $V_{\mathrm{m}} \sin \omega t$ is applied between two circular parallel plates at a distance d apart; the space between the plates is filled with a medium of permittivity ϵ and conductivity γ. If it is assumed that the electric field between the plates is uniform, show that the magnetic flux density between the plates and at a distance r from the central axis is

$$B = \tfrac{1}{2}\mu V_{\mathrm{m}} r (\gamma \sin \omega t + \omega \cos \omega t)/d$$

9.13 A circular wire loop of radius a is placed with its axis parallel to a uniform magnetic field of flux density B. The loop is now rotated with an angular velocity ω about a diametral axis perpendicular to the field. Show that, if the inductance of the coils is neglected, the torque required to maintain the rotation is given by

$$T = \omega \pi^2 a^2 B^2 \sin^2 \omega t / R$$

where R is the resistance of the loop.

9.14 A dielectric cylinder of relative permittivity ϵ_{r} rotates about its axis with a constant angular velocity ω and in a uniform magnetic field of flux density B parallel to the cylinder axis. Show that the polarization of the cylinder at a distance r from its axis is given by

$$\mathbf{P} = \frac{\epsilon_0}{\epsilon_{\mathrm{r}}} (\epsilon_{\mathrm{r}} - 1) B \omega r \mathbf{a}_{\mathrm{r}}$$

Neglect end effects.

9.15 A long non-magnetic conducting tube of inner radius a and outer radius b rotates with a uniform angular velocity ω about its axis and in a uniform magnetic field B parallel to its axis. Show that the potential difference between the inner and outer tube surfaces is $\frac{1}{2}(a^2 - b^2)\omega B$.

CHAPTER TEN

ENERGY AND FORCES IN THE MAGNETIC FIELD

10.1 MAGNETIC ENERGY OF A SINGLE CIRCUIT

According to Faraday's law the emf induced in a circuit is equal to the time rate of change of magnetic flux linking that circuit, and its sense is such as to oppose any current change. Consider an isolated circuit of N closely wound turns carrying a current i. As long as this current is steady there is no change in the magnetic flux produced by it and there is no emf induced in the circuit. If the flux linking each turn of the circuit is now changed by $d\Phi$ in a time dt, the induced emf will be

$$\mathscr{V}' = -N\frac{d\Phi}{dt}$$

and since it opposes the external emf V driving the current i, we may write for the circuit

$$V - N\frac{d\Phi}{dt} = iR$$

$$V = iR + N\frac{d\Phi}{dt}$$

The energy supplied by the external source (battery) during the time interval dt is

$$dW_e = Vi\,dt = i^2R\,dt + iN\,d\Phi \qquad (10\text{--}1)$$

The term $i^2R\,dt$ represents the Joule energy dissipated in the conductor whilst the term $Ni\,d\Phi$ represents an increment in energy associated with the increment in the flux linking the circuit. This increment in energy, which is supplied by the source, is independent of the manner by which the flux increment is produced; it may be produced by a change in the current or by a change in the geometry of the circuit.*

If the medium is everywhere linear (μ constant), the relationship between the current and the magnetic flux is (see equation 7–89b)

$$N\Phi = Li$$

$$N\,d\Phi = L\,di + i\,dL$$

$$= L\,di \quad \text{(constant geometry)}$$

$$= i\,dL \quad \text{(constant current)}$$

---

* A change in geometry must be associated with mechanical work.

If we assume that the circuit is rigid (constant geometry) then the increment in energy supplied by the source will be equal to the increment in magnetic energy stored in the field; thus

$$dW_e = dT = Li\, di$$

and

$$T = \int_0^I Li\, di$$

$$= \tfrac{1}{2}LI^2$$

$$= \tfrac{1}{2}NI\Phi \tag{10-2}$$

Equation 10–2 is the energy supplied by the external source to build up the current in a rigid circuit from $i = 0$ at $t = 0$ to $i = I$ at $t = t$; this energy is stored in the magnetic field. We may thus state that when a rigid coil of inductance L carries a current I, the energy stored in the magnetic field set up by this current is $\tfrac{1}{2}LI^2$; alternatively we may say that if a flux Φ links N turns of a rigid circuit carrying a current I the energy stored in the magnetic field is $\tfrac{1}{2}NI\Phi$.

It has been shown in section 7–15 that the self-inductance of a non-filamentary circuit—i.e., a circuit of finite cross-section—is given by

$$L = \frac{1}{I^2} \int_v \mathbf{A} \cdot \mathbf{j}\, dv \tag{10-3}$$

The self-energy of a volume current distribution may thus be written as

$$T = \tfrac{1}{2} \int_v \mathbf{A} \cdot \mathbf{j}\, dv \tag{10-4}$$

where the integration must be taken throughout the volume in which the current flows.* Equation 10–4 is analogous to equation 5–3 for the self-energy of a static charge distribution

$$W = \tfrac{1}{2} \int_v \rho V\, dv$$

The magnetic energy given by equation 10–4 may be assumed to be distributed throughout the volume v with a density

$$\frac{dT}{dv} = \tfrac{1}{2}\mathbf{A} \cdot \mathbf{j} \tag{10-5}$$

Equation 10–4 defines the magnetic energy as localized in the sources of the magnetic field—i.e., the currents. The energy density is zero at all points where the current density is zero.

* For a surface distribution of current

$$T = \tfrac{1}{2} \int_s \mathbf{A} \cdot \mathbf{K}\, dS$$

Consider now the vector identity

$$\nabla \cdot (\mathbf{B} \times \mathbf{A}) = \mathbf{B} \cdot \nabla \times \mathbf{A} - \mathbf{A} \cdot \nabla \times \mathbf{B}$$

For a steady or quasi-stationary current distribution (displacement current negligible compared with the conduction current) we have, from Ampère's law

$$\nabla \times \mathbf{B} = \mu_0 \mathbf{j}$$

and since, by definition, $\nabla \times \mathbf{A} = \mathbf{B}$ we have that

$$\nabla \cdot (\mathbf{B} \times \mathbf{A}) = B^2 - \mu_0 \mathbf{A} \cdot \mathbf{j}$$

Integration of the above equation over all space gives

$$\int_{\text{all space}} \nabla \cdot (\mathbf{B} \times \mathbf{A}) \, dv = \int_{\text{all space}} B^2 \, dv - \mu_0 \int_v \mathbf{A} \cdot \mathbf{j} \, dv$$

where the last integral is taken over the volume v, since everywhere outside this volume $\mathbf{j} = 0$. To evaluate the left-hand integral let us surround the volume v by a sphere of radius r which completely

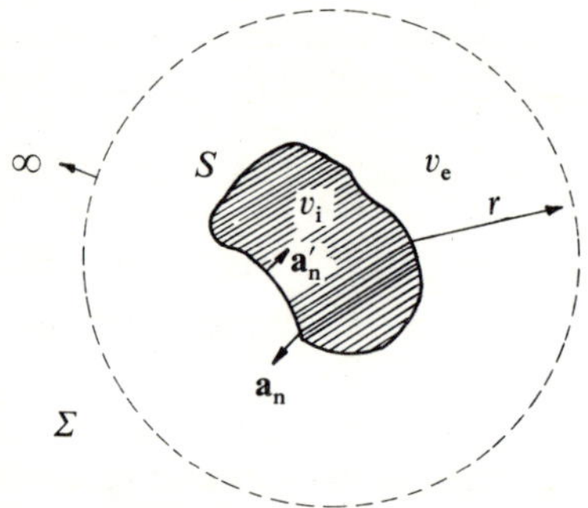

Fig. 10.1 Hypothetical sphere surrounding a finite volume distribution of current

encloses v (Fig. 10.1). Using the divergence theorem and bearing in mind that v_{e} is bounded externally by Σ and internally by S, we may write

$$\int_{v+v_{\mathrm{e}}} \nabla \cdot (\mathbf{B} \times \mathbf{A}) \, dv = \oint_S \mathbf{B} \times \mathbf{A} \cdot d\mathbf{S} + \oint_{S+\Sigma} \mathbf{B} \times \mathbf{A} \cdot d\mathbf{S}$$

$$= \oint_S \mathbf{B} \times \mathbf{A} \cdot dS\mathbf{a}_n + \oint_S \mathbf{B} \times \mathbf{A} \cdot dS\mathbf{a}_{n'} + \oint_\Sigma \mathbf{B} \times \mathbf{A} \cdot d\mathbf{S}$$

$$= \oint_\Sigma \mathbf{B} \times \mathbf{A} \cdot d\mathbf{S}$$

since $\mathbf{a}_n = -\mathbf{a}_n'$. Now to include all space we must carry out the above integration over the surface of a sphere of infinite radius

$(r \to \infty)$. At infinite distances from the current distribution the latter acts as a dipole so that $\mathbf{A}$ falls off as $1/r^2$ and $\mathbf{B}$ as $1/r^3$ and since dS increases as r^2 the integral over Σ becomes zero as $r \to \infty$. It follows that

$$\int_v \mathbf{A} \cdot \mathbf{j}\, dv = \frac{1}{\mu_0} \int_{\text{all space}} B^2\, dv \qquad (10\text{--}6)$$

and the magnetic energy associated with the current distribution is therefore given by

$$T = \frac{1}{2\mu_0} \int_{\text{all space}} B^2\, dv = \tfrac{1}{2}\mu_0 \int_{\text{all space}} H^2\, dv$$

$$= \tfrac{1}{2} \int_{\text{all space}} \mathbf{B} \cdot \mathbf{H}\, dv \qquad (10\text{--}7)$$

This equation shows that the magnetic energy may be regarded as distributed throughout all the space occupied by a magnetic field with a density

$$\frac{dT}{dv} = \tfrac{1}{2}\mathbf{B} \cdot \mathbf{H} \qquad (10\text{--}8)$$

Thus whereas equation 10–4 regards the energy as localized in the sources, equation 10–8 localizes the energy in the field. This dual interpretation of the localization of the magnetic energy is analogous to the one arrived at in electrostatics (Section 5.2). Here too no physical significance can be attached to the concept of energy density and it is to be regarded merely as a convenient quantity for the calculation of the *total* energy stored in the magnetic field.

From equations 10–3, 10–4, and 10–7 we may define the self-inductance of a circuit as

$$L = \frac{2T}{I^2} = \frac{1}{\mu_0 I^2} \int_v B^2\, dv \qquad (10\text{--}9)$$

We shall now give two examples on the application of the above equation.

10.1.1 INTERNAL INDUCTANCE OF AN INFINITELY LONG
       CYLINDRICAL CONDUCTOR

Assume that the conductor carries a current I uniformly distributed over the circular cross-section of the conductor (Fig. 10.2). The flux density at a distance x from the centre is

$$B_x = \frac{\mu_0 I'}{2\pi x} = \frac{\mu_0 I}{2\pi x} \cdot \frac{x^2}{R^2} = \frac{\mu_0 I}{2\pi R^2} x$$

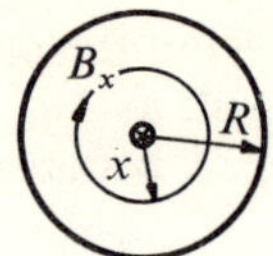

Fig. 10.2 Cylindrical conductor

The total internal magnetic energy is therefore

$$T_{\text{i}} = \frac{1}{2\mu_0} \int_v B^2 \, \mathrm{d}v = \frac{\mu_0 I^2}{8\pi^2 R^4} \int_v x^2 \, \mathrm{d}v$$

Assuming an axial length l, $\mathrm{d}v = 2\pi l x \, \mathrm{d}x$ and

$$T_{\text{i}} = \frac{\mu_0 I^2 l}{16\pi}$$

The internal inductance is therefore

$$L_{\text{i}} = 2T_{\text{i}}/I^2 = \frac{\mu_0}{8\pi} l \tag{10--10}$$

We thus see that in equation 7–114 the first term represents the internal self-inductance of the conductor.

10.1.2 INDUCTANCE OF A COAXIAL CABLE

In the coaxial cable shown in Fig. 10.3 the go and return currents are assumed to be uniformly distributed over their respective conductor cross-section. The magnetic energy per unit length in region 1 is

$$T_1 = \frac{\mu_0 I^2}{16\pi}$$

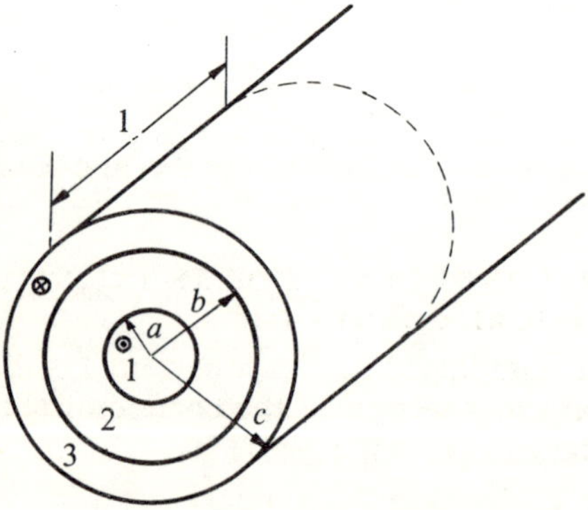

Fig. 10.3 Coaxial cable

The magnetic flux density in region 2 is

$$B_2 = \frac{\mu_0 I}{2\pi x}, \quad (a \le x \le b)$$

and the magnetic energy per unit length in this region is

$$T_2 = \frac{1}{2\mu_0} \int_a^b B_2^2 2\pi x \, dx$$

$$= \frac{\mu_0 I^2}{4\pi} \log\left(\frac{b}{a}\right)$$

In region 3 the magnetic flux density is the algebraic sum of the flux density produced by the whole current flowing in the central conductor and that produced by that part of the external current between $r = b$ and $r = x \, (>b)$. Thus

$$B_3 = \frac{\mu_0 I}{2\pi x}\left[1 - \frac{x^2 - b^2}{c^2 - b^2}\right]$$

$$= \frac{\mu_0 I}{2\pi x}\left[\frac{c^2 - x^2}{c^2 - b^2}\right]$$

The magnetic energy in region 3 is therefore

$$T_3 = \frac{\mu_0 I^2}{4\pi(c^2 - b^2)} \int_b^c \left[c^4 x^{-1} - 2c^2 x + x^3\right] dx$$

$$= \frac{\mu_0 I^2}{4\pi}\left[\frac{c^4}{(c^2 - b^2)^2} \log\left(\frac{c}{b}\right) - \frac{3c^2 - b^2}{4(c^2 - b^2)}\right]$$

The total inductance of a length l of cable is therefore

$$L = \left\{\frac{\mu_0}{8\pi} + \frac{\mu_0}{2\pi}\left[\frac{c^4}{(c^2 - b^2)^2} \log\left(\frac{c}{b}\right) - \frac{3c^2 - b^2}{4(c^2 - b^2)}\right] + \frac{\mu_0}{2\pi} \log\left(\frac{b}{a}\right)\right\} l$$

$$(10\text{–}11)$$

If the thickness of the outer conductor is small compared with b—i.e., $c \simeq b$— the inductance of the cable is given by

$$L = \left[\frac{\mu_0}{8\pi} + \frac{\mu_0}{2\pi} \log\left(\frac{b}{a}\right)\right] l \qquad (10\text{–}12)$$

10.2 MAGNETIC ENERGY OF A SYSTEM OF SEVERAL CIRCUITS

Consider two circuits C_1 and C_2 fixed in position and carrying currents i_1 and i_2 respectively (Fig. 10.4). C_1 has N_1 closely wound

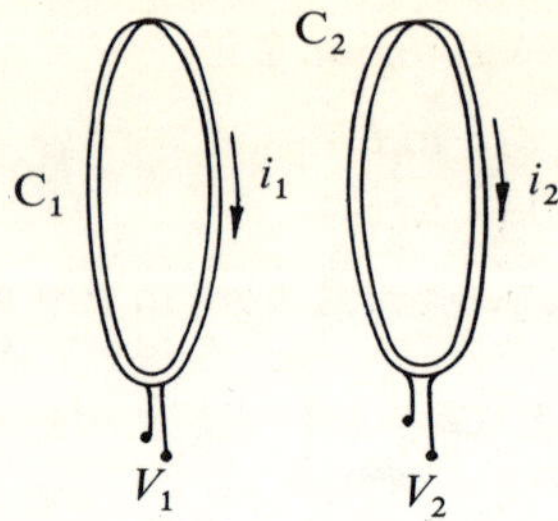

Fig. 10.4 Two neighbouring current circuits

turns and C_2 has N_2 closely wound turns. The following notation for magnetic flux will be used:

$N_1 \Phi_{11}$ = flux linking C_1 due to its own current
$N_1 \Phi_{12}$ = flux linking C_1 due to current i_2 in C_2
$N_2 \Phi_{22}$ = flux linking C_2 due to its own current
$N_2 \Phi_{21}$ = flux linking C_2 due to current i_1 in C_1

The voltage equations for the two circuits are

$$V_1 = i_1 R_1 + N_1 \frac{d\Phi_{11}}{dt} + N_1 \frac{d\Phi_{12}}{dt}$$

$$V_2 = i_2 R_2 + N_2 \frac{d\Phi_{22}}{dt} + N_2 \frac{d\Phi_{21}}{dt}$$

The total energy supplied by the two external sources during the time interval dt is

$$dW = V_1 i_1 \, dt + V_2 i_2 \, dt$$

$$= i_1^2 R_1 \, dt + i_2^2 R_2 \, dt + i_1 N_1 \, d\Phi_{11} + i_2 N_2 \, d\Phi_{22} +$$

$$i_1 N_1 \, d\Phi_{12} + i_2 N_2 \, d\Phi_{21}$$

The electrical work done against the induced emfs is, therefore,

$$dW_e = i_1 N_1 \, d\Phi_{11} + i_2 N_2 \, d\Phi_{22} + i_1 N_1 \, d\Phi_{12} + i_2 N_2 \, d\Phi_{21} \tag{10–13}$$

and since the circuits are fixed, this represents the increment in magnetic energy dT stored in the field. For a linear relationship between current and flux—i.e., constant permeability—

$$N_1 \, d\Phi_{11} = L_1 \, di_1$$

$$N_2 \, d\Phi_{22} = L_2 \, di_2$$

$$N_1 \, d\Phi_{12} = M_{12} \, di_2 = M \, di_2$$

$$N_2 \, d\Phi_{21} = M_{21} \, di_1 = M \, di_1$$

so that

$$\mathrm{d}T = L_1 i_1 \, \mathrm{d}i_1 + L_2 i_2 \, \mathrm{d}i_2 + \mathrm{d}(M i_1 i_2)$$

and

$$T = \tfrac{1}{2} L_1 I_1^2 + \tfrac{1}{2} L_2 I_2^2 + M I_1 I_2 \qquad (10\text{–}14)$$

where I_1 and I_2 are arbitrary final values of the currents flowing in circuits 1 and 2 respectively.

The term $M I_1 I_2$ represents the mutual or interaction energy between the two circuits. The coefficient M may be either positive or negative depending on the direction in which the mutual flux links the respective circuits; M is considered positive if the direction of the mutual flux is the same as that of the self flux.

Equation 10–14 may be written in terms of the self and mutual fluxes as follows

$$T = \tfrac{1}{2} I_1 N_1 \Phi_{11} + \tfrac{1}{2} I_2 N_2 \Phi_{22} + \tfrac{1}{2}[I_1 N_1 \Phi_{12} + I_2 N_2 \Phi_{21}] \qquad (10\text{–}15)$$

where the factor $1/2$ appears in the last two terms because $I_1 N_1 \Phi_{12} = I_2 N_2 \Phi_{21} = I_1 I_2 M$.

If we have two volume current distributions with $\mathbf{j}_1$ in v_1 and $\mathbf{j}_2$ in v_2, then from equation 10–4 we have that

$$T = \tfrac{1}{2} \int_{v_1 + v_2} \mathbf{A} \cdot \mathbf{j} \, \mathrm{d}v$$

$$= \tfrac{1}{2} \int_{v_1} (\mathbf{A}_1 + \mathbf{A}_2) \cdot \mathbf{j}_1 \, \mathrm{d}v_1 + \tfrac{1}{2} \int_{v_2} (\mathbf{A}_1 + \mathbf{A}_2) \cdot \mathbf{j}_2 \, \mathrm{d}v_2$$

$$= \tfrac{1}{2} \int_{v_1} \mathbf{A}_1 \cdot \mathbf{j}_1 \, \mathrm{d}v + \tfrac{1}{2} \int_{v_2} \mathbf{A}_2 \cdot \mathbf{j}_2 \, \mathrm{d}v + \tfrac{1}{2} \int_{v_1} A_2 \cdot \mathbf{j}_1 \, \mathrm{d}v + \tfrac{1}{2} \int_{v_2} \mathbf{A}_1 \cdot \mathbf{j}_2 \, \mathrm{d}v$$

$$(10\text{–}16)$$

The first two terms are the self energies of the distributions and the sum of the last two terms is their mutual or interaction energy; from equation 7–62 we see that the mutual energy may also be written as

$$\int_{v_1} \mathbf{A}_2 \cdot \mathbf{j}_1 \, \mathrm{d}v \quad \text{or} \quad \int_{v_2} \mathbf{A}_1 \cdot \mathbf{j}_2 \, \mathrm{d}v$$

In terms of the flux density the total magnetic energy of the system is given by

$$T = \frac{1}{2\mu_0} \int_{\text{all space}} (\mathbf{B}_1 + \mathbf{B}_2) \cdot (\mathbf{B}_1 + \mathbf{B}_2) \, \mathrm{d}v \qquad (10\text{–}17)$$

Consider now a system of n fixed circuits; let the ith circuit $(i = 1, 2, \ldots n)$ consist of N_i closely wound turns through which a current I_i flows. For such a system the equations corresponding to equations 10–13, 10–14 and 10–15 are respectively,

$$dW_e = \sum_{i=1}^{n} I_i N_i \, d\Phi_{ii} + \sum_{i=1}^{n} I_i \sum_{j \neq i}^{n} N_i \, d\Phi_{ij} \qquad (10\text{--}18)$$

$$T = \tfrac{1}{2} \sum_{i=1}^{n} I_i N_i \Phi_{ii} + \tfrac{1}{2} \sum_{i=1}^{n} I_i \sum_{j \neq i}^{n} N_i \Phi_{ij} \qquad (10\text{--}19)$$

$$T = \tfrac{1}{2} \sum_{i=1}^{n} L_i I_i^2 + \tfrac{1}{2} \sum_{\substack{i,j \\ i \neq j}}^{n} I_i I_j M_{ij} \qquad (10\text{--}20)$$

where the coefficients M_{ij} may be either positive or negative as discussed above.

If we define the total flux linking the N_i turns of the ith circuit by

$$\Phi_i = \sum_{j=1}^{n} N_i \Phi_{ij} \qquad (10\text{--}21)$$

we may rewrite equation 10–19 as

$$T = \tfrac{1}{2} \sum_{i=1}^{n} I_i \Phi_i \qquad (10\text{--}22)$$

This equation represents the total magnetic energy stored in the field of a system of n circuits in which the ith circuit carries a current I_i and is linked by a total flux Φ_i.

Since for a linear medium

$$N_i \Phi_{ij} = I_j L_{ij}$$

then

$$\Phi_i = \sum_{j=1}^{n} I_j L_{ij} \qquad (10\text{--}23)$$

Thus

$$\begin{aligned}
\Phi_1 &= L_{11} I_1 + L_{12} I_2 + L_{13} I_3 + \cdots L_{1n} I_n \\
\Phi_2 &= L_{21} I_1 + L_{22} I_2 + L_{23} I_3 + \cdots L_{2n} I_n \\
&\cdots \cdots \cdots \cdots \cdots \cdots \cdots \cdots \cdots \cdots \cdots \\
\Phi_n &= L_{n1} I_1 + L_{n2} I_2 + L_{n3} I_3 + \cdots L_{nn} I_n
\end{aligned}$$

It follows that the current may be expressed in terms of a linear combination of the fluxes,

$$I_i = \sum_{j=1}^{n} \Lambda_{ij} \Phi_j \qquad (10\text{--}24)$$

The formal analogy between equations 10–23, 10–24 and equations 3–32 and 3–41 is evident. The coefficients L_{ij} and Λ_{ij} (like p_{ij} and C_{ij}) are functions of the geometry of the system only. For a given system of positional coordinates η, referred to a fixed system of axes, we have that in general

$$T(I, \eta) = \tfrac{1}{2} \sum_{i,j}^{n} I_i I_j L_{ij} \qquad (10\text{–}25)$$

$$T(\Phi, \eta) = \tfrac{1}{2} \sum_{i,j}^{n} \Phi_i \Phi_j \Lambda_{ij} \qquad (10\text{–}26)$$

10.3 PHYSICAL INTERPRETATION OF THE MUTUAL ENERGY

Consider a filamentary circuit C carrying a current I and placed in an external magnetic field of density $\mathbf{B}_0$ (Fig. 10.5). Suppose now that

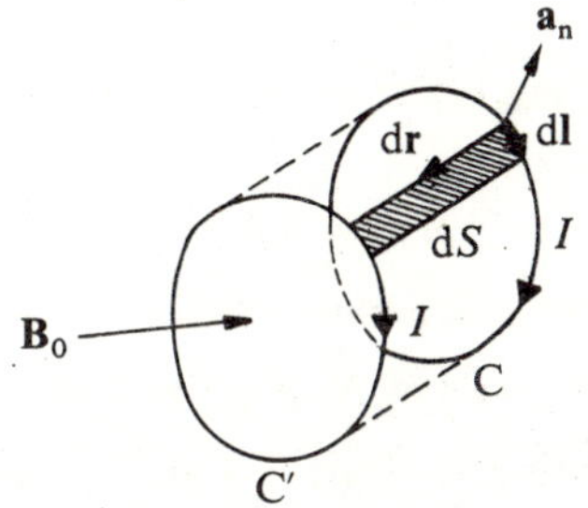

Fig. 10.5 Displacement of a filamentary current circuit in an external magnetic field

the circuit *as a whole* is displaced a distance dr and that during the displacement the current I is kept constant (by an external source such as a battery). The force acting on the current element $I\,\mathrm{d}\mathbf{l}$ is, by equation 7–13

$$\mathrm{d}\mathbf{F} = I\,\mathrm{d}\mathbf{l} \times \mathbf{B}_0$$

The work done by this force is displacing the element a distance dr is therefore

$$\mathrm{d}^2 W_{\mathrm{m}} = \mathrm{d}\mathbf{F} \cdot \mathrm{d}\mathbf{r} = I(\mathrm{d}\mathbf{l} \times \mathbf{B}_0 \cdot \mathrm{d}\mathbf{r})$$

$$= I(\mathrm{d}\mathbf{r} \times \mathrm{d}\mathbf{l} \cdot \mathbf{B}_0) = I(\mathbf{B}_0 \cdot \mathrm{d}\mathbf{S})$$

$$= I\,\mathrm{d}\Phi_0$$

where $\mathrm{d}\Phi_0$ is the net flux crossing the surface element $\mathrm{d}S$ swept out by the displacement of the element $\mathrm{d}l$. The total mechanical work done by the magnetic forces in moving the entire loop contour C is

$$\mathrm{d}W_{\mathrm{m}} = I(\Phi_0' - \Phi_0)$$

If the circuit C was originally at an infinite distance from the flux source such that $\Phi_0 = 0$, then

$$W_\mathrm{m} = I\Phi_0' \qquad (10\text{--}27)$$

Hence we may state that the quantity $I\Phi$ represents the work done by the mechanical forces in bringing the circuit C from an initial position at infinity to a position where the flux (due to sources other than I) linking it is Φ, the current in the circuit remaining constant during the displacement.

Consider now the case of two linear circuits C_1 and C_2 carrying currents I_1 and I_2 respectively, placed a finite distance apart and with no external flux other than that produced by the currents themselves. The mechanical work done to bring C_2 from infinity to its final position in the presence of C_1 is

$$W_\mathrm{m} = I_2 N_2 \Phi_{21}$$

The work done to bring C_1 from infinity to its final position in the presence of C_2 is

$$W_\mathrm{m} = I_1 N_1 \Phi_{12}$$

Since $I_1 N_1 \Phi_{12} = I_2 N_2 \Phi_{21}$ we may write

$$W_\mathrm{m} = \tfrac{1}{2}[I_1 N_1 \Phi_{12} + I_2 N_2 \Phi_{21}]$$

The work done to bring a third circuit C_3 to some final position within the system is

$$(W_\mathrm{m})_{1+2,3} = I_3 N_3 \Phi_{31} + I_3 N_3 \Phi_{32}$$
$$= I_1 N_1 \Phi_{13} + I_2 N_2 \Phi_{23}$$
$$= \tfrac{1}{2}[I_1 N_1 \Phi_{13} + I_2 N_2 \Phi_{23} + I_3 N_3 (\Phi_{31} + \Phi_{32})]$$

The total work done to set up the system is therefore

$$W_\mathrm{m} = \tfrac{1}{2}[I_1 N_1 (\Phi_{12} + \Phi_{13}) + I_2 N_2 (\Phi_{21} + \Phi_{23}) + I_3 N_3 (\Phi_{31} + \Phi_{32})]$$

For a system of n circuits we may write

$$W_\mathrm{m} = \tfrac{1}{2} \sum_{i=1}^{n} I_i \sum_{j \neq i}^{n} N_i \Phi_{ij} \qquad (10\text{--}28)$$

where $\Sigma N_i \Phi_{ij}$ is the flux linking the N_i turns of the ith circuit due to the currents in the other $n-1$ circuits. Equation 10–28 represents the work done by the magnetic forces in bringing the n circuits (all initially an infinite distance apart) from infinity to their specified position within the final configuration of the system, whilst the currents in all the circuits are kept constant as the circuits are moved into position.

A comparison between equations 10–28 and 10–19 shows that equation 10–28 represents the mutual or interaction energy of the system.

10.4 FORCES IN TERMS OF ENERGY CHANGES

The determination of the forces acting on current-carrying circuits in terms of magnetic field energy is based on the principle of conservation of energy. According to this principle, energy can be neither created nor destroyed and thus can only be transformed from one state to another. We have already seen in the preceding section that for a system of fixed circuits the electrical energy expended (against the induced emfs) in establishing constant currents in the circuits is stored as magnetic energy in the field. Let us now assume that in a system of n circuits the ith circuit is allowed to make a virtual incremental displacement $d\mathbf{r}_i$ under the action of the magnetic force $\mathbf{F}_i$ acting on it, subject to the condition that the currents within the system remain constant during the displacement. The principle of conservation of energy now requires that

$$\begin{bmatrix} \text{electrical energy supplied by} \\ \text{the source to maintain the} \\ \text{currents constant} \end{bmatrix} = \begin{bmatrix} \text{increase in} \\ \text{energy} \end{bmatrix} + \begin{bmatrix} \text{mechanical work} \\ \text{done by system} \end{bmatrix}$$

that is,

$$dW_{\mathrm{e}} = dT + dW_{\mathrm{m}} \tag{10–29}$$

Since the currents are kept constant we have from equation 10–22

$$dT = \tfrac{1}{2} \sum I_i \, d\Phi_i$$

The energy supplied by the external sources is, from equations 10–18 and 10–21,

$$dW_{\mathrm{e}} = \sum I_i \, d\Phi_i$$
$$= 2T \tag{10–30}$$

and it follows that

$$dW_{\mathrm{m}} = dT$$
$$\mathbf{F}_i \cdot d\mathbf{r}_i = dT$$

and

$$\mathbf{F}_i = \nabla_i T \tag{10–31}$$

We thus see that whenever mechanical work is done by the system under a constant current constraint, the external electrical sources must supply twice the energy expended, in addition to supplying the

Joule heat losses; the second equal part is stored as magnetic energy in the field. When external work is done against the action of magnetic forces, the energy returned to the external sources is twice the mechanical work done on the system; the additional energy is supplied to the sources by a reduction in the stored magnetic energy. The magnetic field thus serves as an agent for converting electrical energy into mechanical energy or vice versa.

Equation 10–31 shows that, for a constant current constraint, the force is equal to the gradient of the magnetic energy. Now since in mechanics and electrostatics, forces are given by the *negative gradient* of the *potential* energy, we may define a *magnetic potential energy function U* such that

$$U = -T \tag{10–32}$$

In mechanics and electrostatics the negative sign before the gradient indicates that the forces always act in such a direction as to decrease the potential energy; we therefore conclude that the magnetic forces act in such a direction as to increase the magnetic energy of the field.

For a virtual displacement $d\eta_r$ (relative to a fixed system of axes) of the rth circuit of a system of n circuits, the component of the magnetic force acting on this circuit in the direction of the coordinate η_r is

$$\left.\begin{aligned} F_{\eta_r} &= \frac{\partial T}{\partial \eta_r} \\[2mm] &= \tfrac{1}{2} \sum_{i,j}^{n} I_i I_j \frac{\partial L_{ij}}{\partial \eta_r} \end{aligned}\right\} \quad \text{(constant currents)} \tag{10–33}$$

If the rth circuit is rotated by an angle $d\theta_r$ about a given axis, the *torque* on the circuit is given by

$$\tau_{\theta_r} = \tfrac{1}{2} \sum_{i,j}^{n} I_i I_j \frac{\partial L_{ij}}{\partial \theta_r} \tag{10–34}$$

For the case of two circuits, if the coordinate η_1 of the first circuit changes whilst the second circuit remains fixed, the force on circuit 1 in the direction of the coordinate η_1 is

$$F_{\eta_1} = \tfrac{1}{2} I_1^2 \frac{\partial L_{11}}{\partial \eta_1} + I_1 I_2 \frac{\partial M}{\partial \eta_1} \tag{10–35}$$

If the self-inductances of the circuits do not change

$$F_{\eta_1} = F_{\eta_2} = I_1 I_2 \frac{\partial M}{\partial \eta} \tag{10–36}$$

and

$$\tau_{\theta_1} = \tau_{\theta_2} = I_1 I_2 \frac{\partial M}{\partial \theta} \qquad (10\text{--}37)$$

Both the force and torque act in the direction of increasing mutual inductance, that is the circuits tend to move so as to make the flux linkages as large as possible.

If the displacement of one of the circuits of a system takes place such that there is no change in the flux linking any of the circuits, it is evident that in this case there is no additional electrical energy supplied to the system and the work done by the magnetic forces must be equal to the decrease in the magnetic energy of the system; thus

$$\mathbf{F}_n = -\frac{\partial T}{\partial \eta} \quad \text{(constant flux)} \qquad (10\text{--}38)$$

10.4.1 FORCE BETWEEN TWO CURRENT LOOPS

Consider two coaxial circular coils of radii a_1 and a_2 which have N_1 and N_2 closely wound turns and which carry currents I_1 and I_2 respectively. Let z be the distance between the planes of the coils

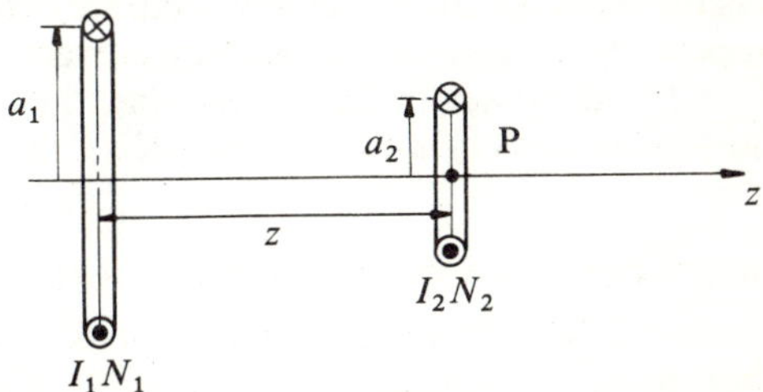

Fig. 10.6 To determine the force between two coaxial circular coils

(Fig. 10.6). It will be assumed that $a_2 \ll a_1$; with this assumption the field over the area of coil 2 may be considered uniform. The magnetic flux density at point P due to the current in coil 1 is, from equation 7–33,

$$B_z = \tfrac{1}{2}\mu_0 N_1 I_1 a_1^2 (z^2 + a_1^2)^{-3/2}$$

hence,

$$\Phi_{21} = \tfrac{1}{2}\mu_0 \pi N_1 I_1 a_1^2 a_2^2 (z^2 + a_1^2)^{-3/2}$$

and

$$M_{12} = N_2 \Phi_{21}/I_1 = \tfrac{1}{2}\mu_0 \pi N_1 N_2 a_1^2 a_2^2 (z^2 + a_1^2)^{-3/2}$$

The force between the coils is given by

$$F_z = I_1 I_2 \frac{\partial M}{\partial z} = -\tfrac{3}{2}\mu_0 \pi I_1 N_1 I_2 N_2 a_1^2 a_2^2 z (z^2 + a_1^2)^{-5/2} \qquad (10\text{--}39)$$

The negative sign indicates that the force is attractive. If a_1 and a_2 are both very small compared with z, M is given by equation 7–110 and the force by

$$F_z = -\frac{3}{2}\frac{\mu_0}{\pi}\frac{(\pi a_1^2 I_1 N_1)(\pi a_2^2 I_2 N_2)}{z^4} \tag{10–40}$$

The exact expression for the force may be obtained from equation 7–109,

$$F_z = I_1 I_2 \frac{\partial M}{\partial k}\frac{\partial k}{\partial z} = -I_1 I_2 \frac{k^3 z}{4a_1 a_2}\frac{\partial M}{\partial k}$$

The derivatives of the complete elliptic integrate F and E are given by

$$\frac{\partial F}{\partial k} = \frac{1}{k}\left[\frac{E}{1-k^2} - F\right]; \quad \frac{\partial E}{\partial k} = \frac{1}{k}[E-F]$$

so that

$$F_z = \frac{\mu_0 I_1 N_1 I_2 N_2 z}{[z^2 + (a_1 + a_2)^2]^{1/2}}\left[\frac{a_1^2 + a_2^2 + z^2}{(a_1 + a_2)^2 + z^2}E - F\right] \tag{10–41}$$

This expression is the basis for the extremely accurate determination of the absolute ampere by means of the Kelvin current balance.* By making $I_1 = I_2 = I$ and by accurately measuring the force between identical precision-made coils of known dimensions, the current I can be determined.

10.4.2 FORCE BETWEEN TWO COAXIAL SOLENOIDS

Consider two coaxial solenoids with the dimensions of the inner one much smaller than those of the outer one (Fig. 10.7). The flux density at any point on the axis of the large solenoid is, from equation 7–34a

$$B_z = \mu_0 I_1 N_1 (\cos \theta_1 - \cos \theta_2)/2l$$

which in terms of the linear dimensions of the coil is,

$$B_z = \frac{\mu_0 I_1 N_1}{2l}\left[\frac{\frac{1}{2}l - z}{[a^2 + (\frac{1}{2}l - z)^2]^{1/2}} + \frac{\frac{1}{2}l + z}{[a^2 + (\frac{1}{2}l + z)^2]^{1/2}}\right]$$

Since the inner solenoid is small we may write

$$\Phi_{21} = B_z S_2$$

where S_2 is the cross-sectional area of the inner solenoid. We have that

$$M_{21} = N_2 \Phi_{21}/I_1 = N_2 S_2 B_z/I_1$$

* E. W. Golding and F. C. Widdis, *Electrical Measurements and Measuring Instruments* (Pitman, 1963).

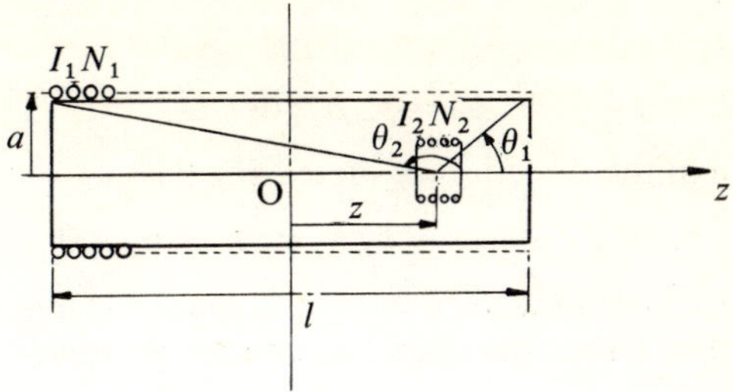

Fig. 10.7 To determine the force between two coaxial solenoids

for the same direction of currents in the two coils M is positive and

$$T = I_1 I_2 M = I_2 N_2 S_2 B_z$$

The force acting on the inner solenoid is therefore

$$F_z = \partial T/\partial z$$

$$= \tfrac{1}{2}(\mu_0 I_1 I_2 N_1 N_2 S_2 a^2/l)\{[a^2 + (\tfrac{1}{2}l + z)^2]^{-3/2} - [a^2 + (\tfrac{1}{2}l - z)^2]^{-3/2}\}$$

$$(10\text{--}42)$$

The force is negative for positive values of z and positive for negative values of z. Thus for the same direction of the currents I_1 and I_2 the inner coil is attracted towards the centre of the outer one. If the currents have opposite directions, M will be negative and the force on the inner coil will be directed away from the centre.

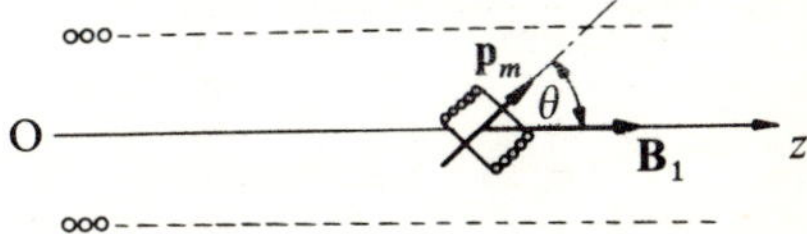

Fig. 10.8 To determine torque on inner solenoid

Suppose that the axis of the inner coil makes an angle θ with Oz (Fig. 10.8). In this case we have

$$\Phi_{21} = S_2 B_1 \cos \theta$$

$$M_{21} = N_2 S_2 B_1 \cos \theta / I_1$$

and

$$T = I_2 N_2 S_2 B_1 \cos \theta$$

Since, by equation 8–76, the equivalent dipole moment of the small solenoid is

$$p_m = I_2 N_2 S_2$$

we may write

$$T = p_m B_1 \cos \theta \qquad (10\text{--}43)$$

and the torque in the direction of increasing θ is

$$\tau = \partial T/\partial \theta = -p_m B_1 \sin \theta = \mathbf{p}_m \times \mathbf{B}_1 \qquad (10\text{--}44)$$

The torque is maximum when $\theta = \pm\pi/2$. Since the magnetic potential energy is equal to minus the magnetic energy, we have

$$U = -p_m B_1 \cos \theta = -\mathbf{p}_m \cdot \mathbf{B}_1 \qquad (10\text{--}45)$$

When $\theta = 0$ the potential energy is a minimum whereas the mutual magnetic energy is a maximum.

10.4.3 LIFTING FORCE OF AN ELECTROMAGNET

Consider the magnetic circuit shown in Fig. 10.9. The two parts (armature and yoke) are made of a soft magnetic material of cross-sectional area S and mean lengths l_1 and l_2 and separated by a small

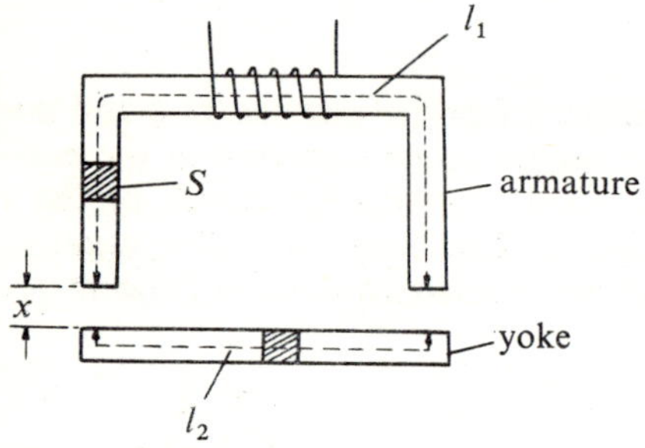

Fig. 10.9 Electromagnet

air gap of length x. Leakage and fringing are neglected and the permeability of the yoke and armature is assumed constant; this is justifiable if the magnetic material is not saturated and is operated in the central portion of its magnetization characteristic.

For a given flux Φ in the circuit the total stored energy is

$$T = \tfrac{1}{2} \int_{\text{material}} \mathbf{B} \cdot \mathbf{H} \, dv + \tfrac{1}{2} \int_{\text{gaps}} \mathbf{B} \cdot \mathbf{H} \, dv$$

The force acting on the armature is given by either

$$F_x = -\partial T/\partial x \text{ (constant flux)}$$

or

$$F_x = \partial T/\partial x \text{ (constant current)}$$

If we assume that the flux in the circuit remains constant, then the magnetic energy in the armature and yoke is constant and

$$F_x = -\frac{\partial}{\partial x} \frac{1}{2} \int_{\text{gaps}} \mathbf{B} \cdot \mathbf{H} \, dv$$

$$= -\frac{\partial}{\partial x} \frac{1}{2} \int_0^x BH2S \, dx = -BHS$$

and since in the gaps $B = \mu_0 H$

$$F_x = -2\left(\frac{B^2 S}{2\mu_0}\right)$$

The negative sign indicates that the force is attractive. *The attractive force per unit area at each air gap* is thus

$$f_x = \frac{B^2}{2\mu_0} \quad \text{newton metre}^{-2} \tag{10-46}$$

For a flux density of 1 tesla this force is $397,887 \, \text{N m}^{-2}$ ($4.352 \, \text{ton ft}^{-2}$). The same result would have been obtained had a constant current constraint been imposed on the system. The total flux in the circuit is (see Section 8.15.1),

$$\Phi = NI/[(l'/\mu S)+(2x/\mu_0 S)]$$

where $l' = l_1 + l_2$. The magnetic energy stored is

$$T = \tfrac{1}{2} NI\Phi = \tfrac{1}{2} N^2 I^2/[(l'/\mu S)+(2x/\mu_0 S)]$$

and the force is given by

$$F_x = \left.\frac{\partial T}{\partial x}\right|_{I=\text{const}} = -\frac{B^2 S}{\mu_0}$$

10.5 FORCES ON LINEAR MAGNETIC BODIES

Let us assume that a linear isotropic magnetic medium whose permeability is everywhere μ_1 extends over all space. Suppose that due to fixed current sources the magnetic flux density at any point in the medium is $\mathbf{B}_0$. The magnetic energy of the system is

$$T_0 = \tfrac{1}{2} \int_{\text{all space}} \mathbf{B}_0 \cdot \mathbf{H}_0 \, dv \tag{10-47}$$

where $\mathbf{B}_0 = \mu_1 \mathbf{H}_0$. Suppose now that in a finite volume v_2 we replace the medium of permeability μ_1 by a linear medium of permeability μ_2 with the condition that the currents remain constant (Fig. 10.10). The

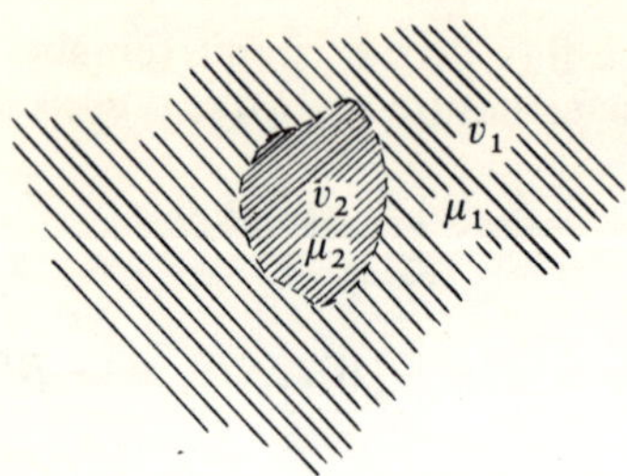

Fig. 10.10 Magnetic body of volume v_2 and permeability μ_2 placed in a medium of infinite extent and permeability μ_1

flux density at any point will be modified such that

$$\mathbf{B} = \mu_1\mathbf{H}_1 \text{ in volume } v_1 \text{ (all space less volume } v_2)$$

$$\mathbf{B} = \mu_2\mathbf{H}_2 \text{ in volume } v_2$$

The total magnetic energy of the system now has the value

$$T = \tfrac{1}{2} \int_{v_1+v_2} \mathbf{B} \cdot \mathbf{H} \, dv$$

The change in energy is given by

$$T - T_0 = \tfrac{1}{2} \int_{v_1+v_2} (\mathbf{B} \cdot \mathbf{H} - \mathbf{B}_0 \cdot \mathbf{H}_0) \, dv$$

$$= \tfrac{1}{2} \int_{v_1+v_2} (\mathbf{B}+\mathbf{B}_0) \cdot (\mathbf{H}-\mathbf{H}_0) \, dv + \tfrac{1}{2} \int_{v_1+v_2} (\mathbf{B} \cdot \mathbf{H}_0 - \mathbf{H} \cdot \mathbf{B}_0) \, dv$$

$$(10\text{--}48)$$

If we let $(\mathbf{B}+\mathbf{B}_0) = \mathbf{B}'$ and $(\mathbf{H}-\mathbf{H}_0) = \mathbf{H}'$, then since we have assumed that no change takes place in the current distribution of the source, we have that

$$\nabla \cdot \mathbf{B}' = 0, \quad \nabla \times \mathbf{H}' = 0$$

At the boundary surface S we have, from equations 8–78 and 8–82,

$$\mathbf{a}_n \cdot (\mathbf{B}'_+ - \mathbf{B}'_-) = 0 \qquad (10\text{--}49)$$

$$\mathbf{a}_n \times (\mathbf{H}'_+ - \mathbf{H}'_-) = 0 \qquad (10\text{--}50)$$

Since the curl of $\mathbf{H}'$ is zero we may write $\mathbf{H}' = \nabla\phi$ where ϕ is a scalar function which, by virtue of equation 10–50, is continuous across the boundary ($\phi_+ = \phi_-$).

We shall now show that

$$\int_{v_1+v_2} \mathbf{B}' \cdot \mathbf{H}' \, dv = 0$$

Since $\mathbf{B}' \cdot \mathbf{H}' = \mathbf{B}' \cdot \nabla\phi = \nabla \cdot (\phi\mathbf{B}') - \phi\nabla \cdot \mathbf{B}'$ and since $\nabla \cdot \mathbf{B}' = 0$ then

$$\int_{v_1+v_2} \mathbf{B}' \cdot \mathbf{H}' \, dv = \int_{v_1} \nabla \cdot (\phi\mathbf{B}') \, dv + \int_{v_2} \nabla \cdot (\phi\mathbf{B}') \, dv$$

If we assume that the volume v'_1 is bounded internally by S and externally by the surface Σ of a sphere of radius r which completely

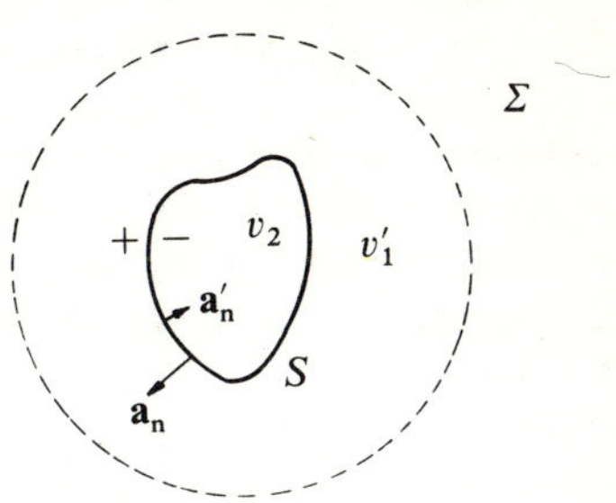

Fig. 10.11 Hypothetical sphere surrounding volume v_2

encloses v_2 (Fig. 10.11) application of the divergence theorem to the above equation gives*

$$\int_{v'_1+v_2} \mathbf{B}' \cdot \mathbf{H}' \, dv = \oint_{S+\Sigma} (\phi\mathbf{B}'_+) \cdot d\mathbf{S}' + \oint_{S} (\phi\mathbf{B}'_-) \cdot d\mathbf{S}$$

where $d\mathbf{S}' = -\mathbf{a}_n \, dS = -d\mathbf{S}$. Since the source of $\mathbf{B}'$ and $\mathbf{H}'$ are located at finite distances from an arbitrary origin, the integral over Σ is zero when $r \to \infty$ and $v'_1 \to v_1$ (see page 000). It follows that

$$\int_{v_1+v_2} \mathbf{B}' \cdot \mathbf{H}' \, dv = -\oint_{S} \phi(\mathbf{B}'_+ - \mathbf{B}'_-) \cdot d\mathbf{S} = 0$$

Equation 10–48 thus reduces to

$$T = T_0 + \tfrac{1}{2} \int_{v_1+v_2} (\mathbf{B} \cdot \mathbf{H}_0 - \mathbf{H} \cdot \mathbf{B}_0) \, dv$$

Over v_1 we have that $\mathbf{B} = \mu_1\mathbf{H}_1$ and $\mathbf{B}_0 = \mu_1\mathbf{H}_0$ so that the integral over v_1 is zero.

Over v_2 we have $\mathbf{B} = \mu_2\mathbf{H}_2$ and $\mathbf{B}_0 = \mu_1\mathbf{H}_0$ so that

$$T = T_0 + \tfrac{1}{2} \int_{v_2} (\mu_2 - \mu_1)\mathbf{H}_2 \cdot \mathbf{H}_0 \, dv \qquad (10\text{–}51)$$

If $\mu_2 > \mu_1$ the introduction of the body of volume v_2 increases the

* In the divergence theorem $\int \nabla \cdot \mathbf{A} \, dv = \oint \mathbf{A} \cdot d\mathbf{S}$, the value of $\mathbf{A}$ in the surface integral is its value on the inner side of S since this alone belongs to the volume v.

magnetic energy of the initial field whereas if $\mu_2 < \mu_1$ the original energy is decreased.

If the medium surrounding v_2 is free space, $\mu_1 = \mu_0$ and since

$$\mu_2 = \mu_0[1+(\mathbf{M}/\mathbf{H}_2)]$$

we may write equation 10–51 in terms of the magnetization $\mathbf{M}$ of the body as

$$T = T_0 + \tfrac{1}{2}\int_{v_2} \mathbf{M} \cdot \mathbf{B}_0 \, dv \qquad (10\text{–}52)$$

Suppose that medium 1 is free space and that a small para- or diamagnetic body is placed in it. Since the relative permeability of the body differs very slightly from 1 we may assume that $H_2 \simeq H_0$; moreover, if the volume of the body is sufficiently small so that variations in H throughout its volume are negligibly small, the magnetic energy may be expressed as

$$T = T_0 + \tfrac{1}{2}v\mu_0(\mu_r - 1)H_0^2 \qquad (10\text{–}53)$$

and the total force acting on the body is therefore

$$\mathbf{F} = \nabla T = \tfrac{1}{2}v\mu_0(\mu_r - 1)\nabla(H_0)^2 \qquad (10\text{–}54)$$

Thus a paramagnetic body will tend to move towards regions of high magnetic field whereas diamagnetic bodies will tend to move away from such regions.

10.6 GENERAL EXPRESSION FOR MAGNETIC ENERGY IN QUASI-STATIONARY CURRENT SYSTEMS

In all the preceding sections it was assumed that the relationship between the current and the flux was linear, that is the permeability of all media involved was constant. We already know from Chapter 8 that for ferromagnetic materials the relationship between the current (and hence $\mathbf{H}$) and the flux (and hence $\mathbf{B}$) is non-linear and depends on the past history of the material. In the presence of such materials, therefore, the magnetic energy can no longer be represented by expressions in the form of equations 10–2, 10–4 to 10–7. To obtain a general expression for the magnetic energy consider a finite volume of a conductor whose material has a conductivity γ. If the current density at any point is $\mathbf{j}$ then the relationship between the electric field intensity and the current density at that point is

$$\mathbf{E} = \mathbf{j}/\gamma \qquad (10\text{–}55)$$

According to equation 9–10 this field consists of two parts: one part is of electric origin ($\mathbf{E}_e = -\nabla V$) and the other part is of magnetic origin ($\mathbf{E}_i = -\partial\mathbf{A}/\partial t$). Since a steady potential gradient can only be maintained inside a conductor by means of an external source (e.g., a battery), $\mathbf{E}_e$ will represent the electric source field. Equation 10–55 may be written as

$$\mathbf{E}_e = (\mathbf{j}/\gamma) + \partial\mathbf{A}/\partial t \qquad (10\text{–}56)$$

or

$$\mathbf{E}_e \cdot \mathbf{j}\,\mathrm{d}t = (\mathbf{j} \cdot \mathbf{j}/\gamma)\,\mathrm{d}t + \mathrm{d}\mathbf{A} \cdot \mathbf{j} \qquad (10\text{–}57)$$

The left side of equation 10–57 represents the energy per unit volume supplied by the electric source; the term j^2/γ represents the Joule heat dissipated per unit volume of the conducting medium whilst the last term is the energy per unit volume which is stored in the magnetic field produced by the current distribution. The increment in magnetic energy associated with the whole conductor volume is thus

$$\mathrm{d}T = \int_v \mathrm{d}\mathbf{A} \cdot \mathbf{j}\,\mathrm{d}v \qquad (10\text{–}58)$$

If the magnetic vector potential changes from $\mathbf{A}_0$ to some value $\mathbf{A}$, the increase or decrease in the magnetic energy stored in the field as a result of this change is given by

$$T = \int \int_{\mathbf{A}_0}^{\mathbf{A}} \mathbf{j} \cdot \mathrm{d}\mathbf{A}\,\mathrm{d}v \qquad (10\text{–}59)$$

Thus to find the magnetic energy the functional relationship between $\mathbf{j}$ and $\mathbf{A}$ must be specified. When the relationship is linear then if the magnetic potential at a given instant is some fraction k of its final value $\mathbf{A}$, the current density at this instant will have the same fraction of its final value $\mathbf{j}$; if the potential and current density are increased by an additional fraction $\mathrm{d}k$ then $\mathrm{d}\mathbf{A} = \mathbf{A}\mathrm{d}k$ and the energy stored in the magnetic field during the whole process of bringing the current distribution to its final value is given by

$$T = \int_v \int_0^1 (\mathbf{j} \cdot \mathbf{A})k\,\mathrm{d}k\,\mathrm{d}v = \tfrac{1}{2}\int \mathbf{j} \cdot \mathbf{A}\,\mathrm{d}v$$

Equation 10–59 localizes the energy to the currents. We may obtain an expression for this energy in terms of the vectors $\mathbf{B}$ and $\mathbf{H}$ as follows. Consider the vector identity

$$\nabla \cdot \mathbf{H} \times \mathrm{d}\mathbf{A} = \mathbf{H} \cdot \nabla \times \mathrm{d}\mathbf{A} - \mathrm{d}\mathbf{A} \cdot \nabla \times \mathbf{H}$$

By definition $\nabla \times \mathrm{d}\mathbf{A} = \mathrm{d}\mathbf{B}$ and from Ampère's law for quasistationary currents $\nabla \times \mathbf{H} = \mathbf{j}$ so that

$$\nabla \cdot (\mathbf{H} \times \mathrm{d}\mathbf{A}) = \mathbf{H} \cdot \mathrm{d}\mathbf{B} - \mathrm{d}\mathbf{A} \cdot \mathbf{j}$$

Integrating over all space and using the arguments given in Section 10.1

we obtain

$$\int_v \mathbf{j} \cdot d\mathbf{A} = \int_{\text{all space}} \mathbf{H} \cdot d\mathbf{B}\, dv$$

We thus have that

$$dT = \int_{\text{all space}} \mathbf{H} \cdot d\mathbf{B}\, dv \tag{10-60}$$

$$T = \int_{\text{all space}} \int_{\mathbf{B}} \mathbf{H} \cdot d\mathbf{B}\, dv \tag{10-61}$$

The magnetic energy density in the field is

$$\frac{dT}{dv} = \int \mathbf{H} \cdot d\mathbf{B} \quad \text{joule metre}^{-3} \tag{10-62}$$

10.7 HYSTERESIS LOSS IN FERROMAGNETIC CORES

We have seen in the preceding sections that in order to create a magnetic field, energy has to be expended by some external source; this energy is stored in the magnetic field. If the magnetic field is destroyed then in a linear medium, the stored energy is wholly returned to the source. In a non-linear ferromagnetic medium, only part of the stored energy is returned to the source; the other part is irreversibly lost in the medium itself in the form of heat. In many practical magnetic devices such as transformers, motors and generators, whose magnetic structures are made of ferromagnetic material, the material is subjected to periodic flux changes in which the flux alternates cyclically between equal positive and negative amplitudes. Under these

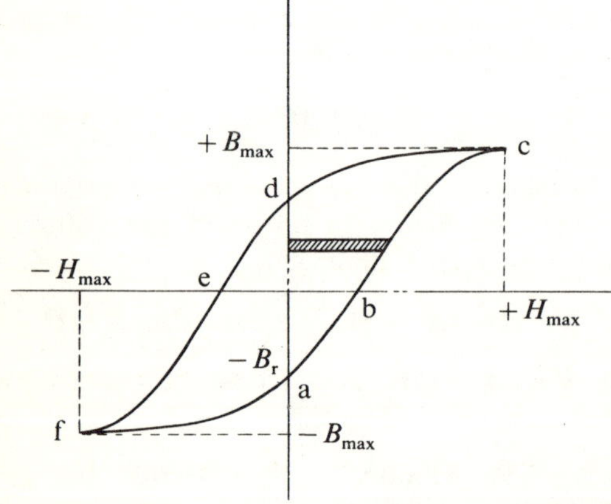

Fig. 10.12 To calculate the energy loss during a hysteresis cycle

conditions the actual relationship between B and H is given by the symmetrical hysteresis loop of the material.

Consider one cyclic variation of flux and let the amplitude of the magnetizing force vary between $+H_{max}$ and $-H_{max}$ and the corresponding variation in flux density be from $+B_{max}$ to $-B_{max}$ (Fig. 10.12). When the flux density varies by dB the corresponding variation in the magnetic energy per unit volume of the material is $\mathbf{H} \cdot d\mathbf{B}$. The energy stored in the magnetic field during the portion ac of the cycle is

$$T_{ac} = \int_{-B_r}^{B_{max}} \mathbf{H} \cdot d\mathbf{B}$$

$$= \text{area} \quad [\mathsf{abc}\, B_{max}]$$

During the portion cd of the cycle energy is returned to the source and

$$T_{cd} = \int_{B_{max}}^{B_r} \mathbf{H} \cdot d\mathbf{B} = \text{area}\,[\mathsf{dc}\, B_{max}]$$

T_{cd} is evidently negative (returned energy) since $B_{max} > B_r$.

During the portion def energy is again absorbed from the source since

$$T_{df} = \int_{B_r}^{-B_{max}} \mathbf{H} \cdot d\mathbf{B} = \text{area}\,[\mathsf{def}(-B_{max})]$$

is positive (H is negative and the upper limit is less than the lower limit).

During the final portion fa of the cycle energy is returned to the source and

$$T_{fa} = \int_{-B_{max}}^{-B_r} \mathbf{H} \cdot d\mathbf{B} = \text{area}\,[\mathsf{fa}(-B_{max})]$$

The net energy input per cycle and per unit volume which is *not* returned to the source is thus given by the area of the hysteresis loop

$$\omega_h = \oint \mathbf{H} \cdot d\mathbf{B} \quad \text{joule metre}^{-3}\ \text{cycle} \tag{10–63}$$

Experimental evidence and solid state theory both indicate that this energy is irreversibly converted into heat and thus represents a loss of energy which is known as the hysteresis loss. Although the heat generated for one cycle is very small, it nevertheless becomes important at the power frequencies 50–60 Hz and above. The area of the hysteresis loop, and hence the hysteresis loss, depends upon the value of B_{max}; hence, in specifying the hysteresis loss of a given material, it is necessary to specify the B_{max} at which the loss was measured. Usually $B_{max} = 1$ tesla; for transformer and motor grade steel (Si–Fe) ω_h varies between 120 and 400 joule metre^{-3} cycle; for permalloy ω_h is about

16 joule metre^{-3} cycle. The hysteresis loss per second is given by

$$p_\text{h} = f \oint \mathbf{H} \cdot d\mathbf{B} \quad \text{watt metre}^{-3} \tag{10-64}$$

where f is the frequency in Hz. Hysteresis loss is also frequently expressed as

$$p_\text{h} = (f/\delta) \oint \mathbf{H} \cdot d\mathbf{B} \quad \text{watt kilogram}^{-1} \tag{10-65}$$

where δ is the density of the material in kilogram metre^{-3}.

In 1892 Steinmetz proposed the following empirical formula for the hysteresis loss per unit volume per cycle (area of loop),

$$\omega_\text{h} = k_\text{h}(B_{\text{max}})^n \tag{10-66}$$

where the values of the coefficient k_h and of the exponent n depend upon the nature of the material. Steinmetz found that $n = 1\cdot6$ for most materials and for maximum flux densities (in tesla) in the range $0\cdot2 < B_\text{m} < 1\cdot2$; however, for many of the new alloys which have come into use in recent years, n may vary from $1\cdot58$ to 2. The constants k_h and n may be found by plotting test results of $\oint \mathbf{H} \cdot d\mathbf{B}$ versus B_{max} on log-log paper; the plot

$$\log\left(\oint \mathbf{H} \cdot d\mathbf{B}\right) = \log k_\text{h} + n \log B_{\text{max}}$$

is a straight line of intercept $\log k_\text{h}$ and slope n.

For low values of the flux density ($B_{\text{max}} < 0\cdot1$ tesla) the area of the hysteresis loop is given by the Raleigh formula

$$\omega_\text{h} = C(B_{\text{max}})^3 \tag{10-67}$$

where C is a material constant.

10.8 EDDY-CURRENT LOSS

We have seen in Chapter 9 that a time-varying magnetic field gives rise to an induced electric field having the property that its integral around any closed contour is equal to the negative time derivation of the magnetic flux enclosed by that contour. When a conducting medium is penetrated by a time-varying magnetic flux, the induced emf around any closed path linking the flux will cause an induced current to circulate around that path; the direction of the current is such that its flux tends to oppose the change in the external flux (Lenz's law). Because the induced currents form loops which resemble eddies in a whirlpool, they are called *eddy currents*. Since the conducting medium has a finite conductivity, the flow of these currents produces an ohmic energy loss which is known as eddy loss.

In order to find an analytical expression for the eddy loss it is necessary to take into consideration the modification in the magnitude and distribution of the magnetic flux which results from the flow of eddy currents. The calculations involved are somewhat complicated but the problem may be simplified considerably if we assume that the applied field remains unchanged throughout the material under consideration.

Consider the metal block of rectangular cross-section shown in Fig. 10.13. We shall assume that the length L of the block is large

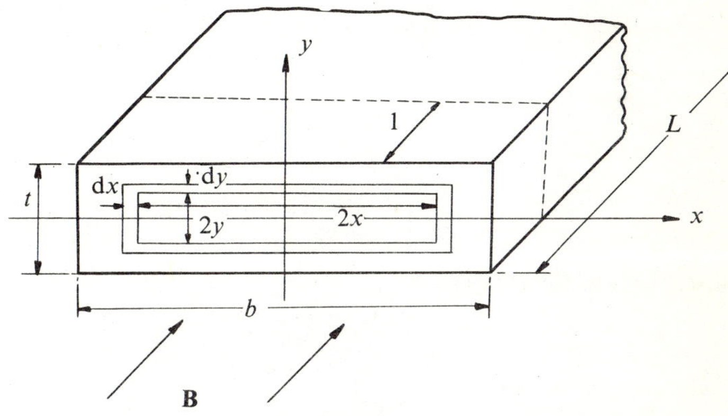

Fig. 10.13 To calculate the eddy loss in a thin rectangular sheet

compared with its other dimensions, that the impressed alternating flux is a sinusoidal function of the time and that this flux is uniformly distributed over the cross-section. The cross-section may be considered as made up of a set of infinitesimal *similar* rectangles of sides $2x$ and $2y$ such that, for similarity,

$$2x/b = 2y/t$$

or

$$\frac{\mathrm{d}x}{b} = \frac{\mathrm{d}y}{t}$$

Consider the elementary path bounded by an inner rectangle of sides $2x$ and $2y$ and an outer rectangle of sides $2(x+\mathrm{d}x)$ and $2(y+\mathrm{d}y)$. The flux linked by that path is

$$\Phi_x(t) = 2x2yB_\mathrm{m}\cos\omega t$$

and the emf induced in the path is therefore

$$v_x = -\,\mathrm{d}\Phi/\mathrm{d}t = 4xy\omega B_\mathrm{m}\sin\omega t$$

The root mean square value of this emf is

$$\mathscr{V}_x = 4xy\omega B_\mathrm{m}/\sqrt{2}$$

Making use of the similarity relationships this may be expressed as

$$\mathscr{V}_x = 4x^2 t\omega B_\mathrm{m}/\sqrt{2} \qquad (b)$$

If the resistivity of the material is ρ, the resistance of the elementary path for 1 metre depth in the L direction is

$$\begin{aligned}
\mathrm{d}R_x &= 2\rho\left[\frac{2x}{\mathrm{d}y}+\frac{2y}{\mathrm{d}x}\right] \\[1em]
&= 2\rho\left[\frac{2x}{\mathrm{d}x(t/b)}+\frac{2x(t/b)}{\mathrm{d}x}\right] \\[1em]
&= \frac{4\rho x}{\mathrm{d}x}\left[\frac{1+(t/b)^2}{(t/b)}\right]
\end{aligned}$$

The power loss in the element is

$$\mathrm{d}P_x = \frac{\mathscr{V}_x^2}{\mathrm{d}R_x} = \frac{2t^2\omega^2 B_\mathrm{m}^2(t/b)}{b^2\rho[1+(t/b)^2]}\,x^3\,\mathrm{d}x$$

The eddy loss per metre length is therefore

$$\frac{P_\mathrm{e}}{L} = \int_0^{b/2}\mathrm{d}P_x = \frac{\omega^2 b^4 B_\mathrm{m}^2}{32\rho}\cdot\frac{(t/b)^3}{1+(t/b)^2} \qquad (10\text{--}68)$$

for $b \gg t$ or $(t/b) \ll 1$ the total eddy loss may be expressed as

$$P_\mathrm{e} = \frac{\omega^2 b t^3 B_\mathrm{m}^2}{32\rho}\,L \quad \text{watt} \qquad (10\text{--}69)$$

This equation shows that for a given frequency and flux density the eddy-current loss is inversely proportional to the resistivity of the metal and directly proportional to the cube of the thickness. For this reason machine and transformer cores are made of thin laminations with each lamination insulated from the next, by the application of a varnish coat or by oxidation of the surfaces in contact, thus restricting the eddy current paths. The laminations are normally made from silicon-steel as its resistivity is greater than that of iron or the low-carbon steels.

The power dissipated per cubic metre of material is

$$p_e = \frac{P_e}{btL} = \frac{\pi^2}{8\rho} \cdot f^2 B_m^2 t^2$$

$$= k_e f^2 B_m^2 t^2 \quad \text{watt metre}^{-3} \tag{10-70}$$

In practice the proportionality constant k_e is always larger than the theoretical value $(\pi^2/8\rho)$. This is partly due to the fact that the insulation between the laminations is not perfect so that the actual current paths are not as simple as those assumed, and partly to the non-uniform distribution of the flux throughout the cross-section of the individual plates.

For very thin laminations (≤ 0.5 mm) and power frequencies the assumption that the amplitude of the flux density is the same throughout the cross-section of the plate is a justifiable approximation. For thick laminations the flux density is not uniformly distributed over the cross-section but tends to be crowded towards the surface or outer region of the lamination.* The reason for this may be explained qualitatively as follows. Consider the core section shown in Fig. 10.14;

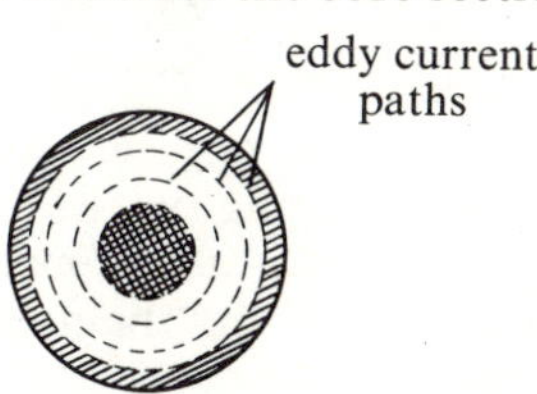

Fig. 10.14 Crowding of flux density towards surface of a conductor

every eddy current produces an mmf which is opposed to that produced by the external source. This demagnetizing mmf increases as the centre of the cross-section is approached from the surface since the cross-hatched area is subjected to the demagnetizing action of eddy currents flowing in all the paths surrounding it. The final result is that the flux density as well as the density of the eddy currents is not uniformly distributed over the cross-section but decreases towards the centre.

The heat generated by eddy-current loss is made use of in the heat treatment of metals (*induction heating*) and for the melting of metals; metal melting furnaces operating on the eddy-current heating principle are known as induction furnaces. Eddy-currents are also used as a means for damping electrical indicating instruments and instrument transducers.

* E. W. Golding, *Electrical Measurements and Measuring Instruments*, p. 501 (Pitman, 1940).

PROBLEMS

Chapter Ten

10.1 A filamentary circular current loop of radius a carries a current I_1 and is placed with its centre at a distance d from an infinitely long coplanar wire which carries a current I_2. Find the mutual inductance between the two circuits and hence show that the force between them is given by

$$F = \mu_0 I_1 I_2 [1 - d(d^2 - a^2)^{-1/2}]$$

10.2 The toroid of Problem 8.5 is cut diametrically into two halves. For a current of 2A in the winding find the force needed to separate the two halves. If the two halves are separated by two 0·5 mm air gaps, determine the force necessary to keep the two halves apart.

10.3 A circular wire loop of radius r carries a current I. Show that the radial pressure acting on the loop is given by

$$f = \frac{I^2}{4\pi r} \frac{\partial L}{\partial r}$$

10.4 A long thin cylindrical tube of mean radius r carries a current I. Show that the pressure acting on the tube is $\mu_0 I^2 / 8\pi^2 r^2$. If the ultimate tensile strength of a tube of mean radius 2 cm and thickness 1 mm is 2000 kg cm^{-2}, find the maximum current which the tube can carry without bursting.

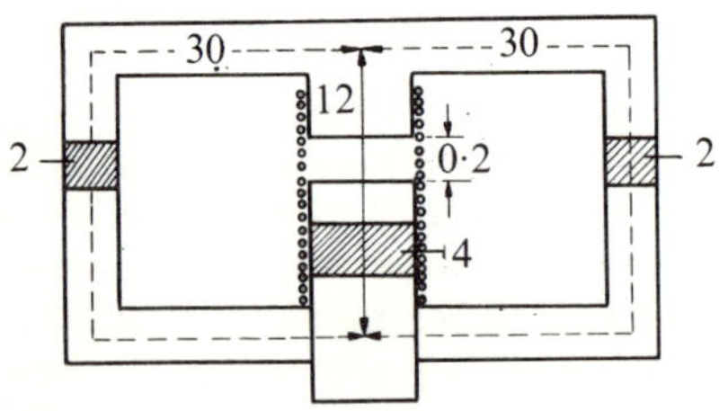

Fig. P. 10.5

10.5 Figure P10.5 shows the magnetic circuit of a relay. Find the ampere-turns required if a force of 1·5 kg is required to lift the plunger. The average lengths and the cross-sectional areas of the circuit are given in cm. The armature and plunger are of cast iron. Neglect leakage and fringing.

10.6 The core shown in Fig. P10.6 has a uniform cross-section and consists of a circular ring of mean diameter D with two diametrically opposite arms separated by an air gap of length g.

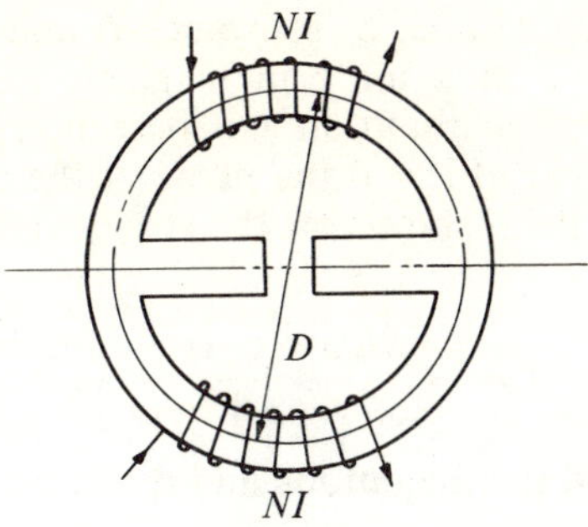

Fig. P. 10.6

The windings on the two halves of the ring are identical and carry equal currents such that their fluxes in the diametral branch are in the same sense. Assuming that the permeability of the core is constant, show that, if leakage is neglected, the magnetic energy stored in the gap will be a maximum when $g = D(\pi+4)/4(\mu_r-1)$.

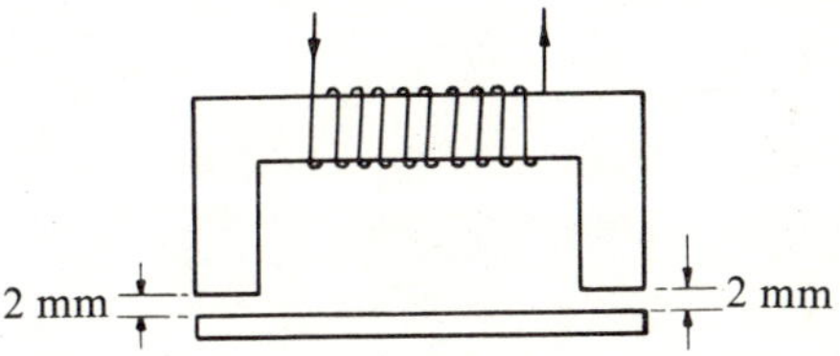

Fig. P. 10.7

10.7 Figure P10.7 shows the yoke and armature of a relay. The reluctance of the armature is negligible and its weight is 100 g. The yoke has a cross-sectional area of 5 cm² and a mean magnetic path length of 50 cm. The coil of 500 turns has a resistance of 200 ohms. If it is connected to a 220 V supply find the time which elapses before the armature begins to move. Neglect leakage and fringing and assume that the permeability of the yoke is constant and equal to 1200.

10.8 Two square plane parallel plates of side l and negligible resistivity are at a distance d apart. At one end a battery of V volts is connected between the plate edges and at the other end a resistance R is connected between the plate edges. Neglecting end effects show that the magnetic force between the plates is $\frac{1}{2}\mu_0 I^2$ and that the electric and magnetic forces acting on the plates become equal when

$$R = \sqrt{\frac{\mu_0}{\epsilon_0}}\left(\frac{d}{l}\right)$$

10.9 A plane circuit of area S, resistance R and self-inductance L is made to rotate in a uniform magnetic field B with constant angular velocity ω about an axis perpendicular to the magnetic field. Given that at $t = 0$ the plane of the circuits is parallel to the magnetic field, show that the steady state current in the coil is given by

$$I = -\frac{SB\omega(R\cos\omega t + \omega L \sin\omega t)}{R^2 + \omega^2 L^2}$$

and hence find the magnitude and direction of the couple acting on the coil.

METHODS OF SOLVING FIELD PROBLEMS WITH SPECIFIED BOUNDARY CONDITIONS

11.1 GENERAL

In homogeneous and isotropic media, stationary and quasi-stationary electric and magnetic fields that are both irrotational and solenoidal satisfy Laplace's equation $\nabla^2\phi = 0$ where ϕ is a function which is continuous and finite and has continuous second-order derivatives. As shown in Section 2.18 the physical significance of this equation is that the function ϕ has the property that its value at the centre of any sphere is equal to the arithmetic mean of its values over the surface of this sphere. Functions which satisfy Laplace's equation are known as *harmonic functions*. As far as we are concerned Laplace's equation is satisfied by the following physical field systems:

(a) *The electrostatic potential* $(\phi = V)$ in a charge-free homogeneous medium. The equipotential surfaces are given by $V = $ constant, and the field lines by $\psi = $ constant; (V- and ψ-lines are orthogonal).

(b) *The electric potential* $(\phi = V)$ in an electric conduction field (equation 6–21). The equipotential surfaces are given by $V = $ constant, and the lines of current flow by $I = $ constant; (V- and I-lines are orthogonal).

(c) *The magnetic scalar potential of* H $(\phi = \Phi_H)$ in a medium in which there is no current flow (equation 8–60). The equipotential surfaces are given by $\Phi_H = $ constant, and the flux lines by $\Phi = $ constant; (Φ_H- and Φ-lines are orthogonal).

The Laplace equation is a second-order partial differential equation and as such it does not express a specific physical problem, but only the general law under consideration. For a specific problem any solution of the equation will not be completely determinate unless it satisfies the requirements dictated by the physical formulation of the problem. In the great majority of field problems these requirements are that the function ϕ, or some of its derivatives, or a combination of the function and its derivatives, shall have certain specified values over the surface or surfaces bounding the space throughout which ϕ is harmonic. Such problems are known as boundary-value problems and are usually classified into three types:

(i) To find ϕ which satisfies $\nabla^2\phi = 0$ throughout a region v when ϕ is given everywhere on the surface (or surfaces) bounding v. This is known as the *Dirichlet problem*.

(ii) To find ϕ which satisfies $\nabla^2\phi = 0$ throughout a region v when the normal derivative $\partial\phi/\partial n$ is given everywhere on the surface (or surfaces) bounding v. This is known as the *Neumann problem*.

(iii) To find ϕ which satisfies $\nabla^2\phi = 0$ throughout a region v when ϕ is specified on part of the boundary surface (or surfaces) and $\partial\phi/\partial n$ on the remaining part. This is known as a *mixed boundary-value problem*.

The solution of Laplace's equation must always be consistent with the conditions at the boundaries between different media:

(a) Electrostatic fields (Sections 3.2 and 4.4):

At the surface of a conductor,

$$V = \text{constant}; \frac{\partial V}{\partial n} = \frac{-\sigma}{\epsilon} \tag{11-1}$$

At the boundary between two dielectrics

$$V_1 = V_2; \epsilon_1 \frac{\partial V_1}{\partial n} = \epsilon_2 \frac{\partial V_2}{\partial n} \tag{11-2}$$

(b) Electric conduction fields (Section 6.6):

At the boundary between two conducting media

$$V_1 = V_2; \gamma_1 \frac{\partial V_1}{\partial n} = \gamma_2 \frac{\partial V_2}{\partial n} \tag{11-3}$$

At the boundary between a conductor and an insulator (no normal flow of current)

$$\frac{\partial V}{\partial n} = 0 \tag{11-4}$$

(c) Magnetostatic fields (Section 8.14):

$$\Phi_{H1} = \Phi_{H2}; \mu_1 \frac{\partial \Phi_{H1}}{\partial n} = \mu_2 \frac{\partial \Phi_{H2}}{\partial n} \tag{11-5}$$

Where Φ_H and $\mathbf{H}$ are related by $\mathbf{H} = -\nabla\Phi_H$. We note that the function Φ_H satisfies the same equations as the electrostatic potential and hence solution of magnetic field problems in regions where there is no current flow can be obtained directly from analogous electrostatic problems. This fact has led many authors to consider Φ_H as the magnetostatic potential.

Solutions to Laplace's equation subject to specified boundary conditions may be obtained by the following methods:

(a) Exact mathematical analysis.
(b) Approximate mathematical methods (numerical methods)—relaxation, iteration, method of moments.
(c) Graphic methods.
(d) Experimental analogue methods—electrolytic tank, conducting Teledeltos paper, resistance networks, photoelastic methods, Moore's fluid mapper technique.

The choice of method depends essentially on the geometry of the boundaries and the boundary conditions involved and on the degree of accuracy required. The most important of the exact mathematical methods of solution are the methods of images, the method of separation of variables and the method of conformal mapping. The first two methods give simple solutions of two- or three-dimensional problems with simple boundaries whereas the third method is limited to two-dimensional problems in which the boundaries are curves of known mathematical shape, or straight lines. An outline of both these methods is given below since by their means it is possible to obtain the solution of many field problems of fundamental importance.

In many practical field problems the boundaries involved are such as to lead to intractable mathematical difficulties if an exact solution is required. In such problems any of the three methods (b), (c) or (d) may be used: the choice of method depends on the nature of the problem. For example if several solutions are required of a design problem in which the choice of the optimum boundary profiles have yet to be determined, the use of approximate mathematical methods may be laborious and expensive (in terms of computer time) whereas if an analogic experiment can be readily set up, then the investigation of the original problem can be carried out quickly and easily. An account of the numerical and experimental methods of field analysis is beyond the scope of this book and there are several excellent textbooks (listed at the end of this chapter) which deal exclusively with these topics. A bibliographic survey of electroanalogic methods, carried out by Higgins in 1956, runs to 480 entries*.

11.2 UNIQUENESS OF SOLUTION OF FIELD PROBLEMS

In Section 3.6.1 we have shown that if all the potentials of a given number of conductors are known, then the potential and field at every point in the space surrounding the conductors are uniquely determined.

* These may be found in the first three of a series of five articles entitled *Electroanalogic Methods*, by T. J. Higgins; these appeared in *Applied Mechanics Reviews*, Vol. 9, No. 1, pp. 1–4, 1956; Vol. 9, No. 2, pp. 49–55, 1956; Vol. 10, No. 2, pp. 49–54, 1957.

Consider now an arbitrary closed surface S which encloses a volume v. Let ψ be a scalar point function which is finite, continuous and twice differentiable within v. In Green's first identity (equation 2–98) if we let $\phi = \psi$ we obtain

$$\oint_S \psi \nabla \psi \cdot d\mathbf{S} = \int_v [\psi \nabla^2 \psi + (\nabla \psi)^2] \, dv \qquad (11\text{–}6)$$

Suppose that there exist two potential functions V_1 and V_2 both of which satisfy the above equation. If ρ is the charge density within v then we must have

$$\nabla^2 V_1 = -\rho/\epsilon_0 \qquad \text{and} \qquad \nabla^2 V_2 = -\rho/\epsilon_0$$

If we let $\psi = V_1 - V_2$ so that $\nabla^2 \psi = 0$ everywhere within v, equation 11–6 becomes

$$\oint_S \psi \nabla \psi \cdot d\mathbf{S} = \oint_S \psi \frac{\partial \psi}{\partial n} \, dS = \int_v (\nabla \psi)^2 \, dv \qquad (11\text{–}7)$$

Suppose that the surface integral is zero; in this case

$$\int_v (\nabla \psi)^2 \, dv = 0$$

and $\psi = V_1 - V_2 = $ constant everywhere in v so that the field is unique and the potential is determined to within an arbitrary additive constant. Now the surface integral can be zero if:

(a) $\psi = 0$ everywhere on S. In this case $V_1 = V_2 = V$ (additive constant is zero) and the condition is fulfilled if V is specified everywhere on S (Dirichlet problem).

(b) $\partial \psi/\partial n = 0$ everywhere on S. In this case $\partial V_1/\partial n = \partial V_2/\partial n = \partial V/\partial n$ and this condition is fulfilled if $\partial V/\partial n$ is specified everywhere on S (Neumann problem).

(c) $\psi = V_1 - V_2 = $ constant everywhere on S (this follows from (b)), for then

$$\oint_S \psi (\partial \psi/\partial n) \, dS = \psi \oint_S (\partial \psi/\partial n) \, dS = \psi \int_v \nabla^2 \psi \, dv = 0$$

In general, therefore, we may state that the criterion for uniqueness is that either

$$V_1 - V_2 = \text{constant, everywhere on } S$$

or

$$\partial \psi/\partial n = 0, \text{ everywhere on } S\text{—i.e.,}$$

$$\frac{\partial V_1}{\partial n} = \frac{\partial V_2}{\partial n}$$

Consider the first condition. If A and B are any two points on the surface S, then we must have that

$$V_{A1} - V_{A2} = V_{B1} - V_{B2} = \text{constant}$$

or

$$V_{A1} - V_{B1} = V_{A2} - V_{B2} = \text{constant}$$

and this condition can only be satisfied if the tangential component of the electric field is specified everywhere on S.

The second condition requires that the normal component of the electric field be specified everywhere on S.

11.3 SOLUTIONS OF ELECTROSTATIC PROBLEMS WITH THE AID OF GREEN'S THEOREM

Let a charge be distributed with a variable density ρ throughout a finite region of space and let S_1 be an arbitrary regular closed surface which encloses a volume v of this region (Fig. 11.1). We wish to find the

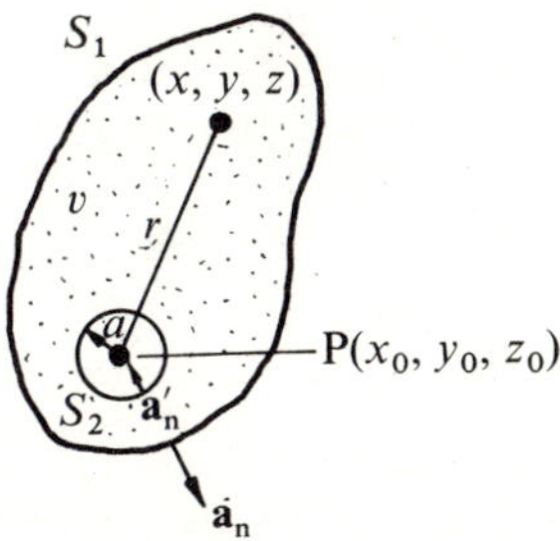

Fig. 11.1 To find the potential at a point P inside a closed surface

potential V_P at some fixed point $P(x_0, y_0, z_0)$ within S_1. In Green's second identity (equation 2–117) let $\phi = V$, where V is the potential at some variable point within v, and $\psi = 1/r$, where r is the distance from the fixed point P to the variable point; the point P is chosen as the origin of the coordinate r; we thus have that

$$\oint_{S_1} [(1/r)\nabla V - V\nabla(1/r)] \cdot d\mathbf{S} = \int_v [(1/r)\nabla^2 V - V\nabla^2(1/r)] \, dv$$

The function $1/r$ is clearly discontinuous at $r = 0$. In order to exclude this singularity we modify the volume of integration by constructing a small sphere S_2 of radius a and volume τ whose centre is the point P. The volume $v - \tau$ is now bounded externally by S_1 and internally by S_2. Since $\nabla^2(1/r) = 0$ everywhere within $v - \tau$, the above identity may be written as

$$\oint_{S_1+S_2} [(1/r)\nabla V - V\nabla(1/r)] \cdot d\mathbf{S} = \int_{v-\tau} (1/r)\nabla^2 V\, dv \qquad (11\text{--}8)$$

The left-hand side consists of two integrations, one over S_1 and the other over S_2. Consider the integral over S_2. On this surface $r = a$ and $\mathbf{a}'_n = -\mathbf{a}_r$ since in this case $\mathbf{a}'_n$ is the *outward* unit normal and is therefore directed towards the origin P, thus we have that

$$\left[\left(\frac{1}{r}\right)\nabla V \cdot d\mathbf{S}_2\right]_{r=a} = \frac{1}{a}\left(\frac{\partial V}{\partial r}\mathbf{a}_r\right) \cdot dS_2\mathbf{a}_n = -\left|\frac{1}{a}\frac{\partial V}{\partial r}\right|_{r=a} dS_2$$

$$\left[V\nabla\left(\frac{1}{r}\right) \cdot d\mathbf{S}_2\right]_{r=a} = V(-r^{-2}\mathbf{a}_r \cdot dS_2\mathbf{a}_n) = \frac{1}{a^2}V\,dS_2$$

Now let $\langle \partial V/\partial r\rangle$ and $\langle V\rangle$ be the average values of $\partial V/\partial r$ and of V over the surface S_2; by definition we have that

$$\left\langle\frac{\partial V}{\partial r}\right\rangle = \frac{1}{4\pi a^2}\oint_{S_2}\left(\frac{\partial V}{\partial r}\right)dS_2$$

$$\langle V\rangle = \frac{1}{4\pi a^2}\oint_{S_2} V\,dS_2$$

and we may thus write

$$\oint_{S_2}\left[\left(\frac{1}{r}\right)\nabla V - V\nabla\left(\frac{1}{r}\right)\right] \cdot d\mathbf{S}_2 = -\frac{1}{a}4\pi a^2\left\langle\frac{\partial V}{\partial r}\right\rangle - \frac{1}{a^2}4\pi a^2\langle V\rangle$$

and if we now take the limit as $a \to 0$, that is allow the surface S_2 to shrink to the point P, the first term on the right-hand side goes to zero; the average value of V in the second term becomes simply the potential V_P at the point P so that this term becomes equal to $-4\pi V_P$. Since in a homogeneous medium V satisfies Poisson's equation $\nabla^2 V = -\rho/\epsilon$, we may now write equation 11–8 as

$$V_P = \frac{1}{4\pi\epsilon}\int_v \left(\frac{\rho}{r}\right)dv + \frac{1}{4\pi}\oint_{S_1}\left[\left(\frac{1}{r}\right)\nabla V - V\nabla\left(\frac{1}{r}\right)\right] \cdot dS\mathbf{a}_n \qquad (11\text{--}9a)$$

which may also be written as (since $\mathbf{a}_n \cdot \nabla = \partial/\partial n$),

$$V_P = \frac{1}{4\pi\epsilon}\int_v \left(\frac{\rho}{r}\right)dv + \frac{1}{4\pi}\oint_{S_1}\left[\frac{1}{r}\frac{\partial V}{\partial n} - V\frac{\partial}{\partial n}\left(\frac{1}{r}\right)\right]dS \qquad (11\text{--}9b)$$

This equation shows that the potential at any point P within a volume v bounded by a surface S can be determined if we know

(a) the charge density at every point in v

(b) the normal component of the gradient of the potential at every point of S

(c) the value of V at every point of the surface S.

Let us now consider the physical meaning of equation 11–9. The volume integral represents the contribution to the potential at any point within S due to the charge distribution inside S. If there are no charges whatsoever throughout the volume v, this integral is zero and everywhere within S the potential V satisfies Laplace's equation $\nabla^2 V = 0$; the potential at any interior point is given by

$$V_P = \frac{1}{4\pi} \oint_S \left[\frac{1}{r} \frac{\partial V}{\partial n} - V \frac{\partial}{\partial n}\left(\frac{1}{r}\right) \right] dS \qquad (11\text{–}10)$$

Thus the surface integral represents the contribution to the potential at any point within S from all charges which lie outside S. Equation 11–10 shows that if the potential satisfies $\nabla^2 V = 0$ everywhere within a closed surface S, then the potential at any point within S may be expressed in terms of the values of V and $\partial V/\partial n$ on the surface S. It should be noted that both V and $\partial V/\partial n$ must be calculated on the inner or negative side of S since this alone belongs to v. The first term in equation 11–10 is the potential at P produced by a surface charge distribution whose density σ at any point on S is given by

$$\sigma = \epsilon \frac{\partial V}{\partial n} \qquad (11\text{–}11)$$

and the second term is the potential at P produced by a surface distribution of dipoles of density (see equation 2–99)*:

$$\tau = -\epsilon V \mathbf{a_n} \qquad (11\text{–}12)$$

Thus we may rewrite equation 11–10 as

$$V_P = \frac{1}{4\pi\epsilon} \oint_S \left[\left(\frac{\sigma}{r}\right) + \tau \cdot \nabla\left(\frac{1}{r}\right) \right] dS \qquad (11\text{–}13)$$

Equation 11–13 shows that the potential produced at any point inside a closed surface S by a charge distribution lying outside S is the same as that obtained by replacing the external charge distribution by a single-layer distribution and a double-layer distribution on the bounding surface S. It must be stressed that the equivalence of the actual external charge distribution and the fictitious surface layers is only valid for points inside S; for points outside S this equivalence does not apply.

* The negative sign in equation 11–12 is necessary because in equation 2–99 the term $\nabla(1/r) = -\mathbf{a}_{rQP}/r^2$ whereas in equation 11–12 the term $\nabla(1/r) = -\mathbf{a}_{rPQ}/r^2 = \mathbf{a}_{rQP}/r^2$, where Q is the source point and P the field point.

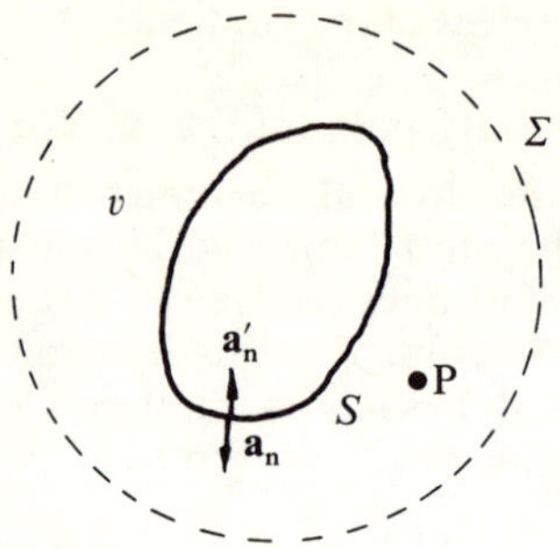

Fig. 11.2 To find the potential at a point P outside a domed surface

Suppose now that the point P lies outside S (Fig. 11.2); consider the volume v bounded internally by S and externally by the surface Σ of a sphere of radius R such that S and all charges are completely inside Σ. By surrounding P with a small sphere of radius a and then taking the limit as $a \to 0$ we arrive, by a procedure similar to that given above, at the result

$$V_P = \frac{1}{4\pi\epsilon} \int_v \frac{\rho}{r}\, dv - \frac{1}{4\pi} \oint_{S+\Sigma} \left[\frac{1}{r}\frac{\partial V}{\partial n} - V\frac{\partial}{\partial n}\left(\frac{1}{r}\right) \right] dS \quad (11\text{--}14)$$

If we now let Σ expand to infinity and assume that V is at least of the order of $1/r$ (which is equivalent to assuming that infinity is the zero reference level of potential) then both $r^{-1}\,\partial V/\partial n$ and $V\,\partial/\partial n(r^{-1})$ will decrease at least as $1/r^3$. Since dS will increase as r^2 the integrand over Σ will be of the order of $1/r$ and the integral will be zero in the limit as $R \to \infty$. We may thus write

$$V_P = \frac{1}{4\pi\epsilon} \int_v \frac{\rho}{r}\, dv - \frac{1}{4\pi} \oint_{S} \left[\frac{1}{r}\frac{\partial V}{\partial n} - V\frac{\partial}{\partial n}\left(\frac{1}{r}\right) \right] dS \quad (11\text{--}15a)$$

or alternatively

$$V_P = \frac{1}{4\pi\epsilon} \int_v \frac{\rho}{r}\, dv - \frac{1}{4\pi} \oint_{S} \left[\left(\frac{1}{r}\right)\nabla V - V\nabla\left(\frac{1}{r}\right) \right] \cdot d\mathbf{S} \quad (11\text{--}15b)$$

In equations 11–14 and 11–15 the negative sign appears before the surface integral because the unit outward normal is now $\mathbf{a}'_n$ instead of $\mathbf{a}_n (d\mathbf{S} = dS\mathbf{a}'_n = -dS\mathbf{a}_n)$. We thus see that the potential produced at any point outside S by the charges inside S is unchanged if we remove these charges and replace them by single- and double-layer distributions of densities

$$\sigma = -\epsilon\frac{\partial V}{\partial n}, \quad \tau = \epsilon V\mathbf{a}_n \quad (11\text{--}16)$$

respectively on the bounding surface.

If the surface S is an equipotential surface, V is constant over it; τ is therefore constant over S and the potential at the point P produced by the double layer is zero (equation 2–101). The potential at P due to charges inside S is the same as that produced by a single layer of charge distributed over S with a density $\sigma = -\epsilon \partial V/\partial n$. This layer is known as *Green's equivalent stratum.*

We have seen from the above discussion that at any point in a charge-free region the potential is given by

$$V_{\mathrm{P}} = \pm \oint_S \left[\frac{1}{r}\frac{\partial V}{\partial n} - V\frac{\partial}{\partial n}\left(\frac{1}{r}\right) \right] \mathrm{d}S \tag{11–17}$$

and hence can be calculated provided that we know V and $\partial V/\partial n$ on the surface S. However, these two quantities are not independent; that is both V and $\partial V/\partial n$ may not be arbitrarily given. If V alone is given over the boundary surface the potential at all points inside S (or outside S in the case of the exterior problem) is uniquely determined (see Section 11.2) and hence the gradient is determined at every point and with it $\partial V/\partial n$. If $\partial V/\partial n$ alone is given then the potential at all points is completely determined to within an arbitrary additive constant.

11.4 SOLUTION OF ELECTROSTATIC FIELD PROBLEMS BY THE METHOD OF IMAGES

Consider a charge distribution ρ in a region v bounded externally by a sphere at infinity and internally by an equipotential surface such as S (Fig. 11.3) consisting of an uncharged conductor. The potential at any

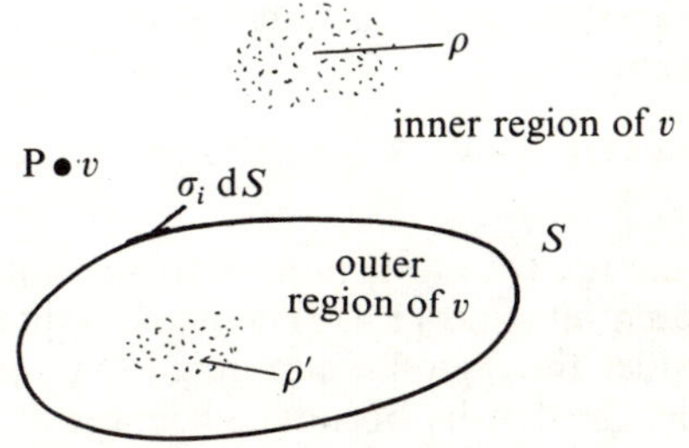

Fig. 11.3 Principle of the method of images

point P in v (and hence outside the conductor) is the sum of the potentials produced at P by the charge distribution and by the surface charges induced on S. The problem is solved if we know σ_i on S; this is equivalent to knowing $\partial V/\partial n$ everywhere on S. σ_i represents the Green equivalent stratum and may be replaced by a fictitious equivalent charge distribution ρ' outside v which produces the same potential at

P as σ_i. This equivalent charge distribution is said to be the electrical image of the charge distribution ρ with respect to the surface S.

If $\partial V/\partial n$ is prescribed everywhere on the boundary surface S, the field at P is unique. However, we have seen in Section 11.2 that the condition for uniqueness is also satisfied if the tangential component of the electric field is specified everywhere on S. This means that there can be fields tangential to the surface S so that S need not necessarily be an equipotential surface but may be a dielectric boundary. The method of images is thus not necessarily limited to problems in which the boundary surfaces are equipotentials. The image problem in electrostatics may be formulated as follows.*

Consider a closed surface S which divides all space into two distinct regions (either region may be taken as the interior of S). Determine the normal or tangential component of the electric field on S due to charges (free or bound) induced thereon by charges inside S. Find a distribution of charges outside S to give on S the same normal or tangential field; this distribution will give inside S the same field as that given by the induced charges on S and is the image distribution required.

It must be emphasized that the image charges or charge distribution must always lie in the region which is considered to be external with respect to the boundary surface S. In essence the method of electrical images converts a given electrostatic field problem into another equivalent problem and is only useful, therefore, if this new problem is simpler to solve than the original one. This is usually the case whenever the conductor or dielectric boundaries have a simple geometrical form such as plane, spherical or cylindrical. The examples which follow will help to clarify the application of the method to a number of important electrostatic problems.

11.4.1 POINT CHARGE NEAR A GROUNDED CONDUCTING PLANE

Consider a point charge Q at a distance h from an infinite conducting plane S at zero potential (Fig. 11.4). The plane S divides all space into two regions. Consider the region containing the point charge Q as the interior of S and let the density of induced charge at a point P on S be σ. The normal component of the field at P due to the charge Q alone is

$$E_n = E\cos\theta = \frac{Q\cos\theta}{4\pi\epsilon_0 r^2} = \frac{Qh}{4\pi\epsilon_0 r^3} \qquad (11\text{–}18)$$

The normal component of the resultant field intensity on either side of

* See P. Hammond, 'Electric and Magnetic Images', *Proc. Instn elec. Engrs* **107C**, 306–11 (1960).

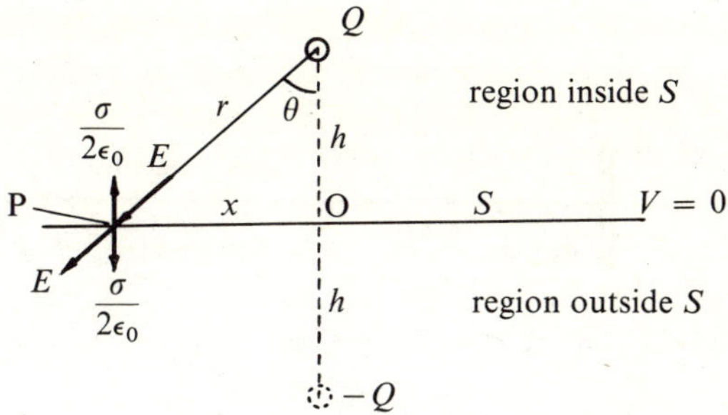

Fig. 11.4 Point charge near a grounded conducting plane

S at P must satisfy the boundary condition (see equation 2–55),

$$\left(\frac{\sigma}{2\epsilon_0} - E_\mathrm{n}\right) - \left(E_\mathrm{n} + \frac{\sigma}{2\epsilon_0}\right) = \frac{\sigma}{\epsilon_0}$$

so that

$$\sigma = -2\epsilon_0 E_\mathrm{n} \qquad (11\text{--}19)$$

The normal component of the electric field on the inner side of S, due to σ, is thus

$$\frac{\sigma}{2\epsilon_0} = -E_\mathrm{n} = -\frac{Qh}{4\pi\epsilon_0 r^3} \qquad (11\text{--}20)$$

This field would be produced by a point charge $-Q$ placed at a vertical distance h from O outside S. This is the required image charge. The surface charge density on S is given by

$$\sigma = -\frac{Qh}{2\pi r^3} \qquad (11\text{--}21)$$

and the total induced charge on S is

$$Q_i = \int_0^\infty \sigma 2\pi x \,\mathrm{d}x = -Q$$

as expected.

11.4.2 POINT CHARGE NEAR A PLANE DIELECTRIC INTERFACE

Let the density of induced charge at P (Fig. 11.5) due to the charge Q alone be σ_b, which is evidently a surface density of bound charges. The normal component of the field at P due to the charge Q alone is

$$E_\mathrm{n} = \frac{Qh}{4\pi\epsilon_0 r^3}$$

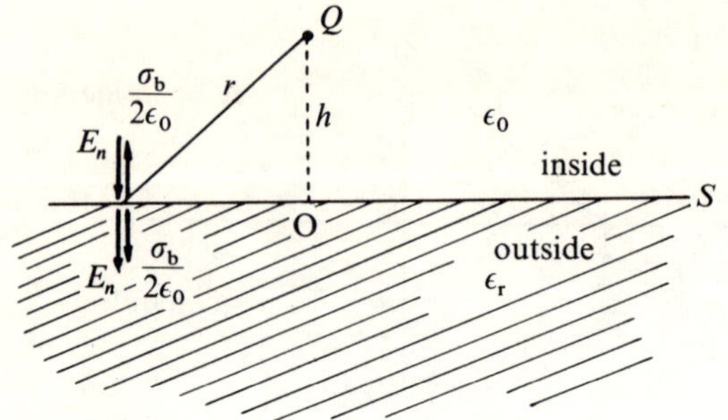

Fig. 11.5 Point charge near a plane dielectric interface

The normal component of the resultant field intensity on either side of
S at P must satisfy the boundary condition (equation 4–24a)

$$\epsilon_0\left(\frac{\sigma_b}{2\epsilon_0}-E_n\right) = -\epsilon_0\epsilon_r\left(E_n+\frac{\sigma_b}{2\epsilon_0}\right)$$

so that

$$\sigma_b = -\frac{\epsilon_r-1}{\epsilon_r+1}2\epsilon_0 E_n \qquad\qquad (11\text{--}22)$$

The normal component of the electric field on the inner side of S due to
σ_b is,

$$\frac{\sigma_b}{2\epsilon_0} = -\frac{\epsilon_r-1}{\epsilon_r+1}\frac{Qh}{4\pi\epsilon_0 r^3} \qquad\qquad (11\text{--}23)$$

This field would be produced by a point charge

$$Q_1 = -\frac{\epsilon_r-1}{\epsilon_r+1}Q \qquad\qquad (11\text{--}24)$$

located at a distance h from O inside the dielectric. The image charge
Q_1 together with the point charge Q gives the correct field for all
points outside the dielectric—i.e., inside S. If $\epsilon_r \rightarrow \infty$ the problem
becomes that of a point charge opposite a conducting plane.

To find the field at points inside the dielectric we note that the normal
component of the electric field, due to σ_b, on what is now the inner side
of S is given by

$$\frac{\sigma_b}{2\epsilon_0} = -\frac{\epsilon_r-1}{\epsilon_r+1}\frac{Qh}{4\pi\epsilon_0 r^3}$$

This normal component would be produced by a point charge Q_1
located at a distance h from O outside the dielectric or by a charge $-Q_1$

at a distance h from O inside the dielectric. Since, however, the image charge must always lie outside the region in which the field is to be determined, it follows that in this case the position of the image charge must coincide with that of Q. Hence the field at any point in the dielectric is obtained by replacing both Q and the dielectric by a single charge:

$$Q_2 = Q + Q_1 = \frac{2Q}{\epsilon_r + 1} \qquad (11\text{--}25)$$

If the surface S separates two media of permittivities ϵ_1 and ϵ_2 then it can be readily shown that the field in medium 1 is the same as if the whole space were occupied by medium 1 and we had an additional charge

$$Q_1' = -\frac{\epsilon_2 - \epsilon_1}{\epsilon_2 + \epsilon_1} Q \qquad (11\text{--}26)$$

at the image position; the field in medium 2 is the same as if the whole space were occupied by medium 2 and we had a single charge

$$Q_2' = \frac{2\epsilon_2}{\epsilon_2 + \epsilon_1} Q \qquad (11\text{--}27)$$

in place of the charge Q. Figures 11.6 and 11.7 show the field lines when $\epsilon_2 > \epsilon_1$ and $\epsilon_2 < \epsilon_1$ respectively.

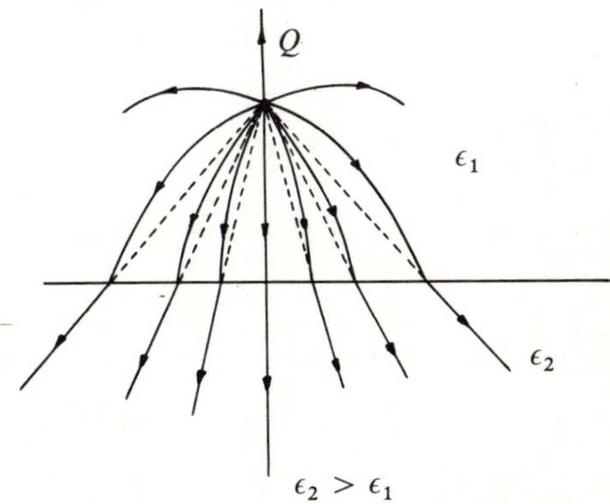

Fig. 11.6 Field lines due to point charge near boundary between two dielectrics

11.4.3 POINT CHARGE BETWEEN TWO SEMI-INFINITE CONDUCTING PLANES INTERSECTING AT RIGHT ANGLES

It is evident from Fig. 11.8 that the conducting planes may be replaced by the three image charges shown. The potential at any point P is

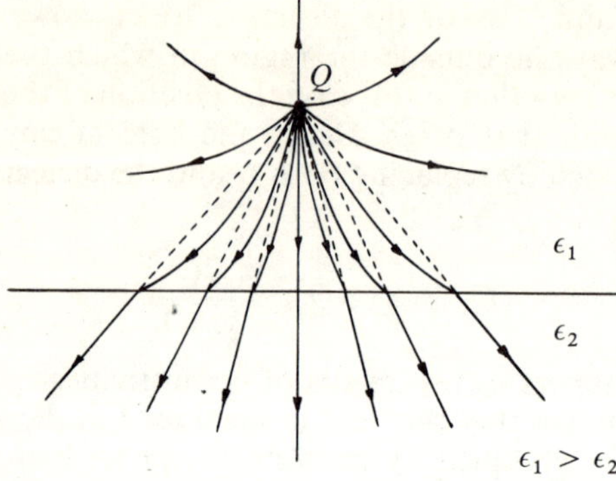

Fig. 11.7 Field of point charge near boundary between two dielectrics

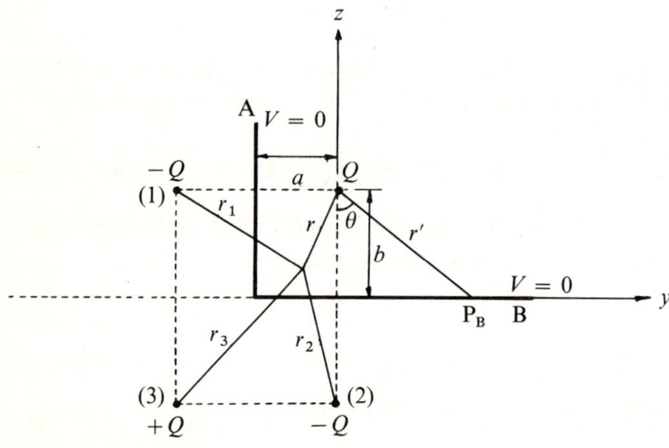

Fig. 11.8 Point charge between two semi-infinite conducting planes

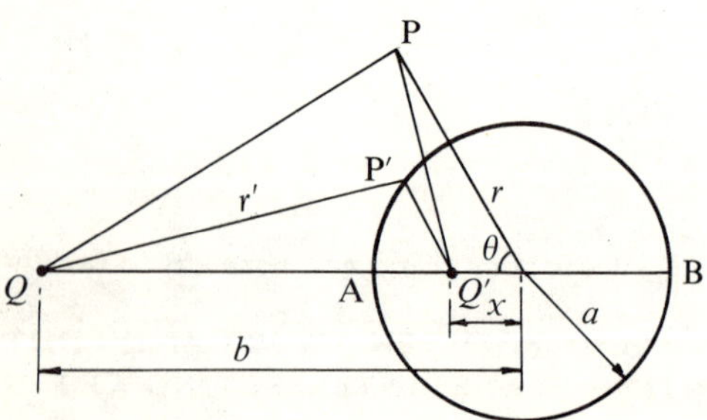

Fig. 11.9 Point charge outside a conducting sphere

given by

$$V = \frac{Q}{4\pi\epsilon_0}\left(\frac{1}{r} - \frac{1}{r_1} - \frac{1}{r_2} + \frac{1}{r_3}\right) \tag{11-28}$$

The surface charges induced on the planes may be determined as follows. The surface charge density at any point on plane B due to the charge Q and the image charge $-Q$ at 2 is

$$\sigma_B = -\frac{Qb}{2\pi r'^3}$$

The induced charge on plane B due to these two charges is

$$Q_B' = -(Qb/2\pi) \int_{-a}^{\infty} \int_{-\infty}^{\infty} (x^2+y^2+b^2)^{-3/2}\, dx\, dy$$

$$= -(Qb/\pi) \int_{-a}^{\infty} (y^2+b^2)^{-1}\, dy$$

$$= -\frac{Q}{\pi}\left[\tfrac{1}{2}\pi + \tan^{-1}\left(\frac{a}{b}\right)\right]$$

Similarly the charge induced on plane B due to the image charges at 1 and 3 is given by

$$Q_B'' = \frac{Q}{\pi}\left[\tfrac{1}{2}\pi - \tan^{-1}\left(\frac{a}{b}\right)\right]$$

The total charge induced on plane B is therefore

$$Q_B = Q_B' + Q_B'' = -\frac{2Q}{\pi}\tan^{-1}\left(\frac{a}{b}\right) \tag{11-29}$$

and it follows that

$$Q_A = Q - \frac{2Q}{\pi}\tan^{-1}\left(\frac{a}{b}\right) \tag{11-30}$$

11.4.4 POINT CHARGE AND A CONDUCTING SPHERE

Consider a point charge Q placed at a distance b from the centre of a conducting sphere of radius a such that $b > a$ (Fig. 11.9).

Case (i) Sphere is at zero potential. If there is an image charge it must, by symmetry, lie on a diameter through Q. Since the potential of the sphere is zero, we must have that

$$\frac{Q}{|QP'|} + \frac{Q'}{|Q'P'|} = 0$$

at all points on the sphere. At points A and B,

$$-\frac{Q}{Q'} = \frac{b-a}{a-x} = \frac{b+a}{b+x} = \frac{b}{a} = \frac{a}{x}$$

Hence it follows that

$$Q' = -Q(a/b) \tag{11-31}$$

$$x = a^2/b \tag{11-32}$$

Thus the potential at an arbitrary point P outside the sphere is given by

$$V = \frac{1}{4\pi\epsilon_0}\left[\frac{Q}{|QP|} + \frac{Q'}{|Q'P|}\right]$$

$$= \frac{Q}{4\pi\epsilon_0}[(r^2+b^2-2rb\cos\theta)^{-1/2} - (a/b)(r^2+x^2-2rx\cos\theta)^{-1/2}]$$

and

$$V(r, \theta) = \frac{Q}{4\pi\epsilon_0}\{(r^2+b^2-2rb\cos\theta)^{-1/2} - (a/b)[r^2+(a^4/b^2)-$$

$$(2ra^2/b)\cos\theta]^{-1/2}\} \tag{11-33}$$

The electric field at P is

$$\mathbf{E} = -\frac{\partial V}{\partial r}\mathbf{a_r} - \frac{1}{r}\frac{\partial V}{\partial\theta}\mathbf{a_\theta}$$

$$E_r = \frac{Q}{4\pi\epsilon_0}\left\{\frac{r-b\cos\theta}{[r^2+b^2-2rb\cos\theta]^{3/2}} - \right.$$

$$\left.\frac{(a/b)[r-(a^2/b)\cos\theta]}{[r^2+(a^2/b)^2-(2a^2r/b)\cos\theta]^{3/2}}\right\} \tag{11-34}$$

$$E_\theta = \frac{Q}{4\pi\epsilon_0}\left\{\frac{b}{[r^2+b^2-2rb\cos\theta]^{3/2}} - \right.$$

$$\left.\frac{a^3/b^2}{[r^2+(a^2/b)^2-(2a^2r/b)\cos\theta]^{3/2}}\right\}\sin\theta \tag{11-35}$$

For $r = a$ we have

$$V(a, \theta) = 0$$

$$E_\theta(a, \theta) = 0$$

$$E_r(a, \theta) = -\frac{Q}{4\pi\epsilon_0 a}\frac{b^2-a^2}{r'^3} \tag{11-36}$$

The induced surface charge density is given by $\sigma = \epsilon_0 E_r(a, \theta)$. The force of attraction between the charge and the sphere is equal to that between the charges Q and Q'; thus

$$F = \frac{QQ'}{4\pi\epsilon_0(b-x)^2} = -\frac{abQ^2}{4\pi\epsilon_0(b^2-a^2)^2} \tag{11-37}$$

It is evident that if a point charge Q is placed at a point *inside* an earthed conducting sphere at a distance b from its centre, then its image will lie outside the sphere at a distance $x = a^2/b$ from the centre and its magnitude will be $-Qa/b$ as before. The charge density induced on the inner surface of the sphere is in this case

$$\sigma = -\frac{Q}{4\pi a}\frac{a^2-b^2}{r'^3} \tag{11-38}$$

Case (ii) Sphere insulated and uncharged. In this case it is evident that in addition to the induced surface charge Q' there must be a charge $-Q'$ superposed on it in order to make the total charge on the sphere zero. Hence a second image charge $Q'' = -Q'$ must be placed within the sphere; its position must be at the centre of the sphere to ensure that the surface of the sphere remains an equipotential. The potential at any point outside the sphere is the sum of the potentials due to the three charges (superposition of two equilibrium states). The charge Q'' raises the sphere to a potential

$$V'' = \frac{Q''}{4\pi\epsilon_0 a} = \frac{Q}{4\pi\epsilon_0 b}$$

Thus the potential of the uncharged sphere due to the original point charge Q is $Q/4\pi\epsilon_0 b$; this result was also obtained in Section 3.6.4 by a different method.

Case (iii) Sphere carries a net charge. If the sphere carries a net charge Q_0 it is evident that the potential outside the sphere is given by the charges,

$$Q \qquad\qquad \text{(at a distance } b \text{ from the sphere centre)}$$

$$Q' = -Qa/b \qquad \text{(at a distance } a^2/b \text{ from the sphere centre)}$$

$$Q'' + Q_0 = Qa/b + Q_0 \quad \text{(at sphere centre)}$$

11.4.5 PROBLEM OF TWO SPHERES

Consider two conducting spheres of radii a and b respectively with their centres at a distance c apart (Fig. 11.10). Suppose that sphere 1 is at a potential V_1 and that sphere 2 is kept at zero potential. The field at any point outside the two spheres may be found by replacing the

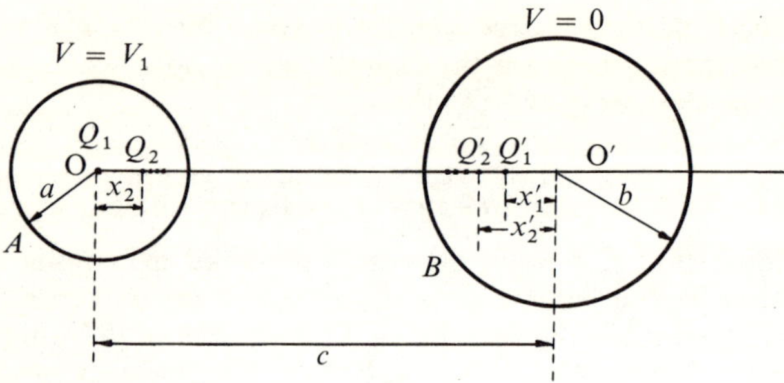

Fig. 11.10 Problem of two conducting spheres

spheres by an infinite series of image point charges. The magnitude and position of the image charges are determined by the method already described. The procedure of establishing the necessary image charges is as follows.

(1) The sphere whose potential is V_1 may be replaced by a charge $Q_1 = 4\pi\epsilon_0 a V_1$ placed at its centre O.

(2) To maintain the second sphere at zero potential we require an image charge

$$Q'_1 = -Q_1 b/c$$

at a distance

$$x'_1 = b^2/c$$

from O'.

(3) In order to cancel the effect which the charge Q'_1 has on the potential of sphere 1, it is necessary to place an image charge

$$Q_2 = -\frac{Q'_1 a}{(c-x'_1)} = \frac{Q_1 ab}{c(c-x'_1)}$$

at a distance

$$x_2 = \frac{a^2}{c-x'_1}$$

from O.

(4) To keep sphere 2 at zero potential in the presence of Q_2 it is necessary to place an image charge

$$Q'_2 = -\frac{Q_2 b}{c-x_2} = -\frac{Q_1 ab^2}{c(x-x'_1)(x-x_2)}$$

at a distance

$$x_2' = \frac{b^2}{c - x_2}$$

from O'.

(5) It is evident that this procedure of successive imaging must be continued indefinitely. However, each new image charge is smaller than the one before and with reasonable values of a, b and c the effect of additional image charges rapidly becomes negligibly small and a limited number of images will suffice.

It can be readily shown that the image charges are given by

$$Q_{n+1} = Q_1 a^n b^n \prod_{k=1}^{n} \frac{1}{(c - x_k)(c - x_k')} \qquad (11\text{--}39)$$

$$Q_n' = -\frac{Q_n b}{c - x_n} \qquad (11\text{--}40)$$

where

$$\left.\begin{aligned}
x_1 &= 0 \quad ; \quad x_1' = \frac{b^2}{c - x_1} \\[2mm]
x_2 &= \frac{a^2}{c - x_1'}; \quad x_2' = \frac{b^2}{c - x_2} \\[2mm]
x_{n+1} &= \frac{a^2}{c - x_n'}; \quad x_n' = \frac{b^2}{c - x_n}
\end{aligned}\right\} \qquad (11\text{--}41)$$

The total charge on sphere A and the total charge on sphere B is obtained by summing the charges inside each sphere; thus

$$Q_A = Q_1 + \sum_{n=1}^{\infty} Q_{n+1} \qquad (11\text{--}42)$$

$$Q_B = \sum_{n=1}^{\infty} Q_n' \qquad (11\text{--}43)$$

If the potential of sphere B were kept at V_2 instead of zero, then starting with a charge $4\pi\epsilon_0 b V_2$ at O', a second set of image charges can be found which will maintain A at zero potential and B at a potential V_2. By superposing the two sets of images we obtain a solution to the problem of two spheres, one raised to a potential V_1 and the other to a potential V_2.

If the spheres A and B have equal radii a, and are raised to equal but opposite potentials ($V_1 = -V_2$) it can be shown that

$$Q_A = -Q_B = Q_1(1 + m + m^2 + m^3 + 2m^4 + 3m^5 + \cdots) \qquad (11\text{--}44)$$

where $m = a/c$. The capacitance of the two spheres is

$$C = Q_A/2V = 2\pi\epsilon_0 a(1 + m + m^2 + m^3 + 2m^4 + 3m^5 + \cdots) \qquad (11\text{--}45)$$

11.4.6 LINE CHARGE OUTSIDE A CIRCULAR CONDUCTING CYLINDER

In Section 2.9 (page 48) we showed that for two infinite, parallel line charges of densities $+\lambda$ and $-\lambda$, the equipotential surfaces are a set of circular cylinders having their axes parallel to and collinear with the line charges.

Consider now the problem of a line charge of density $+\lambda$ at a distance d from the centre of a conducting cylinder of radius R (Fig. 11.11).

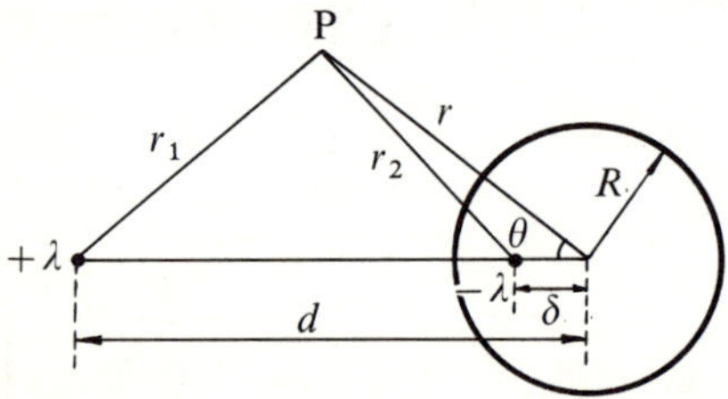

Fig. 11.11 Line charge outside a circular conducting cylinder

From equation 2–71 we see that if we place an image line charge $-\lambda$ at a distance

$$\delta = R^2/d$$

from the centre of the cylinder, we satisfy the condition that the cylinder is an equipotential surface. In this case the total charge per unit axial length of the cylinder surface is $-\lambda$. The potential at any point outside the cylinder is given by

$$V = \frac{\lambda}{2\pi\epsilon_0} \log\frac{r_2}{r_1} + C$$

where C is an arbitrary constant. If the cylinder is at zero potential ($V = 0$ when $r_2/r_1 = R/d$) then

$$V = \frac{\lambda}{2\pi\epsilon_0} \log\frac{r_2}{r_1} \cdot \frac{d}{R} \qquad (11\text{--}46)$$

The surface charge density is in this case given by

$$\sigma(R, \theta) = -\frac{\lambda}{2\pi R} \frac{d^2 - R^2}{(d^2 + R^2 - 2dR\cos\theta)} \qquad (11\text{--}47)$$

If the cylinder is uncharged and insulated then we require an additional image line charge of density $+\lambda$ placed at the centre of the cylinder. The potential at any point P is then given by

$$V = \frac{\lambda}{2\pi\epsilon_0}\log\frac{r_2}{r_1} + C + \frac{\lambda}{2\pi\epsilon_0}\log\frac{r_0}{r} + V_0$$

$$= \frac{\lambda}{2\pi\epsilon_0}\log\frac{r_2}{r_1}\cdot\frac{1}{r} + C' \qquad (11\text{-}48)$$

where C' is an arbitrary constant. If the cylinder is to be chosen as zero potential reference then,

$$V = \frac{\lambda}{2\pi\epsilon_0}\log\frac{r_2}{r_1}\cdot\frac{d}{r} \qquad (11\text{-}49)$$

The surface charge density distribution for the uncharged insulated cylinder is given by

$$\sigma(R,\theta) = \frac{\lambda}{\pi}\frac{R - d\cos\theta}{(d^2 + R^2 - 2dR\cos\theta)} \qquad (11\text{-}50)$$

11.5 SOLUTION OF MAGNETOSTATIC FIELD PROBLEMS BY THE METHOD OF IMAGES

In the previous section we discussed the application of the method of images to electrostatic field problems. The method is equally applicable to magnetostatic field problems*. The source of a magnetostatic field may be either a steady current or a permanent magnet; in the first case the current may be replaced by a magnetic dipole shell bounded by the current circuit (Section 7.13) and in the second the magnetized body may be replaced by a volume and surface distribution of magnetic poles (Section 8.3). In either case the magnetizing force $\mathbf{H}$ can be derived from a scalar potential function Φ_H and the formulation of the image problem becomes analogous to that in electrostatics.

Consider a closed surface S which divides all space into two distinct regions. Determine the normal or tangential component of the magnetic field $\mathbf{H}$ on S due to the magnetic poles induced thereon by magnetic poles inside S. Find a distribution of poles outside S to give on S the same normal or tangential $\mathbf{H}$; this distribution will give inside S the same $\mathbf{H}$ as that given by the induced poles on S. It is the image distribution required.

As an example consider the infinite line current I at a distance h from a semi-infinite magnetic slab of permeability μ (Fig. 11.12). Let P be a

* For an excellent and extensive discussion of this topic see *The Principles of Electromagnetism Applied to Electrical Machines* by B. Hague (Dover, New York, 1962).

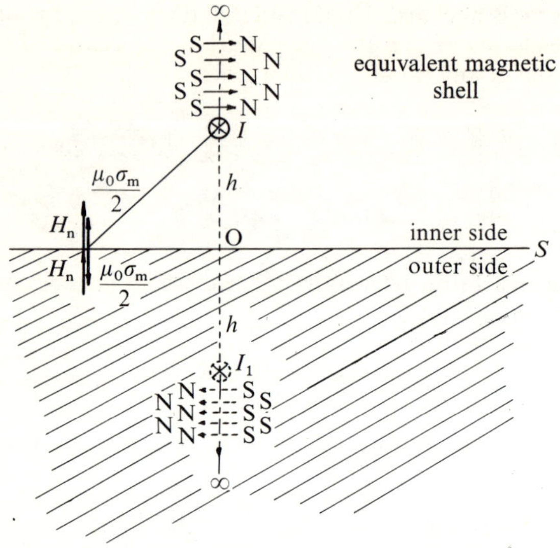

Fig. 11.12 Infinite line current above a semi-infinite magnetic slab of permeability μ

point on the boundary surface S, σ_m be the induced magnetic pole density at P and H_n be the normal component of the magnetizing force at P, due to the magnetic shell bounded at I. The normal component of the resultant H on either side of S at P must satisfy the boundary condition $\mu_1 H_{n1} = \mu_2 H_{n2}$ so that

$$\mu_0(\tfrac{1}{2}\sigma_m + H_n) = \mu_0\mu_r(H_n - \tfrac{1}{2}\sigma_m)$$

or

$$\tfrac{1}{2}\sigma_m = \left(\frac{\mu_r - 1}{\mu_r + 1}\right)H_n \tag{11–51}$$

where H_n is the normal component of the magnetizing force at P due to the current I alone. It is evident that the field given by equation 11–51 would be produced by an infinite line-current carrying a current

$$I_1 = \left(\frac{\mu_r - 1}{\mu_r + 1}\right)I \tag{11–52}$$

and placed at a distance h below O. Since $\mu_r > 1$, I_1 and I have the same direction. We thus see that the field at any point P in the region containing the line current (Fig. 11.13a) is the same as that produced by line currents I and I_1 at a distance $2h$ apart placed in air (same medium as that in which the original current is situated).

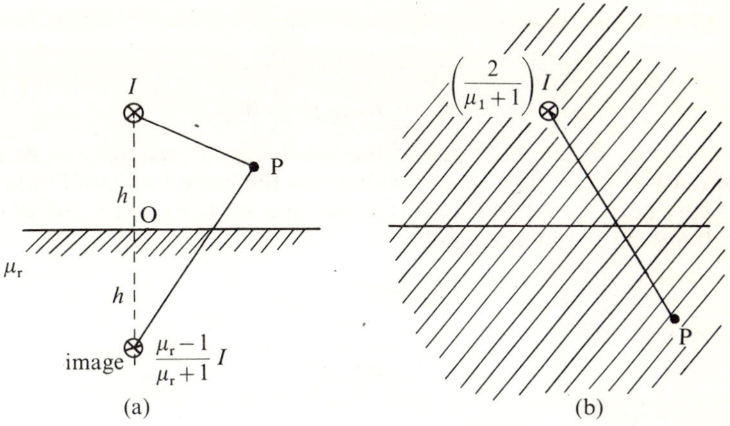

Fig. 11.13 (a) Equivalent currents for obtaining field at point P in region containing original line current. (b) Equivalent current for obtaining field at point P inside magnetic medium

To find **H** at a point *inside* the magnetic medium, we note that the normal component of the field **H**, due to σ_m, on what is now the inside of S, is given by

$$\tfrac{1}{2}\sigma_m = \left(\frac{\mu_r-1}{\mu_r+1}\right)H_n \tag{11-53}$$

This field would be produced by a line current $-I[(\mu_r-1)/(\mu_r+1)]$ at a distance h above O; the negative sign indicates that the direction of this current is into the plane of the paper. The field in the region that does not contain the original current is thus due to a single line current

$$I_2 = I - \left(\frac{\mu_r-1}{\mu_r+1}\right)I = \frac{2I}{\mu_r+1} \tag{11-54}$$

having the same direction as and at the position of the original current and placed in an unbounded medium of permeability μ_r (Fig. 11.13b). The flux lines in the iron are thus circles with centres at I_2.

Let us now examine the limiting case in which $\mu_r \to \infty$. The image current $I_1 \to I$ and the field in air is that due to two equal line-currents (Fig. 7.43); it is evident that the flux lines at the boundary surface S are everywhere perpendicular to S. The image current $I_2 \to 0$ and since the flux density in the iron is given by

$$B_1 = \frac{\mu_0 I}{\pi r} \cdot \frac{\mu_r}{\mu_r+1}$$

we have that as $\mu_r \to \infty$,

$$\left.\begin{array}{l} B_1 \to \mu_0 I/\pi r \\[4pt] H_1 = B_1/\mu_0\mu_r \to 0 \end{array}\right\} \tag{11–55}$$

Thus B_1 and B_2 are finite and of the same order of magnitude. At the boundary S, $\theta_2 = 0$, $\theta_1 = $ any value and the boundary conditions in this case correspond to Case (a) of Section 8.14. Figure 11.14 shows the *B-lines* in the two regions.

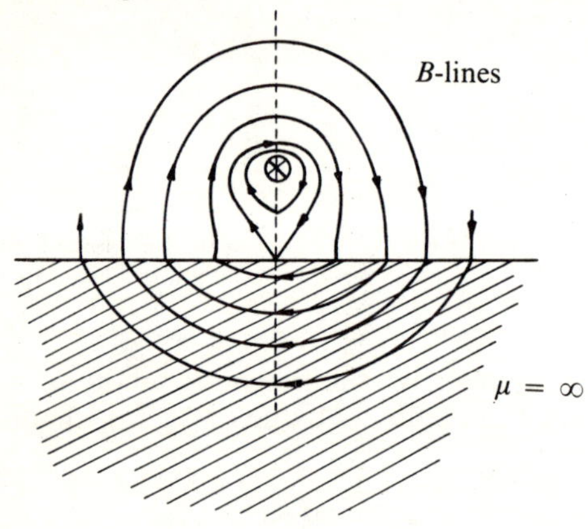

Fig. 11.14 *B*-lines for configuration shown in Fig. 11.12 with $\mu = \infty$

If the line current were embedded in the ferromagnetic semi-infinite slab it can be readily shown that the field in the region containing the current is the same as that produced by the line current I and an image line current

$$I_1 = -\left(\frac{\mu_r-1}{\mu_r+1}\right)I \tag{11–56}$$

at a distance h below O, the whole system being situated in a homogeneous medium of permeability μ_r (Fig. 11.15); the negative sign indicates that I_1 and I are in opposite directions. The field in the air region is the same as that produced by a single line current

$$I_2 = \left(\frac{2\mu_r}{\mu_r+1}\right)I \tag{11–57}$$

having the same direction as and at the position of the original current I_1 the whole medium being air. The flux lines in the air are circles centred at I_2.

In the limiting case in which $\mu_r \to \infty$, the image current $I_1 \to -I$ and the field in the iron is that due to two equal but opposite line currents.

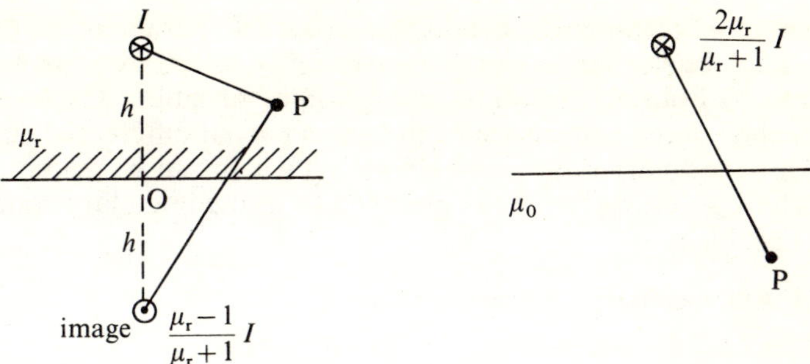

Fig. 11.15 Equivalent currents for obtaining field at points inside and outside the region in which original current is embedded

The flux lines at the boundary are everywhere tangential to S. The image current $I_2 \to 2I$. The flux density B_2 in the air is everywhere finite and given by

$$B_2 = \mu_0 I/\pi r \qquad (11\text{--}58)$$

The flux density B_1 in the iron is evidently infinite but H_1 is finite. At the boundary $\theta_1 = \pi/2$, $\theta_2 =$ any value; in this case the boundary conditions correspond to Case (b) of Section 8.14. Figure 11.16 shows the *H-lines* in the two regions.

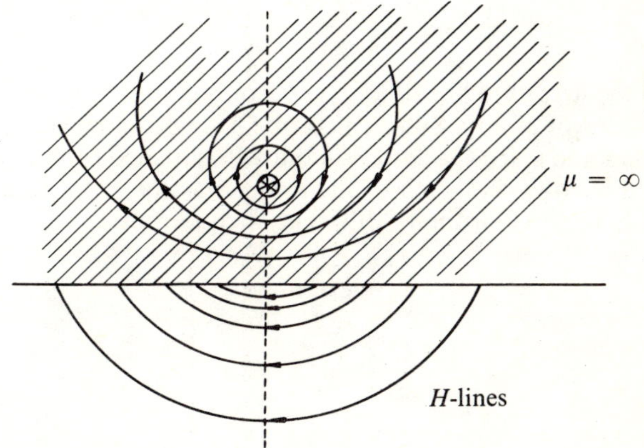

Fig. 11.16 *H*-lines for configuration shown in Fig. 11.15 and $\mu = \infty$.

11.6 ANALYTICAL SOLUTION OF LAPLACE'S EQUATION BY SEPARATION OF VARIABLES

When the boundaries, over which the potential or its normal derivative are known, can be made to coincide with an orthogonal coordinate system—e.g., rectangular, cylindrical, spherical, parabolic, ellipsoidal, etc.—Laplace's equation can be conveniently solved by a method, put forward by Fourier, known as separation of variables. The essence of the method is the conversion of the given partial differential equation into three ordinary differential equations which can be readily solved and whose solutions are then combined to fit the boundary conditions of the problem.

11.6.1 RECTANGULAR COORDINATES

Laplace's equation in rectangular coordinates is

$$\frac{\partial^2 \Phi}{\partial x^2} + \frac{\partial^2 \Phi}{\partial y^2} + \frac{\partial^2 \Phi}{\partial z^2} = 0$$

Suppose that solutions for Φ exist as products of a function of x alone, a function of y alone and a function of z alone, then

$$\Phi = X(x)Y(y)Z(z) \tag{11–59}$$

Substituting this form into Laplace's equation gives

$$YZ\frac{d^2 X}{dx^2} + ZX\frac{d^2 Y}{dy^2} + XY\frac{d^2 Z}{dz^2} = 0$$

and on dividing throughout by XYZ we obtain

$$\frac{1}{X}\frac{d^2 X}{dx^2} + \frac{1}{Y}\frac{d^2 Y}{dy^2} + \frac{1}{Z}\frac{d^2 Z}{dz^2} = 0$$

in which the first term is a function of x only, the second term of y only and the third of z only. Since the terms have no independent variable in common and since the equation must be satisfied for all values of x, y and z, it follows that each of the terms must be equal to a constant. We thus have three ordinary differential equations

$$\left.\begin{aligned}
\frac{d^2 X}{dx^2} + a_x^2 X &= 0 \\[2ex]
\frac{d^2 Y}{dy^2} + a_y^2 Y &= 0 \\[2ex]
\frac{d^2 Z}{dz^2} + a_z^2 Z &= 0
\end{aligned}\right\} \tag{11–60}$$

where a_x, a_y and a_z are arbitrary constants subject to the condition

$$a_x^2 + a_y^2 + a_z^2 = 0$$

so that at least one of the a^2s will be negative—i.e., one of the characteristic equations will have real roots. Each of the equations in (11–60)

has characteristic function solutions which can be expressed in the
following forms

$$N = A_n n + B_n \qquad\qquad\qquad a_n^2 = 0 \qquad (11\text{–}61)$$

$$\left.\begin{aligned} N &= A_n' \exp(a_n n) + B_n' \exp(-a_n n) \\ &= C_n \sinh a_n n + D_n \cosh a_n n \end{aligned}\right\} a_n^2 < 0 \qquad (11\text{–}62)$$

$$\left.\begin{aligned} N &= A_n'' \exp(i a_n n) + B_n'' \exp(-i a_n n) \\ &= C_n' \cos a_n n + D_n' \sin a_n n \end{aligned}\right\} a_n^2 > 0 \qquad (11\text{–}63)$$

where $N = X$, Y or Z and $n = x$, y or z. The values of a_n and of the
arbitrary constants A_n, B_n, . . ., are determined by the boundary conditions of the problem. If all a_ns are zero, the solution is of the form

$$\Phi = (A_x x + B_x)(A_y y + B_y)(A_z z + B_z) \qquad (11\text{–}64)$$

If both a_x^2 and a_y^2 are positive constants then a_z^2 will be negative and
the solution will have the form

$$\Phi = (C_x' \cos a_x x + D_x' \sin a_x x)(C_y' \cos a_y y + D_y' \sin a_y y)(C_z \sinh a_z z$$
$$+ D_z \cosh a_z z) \qquad (11\text{–}65)$$

Since the a_ns may take on more than one value, the complete formal
solution is obtained by summing* over all values of a_x, a_y and a_z,

$$\Phi = \sum_{a_z} \sum_{a_y} \sum_{a_x} (C_x' \cos a_x x + D_x' \sin a_x x)(C_y' \cos a_y y$$
$$+ D_y' \sin a_y y)(C_z \sinh a_z z + D_z \cosh a_z z) \qquad (11\text{–}66)$$

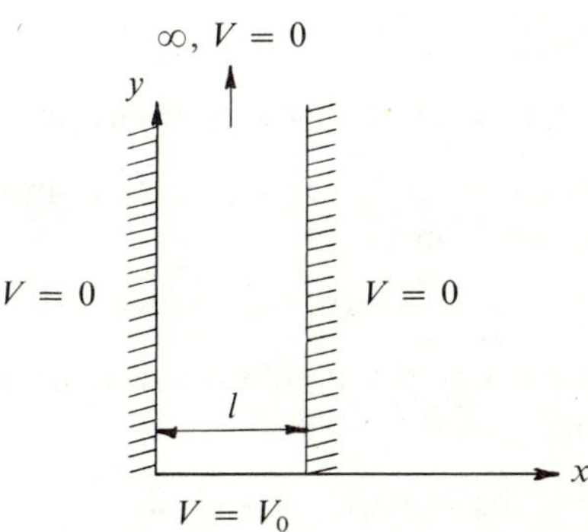

Fig. 11.17 Boundary conditions for a semi-infinite metal plate

* If Φ_1, Φ_2, . . . Φ_n are solutions of Laplace's equation, then

$$\Phi = A_1 \Phi_1 + A_2 \Phi_2 + \cdots A_n \Phi_n$$

is also a solution. This can be proved by direct substitution.

In a two-dimensional problem $(\partial^2\Phi/\partial x^2 + \partial^2\Phi/\partial y^2)$ we have that $a_x^2 + a_y^2 = 0$ and the solution can be of the form

$$\Phi = (C_x' \cos a_x x + D_x' \sin a_x x)[A_y' \exp(a_y y) + B_y' \exp(-a_y y)] \qquad (11\text{–}67)$$

if a_x^2 is positive, or of the form

$$\Phi = [A_x' \exp(a_x x) + B_x' \exp(-a_x x)](C_y' \cos a_y y + D_y' \sin a_y y) \qquad (11\text{–}68)$$

if a_y^2 is positive. The choice is determined by the boundary conditions.
Example. It is required to find the potential distribution throughout a semi-infinite metal plate of uniform thickness and width l subject to the boundary conditions (Fig. 11.17).

$$V = 0, \quad x = 0$$
$$V = 0, \quad x = l$$
$$V = 0, \quad y \to \infty$$
$$V = V_0, \quad y = 0$$

It is evident from the first three boundary conditions that in this case the solution must be of the form equation 11–67 since a solution of the form equation 11–68 does not tend to zero as $y \to \infty$ and a linear combination of $\exp(a_x x)$ and $\exp(-a_x x)$ cannot be zero at both $x = 0$ and $x = l$.

$$\text{Since } V = 0 \quad \text{as } y \to \infty \text{ we must have } A_y' = 0$$
$$\text{Since } V = 0 \quad \text{when } x = 0 \text{ we must have } C_x' = 0$$

Thus the solution reduces to

$$V = D_x' B_y' \exp(-a_y y) \sin a_x x$$

Introducing a new constant b_n for $D_x' B_y'$, and since $|a_x| = |a_y| = a_n$, we may rewrite the solution as

$$V = b_n \exp(-a_n y) \sin a_n x \qquad (11\text{–}69)$$

with two remaining boundary conditions for determining b_n and a_n. Since $V = 0$ when $x = l$, it follows that

$$a_n = n\pi/l, \quad n = 1, 2, \ldots$$

and the complete formal solution will be of the form

$$V = \sum_{n=1}^{\infty} b_n \exp(-n\pi y/l) \sin(n\pi x/l) \qquad (11\text{–}70)$$

The coefficients b_n must be such as to satisfy the condition $V = V_0$

when $y = 0$, that is,

$$V(x, 0) = \sum_{n=1}^{\infty} b_n \sin\frac{n\pi x}{l} \qquad (11\text{--}71)$$

Equation 11–71 is just the Fourier sine series* for $f(x) = V_0$, so that

$$b_n = \frac{2}{l}\int_0^l V_0 \sin(n\pi x/l)\,\mathrm{d}x = \begin{cases} 4V_0/n\pi & \text{if } n \text{ is odd} \\ 0 & \text{if } n \text{ is even} \end{cases}$$

The final solution is thus

$$V(x, y) = \frac{4V_0}{\pi}\sum_{n=1}^{\infty}\frac{1}{n}\exp\left(-\frac{n\pi y}{l}\right)\sin\frac{n\pi x}{l} \qquad (11\text{--}72)$$

11.6.2 CYLINDRICAL COORDINATES

Laplace's equation in circular cylindrical coordinates is

$$\nabla^2 V = \frac{1}{r}\frac{\partial}{\partial r}\left(r\frac{\partial V}{\partial r}\right) + \frac{1}{r^2}\frac{\partial^2 V}{\partial\phi^2} + \frac{\partial^2 V}{\partial z^2}$$

To separate variables let

$$V = R(r)\Phi(\phi)Z(z) \qquad (11\text{--}73)$$

and when this is substituted in the Laplace equation and the resulting expression is divided throughout by $R\Phi Z$ we obtain

$$\frac{1}{rR}\frac{\partial}{\partial r}\left(r\frac{\partial R}{\partial r}\right) + \frac{1}{r^2\Phi}\frac{\partial^2\Phi}{\partial\phi^2} + \frac{1}{Z}\frac{\partial^2 Z}{\partial z^2} = 0 \qquad (11\text{--}74)$$

The last term is a function of z only while the other two terms do not contain z. The last term can therefore be equated to a constant and the sum of the first two terms must be equal to minus the same constant. We thus have

$$\frac{1}{Z}\frac{\mathrm{d}^2 Z}{\mathrm{d}z^2} = a_z^2$$

so that

$$Z = A_z z + B_z, \quad (a_z = 0)$$

$$Z = A_z' \exp(a_z z) + B_z' \exp(-a_z z), \quad (a_z^2 > 0)$$

Substituting a_z^2 for the last term in equation 11–74 and multiplying

* See I. N. Sneddon, *Fourier Series* (Routledge & Kegan-Paul, London, 1961).

through by r^2 gives

$$\frac{r}{R}\frac{\partial}{\partial r}\left(r\frac{\partial R}{\partial r}\right)+\frac{1}{\Phi}\frac{\partial^2\Phi}{\partial\phi^2}+a_z^2 r^2 = 0 \qquad (11\text{–}75)$$

It is evident the second term may be equated to a constant so that

$$\frac{1}{\Phi}\frac{\partial^2\Phi}{\partial\phi^2} = -a_\phi^2$$

and

$$\Phi = C_\phi\cos a_\phi\phi + D_\phi\sin a_\phi\phi \qquad (11\text{–}76)$$

This is the only possible form of solution for $a_\phi^2 \neq 0$, that is we must use $-a_\phi^2$ as the separation constant. This is because the location of a point by means of cylindrical coordinates is not altered if we choose the angle as ϕ or as $\phi+2\pi m$ where m is any integer; thus the solution for V must, at a given point, have the same value at ϕ and at $\phi+2\pi m$; V must therefore be a periodic function of ϕ with period 2π. Hence the solution must be in terms of sines and cosines and not exponentials and a_ϕ *must be an integer*. It is more convenient to use n instead of a_ϕ. To obtain the solution for the radial function, we replace the second term in equation 11–75 by $-n^2$; this gives

$$\frac{d^2R}{dr^2}+\frac{1}{r}\frac{dR}{dr}+\left(a_z^2-\frac{n^2}{r^2}\right)R = 0 \qquad (11\text{–}77)$$

Equation 11–77 is Bessel's differential equation. If we let $x = a_z r$,

$$\frac{dR}{dr} = a_z\frac{dR}{dx}, \quad \frac{d^2R}{dr^2} = a_z^2\frac{d^2R}{dx^2}$$

and it is possible to rewrite equation 11–77 in the form

$$\frac{d^2R}{dx^2}+\frac{1}{x}\frac{dR}{dx}+\left(1-\frac{n^2}{x^2}\right)R = 0 \qquad (11\text{–}78)$$

which is another form of Bessel's equation.

The complete solution of equation 11–78 in standard form is

$$\left.\begin{array}{c} R = A_1 J_n(x) + B_1 Y_n(x) \\[4pt] = A_1 J_n(a_z r) + B_1 Y_n(a_z r) \end{array}\right\} \qquad (11\text{–}79)$$

where $J_n(x)$ is the Bessel function of the first kind of order n and $Y_n(x)$ is the Bessel function of the second kind of order n; $Y_n(x)$ has a singularity on the axis $r = 0$ and is therefore discarded in problems where $R(r)$ must remain finite on the axis. The Bessel equation and Bessel functions will not be considered any further because problems involving

Bessel functions will not be dealt with in this book. Extensive treatment of this subject may be found in mathematics books* and tables† of $J_n(x)$ and $Y_n(x)$ are available for commonly occurring values of n and x so that it is as easy to evaluate an expression containing a Bessel function as one containing a trigonometric function.

If the potential function V is not a function of z, then let $a_z = 0$ in equation 11–77 so that

$$\frac{d^2R}{dr^2} + \frac{1}{r}\frac{dR}{dr} = n^2\frac{R}{r^2} \tag{11-80}$$

and it is easy to verify that the solution of this equation is

$$R = A_n r^n + B_n r^{-n} \tag{11-81}$$

so that the complete formal solution of Laplace's equation in this case is given by

$$V = \sum_{n=1}^{\infty} (A_n r^n + B_n r^{-n})(C_n \cos n\phi + D_n \sin n\phi) \tag{11-82}$$

If $n = 0$, it is evident that the solution must be of the form

$$V = (A \log r + B)(C\phi + D) \tag{11-83}$$

(a) Dielectric cylinder in a uniform electric field. Consider an infinitely long right-circular dielectric cylinder of radius a and permittivity ϵ placed with its axis (chosen as the z-axis) perpendicular to an originally uniform electric field of intensity E_0 (Fig. 11.18). Since there will be a

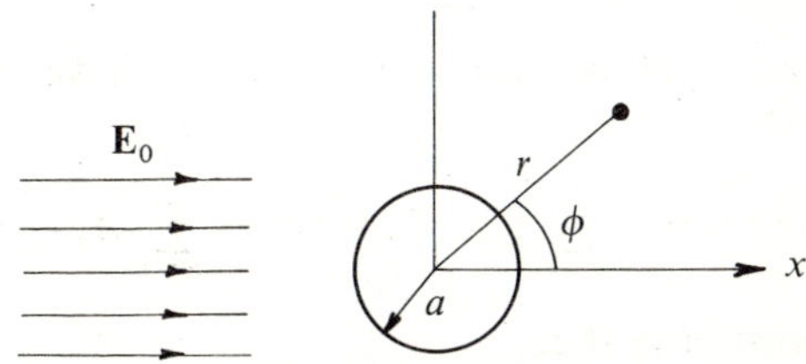

Fig. 11.18 Dielectric cylinder in a uniform electric field

field both inside and outside the cylinder, the electrostatic potential both inside and outside the cylinder must satisfy Laplace's equation,

$$\nabla^2 V_i = 0 \quad (0 \leq r \leq a)$$

$$\nabla^2 V_e = 0 \quad (a \leq r \leq \infty)$$

* N. W. McLachlan, *Bessel Functions for Engineers* (Oxford University Press, London, 1941).
† E. Jahnke and F. Emde, *Tables of Functions* (Dover, New York, 1945).

The boundary conditions which must be satisfied are:

(*i*) At points sufficiently far from the cylinder ($r \to \infty$) the field will retain its original uniform character so that (arbitrarily taking the potential as zero at $x = 0$),

$$V_0 = -E_0 x = -E_0 r \cos \phi$$

(*ii*) At the surface of the cylinder ($r = a$),

$$\epsilon \frac{\partial V_\mathrm{i}}{\partial r} = \epsilon_0 \frac{\partial V_\mathrm{e}}{\partial r} \quad \text{(continuity of the normal component of flux density)}$$

$$V_\mathrm{i} = V_\mathrm{e} \quad \text{(continuity of electrostatic potential)}$$

For points outside the cylinder, it is evident that in order to satisfy the boundary condition at infinity, no terms of the form r^n ($n > 1$) can appear in the expression for the potential; also, because the potential must be symmetrical about the x-axis, no terms involving $\sin n\phi$ can occur. Equation 11–82 thus reduces to

$$V_\mathrm{e} = -E_0 r \cos \phi + \sum_{n=1}^{\infty} C'_n r^{-n} \cos n\phi \tag{11–84}$$

For points inside the cylinder no terms of the form r^{-n} can occur in the expression for the potential since V_i is finite for $r = 0$; thus $B_n = 0$ and since $D_n = 0$ because of symmetry, equation 11–82 becomes

$$V_\mathrm{i} = \sum A'_n r^n \cos n\phi \tag{11–85}$$

For the continuity of potential at $r = a$ we have from equations 11–84 and 11–85,

$$-E_0 a \cos \phi + \sum_{n=1}^{\infty} C'_n a^{-n} \cos n\phi = \sum_{n=1}^{\infty} A'_n a^n \cos n\phi$$

If $n \neq 1$, we must have that

$$a^{-n} C'_n = a^n A'_n$$

a condition that can only be satisfied if $A'_n = C'_n = 0$. If $n = 1$ then

$$-E_0 a + \frac{C'_1}{a} = a A'_1 \tag{11–86}$$

For the continuity of the normal component of the flux density, we have

$$\epsilon_0 E_0 + \frac{\epsilon_0 C'_1}{a^2} = -\epsilon A_1 \tag{11–87}$$

Solving equations 11–86 and 11–87 for A'_1 and C'_1, we obtain

$$A'_1 = -\frac{2\epsilon_0}{\epsilon+\epsilon_0}E_0 \tag{11-88}$$

$$C'_1 = \frac{\epsilon-\epsilon_0}{\epsilon+\epsilon_0}a^2E_0 \tag{11-89}$$

The potential outside and inside the cylinder is thus given by

$$V_e = -E_0 r\cos\phi+\frac{\epsilon-\epsilon_0}{\epsilon+\epsilon_0}\frac{a^2}{r}E_0\cos\phi \tag{11-90}$$

$$V_i = -\frac{2\epsilon_0}{\epsilon+\epsilon_0}rE_0\cos\phi \tag{11-91}$$

The external and internal fields are given by the negative gradient of the respective potentials

$$\mathbf{E}_e = E_0\mathbf{a}_x+E_0\frac{\epsilon-\epsilon_0}{\epsilon+\epsilon_0}\frac{a^2}{r^2}(\cos\phi\,\mathbf{a}_r+\sin\phi\,\mathbf{a}_\phi) \tag{11-92}$$

$$\mathbf{E}_i = \frac{2\epsilon_0}{\epsilon+\epsilon_0}(\cos\phi\,\mathbf{a}_r+\sin\phi\,\mathbf{a}_\phi)E_0$$

$$= \frac{2\epsilon_0}{\epsilon+\epsilon_0}E_0\mathbf{a}_x \tag{11-93}$$

Equation 11–93 shows that inside the cylinder the field is uniform and parallel to the original field E_0; its magnitude, however, is smaller than E_0 since $\epsilon > \epsilon_0$ (Fig. 11.19). This reduction is due to the depolarizing effect of the bound charges induced on the surface of the cylinder. The

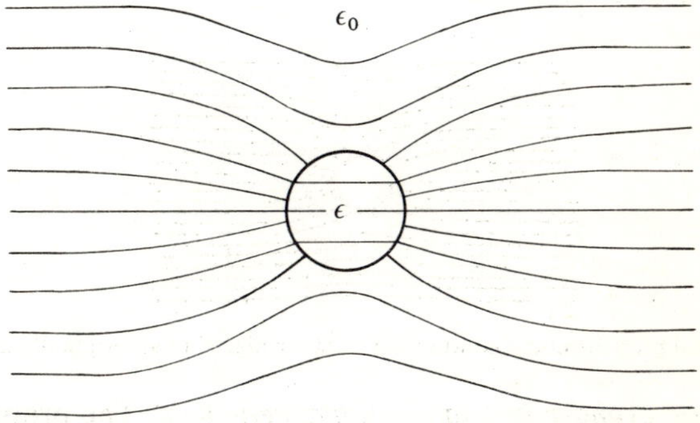

Fig. 11.19 Electric field inside and around dielectric cylinder placed in an originally uniform field

depolarizing factor, as defined in Section 4.4, is given by

$$L = \frac{\epsilon_0(\mathbf{E}_0 - \mathbf{E}_i)}{\mathbf{P}}$$

From equation 11–93 we have that

$$\mathbf{E}_0 - \mathbf{E}_i = \frac{\epsilon - \epsilon_0}{\epsilon + \epsilon_0}\mathbf{E}_0$$

Also,

$$\mathbf{P} = \epsilon_0(\epsilon_r - 1)\mathbf{E}_i = \frac{2\epsilon_0(\epsilon - \epsilon_0)}{\epsilon + \epsilon_0}\mathbf{E}_0$$

so that

$$L = 1/2$$

(b) Conducting cylinder in a uniform field. Since in purely static fields conductors behave like dielectrics with an infinite permittivity, the potential and field relations for the case of an infinitely long uncharged conducting cylinder placed in a uniform external field may be obtained by letting $\epsilon \to \infty$ in the foregoing results. Thus,

$$V_i = 0$$

$$V_e = -E_0 r\left(1 - \frac{a^2}{r^2}\right)\cos\phi \tag{11–94}$$

$$E_e = E_0\mathbf{a}_x + E_0\frac{a^2}{r^2}(\cos\phi\,\mathbf{a}_r + \sin\phi\,\mathbf{a}_\phi) \tag{11–95}$$

The field lines are shown in Fig. 11.20. On the surface of the cylinder the field has a maximum value of $2E_0$ at $\phi = 0$ and $\phi = \pi$, and a minimum value of zero at $\phi = \pm\pi/2$.

Fig. 11.20 Field lines around a conducting cylinder placed in an originally uniform field

(c) Magnetic cylinder in a uniform magnetic field. The problem of an infinitely long cylinder of constant magnetic permeability μ placed with its axis normal to an originally uniform magnetic field $B_0 = \mu_0 H_0$,

is mathematically identical to that of the dielectric cylinder in an electric field. The solution is obtained by changing V to Φ_H, $\mathbf{E}$ to $\mathbf{H}$, $\mathbf{D}$ to $\mathbf{B}$ and ϵ to μ; thus

$$\Phi_{He} = -H_0 r \cos\phi + \frac{\mu-\mu_0}{\mu+\mu_0}\frac{a^2}{r}H_0 \cos\phi \qquad (11\text{–}96)$$

$$\Phi_{Hi} = -\frac{2\mu_0}{\mu+\mu_0}r H_0 \cos\phi \qquad (11\text{–}97)$$

and

$$\mathbf{H}_e = H_0\mathbf{a}_x + H_0\frac{\mu-\mu_0}{\mu+\mu_0}\frac{a^2}{r^2}(\cos\phi\,\mathbf{a}_r + \sin\phi\,\mathbf{a}_\phi) \qquad (11\text{–}98)$$

$$\mathbf{H}_i = \frac{2\mu_0}{\mu+\mu_0}H_0\mathbf{a}_x \qquad (11\text{–}99)$$

The reader may readily verify that the demagnetizing factor (see Section 8.3)

$$D = \frac{\mathbf{H}_0-\mathbf{H}_i}{\mathbf{M}}$$

where

$$\mathbf{M} = \frac{1}{\mu_0}(\mu-\mu_0)\mathbf{H}_i$$

is equal to 1/2.

11.6.3 SPHERICAL COORDINATES

Laplace's equation in spherical coordinates is

$$\frac{1}{r^2}\frac{\partial}{\partial r}\left(r^2\frac{\partial V}{\partial r}\right) + \frac{1}{r^2\sin\theta}\frac{\partial}{\partial\theta}\left(\sin\theta\frac{\partial V}{\partial\theta}\right) + \frac{1}{r^2\sin^2\theta}\frac{\partial^2 V}{\partial\phi^2} = 0$$

We assume a solution of the form

$$V = R(r)\Theta(\theta)\Phi(\phi) \qquad (11\text{–}100)$$

Substituting this in Laplace's equation and multiplying by $r^2/R\Theta\Phi$ we obtain

$$\frac{1}{R}\frac{d}{dr}\left(r^2\frac{dR}{dr}\right) + \frac{1}{\Theta\sin\theta}\frac{d}{d\theta}\left(\sin\theta\frac{d\Theta}{d\theta}\right) + \frac{1}{\Phi\sin^2\theta}\frac{d^2\Phi}{d\phi^2} = 0 \qquad (11\text{–}101)$$

Only the first term of this equation depends on r so that we may write

$$\frac{1}{R}\frac{d}{dr}\left(r^2\frac{dR}{dr}\right) = a_r^2 = n(n+1) \qquad (11\text{–}102)$$

where for convenience the separation constant has been chosen as $n(n+1)$. The solution of equation 11–102 is

$$R = A_n'' r^n + B_n'' r^{-(n+1)} \qquad (11\text{–}103)$$

where A_n'' and B_n'' are constants.

Substituting $n(n+1)$ for the first term in equation 11–101 and multiplying through by $\sin^2 \theta$ we obtain

$$n(n+1)\sin^2\theta + \frac{\sin\theta}{\Theta}\frac{\mathrm{d}}{\mathrm{d}\theta}\left(\sin\theta\frac{\mathrm{d}\Theta}{\mathrm{d}\theta}\right) + \frac{1}{\Phi}\frac{\mathrm{d}^2\Phi}{\mathrm{d}\phi^2} = 0 \quad (11\text{–}104)$$

We now let

$$\frac{1}{\Phi}\frac{\mathrm{d}^2\Phi}{\mathrm{d}\phi^2} = -m^2 \qquad (11\text{–}105)$$

so that

$$\Phi = C_m \cos m\phi + D_m \sin m\phi \qquad (11\text{–}106)$$

or, for $m = 0$,

$$\Phi = E\phi + F \qquad (11\text{–}107)$$

The separation constant must be negative and m an integer in order to make Φ a periodic single-valued function of ϕ $[V(r, \theta, \phi) = V(r, \theta, \phi + 2m\pi)]$.

For $m \neq 0$ the resulting equation for Θ is

$$\sin^2\theta\frac{\mathrm{d}^2\Theta}{\mathrm{d}\theta^2} + \sin\theta\cos\theta\frac{\mathrm{d}\Theta}{\mathrm{d}\theta} + [n(n+1)\sin^2\theta - m^2]\Theta = 0 \quad (11\text{–}108)$$

If we let $u = \cos\theta$, then equation 11–108 may be expressed as

$$(1-u^2)\frac{\mathrm{d}^2\Theta}{\mathrm{d}u^2} - 2u\frac{\mathrm{d}\Theta}{\mathrm{d}u} + \left[n(n+1) - \frac{m^2}{1-u^2}\right]\Theta = 0 \quad (11\text{–}109)$$

This equation is known as the associated Legendre equation.*

If $m = 0$, that is if the solution of the original problem is independent of the longitude angle ϕ, equation 11–109 reduces to

$$(1-u^2)\frac{\mathrm{d}^2\Theta}{\mathrm{d}u^2} - 2u\frac{\mathrm{d}\Theta}{\mathrm{d}u} + n(n+1)\Theta = 0 \qquad (11\text{–}110)$$

which is the *Legendre equation*.

---

* H. Margenau and G. M. Murphy, *The Mathematics of Physics and Chemistry* (Van Nostrand, New York, 1956).

  I. Sneddon, *Special Functions of Mathematical Physics and Chemistry* (Oliver & Boyd, Edinburgh, 1956).

If n is not an integer all solutions of equations 11–109 and 11–110 become infinite when $u = \pm 1$, that is when $\theta = 0$ or π. If n is an integer $(0, 1, 2, \ldots)$ the general solution of the Legendre equation is given by

$$\Theta = A'_n P_n(u) + B'_n Q_n(u) \tag{11–111}$$

where A'_n and B'_n are constants determined by the boundary conditions and

$$P_n(u) \equiv \text{Legendre polynomial of degree } n.$$

$$Q_n(u) \equiv \text{Legendre function of the second kind of degree } n.$$

The function $Q_n(u)$ is infinite when $u = \pm 1$ and thus cannot represent a finite function in a region which includes either $\theta = 0$ or $\theta = \pi$. The Legendre polynomials on the other hand are bounded in any finite interval, and since in most physical problems we know that the function Θ remains finite along the polar axis $\theta = 0$, it follows that the only solution of Legendre's equation which is of practical interest is

$$\Theta = A'_n P_n(u) \tag{11–112}$$

in which $P_n(u)$ may be obtained* from Roderigue's formula (equation 2–89). The first few values of $P_n(\cos \theta)$ are

$$P_0(\cos \theta) = 1$$

$$P_1(\cos \theta) = \cos \theta$$

$$P_2(\cos \theta) = \tfrac{1}{2}(3\cos^2 \theta - 1) = \tfrac{1}{4}(3\cos 2\theta + 1)$$

$$P_3(\cos \theta) = \tfrac{1}{2}(5\cos^3 \theta - 3\cos \theta) = \tfrac{1}{8}(5\cos 3\theta + 3\cos \theta)$$

The most general solution of Laplace's equation in spherical polar coordinates, assuming axial symmetry, is given by

$$V = \sum_{n=0}^{\infty} \left[A_n r^n P_n(\cos \theta) + B_n r^{-(n+1)} P_n(\cos \theta) \right] \tag{11–113}$$

Sphere in a uniform electric field. Consider a dielectric sphere of radius a and permittivity ϵ_2 placed in an originally uniform electrostatic field in a region of space of permittivity ϵ_1. The potential both inside and outside the sphere must satisfy Laplace's equation

$$\nabla^2 V_{\text{i}} = 0, \quad 0 \leq r \leq a$$

$$\nabla^2 V_{\text{e}} = 0, \quad a \leq r \leq \infty$$

Choose the centre of the sphere as the origin of a system of coordinates with the field pointing in the direction of the positive z-axis. The

* Numerical values of Legendre polynomials and functions may be obtained from *Tables of Associated Legendre Functions* (Columbia University Press, New York, 1945).

boundary conditions which must be satisfied are

$$V_e = -E_0 z = -E_0 r \cos\theta \qquad (r \to \infty)$$

$$\left. \begin{array}{c} \epsilon_2 \dfrac{\partial V_i}{\partial r} = \epsilon_1 \dfrac{\partial V_e}{\partial r} \\[3mm] V_i = V_e \end{array} \right\} \qquad (r = a)$$

For points outside the sphere it is evident that the potential must be of the form

$$V_e = -E_0 r \cos\theta + \sum_{n=0}^{\infty} B_n r^{-(n+1)} P_n(\cos\theta) \qquad (11\text{--}114)$$

For points inside the sphere no terms of the form r^{-n} can occur in the potential since V_i is finite for $r = 0$; thus $B_n = 0$ and

$$V_i = \sum_{n=0}^{\infty} A_n r^n P_n(\cos\theta) \qquad (11\text{--}115)$$

From the boundary condition $V_i = V_e$ at $r = a$ we have

$$-E_0 a \cos\theta + \sum_{n=0}^{\infty} B_n a^{-(n+1)} P_n(\cos\theta) = \sum_{n=0}^{\infty} A_n a^n P_n(\cos\theta)$$

or

$$-E_0 a \cos\theta + \frac{B_0}{a} + \frac{B_1}{a^2}\cos\theta + \frac{B_2}{a^3} P_2(\cos\theta) + \cdots$$

$$= A_0 + A_1 a \cos\theta + A_2 a^2 P_2(\cos\theta) + \cdots$$

We must therefore have that

$$\frac{B_0}{a} = A_0 \qquad (11\text{--}116)$$

$$\frac{B_n}{a^{n+1}} = A_n a^n \qquad (11\text{--}117)$$

$$-E_0 a + \frac{B_1}{a^2} = A_1 a \qquad (11\text{--}118)$$

It is clear that equations 11–116 and 11–117 can only be satisfied if $A_0 = B_0 = 0$, and $A_n = B_n = 0 \ (n \neq 1)$.

The continuity of the normal component of the flux density requires that

$$-\epsilon_1 E_0 - \frac{2\epsilon_1}{a^3} B_1 = \epsilon_2 A_1 \qquad (11\text{--}119)$$

Solving 11–118 and 11–119 for A_1 and B_1 we find that

$$A_1 = -\frac{3\epsilon_1}{\epsilon_2+2\epsilon_1}E_0 \tag{11–120}$$

$$B_1 = -\frac{\epsilon_2-\epsilon_1}{\epsilon_2+2\epsilon_1}a^3E_0 \tag{11–121}$$

The two potentials are thus given by

$$V_e = -E_0 r\cos\theta + \frac{\epsilon_2-\epsilon_1}{\epsilon_2+2\epsilon_1}\frac{a^3}{r^2}E_0\cos\theta \tag{11–122}$$

$$V_i = -\frac{3\epsilon_1}{\epsilon_2+2\epsilon_1}E_0 r\cos\theta \tag{11–123}$$

The corresponding values of the field intensities are

$$\mathbf{E}_e = E_0\mathbf{a}_z + E_0\frac{\epsilon_2-\epsilon_1}{\epsilon_2+2\epsilon_1}\cdot\frac{a^3}{r^3}(2\cos\theta\,\mathbf{a}_r+\sin\theta\,\mathbf{a}_\theta) \tag{11–124}$$

$$\mathbf{E}_i = \frac{3\epsilon_1}{\epsilon_2+2\epsilon_1}E_0\mathbf{a}_z \tag{11–125}$$

Equations 11–122 and 11–124 show that at points outside the sphere the potential or the field is the sum of the original potential or field and the potential or field produced by a point dipole of moment

$$p = 4\pi\epsilon_1\left(\frac{\epsilon_2-\epsilon_1}{\epsilon_2+2\epsilon_1}\right)a^3E_0 \tag{11–126}$$

placed at the centre of the sphere.

Equation 11–125 shows that the field inside the sphere is uniform and in the same direction as the original field. If $\epsilon_2 > \epsilon_1$ the field inside the sphere is smaller than E_0 whereas if $\epsilon_2 < \epsilon_1$ (e.g., a spherical cavity inside a dielectric) then E_i will be larger than E_0 (Fig. 11.21).

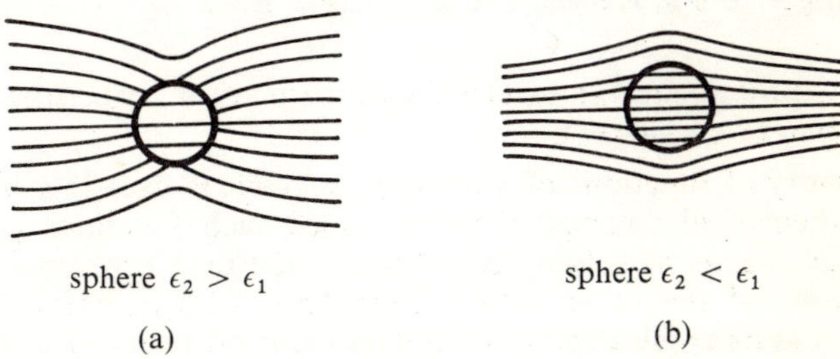

Fig. 11.21 Field inside and around a dielectric sphere placed in an originally uniform field

However, $D_i > D_0$ for $\epsilon_2 > \epsilon_1$ whilst $D_i < D_0$ for $\epsilon_2 < \epsilon_1$; the reason for this is that whereas lines of force may begin and end either at free or at bound charges, flux lines begin and end only at free charges.

The depolarizing factor for the sphere is found as follows.

$$\mathbf{E}_0 - \mathbf{E}_i = \frac{\epsilon_2 - \epsilon_1}{\epsilon_2 + 2\epsilon_1}\mathbf{E}_0$$

$$\mathbf{P} = (\epsilon_2 - \epsilon_0)\mathbf{E}_i = \frac{3\epsilon_1(\epsilon_2 - \epsilon_0)}{\epsilon_2 + 2\epsilon_1}\mathbf{E}_0$$

Hence

$$L = \frac{\epsilon_0(\mathbf{E}_0 - \mathbf{E}_i)}{\mathbf{P}} = \frac{1}{3}\frac{\epsilon_0(\epsilon_2 - \epsilon_1)}{\epsilon_1(\epsilon_2 - \epsilon_0)}$$

For $\epsilon_1 = \epsilon_0$,

$$L = \frac{1}{3}$$

If an *uncharged conducting sphere* is placed in an originally uniform electric field, the expression for the potential and field may be obtained by letting $\epsilon_2 \to \infty$ in equations 11–122/125. We find that

$$V_i = E_i = 0$$

$$V_e = -E_0 r \cos\theta + \frac{a^3}{r^2}E_0 \cos\theta \tag{11–127}$$

$$E_e = E_0\mathbf{a}_z + E_0\frac{a^3}{r^3}(2\cos\theta\,\mathbf{a}_r + \sin\theta\,\mathbf{a}_\theta) \tag{11–128}$$

The problem of a sphere of radius a and constant magnetic permeability placed in a uniform magnetic field of strength $B_0 = \mu_0 H_0$ is mathematically identical to that of a dielectric sphere in an electric field. The solution can be obtained directly from equations 11–122/125 by changing V to Φ_H, $\mathbf{E}$ to $\mathbf{H}$, ϵ to μ and $\mathbf{D}$ to $\mathbf{B}$.

11.7 TWO-DIMENSIONAL SOLUTION OF LAPLACE'S EQUATION BY THE COMPLEX VARIABLE

The theory of functions of a complex variable is usually a basic part of mathematical courses. However since such functions provide a powerful means of solving two-dimensional field problems, we give here some of the basic notions associated with complex functions theory and its application to boundary value problems. Those readers who desire to extend their knowledge beyond the contents of the present section should consult the references given at the end of this chapter.

When a real variable such as x takes on a sequence of values, it can be represented by a series of points along a straight line. When a complex variable $z = x + iy$ takes on a sequence of values, it can be represented by a series of points in a *complex plane*. This plane is the ordinary xy-plane used to plot complex numbers; its x-axis is the real axis and the y-axis is the imaginary axis. To each complex number there corresponds one and only one point in the plane, and inversely to each point in the plane there corresponds one and only one complex number.

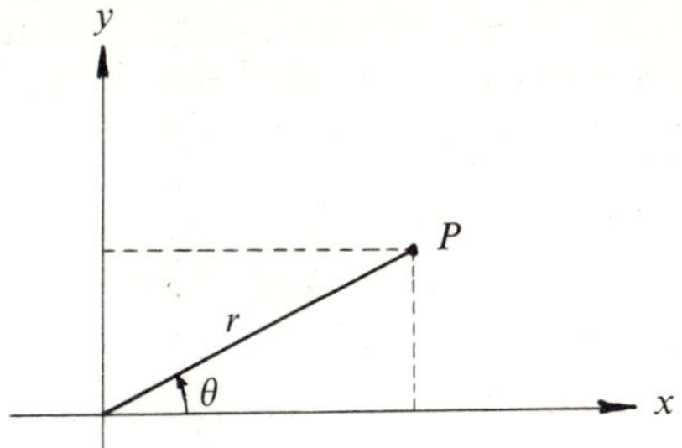

Fig. 11.22 Polar representation of a complex number

A complex number can also be written in the *polar form* (Fig. 11.22)

$$z = r(\cos\theta + i\sin\theta) = re^{i\theta} \qquad (11\text{--}129)$$

We can always add an integral multiple of 2π to the angle θ without changing the position of the point described by the polar coordinates. This multi-valuedness of the polar coordinates is eliminated by restricting the value of θ to the interval $-\pi < \theta \leq \pi$; this value of θ is called the *principal value*.

In general a function describes the pattern of points followed by one variable as another variable takes on a sequence of values. A function of a real variable x is given by $y = f(x)$, where y is a real variable, whereas a function of a complex variable z is given by $w = f(z)$ where w is a complex variable. Consider the simple function of z, $w = z^2$; we may write

$$w = z^2 = (x+iy)^2 = x^2 - y^2 - i2xy$$

$$= u(x, y) + iv(x, y)$$

where $u = x^2 - y^2$ and $v = 2xy$. In general any function of the complex variable z is equivalent to a pair of real functions $u(x, y)$ and $v(x, y)$ of the real variables x and y; thus

$$w = f(z) = u(x, y) + iv(x, y) \qquad (11\text{--}130)$$

u and v are called the *conjugate functions* of x and y. In physical applications the function $w = f(z)$ is called the *complex potential*.

A function $w = f(z)$ is said to be *analytic* in a given region if it is single-valued and has a unique derivative at every point in that region. The derivative of an analytic function is defined by

$$f'(z) = \frac{dw}{dz} = \lim_{\Delta z \to 0} \frac{\Delta w}{\Delta z} \qquad (11\text{--}131)$$

provided that the value of the limit is independent of the way in which Δz tends to zero. If at a given point z_0 a unique limit does not exist, then the function $f(z)$ is not differentiable at z_0 and hence is not analytic at z_0. Points such as z_0 are called singular points.

If the function $f(z)$ is to have a unique limit then

$$\frac{dw}{dz} = \lim_{\Delta x \to 0} \left(\frac{\Delta u}{\Delta x} + i\frac{\Delta v}{\Delta x} \right) = \frac{\partial u}{\partial x} + i\frac{\partial v}{\partial x}$$

$$= \lim_{i\Delta y \to 0} \left(\frac{\Delta u}{i\Delta y} + \frac{i\Delta v}{i\Delta y} \right) = \frac{\partial v}{\partial y} - i\frac{\partial u}{\partial y}$$

Equating the real and imaginary terms, we obtain the *Cauchy–Riemann* equations,

$$\frac{\partial u}{\partial x} = \frac{\partial v}{\partial y}; \quad \frac{\partial v}{\partial x} = -\frac{\partial u}{\partial y} \qquad (11\text{--}132)$$

The necessary and sufficient conditions for a function to be analytic is that u and v satisfy the Cauchy–Riemann equations.

If we take the derivative of equation 11–132 with respect to x and y in turn we obtain

$$\frac{\partial^2 u}{\partial x^2} = \frac{\partial^2 v}{\partial x \partial y}; \quad \frac{\partial^2 v}{\partial x^2} = -\frac{\partial^2 u}{\partial x \partial y}$$

$$\frac{\partial^2 u}{\partial x \partial y} = \frac{\partial^2 v}{\partial y^2}; \quad \frac{\partial^2 v}{\partial x \partial y} = -\frac{\partial^2 u}{\partial y^2}$$

from which it follows that

$$\frac{\partial^2 u}{\partial x^2} + \frac{\partial^2 u}{\partial y^2} = 0 \quad \text{and} \quad \frac{\partial^2 v}{\partial x^2} + \frac{\partial^2 v}{\partial y^2} = 0 \qquad (11\text{--}133)$$

Thus the real and imaginary parts of an analytic function satisfy Laplace's equation in two dimensions. Since $f(z)$ is not analytic at singular points, equation 11–133 fails to hold at such points; in physical applications singular points are those points at which there are sources or sinks.

Consider now the two families of curves:

$$u(x, y) = \text{constant}; \quad v(x, y) = \text{constant}$$

Along the curve $u = C_1$

$$\frac{\partial u}{\partial x}\,dx + \frac{\partial u}{\partial y}\,dy = 0$$

and the slope of the tangent to this curve at any point (x, y) is

$$\left(\frac{dy}{dx}\right)_u = -\frac{\partial u/\partial x}{\partial u/\partial y}$$

Likewise for the curve $v = C_2$ the slope of the tangent at (x, y) is

$$\left(\frac{dy}{dx}\right)_v = -\frac{\partial v/\partial x}{\partial v/\partial y}$$

For an analytic function the Cauchy–Riemann equations must be satisfied so that

$$\left(\frac{dy}{dx}\right)_v = \frac{\partial u/\partial y}{\partial u/\partial x} = -\frac{1}{(dy/dx)_u} \tag{11–134}$$

Since the slopes of the curves are negative reciprocals it follows that the curves $u(x, y) = C_1$ and $v(x, y) = C_2$ intersect orthogonally in the z-plane. Hence the curves $u = \text{constant}$, $v = \text{constant}$ represent two families of orthogonal curves in the z-plane, that is they intersect at right angles at all their points of intersection.

Since both u and v independently satisfy Laplace's equation, either one can be identified with the potential (electrostatic, electric or magneto-static) in a source free region. If u is chosen as the potential function, then the curves $u = \text{constant}$ represent the equipotentials. Since the curves $v = \text{constant}$ are orthogonal to these equipotentials they represent the flux on current-flow lines; the function v is called the *stream function*. The roles of u and v may of course be interchanged; the choice between the two possibilities depends on the boundary con-ditions of the problem to be solved. If u is chosen to represent the electrostatic potential, for example, then since $\mathbf{E} = -\nabla u$, we have, making use of the Cauchy–Riemann equations,

$$E_x = -\frac{\partial u}{\partial x} = -\frac{\partial v}{\partial y} \tag{11–135}$$

$$E_y = -\frac{\partial u}{\partial y} = \frac{\partial v}{\partial x} \tag{11–136}$$

so that

$$\mathbf{E} = E_x + iE_y = -\left(\frac{\partial u}{\partial x} - i\frac{\partial v}{\partial x}\right) = -\left(\frac{\partial v}{\partial y} + i\frac{\partial v}{\partial x}\right)$$

$$= -\left(\frac{\partial w^*}{\partial z}\right) \qquad (11\text{--}137)$$

where $w^* = u - iv$ is the complex conjugate of w. If v were chosen to represent the potential, it can readily be shown that

$$\mathbf{E} = -i\left(\frac{\partial w^*}{\partial z}\right) \qquad (11\text{--}138)$$

In either case it is evident that the magnitude of the field intensity is given by

$$E = \left|\frac{dw}{dz}\right| \qquad (11\text{--}139)$$

Suppose that we wish to find the flux passing two points 1 and 2 on an equipotential (Fig. 11.23). Since the problem is two-dimensional we

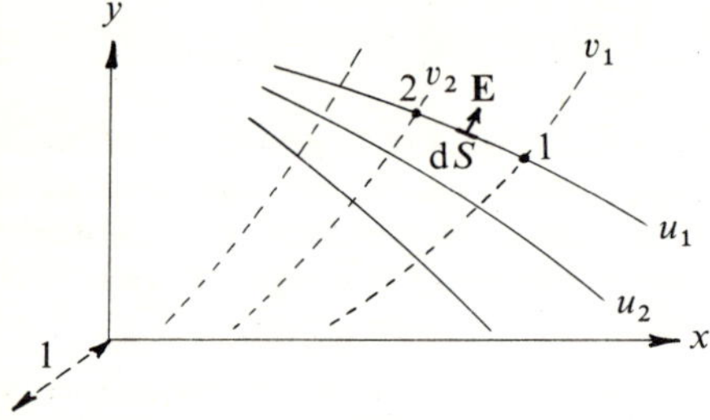

Fig. 11.23 To determine the flux passing between points 1 and 2 on the equipotential u_1

shall assume a unit length of one metre measured in a direction perpendicular to the paper. If dl is an element of length along an equipotential then the flux crossing this element is

$$d\psi = |D|\,|dl| = \epsilon|E|\,|dl|$$

since D or E is everywhere normal to the equipotential surface. We may write $|E|\,|dl|$ as the cross product† of E and dl,

$$E \times dl = |E|\,|dl|\sin 90° = |E|\,|dl|$$

$$= E_x\,dy - E_y\,dx \qquad (11\text{--}140)$$

† If $z_1 = x_1 + iy_1$ and $z_2 = x_2 + iy_2$ the cross product of z_1 and z_2 is defined by

$$z_1 \times z_2 = |z_1||z_2|\sin\theta = x_1y_2 - x_2y_1$$

Thus the flux passing between the points 1 and 2 in the z-plane is given by

$$\psi = \epsilon \int_1^2 [E_x \, \mathrm{d}y - E_y \, \mathrm{d}x]$$

and substituting for E_x and E_y from equations 11–135 and 11–136 gives

$$\psi = -\epsilon \int_1^2 \left(\frac{\partial v}{\partial y} \mathrm{d}y + \frac{\partial v}{\partial x} \mathrm{d}x \right) = -\epsilon \int_1^2 \mathrm{d}v$$

$$= -\epsilon(v_1 - v_2) \qquad (11\text{–}141)$$

Thus to find the electric flux per unit length passing between any two points we need to evaluate v at these points, take the difference between these numerical values and multiply it by the permittivity of the medium.

11.7.1 CONFORMAL TRANSFORMATIONS

A function of a real variable such as $y = f(x) = \sin x$ can be represented graphically as a curve in the xy-plane. To represent an analytic function $w = f(z) = u(x, y) + iv(x, y)$ graphically, we need a four-dimensional space since four variables (x, y, u, v) are involved. Instead, we make use of two complex planes one for the independent variable $z = x + iy$ and the other for the dependent variable $w = u + iv$. Corresponding to any given point $P(x, y)$ in the z-plane there is a particular point $P'(u, v)$ in the w-plane (or vice versa). The pair of points P and P' are called images of each other. Thus a set of points (such as a curve or a region) in the z-plane has a corresponding set of image points (curve or region) in the w-plane (Fig. 11.24). We say that

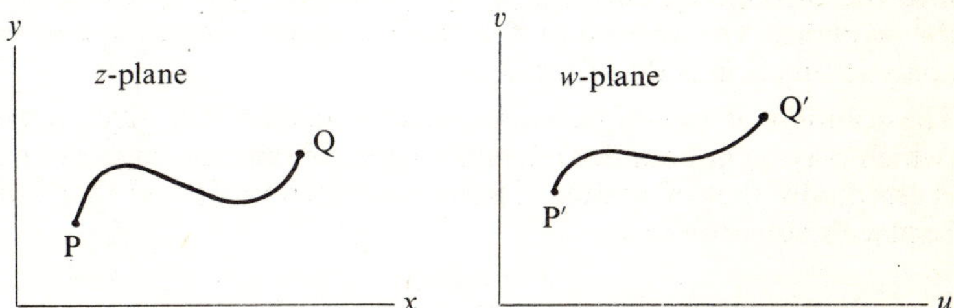

Fig. 11.24 Mapping of curve PQ in z-plane into P'Q' in w-plane

a point, curve or region in the z-plane is *mapped* or *transformed* into the corresponding point, curve, or region in the w-plane. The function $w = f(z)$ is called a mapping or transformation function.

If the mapping function $f(z)$ is analytic (and here we are only interested in analytic functions) and $f'(z) \neq 0$, the mapping or transformation is said to be *conformal* (same form or shape). Points where $f'(z) = 0$ are called critical points and at these points mapping by the function $w = f(z)$ ceases to be conformal. A conformal transformation has the following properties:

(a) If curves C_1 and C_2 which intersect at a point (x_0, y_0) in the z-plane are mapped into curves C_1' and C_2' respectively which intersect at the point (u_0, v_0) in the w-plane, then the angle at (x_0, y_0) between C_1 and C_2 is equal to the angle at (u_0, v_0) between C_1' and C_2', both in magnitude and sense (Fig. 11.25).

(b) Infinitesimal distances drawn from a point z_0 in the z-plane are magnified by a factor $|f'(z_0)|$ when transformed to the image point w_0 in the w-plane. Infinitesimal areas are thus magnified by the factor $|f'(z_0)|^2$. It follows that infinitesimal figures in the z-plane map into similar infinitesimal figures in the w-plane. However, since in general the derivative of a complex function changes from point to point, large figures in the z-plane usually map into figures in the w-plane which are far from similar.

We have already shown that if

$$w = f(z) = f(x + iy) = u(x, y) + iv(x, y)$$

is analytic, then $u(x, y) = \text{constant}$ and $v(x, y) = \text{constant}$ represent two families of orthogonal curves in the z-plane. The transformation $w = f(z)$ maps these curves into the orthogonal lines $u = \text{constant}$, $v = \text{constant}$ in the w-plane (Fig. 11.26). If the inverses $x(u, v)$ and $y(u, v)$ of the two functions $u(x, y)$ and $v(x, y)$ exist, then there exists a unique function $z = g(w) = g(u + iv) = x(u, v) + iy(u, v)$ that will conformally map the lines $x = \text{constant}$, $y = \text{constant}$ in the z-plane into the orthogonal curves $x(u, v) = \text{constant}$, $y(u, v) = \text{constant}$ in the w-plane. The condition for the existence of the inverse of an analytic function is that $f'(z) \neq 0$.

The solution of two-dimensional field problems in a given region R (which may be unbounded) that does not contain any sources or sinks is essentially that of finding a potential function $\Phi(x, y)$ that satisfies Laplace's equation in R,

$$\frac{\partial^2 \Phi}{\partial x^2} + \frac{\partial^2 \Phi}{\partial y^2} = 0$$

and also satisfies prescribed conditions on the given boundaries enclosing the region R (the value of Φ or its normal derivative specified at all points of the boundaries).

It can be proved* that if a function $\Phi(x, y)$ satisfies Laplace's equation

* See C. R. Wylie, *Advanced Engineering Mathematics* (McGraw-Hill, New York, 1966).

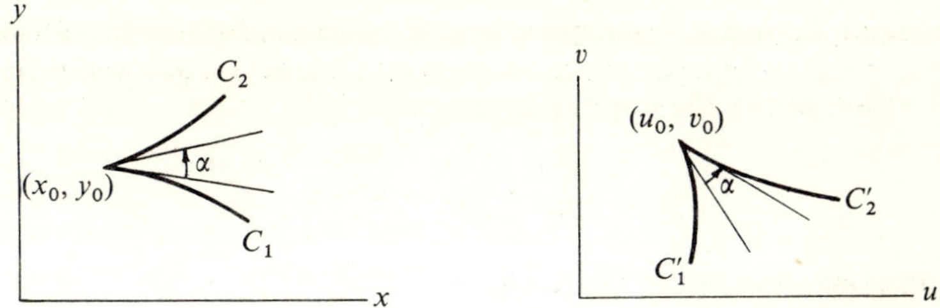

Fig. 11.25 Conformal mapping conserves the angle α both in magnitude and sense

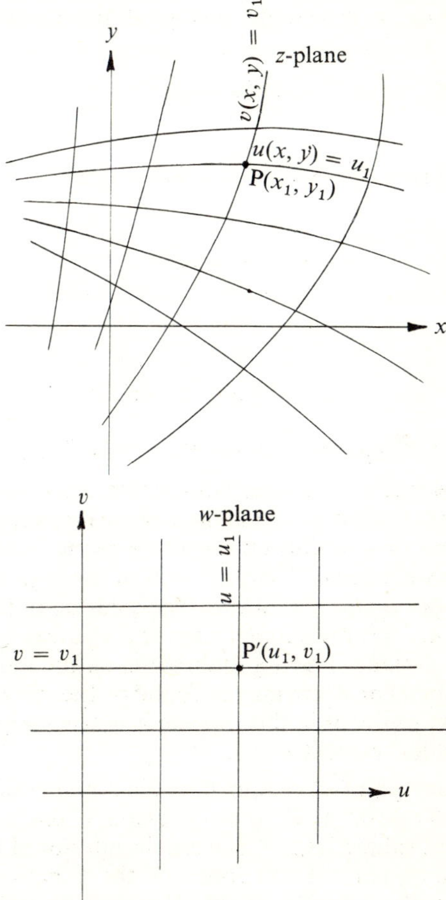

Fig. 11.26 The orthogonal curves $u = $ constant, $v = $ constant in the z-plane, are mapped into orthogonal lines in the w-plane

in a region R, then if R is mapped into R' by the conformal transformation $w = f(z)$ which transforms $\Phi(x, y)$ into a function of u and v, then $\Phi(u, v)$ will satisfy the Laplace equation

$$\frac{\partial^2 \Phi}{\partial u^2} + \frac{\partial^2 \Phi}{\partial v^2} = 0$$

As an example consider the function

$$\Phi(x, y) = \sin x \cosh y$$

which is harmonic in any finite region of the z-plane. Under the transformation

$$z = w^3$$

we have that

$$x + iy = (u + iv)^3 = u^3 - 3uv + i(3u^2v - v^3)$$

so that

$$x = u^3 - 3uv$$

$$y = 3u^2v - v^3$$

Hence,

$$\Phi(u, v) = \sin(u^3 - 3uv^2)\cosh(3u^2v - v^3)$$

straightforward differentiation shows that $\nabla^2 \Phi(u, v) = 0$.

If it is possible to find a conformal transformation that will transform the given boundaries of a region R in the z-plane into simple rectangular lines $u = $ constant or $v = $ constant in the w-plane, it is then possible to solve Laplace's equation $\nabla^2 \Phi(u, v) = 0$ in the region R' subject to the transformed boundary conditions. The solution of this equation in the w-plane is considerably simpler than the solution of the equation $\nabla^2 \Phi(x, y)$ in the z-plane. Having found $\Phi(u, v)$, the required solution $\Phi(x, y)$ of the original problem may be found by the inverse transformation. To solve the problem in the w-plane it is necessary to know how the boundary values transform.

If z_0 is a boundary point of region R in the z-plane, then $w_0 = f(z_0)$ will be a boundary point of R' in the w-plane. If the function $\Phi(x, y)$ tends to a constant value k as $z \to z_0$ from the interior of R, the function $\Phi(u, v) = \Phi[x(u, v), y(u, v)]$ will approach the same constant value k as $w \to w_0$ from the interior of R'. Hence a boundary condition $\Phi(x, y) = $ constant in the z-plane remains invariant under a conformal transformation $w = f(z)$. On the other hand boundary values of the normal derivatives transform according to the relationship

$$\frac{d\Phi(x,\,y)}{dn_z} = |f'(z_0)|\frac{d\Phi(u,\,v)}{dn_w} \qquad (11\text{–}142)$$

where $d\Phi/dn_z$ and $d\Phi/dn_w$ are the normal derivatives at the points z_0 and w_0 respectively.

The major problem involved in the use of conformal mapping to solve field problems is that of finding the appropriate transformation that can be fitted to the boundary conditions. There is no standard method of finding the appropriate transformation function but over the years an extensive number of transformations have been investigated; an excellent source of transformations that can be adopted to solve practical problems may be found in Kober's *Dictionary of Conformal Representations* (Dover, New York, 1952). In the present brief treatment of the subject we will confine ourselves to a few examples of some simple transformations to illustrate the scope of the method.

Example 1. The transformation $w = k\log z$, where k is a constant. Using polar coordinates, we may write

$$u+iv = k\log(re^{i\theta}) = k\log r + ik\theta \quad (-\pi < \theta \leq \pi)$$

$$u = k\log r \qquad (11\text{–}143)$$

$$v = k\theta \qquad (11\text{–}144)$$

The $u = $ constant curves in the z-plane are a family of concentric circles with centre at the origin and radii $e^{u/k}$. The $v = $ constant curves are straight lines passing through the origin (Fig. 11.27).

Suppose now that we have two infinitely long concentric cylindrical conductors of radii r_1 and r_2 which are kept at potentials V_1 and V_2 respectively. It is required to find the potential at any point between these two cylinders. The above transformation maps the circles of radii r_1 and r_2 in the z-plane into the infinite parallel lines u_1 and u_2 in the w-plane. The problem is thus reduced to that of finding the potential at a point between the line u_1 at potential V_1 and the line u_2 at potential V_2; the solution in this case is elementary and is given by

$$\Phi(u) = V(u) = V_1 - \frac{V_1-V_2}{u_2-u_1}(u-u_1)$$

transforming this solution to the z-plane we obtain

$$V(r) = V_1 - \frac{V_1-V_2}{\log(r_2/r_1)}\log(r/r_1)$$

The electric field intensity, found by taking the negative gradient of

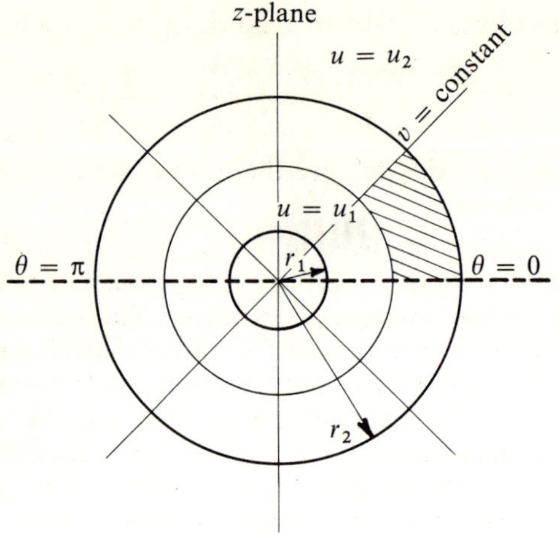

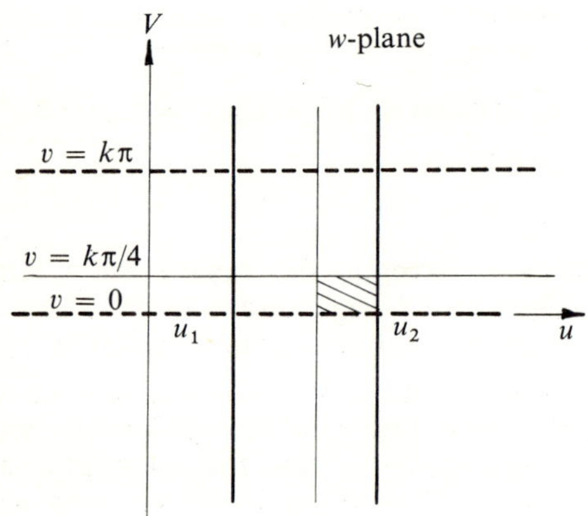

Fig. 11.27 The transformation $w = k \log z$

the above potential, is

$$E(r) = \frac{V_1 - V_2}{r \log(r_2/r_1)}$$

We also know from equation 11–139 that

$$E = \left| \frac{dw}{dz} \right| = \frac{k}{r}$$

so that

$$k = \frac{V_1 - V_2}{\log(r_2/r_1)}$$

The total electric flux leaving the cylinder, that is the flux between the lines $\theta = 0$ and $\theta = 2\pi$, may be found from equation 11–141 and is given by

$$\psi = -\epsilon(0 - 2\pi k) = 2\pi\epsilon k$$

Since by Gauss's theorem this is equal to the total charge on the inner cylinder, we have that

$$k = Q_1/2\pi\epsilon$$

Suppose that, instead of the coaxial cylinders, we have two semi-infinite conducting sheets one extending from $x = 0$ to $x = +\infty$ and kept at a potential V_0 and the other extending from $x = 0$ to $x = -\infty$ and kept at a potential $-V_0$. The above transformation can again be used to find the potential and field at any point in the z-plane. Consider the upper half of the z-plane bounded by the x-axis; this is transformed into the region between the lines $v = 0$ and $v = k\pi$ in the w-plane. It is evident that the potential at any point in this region is given by

$$V(v) = V_0 - \frac{V_0 - (-V_0)}{k\pi - 0}(v - 0)$$

$$= V_0 - \frac{2V_0}{k\pi}v$$

and

$$V(\theta) = V_0\left(1 - \frac{2}{\pi}\tan^{-1}\frac{y}{x}\right)$$

The potential in the lower half of the z-plane is obtained by symmetry. The equipotentials are thus the straight lines $\theta = $ constant passing through the origin. The flux lines are circles with their centres at the origin.

Example 2. The transformation $w = c\sin^{-1}z$, where c is a constant.

$$z = k\sin w = k\sin(u + iv)$$

$$x + iy = k(\sin u \cosh v + i\cos u \sinh v)$$

$$x = k\sin u \cosh v \tag{11–145}$$

$$y = k\cos u \sinh v \tag{11–146}$$

To find the curves of constant u or v in the z-plane we eliminate u and v in turn from the above equations; thus

$$\frac{x^2}{k^2 \cosh^2 v} + \frac{y^2}{k^2 \sinh^2 v} = 1$$

$$\frac{x^2}{k^2 \sin^2 u} - \frac{y^2}{k^2 \cos^2 u} = 1$$

The curves of constant v are a family of confocal ellipses with foci at $\pm k$, whilst curves of constant u are a family of confocal hyperbolas with foci at $\pm k$; the two families of curves form an orthogonal set (Fig. 11.28).

If $v = 0$ we see from equation 11–146 that $y = 0$ for all values of u. Thus $x = k \sin u$ and

$$x = 0, \qquad u = 0$$

$$x = +k, \qquad u = \pi/2$$

$$x = -k, \qquad u = -\pi/2$$

Thus for $v = 0$ the ellipse degenerates into the straight line between $x = k$ and $x = -k$.

If $u = 0$ we see from equation 11–145 that $x = 0$ for all values of v and the hyperbolas degenerate into the y-axis. If $u = \pm\pi/2$ then $x = \pm k \cosh v$ and $y = 0$ for all v, so that the hyperbolas now degenerate into the two semi-infinite lines extending from $+k$ to ∞ and from $-k$ to $-\infty$.

The shaded region (Fig. 11.28) bounded by the curves $u = a_1, u = a_2, v = b_1, v = b_2$ in the z-plane corresponds to the region bounded by the corresponding lines in the w-plane. The region between two ellipses for which $y > 0$ in the z-plane corresponds to the region bounded by the lines $v = b_1, v = b_2$ and $u = \pm\pi/2$ in the w-plane.

The above transformation can be used to solve the following problems:

(a) potential and field due to a charged elliptic cylinder,

(b) potential and field between two elliptic cylinders at different potentials,

(c) potential and field due to a charged flat plate extending from $z = k$ to $z = -k$,

(d) potential and field due to two semi-infinite coplanar sheets at a distance $2k$ apart and whose potentials differ by π.

A special class of problems consisting of figures with rectilinear boundaries may be solved by the Schwartz–Christoffel transformation.

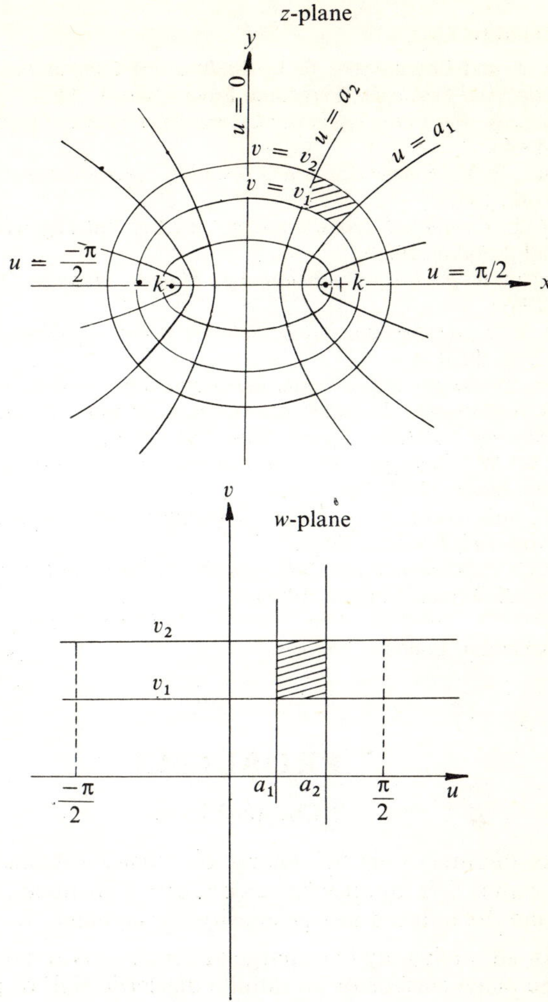

Fig. 11.28 The transformation $w = c \sin^{-1} z$

This transformation will map the interior of a polygon in the z-plane into the upper half of the w-plane in such a way that the sides of the polygon are transformed into the real axis of the w-plane. A discussion of this transformation lies beyond the scope of this book but the reader will find an excellent discussion of this transformation in *The Schwartz–Christoffel Transformation and its Applications*, by M. Walker (Dover, New York, 1964).

11.8 REFERENCES

BINNS, K. J. and LAWRENSON, P. J., *Analysis and Computation of Electric and Magnetic Field Problems*, (Pergamon Press, Oxford, 1963).

CHURCHILL, R. V., *Complex Variables and Applications*, (McGraw-Hill, New York, 1960).

CHURCHILL, R. V., *Fourier Series and Boundary-value Problems*, (McGraw-Hill, New York, 1941).

GIBBS, W. J., *Conformal Transformations in Electrical Engineering*, (Chapman and Hall, London, 1958).

HARRINGTON, R. F., *Field Computation by Moment Methods*, (Macmillan, New York, 1968).

KARPLUS, W. J., *Analog Simulation: Solution of Field Problems*, (McGraw-Hill, New York, 1958).

KOBER, H., *Dictionary of Conformal Representations*, (Dover, New York, 1952).

ROTHE, R., OLLENDORF, F. and POHLHAUSEN, K., *Theory of Functions as Applied to Engineering Problems*, (Dover, New York, 1964).

SOROKA, W. W., *Analog Methods in Computation and Simulation*, (McGraw-Hill, New York, 1954).

THOM, A., and APELT, C. J., *Field Computations in Engineering and Physics*, (Van Nostrand, London, 1961).

TOZZONI, O. V., *Mathematical Models for the Evaluation of Electric and Magnetic Fields*, (Iliffe Books, London, 1968).

VITKOVITCH, D., *Field Analysis: Experimental and Computational Methods*, (Van Nostrand, London, 1966).

PROBLEMS

Chapter Eleven

11.1   An infinitely long line charge of density λ is placed in air at a distance h from an infinite conducting plane at zero potential. Find the induced charge density on the plane.

11.2   An infinitely long line charge of density λ is at a distance h from the plane surface of an infinite dielectric slab of permittivity ϵ. Show that the field at any point on the same side of the dielectric as λ may be obtained by replacing the dielectric by an image line charge $-\lambda(\epsilon_r - 1)/(\epsilon_r + 1)$ at a distance $2h$ vertically below λ, and the field at any point in the dielectric is obtained by replacing both λ and the dielectric by a single line charge of density $2\lambda/(\epsilon_r + 1)$ at the position occupied by λ. Hence find the force per unit length acting on the line charge.

11.3   Show that for a line charge of density λ placed outside an earthed conducting cylinder the charge density on the surface of the cylinder is given by equation 11–47.

11.4   Show that for a line charge of density λ placed outside an uncharged and insulated conducting cylinder the charge density on the surface of the cylinder is given by equation 11–50.

11.5   An electron is at a distance x in vacuum from an infinite conducting plane at zero potential. Plot the potential energy of the electron as a function of x. If there is a uniform electric field in the region outside the plane and terminating at the plane, plot the energy of the electron as a function of distance and find the distance from the plane at which this energy is a minimum.

11.6   Repeat Problem 11.5 for the case in which the region outside the the plane is a dielectric of permittivity ϵ.

11.7   Two long thin parallel wires with charge densities $\pm\lambda$ are at a distance x apart. They are symmetrically enclosed by a thin hollow earthed conducting cylinder of radius R. Show that the electric force between the wires is zero if $x = 0\cdot98R$.

11.8   A point charge Q is placed at a distance b from the centre of an uncharged sphere of radius a $(b > a)$. Show that the part of the sphere which is positively charged is separated from that which is negatively charged by a circle of points at a distance r from Q given by

$$r^3 = b(b-a)(b+a)$$

11.9   A line charge of length $2l$ and density λ is placed such that its midpoint is at a distance b from the centre of an earthed conducting sphere of radius a. Show that the charge induced on the sphere is given by

$$Q = -2\lambda a \log\{[l+(l^2+b^2)^{1/2}]/b\}$$

11.10 An infinite grounded conducting plane has a hemispherical boss of radius a on its surface. If a point charge Q is placed at a distance h from the centre of the hemisphere along a line perpendicular to the plane, find the system of image charges which will give the potential at any point outside the plane. Hence find the force on the point charge and the charge induced on the boss.

11.11 A magnetic dipole of moment m is placed at a distance h from an infinite slab of relative permeability μ_r. If the dipole is parallel to the plane face of the slab, show that the force acting on the dipole is

$$F = \frac{\mu_r-1}{\mu_r+1} \frac{\mu_0 3m^2}{64\pi\mu_0 h^4}$$

11.12 A long hollow cylinder of relative permeability μ_r and radii a and

b $(b > a)$ is placed with its axis perpendicular to an initially uniform magnetic field. Show that inside the cylinder the original field is reduced by the ratio $4\mu_r/[4\mu_r-(\mu_r-1)^2(a^2/b^2-1)]$.

11.13 A hollow dielectric sphere of radii a and b $(b > a)$ and relative permittivity ϵ_r is placed in an originally uniform electric field. Show that inside the sphere the original field is reduced by the ratio $9\epsilon_r/[9\epsilon_r-2(\epsilon_r-1)^2(a^3/b^3-1)]$.

11.14 Two point charges are at a distance r apart in a dielectric medium of infinite extent and relative permittivity ϵ_r. If each charge is now surrounded by a very small spherical cavity, show that the force acting on each charge is increased by the ratio $3\epsilon_r/(2\epsilon_r+1)$.

11.15 An infinite plane boundary separates two media of permeabilities μ_1 and μ_2. In medium 1 and at a distance h from the boundary there is a long wire which carries a current I. Show that the wire experiences a force per unit length given by

$$F = \frac{\mu_1(\mu_2-\mu_1)I^2}{4\pi a(\mu_2+\mu_1)}$$

11.16 A spherical air bubble of radius a is surrounded by a liquid of infinite extent and relative permeability ϵ_r. Show that if a point charge Q is placed at a distance $b \gg a$ from the centre of the bubble, the bubble will experience a force

$$F = \frac{(\epsilon_r-1)Q^2a^3}{2\pi\epsilon_0\epsilon_r(2\epsilon_r+1)b^5}$$

along the line joining the charge and the bubble centre.

11.17 Show that the capacitance per unit length between a flat strip of length $2k$ and an elliptic cylinder with major axis of length $2a$ and whose foci coincide with the ends of the strip is given by

$$C = 2\pi\epsilon/\cosh^{-1}(a/k)$$

BIBLIOGRAPHY

ABRAHAM, M., and BECKER, R., *The Classical Theory of Electricity and Magnetism* (Blackie, London, 1950).

ATTWOOD, S. S., *Electric and Magnetic Fields* (Wiley, New York, 1958).

BEWLEY, L. V., *Flux Linkages and Electromagnetic Induction* (Dover, New York, 1964).

BEWLEY, L. V., *Two-Dimensional Fields in Electrical Engineering* (Dover, New York, 1963).

BLEANEY, B. I., and BLEANEY, B., *Electricity and Magnetism* (Oxford University Press, London, 1965).

BOAST, W. B., *Vector Fields* (Harper and Row, New York, 1964).

BOHN, E. V., *Introduction to Electromagnetic Fields and Waves* (Addison-Wesley, Reading, Mass., 1968).

BOOKER, H. G., *An Approach to Electrical Science* (McGraw-Hill, New York, 1959).

BRADSHAW, M. D. and BYATT, W. J., *Introductory Engineering Field Theory* (Prentice-Hall, Englewood Cliffs, N.J., 1967).

CARTER, G. W., *The Electromagnetic Field in its Engineering Aspects* (Longmans, Green, London, 1967).

CHESTON, G. W., *Elementary Theory of Electric and Magnetic Fields* (Wiley, New York, 1964).

CORSON, D. R., and LORRAIN, P., *Introduction to Electromagnetic Fields and Waves* (Freeman, San Francisco, 1962).

COULSON, C. A., *Electricity* (Oliver and Boyd, Edinburgh, 1958).

CULLWICK, E. G., *The Fundamentals of Electromagnetism* (Cambridge University Press, London, 1966).

FANO, R. M., CHU, L. J., and ADLER, R. B., *Electromagnetic Fields, Energy and Forces* (Wiley, New York, 1960).

FERRARO, V. C. A., *Electromagnetic Theory* (Athlone Press, London, 1967).

HAGUE, B., *The Principles of Electromagnetism Applied to Electrical Machines* (Dover, New York, 1962).

HALLÉN, E., *Electromagnetic Theory* (Wiley, New York, 1962).

HARNWELL, G. P., *Principles of Electricity and Electromagnetism* (McGraw-Hill, New York, 1949).

HARRINGTON, R. F., *Introduction to Electromagnetic Engineering*, (McGraw-Hill, New York, 1958).

HAYT, W. H. JR., *Engineering Electromagnetics* (McGraw-Hill, New York, 1967).

JAVID, M., and BROWN, P. M., *Field Analysis and Electromagnetics* (McGraw-Hill, New York, 1963).

KRAUS, J. D., *Electromagnetics* (McGraw-Hill, New York, 1953).

MASON, M., and WEAVER, W., *The Electromagnetic Field* (Dover, New York, 1929).

MAXWELL, J. C., *A Treatise on Electricity and Magnetism*, Vols. 1 and 2 (Dover, New York, 1954).

MOULLIN, E. B., *The Principles of Electromagnetism* (Oxford University Press, London, 1950).

PAGE, L., and ADAMS, N. I. JR., *Principles of Electricity* (Van Nostrand, Princeton, N.J., 1958).

PLONSEY, R., and COLLIN, R. E., *Principles and Applications of Electromagnetic Fields* (McGraw-Hill, New York, 1961).

PUGH, E. M., and PUGH, E. W., *Principles of Electricity and Magnetism* (Addison-Wesley, Reading, Mass., 1960).

PURCELL, E. M., *Electricity and Magnetism* (McGraw-Hill, New York, 1965).

REITZ, J. R., and MILFORD, F. J., *Foundations of Electromagnetic Theory* (Addison-Wesley, Reading, Mass., 1967).

ROGERS, W. E., *Introduction to Electric Fields* (McGraw-Hill, New York, 1954).

SCHELKUNOFF, S. A., *Electromagnetic Fields* (Blaisdell, New York, 1963).

SCHWARTZ, W. M., *Intermediate Electromagnetic Theory* (Wiley, New York, 1964).

SCOTT, W. T., *The Physics of Electricity and Magnetism,* (Wiley, New York, 1966).

SEELEY, S., *Introduction to Electromagnetic Fields* (McGraw-Hill, New York, 1958).

SILVESTER, P., *Modern Electromagnetic Fields* (Prentice-Hall, Englewood Cliffs, N.J., 1968).

WALSH, J. B., *Electromagnetic Theory and Engineering Applications* (The Ronald Press, New York, 1960).

WEBER, E., *Electromagnetic Fields, Theory and Applications*, Vol. 1—*Mapping of Fields* (Wiley, New York, 1957).

WOODSON, H. H., and MELCHER, J. R., *Electromechanical Dynamics*, Parts I and II (Wiley, New York, 1968).

Advanced Level

BIRSS, R. R., *Electric and Magnetic Forces* (Elsevier, New York, 1968).

BUCHHOLZ, H., *Elektrische und Magnetische Potentialfelder* (Springer-Verlag, Berlin, 1957).

DURAND, E., *Electrostatique*, Vol. I—*Les Distributions* (Masson, Paris, 1964).

DURAND, E., *Electrostatique*, Vol. II—*Problèmes Généraux, Conducteurs* (Masson, Paris, 1966).

DURAND, E., *Electrostatique*, Vol. III—*Méthodes de Calcul, Diélectriques* (Masson, Paris, 1966).

DURAND, E., *Magnétostatique* (Masson, Paris, 1968).

ELLIOT, R. S., *Electromagnetics* (McGraw-Hill, New York, 1966).

JACKSON, J. D., *Classical Electrodynamics* (Wiley, New York, 1962).

JEANS, J., *The Mathematical Theory of Electricity and Magnetism* (Cambridge University Press, London, 1951).

JONES, D. S., *The Theory of Electromagnetism* (Pergamon Press, New York, 1964).

KING, R. W. P., *Fundamental Electromagnetic Theory* (Dover, New York, 1963).

LANDAU, L., and LIFSHITZ, E., *The Classical Theory of Fields*, Addison-Wesley, Reading, Mass., 1951).

MOON, P., and SPENCER, D. E., *Field Theory for Engineers* (Van Nostrand, Princeton, N.J., 1961).

PANOFSKY, W. K. H., and PHILLIPS, *Classical Electricity and Magnetism* (Addison-Wesley, Reading, Mass., 1962).

ROSSER, W. G. V., *Classical Electromagnetism Via Relativity* (Butterworths, London, 1968).

SMYTHE, W. R., *Static and Dynamic Electricity* (McGraw-Hill, New York, 1968).

SOMMERFELD, A., *Electrodynamics* (Academic Press, New York, 1952).

STRATTON, J. A., *Electromagnetic Theory* (McGraw-Hill, New York, 1941).

VAN BLADEL, J., *Electromagnetic Fields* (McGraw-Hill, New York, 1964).

WEEKS, W. L., *Electromagnetic Theory for Engineering Applications* (Wiley, New York, 1964).

APPENDIX I

VECTOR OPERATIONS

1   VECTOR IDENTITIES

$$\mathbf{A} \cdot \mathbf{B} = AB\cos\theta \quad (\theta \text{ is the angle between } \mathbf{A} \text{ and } \mathbf{B})$$

$$\mathbf{A} \cdot \mathbf{B} = \mathbf{B} \cdot \mathbf{A}$$

$$\mathbf{A} \cdot \mathbf{A} = \mathbf{A}^2 = A^2$$

$$\mathbf{A} \cdot (\mathbf{B}+\mathbf{C}) = \mathbf{A} \cdot \mathbf{B} + \mathbf{A} \cdot \mathbf{C}$$

$$\mathbf{A} \times \mathbf{B} = AB\sin\theta\,\mathbf{a_n} \quad (\mathbf{a_n} \text{ is unit vector normal to plane of}$$
$$\mathbf{A} \text{ and } \mathbf{B}, \text{ direction given by right-hand screw rule)}$$

$$\mathbf{B} \times \mathbf{A} = -\mathbf{A} \times \mathbf{B}$$

$$A \times (\mathbf{B}+\mathbf{C}) = \mathbf{A} \times \mathbf{B} + \mathbf{A} \times \mathbf{C}$$

$$\mathbf{A} \cdot (\mathbf{B} \times \mathbf{C}) = \mathbf{C} \cdot (\mathbf{A} \times \mathbf{B}) = \mathbf{B} \cdot (\mathbf{C} \times \mathbf{A})$$

$$\mathbf{A} \times (\mathbf{B} \times \mathbf{C}) = (\mathbf{A} \cdot \mathbf{C})\mathbf{B} - (\mathbf{A} \cdot \mathbf{B})\mathbf{C}$$

$$(\mathbf{A} \times \mathbf{B}) \times \mathbf{C} = (\mathbf{A} \cdot \mathbf{C})\mathbf{B} - (\mathbf{B} \cdot \mathbf{C})\mathbf{A}$$

$$\mathbf{A} \times \mathbf{B} \cdot \mathbf{C} \times \mathbf{D} = (\mathbf{A} \cdot \mathbf{C})(\mathbf{B} \cdot \mathbf{D}) - (\mathbf{A} \cdot \mathbf{D})(\mathbf{B} \cdot \mathbf{C})$$

$$(\mathbf{A} \times \mathbf{B})^2 = \mathbf{A}^2\mathbf{B}^2 - (A \cdot B)^2$$

$$(\mathbf{A} \times \mathbf{B}) \times (\mathbf{C} \times \mathbf{D}) = (\mathbf{A} \cdot \mathbf{B} \times \mathbf{D})\mathbf{C} - (\mathbf{A} \cdot \mathbf{B} \times \mathbf{C})\mathbf{D}$$

2   VECTOR IDENTITIES INVOLVING ∇

(a) Identities independent of coordinate system

$$\nabla \cdot (\phi\mathbf{A}) = \phi\nabla \cdot \mathbf{A} + \mathbf{A} \cdot \nabla\phi$$

$$\nabla \times (\phi\mathbf{A}) = \phi\nabla \times \mathbf{A} + \nabla\phi \times \mathbf{A}$$

$$\nabla \cdot (\mathbf{A} \times \mathbf{B}) = \mathbf{B} \cdot \nabla \times \mathbf{A} - \mathbf{A} \cdot \nabla \times \mathbf{B}$$

$$\nabla \cdot \mathbf{r} = 3 \quad (r = \text{radius vector from origin of coordinates})$$

$$\nabla \times \mathbf{r} = 0$$

(b) Identities valid for rectangular coordinates

$$\nabla \times (\mathbf{A} \times \mathbf{B}) = (A\nabla \cdot - \mathbf{A} \cdot \nabla)\mathbf{B} - (B\nabla \cdot - \mathbf{B} \cdot \nabla)\mathbf{A}$$

$$\nabla(\mathbf{A} \cdot \mathbf{B}) = \mathbf{A} \cdot \nabla\mathbf{B} + \mathbf{B} \cdot \nabla\mathbf{A} + \mathbf{A} \times (\nabla \times \mathbf{B}) + \mathbf{B} \times (\nabla \times \mathbf{A})$$

$$\mathbf{A} \cdot \nabla \mathbf{B} = \left(A_x \frac{\partial B_x}{\partial x} + A_y \frac{\partial B_x}{\partial y} + A_z \frac{\partial B_x}{\partial z} \right) \mathbf{a}_x$$

$$+ \left(A_x \frac{\partial B_y}{\partial x} + A_y \frac{\partial B_y}{\partial y} + A_z \frac{\partial B_y}{\partial z} \right) \mathbf{a}_y$$

$$+ \left(A_x \frac{\partial B_z}{\partial x} + A_y \frac{\partial B_z}{\partial y} + A_z \frac{\partial B_z}{\partial z} \right) \mathbf{a}_z$$

$$\nabla \times (\nabla \times \mathbf{A}) = \nabla(\nabla \cdot \mathbf{A}) - \nabla^2 \mathbf{A}$$

3   SURFACE DIVERGENCE AND SURFACE CURL

$${}^S\nabla \cdot \mathbf{F} = (\mathbf{F}_2 - \mathbf{F}_1) \cdot \mathbf{a}_{n12}$$

$${}^S\nabla \times \mathbf{F} = \mathbf{a}_{n12} \times (\mathbf{F}_2 - \mathbf{F}_1)$$

4   RELATIONSHIPS BASED ON GAUSS AND STOKES'S THEOREMS

$$\int_v \nabla \cdot \mathbf{F} \, dv = \oint_S \mathbf{F} \cdot d\mathbf{S} \quad \text{(Gauss)}$$

$$\int_S (\nabla \times \mathbf{F}) \cdot d\mathbf{S} = \oint_C \mathbf{F} \cdot d\mathbf{l} \quad \text{(Stokes)}$$

$$\int_v \nabla \phi \, dv = \int_S \phi \, d\mathbf{S}$$

$$\int_v (\nabla \times \mathbf{F}) \, dv = \int_S d\mathbf{S} \times \mathbf{F}$$

$$\int_S d\mathbf{S} \times \nabla \phi = \oint_C \phi \, d\mathbf{l}$$

5   GRADIENT, DIVERGENCE, CURL AND LAPLACIAN IN RECTANGULAR, CYLINDRICAL AND SPHERICAL COORDINATES

Rectangular coordinates

$$\nabla V = \frac{\partial V}{\partial x} \mathbf{a}_x + \frac{\partial V}{\partial y} \mathbf{a}_y + \frac{\partial V}{\partial z} \mathbf{a}_z$$

$$\nabla \cdot \mathbf{F} = \frac{\partial F_x}{\partial x} + \frac{\partial F_y}{\partial y} + \frac{\partial F_z}{\partial z}$$

$$\nabla \times \mathbf{F} = \left(\frac{\partial F_z}{\partial y} - \frac{\partial F_y}{\partial z} \right) \mathbf{a}_x + \left(\frac{\partial F_x}{\partial z} - \frac{\partial F_z}{\partial x} \right) \mathbf{a}_y + \left(\frac{\partial F_y}{\partial x} - \frac{\partial F_x}{\partial y} \right) \mathbf{a}_z$$

$$\nabla^2 V = \frac{\partial^2 V}{\partial x^2} + \frac{\partial^2 V}{\partial y^2} + \frac{\partial^2 V}{\partial z^2}$$

Cylindrical coordinates

$$\nabla V = \frac{\partial V}{\partial r}\mathbf{a_r} + \frac{1}{r}\frac{\partial V}{\partial \phi}\mathbf{a_\phi} + \frac{\partial V}{\partial z}\mathbf{a_z}$$

$$\nabla \cdot \mathbf{F} = \frac{1}{r}\frac{\partial}{\partial r}(rF_r) + \frac{1}{r}\frac{\partial F_\phi}{\partial \phi} + \frac{\partial F_z}{\partial z}$$

$$\nabla \times \mathbf{F} = \left(\frac{1}{r}\frac{\partial F_z}{\partial \phi} - \frac{\partial F_\phi}{\partial z}\right)\mathbf{a_r} + \left(\frac{\partial F_r}{\partial z} - \frac{\partial F_z}{\partial r}\right)\mathbf{a_\phi} + \frac{1}{r}\left[\frac{\partial}{\partial r}(rF_\phi) - \frac{\partial F_r}{\partial \phi}\right]\mathbf{a_z}$$

$$\nabla^2 V = \frac{1}{r}\frac{\partial}{\partial r}\left(r\frac{\partial V}{\partial r}\right) + \frac{1}{r^2}\frac{\partial^2 V}{\partial \phi^2} + \frac{\partial^2 V}{\partial z^2}$$

Spherical coordinates

$$\nabla V = \frac{\partial V}{\partial r}\mathbf{a_r} + \frac{1}{r}\frac{\partial V}{\partial \theta}\mathbf{a_\theta} + \frac{1}{r\sin\theta}\frac{\partial V}{\partial \phi}\mathbf{a_\phi}$$

$$\nabla \cdot \mathbf{F} = \frac{1}{r^2}\frac{\partial}{\partial r}(r^2 F_r) + \frac{1}{r\sin\theta}\frac{\partial}{\partial \theta}(F_\theta \sin\theta) + \frac{1}{r\sin\theta}\frac{\partial F_\phi}{\partial \phi}$$

$$\nabla \times \mathbf{F} = \frac{1}{r\sin\theta}\left[\frac{\partial}{\partial \theta}(F_\phi \sin\theta) - \frac{\partial F_\theta}{\partial \phi}\right]\mathbf{a_r} + \frac{1}{r}\left[\frac{1}{\sin\theta}\frac{\partial F_r}{\partial \phi} - \frac{\partial}{\partial r}(rF_\phi)\right]\mathbf{a_\theta}$$

$$+ \frac{1}{r}\left[\frac{\partial}{\partial r}(rF_\theta) - \frac{\partial F_r}{\partial \theta}\right]\mathbf{a_\phi}$$

$$\nabla^2 V = \frac{1}{r^2}\frac{\partial}{\partial r}\left(r^2\frac{\partial V}{\partial r}\right) + \frac{1}{r^2\sin\theta}\frac{\partial}{\partial \theta}\left(\sin\theta\frac{\partial V}{\partial \theta}\right) + \frac{1}{r^2\sin^2\theta}\frac{\partial^2 V}{\partial \phi^2}$$

APPENDIX II

SOLID ANGLES

The plane angle (measured in radians) subtended at the point O by an
arbitrary curve C (Fig. A1) is defined as the length of the arc cut off by
the lines OA and OB on the circle of unit radius and centre O. If in
going from A to B the angle θ is swept out anticlockwise, it is considered positive; otherwise it is considered negative—e.g. at O'. For a

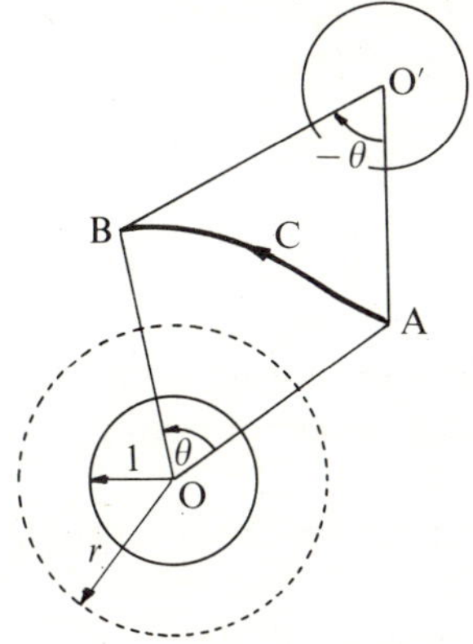

Fig. A1

circle of radius r the angle θ is given by the ratio l/r where l is the length
of the arc; this ratio is independent of r. If the curve C forms a closed
path then the plane angle subtended by C at a point O is 2π if O is
within C and zero if O is outside C (Fig. A2).

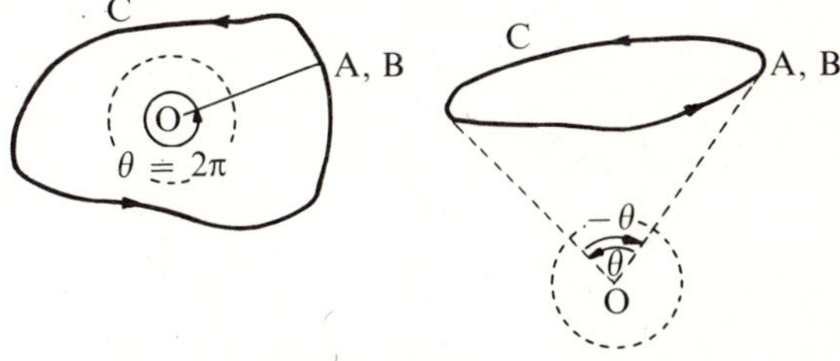

Fig. A2

The solid angle (measured in steradians) subtended at a point O by
any area S is defined as the projection S_u of that area onto the surface

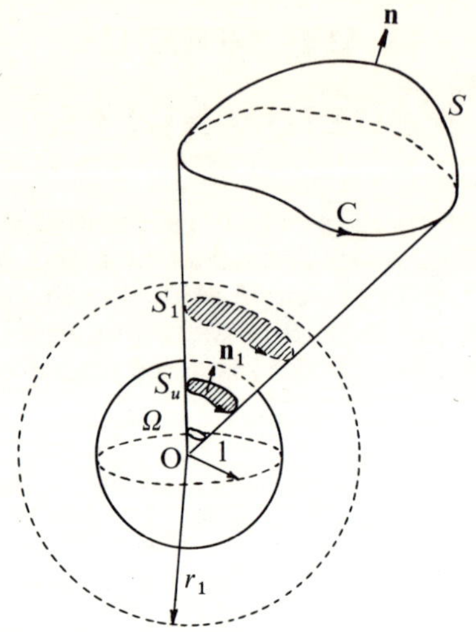

Fig. A3

of a sphere of unit radius and centre O (Fig. A3). If the area S is projected onto the surface of a sphere of radius r_1 the solid angle is given by $\Omega = S_1/r_1^2$ and this ratio is independent of the size of the sphere.

If the surface S is a closed surface the solid angle subtended by S at O is 4π if O is inside S and zero if O is outside S.

If S is an open surface, the angle Ω may be either positive or negative. Consider an open surface S bounded by the curve C (Fig. A3). The positive direction in which C is traversed is that direction which would

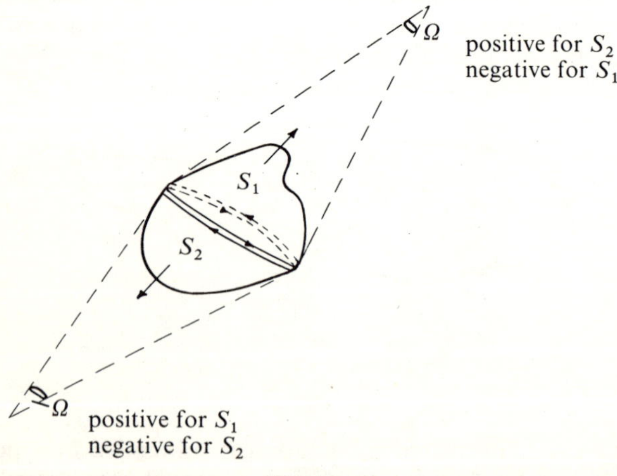

Fig. A4

advance a right-hand screw in the same sense as that of the outward normal at any point on S. If the direction of the normal to the area projected on the sphere points away from O, then Ω is considered positive; otherwise it is negative (Fig. A4).

The solid angle subtended at the vertex of a right circular cone of apex angle 2θ is $2\pi(1 - \cos\theta)$. This is the area of the spherical cap cut off by the cone on a sphere of unit radius and centre at the apex of the cone.

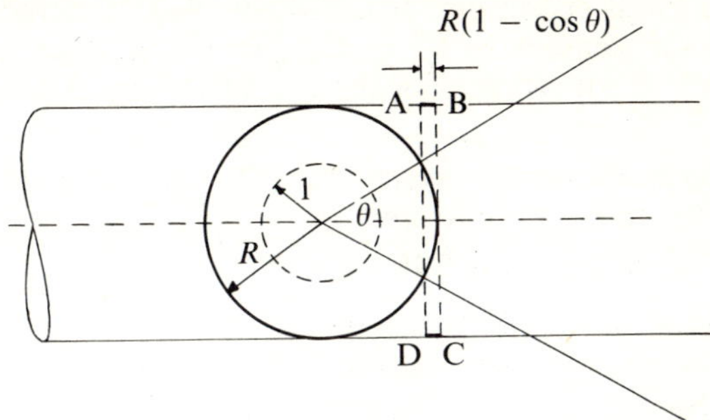

Fig. A5

It is easily shown that the area cut off from a sphere of radius R is equal to the area of the annulus ABCD of the cylindrical surface shown in Fig. A5.

APPENDIX III

SI UNITS

SI is the international abbreviation for Système International d'Unités (International System of Units) adopted by the 11th Conférence Générale des Poids et Mesures (CGPM) in 1960. It includes the base-units of SI, supplementary SI units, derived SI units and decimal multiples and sub-multiples of these units, formed by the use of prefixes.

The SI system is a coherent* system of units and the name 'SI units' is reserved for the coherent units only.

The six base SI units are the following

metre	m	length
kilogram	kg	mass
second	s	time
ampere	A	electric current
kelvin	K	thermodynamic temperature
candela	cd	luminous intensity

The SI units for plane angle and solid angle, the radian (rad) and the steradian (sr) respectively, are called supplementary units in the International System of Units.

The expressions for the derived SI units are stated in terms of base units; for example, the SI unit for velocity is metre per second ($\mathrm{m\,s^{-1}}$).

The CGPM has approved the following special names and symbols for some of the derived SI units:

Quantity	Name of SI unit	Symbol	Expressed in terms of base SI units or derived units
frequency	hertz	Hz	$1\,\mathrm{Hz} = 1\,\mathrm{s^{-1}}$
force	newton	N	$1\,\mathrm{N} = 1\,\mathrm{kg\,m\,s^{-2}}$
work, energy, quantity of heat	joule	J	$1\,\mathrm{J} = 1\,\mathrm{N\,m}$
power	watt	W	$1\,\mathrm{W} = 1\,\mathrm{J\,s^{-1}}$
quantity of electricity	coulomb	C	$1\,\mathrm{C} = 1\,\mathrm{A\,s}$

---

* A system of units is coherent if the product or quotient of any two unit quantities of the system is the unit of the resultant quantity. For example, newton, kilogram, metre, second and $F = M \cdot L \cdot T^{-2}$ form a coherent system; on the other hand newton, gram, centimetre, second and $F = M \cdot L \cdot T^{-2}$ do not form a coherent system. The units of a coherent system are said to be coherent units.

Quantity	Name of SI unit	Symbol	Expressed in terms of base SI units or derived units
electric potential, potential difference, tension, electromotive force	volt	V	$1\,V = 1\,W\,A^{-1}$
electric capacitance	farad	F	$1\,F = 1\,A\,s\,V^{-1}$
electric resistance	ohm	Ω	$1\,\Omega = 1\,V\,A^{-1}$
flux of magnetic induction, magnetic flux	weber	Wb	$1\,Wb = 1\,V\,s$
magnetic flux density, magnetic induction	tesla	T	$1\,T = 1\,Wb\,m^{-2}$
inductance	henry	H	$1\,H = 1\,V\,s\,A^{-1}$
luminous flux	lumen	lm	$1\,lm = 1\,cd\,sr$
illumination	lux	lx	$1\,lx = 1\,lm\,m^{-2}$

Decimal multiples and submultiples of the SI units are formed by means of prefixes. The preferred prefixes are the following:

Multiplication factor	Prefix	Symbol
10^{12}	tera	T
10^{9}	giga	G
10^{6}	mega	M
10^{3}	kilo	k
10^{-2}	centi	c
10^{-3}	milli	m
10^{-6}	micro	μ
10^{-9}	nano	n
10^{-12}	pico	p
10^{-15}	femto	f
10^{-18}	atto	a

The symbol of a prefix is considered to be combined with the symbol to which it is directly attached, forming with it a new unit symbol which can be raised to a positive or negative power and which can be combined with other unit symbols to form symbols for compound units. Examples are

$$1\,cm^3 = (10^{-2}\,m)^3 = 10^{-6}\,m^3$$

$$1\,\mu s^{-1} = (10^{-6}\,s)^{-1} = 10^{6}\,s^{-1}$$

$$1\,mm^2\,s^{-1} = (10^{-3}\,m)^2\,s^{-1} = 10^{-6}\,m^2\,s^{-1}$$

Compound prefixes should not be used; for example, write nm (nanometer) instead of $m\mu m$.

Rules for the use of SI units and their decimal multiples and sub-multiples

The SI units are *preferred* but it will not be practical to limit usage to these; in addition, therefore, their decimal multiples and sub-multiples, formed by using the prefixes, are required.

In order to avoid errors in calculation it is essential to use coherent units. Therefore, it is strongly recommended that in calculations only SI units themselves be used, and not their decimal multiples and sub-multiples.

The use of prefixes representing 10 raised to a power which is a multiple of 3 is especially recommended. In certain cases, to ensure convenience in the use of the units, this recommendation cannot be followed.

It is recommended that only one prefix be used in forming the decimal multiples or sub-multiples of a derived SI unit, and that this prefix be attached to a unit in the numerator. In certain cases convenience in the use requires attachment of prefix to both the numerator and the denominator at the same time, and sometimes only to the denominator.

Numerical values

When expressing a quantity by a numerical value and certain unit it has been found suitable in most applications to use units resulting in numerical values between 0·1 and 1000.

The units which are decimal multiples and sub-multiples of the SI units should therefore be chosen to provide values in this range; for example

Observed or calculated values	Can be expressed as
12 000 N	12 kN
0·00394 m	3·94 mm
14 010 N m^2	14·01 KN m^{-2}
0·0003 s	0·3 ms

The above rule cannot, however, be consistently applied. In one and the same context the numerical values expressed in a certain unit can extend over a considerable range; this applies especially to tabulated numerical values. In such cases it is often appropriate to use the same unit, even when this means exceeding the preferred value range 0·1 to 1000.

Definitions of some important derived SI units used in electricity and magnetism

Force	The newton is that force which, when applied to a body having a mass of one kilogram, gives it an acceleration of one metre per second squared.
Energy	The joule is the work done when the point of application of a force of one newton is displaced through a distance of one metre in the direction of the force.
Power	The watt is the power which gives rise to the production of energy at the rate of one joule per second.
Charge	The coulomb is the quantity of electricity transported in one second by a current of one ampere.
Potential difference and emf	The volt is the difference of potential between two parts of a conducting wire carrying a constant current of one ampere, when the power dissipated between these points is equal to one watt.
Resistance	The ohm is the resistance between two points of a conductor when a constant difference of potential of one volt, applied between these two points, produces in this conductor a current of one ampere, this conductor not being the source of any electromotive force.
Capacitance	The farad is the capacitance of a capacitor between the plates of which there appears a difference of potential of one volt when it is charged by a quantity of electricity equal to one coulomb.
Inductance	The henry is the inductance of a closed circuit in which an electromotive force of one volt is produced when the electric current in the circuit varies uniformly at the rate of one ampere per second.
Magnetic flux	The weber is the flux which, linking a circuit of one turn, produces in it an electromotive force of one volt as it is reduced to zero at a uniform rate in one second.

References

'Metrication, an apology for SI units', by D. S. Margolis (*Electronics and Power*, December 1969, pp. 431–3).

'Recommendations in the field of quantities and units used in electricity', IEC publication No. 164 (1964).

'The use of SI units', BSI publication PD5686 (January 1969).

'IEEE recommended practice: rules for the use of units of the International System of Units', *Inst. elec. electron. Engrs Spectrum*, March 1971, pp. 77–8.

System of Units in Electricity and Magnetism, by L. Young (Oliver & Boyd, Edinburgh, 1969).

Engineering Units and Physical Quantities, by H. S. Hvistendahl (Macmillan, London, 1964).

Units, Dimensional Analysis and Physical Similarity, by B. S. Massey (Van Nostrand, Princeton, N.J., 1971).

APPENDIX IV

SOME PHYSICAL CONSTANTS

Charge of electron, e $\qquad = -1{\cdot}602 \times 10^{-19}\,\text{C}$

Mass of electron, m $\qquad = 9{\cdot}108 \times 10^{-31}\,\text{kg}$

e/m for electron $\qquad = 1{\cdot}759 \times 10^{11}\,\text{C kg}^{-1}$

Mass of proton, m_p $\qquad = 1{\cdot}672 \times 10^{-27}\,\text{kg}$

Boltzmann's constant, k $\qquad = 1{\cdot}380 \times 10^{-23}\,\text{J degree}^{-1}$

Planck's constant, h $\qquad = 6{\cdot}624 \times 10^{-34}\,\text{J s}$

Avogadro's number, N_0 $\qquad = 6{\cdot}0254 \times 10^{23}$ per gram molecule

1 Bohr magneton, β $\qquad = 9{\cdot}27 \times 10^{-24}\,\text{A m}^2$

1 Debye unit $\qquad = 3{\cdot}33 \times 10^{-30}\,\text{C m}$

Electric constant, ϵ_0 $\qquad = 1/36\pi \times 10^9 = 8{\cdot}854 \times 10^{-12}\,\text{F m}^{-1}$

Magnetic constant, μ_0 $\qquad = 4\pi \times 10^{-7}\,\text{H m}^{-1}$

Velocity of light in vacuum, c_0 $= 2{\cdot}9979 \times 10^8 \simeq 3 \times 10^8\,\text{m s}^{-1}$

Universal gravitational
constant, G $\qquad = 6{\cdot}673 \times 10^{-11}\,\text{N m}^2\,\text{kg}^{-2}$

APPENDIX V

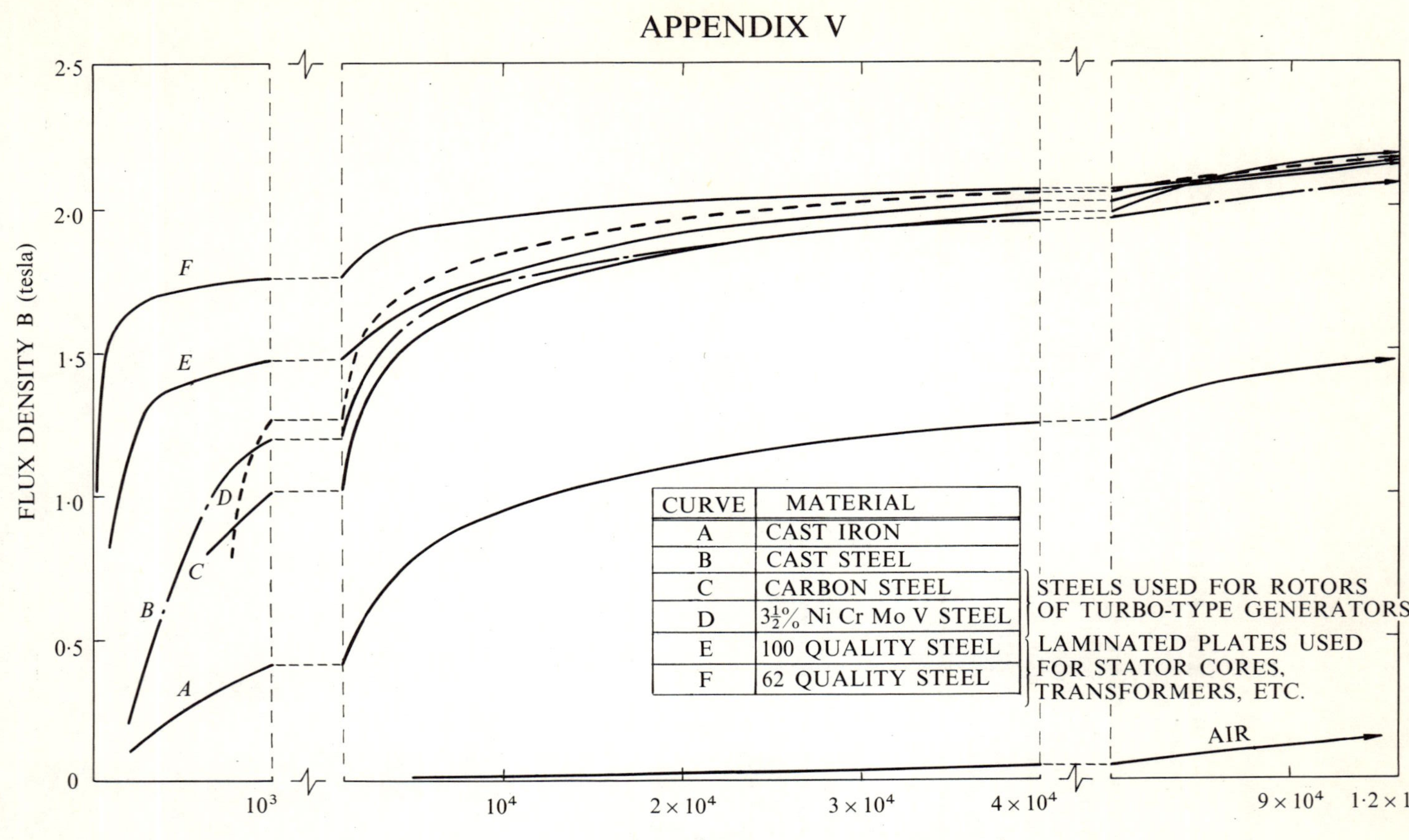

CURVE	MATERIAL
A	CAST IRON
B	CAST STEEL
C	CARBON STEEL
D	$3\frac{1}{2}\%$ Ni Cr Mo V STEEL
E	100 QUALITY STEEL
F	62 QUALITY STEEL

ANSWERS TO PROBLEMS

Chapter 1

1.1 $7\mathbf{a}_x - 6\mathbf{a}_y,\ \mathbf{a}_x + 9\mathbf{a}_y,\ 2\cdot5\mathbf{a}_x - 0\cdot5\mathbf{a}_y$

1.3 straight line

1.4 sphere of radius **r**

1.5 (a) $\mathbf{a}_A = 0\cdot8\mathbf{a}_x - 0\cdot6\mathbf{a}_y,\ \mathbf{a}_B = \frac{1}{3}(2\mathbf{a}_x - 2\mathbf{a}_y + \mathbf{a}_z)$ (b) $\frac{5}{3}$ (c) $70\cdot5°$

1.6 (a) 11   (b) $7\mathbf{a}_x - 5\mathbf{a}_y + \mathbf{a}_z$ (c) $\left(\dfrac{2}{\sqrt{14}}, \dfrac{3}{\sqrt{14}}, \dfrac{1}{\sqrt{14}}\right);$

$\left(\dfrac{1}{\sqrt{14}}, \dfrac{2}{\sqrt{14}}, \dfrac{3}{\sqrt{14}}\right); \left(\dfrac{7}{5\sqrt{3}}, -\dfrac{1}{\sqrt{3}}, \dfrac{1}{5\sqrt{3}}\right)$

1.7 -6

1.8 $\frac{1}{17}(48\mathbf{a}_x - 24\mathbf{a}_y + 48\mathbf{a}_z)$

1.9 100

1.10 $115\cdot2°$

1.11 $7, 7, 5\mathbf{a}_x - 13\mathbf{a}_y + 11\mathbf{a}_z, 14\mathbf{a}_z$

1.12 (a) $\mathbf{A} = \sqrt{5}\mathbf{a}_r - 2\mathbf{a}_z,\ \mathbf{B} = \sqrt{10}\mathbf{a}_r + \mathbf{a}_z$ (b) $\mathbf{A} = 3\mathbf{a}_r,\ \mathbf{B} = \sqrt{11}\mathbf{a}_r$

1.14 $\dfrac{1}{(x^2+y^2)^{1/2}}\left[\dfrac{4x(x^2+y^2)^{1/2}+3xz}{r}+2y\right]\mathbf{a}_x$

$+\dfrac{1}{(x^2+y^2)^{1/2}}\left[\dfrac{4y(x^2+y^2)^{1/2}+yz}{r}+2x\right]\mathbf{a}_y$

$+\left[\dfrac{4z-3(x^2+y^2)^{1/2}}{r}\right]\mathbf{a}_z,\quad \text{where } r = (x^2+y^2+z^2)^{1/2}$

Chapter 2

2.1 $5\times10^{-3}\,\text{N},\ \pm(\mathbf{a}_x + 2\mathbf{a}_y + 2\mathbf{a}_z)$

2.2 $0\cdot0232\,\text{m}$ from 1µC charge

2.3 $-Q/2\sqrt{2}$

2.5 $2\cdot02\times10^5\,\text{V},\ -4\cdot08\times10^5\,\text{V},\ -2\cdot56\times10^5\,\text{V}; 4\cdot58\times10^5\,\text{V},$
$6\cdot1\times10^5\,\text{V},\ 1\cdot52\times10^5\,\text{V}$

2.6 $22\cdot2\times10^{-12}\,\text{C}$

2.7 $-6xy\mathbf{a}_x - 3(x^2-y^2z)\mathbf{a}_y + 2y^3z\mathbf{a}_z;\ -\left(1+\dfrac{b}{r^2}\right)\cos\phi\,\mathbf{a}_r$

$+\left(1-\dfrac{b}{r^2}\right)\sin\phi\,\mathbf{a}_\phi;\ -\left(a-\dfrac{b^3}{r^2}\right)\cos\theta\,\mathbf{a}_r+\left(a+\dfrac{b^3}{r^2}\right)\sin\theta\,\mathbf{a}_\theta$

2.9 $2xa_x - za_y - ya_z$; $0{\cdot}825a_x - 0{\cdot}137a_y - 0{\cdot}55a_z$

2.10 $0, 0$

2.13 $60°$; $\sqrt{3}a$

2.15 0

2.16 $Q/4\pi\epsilon_0 a$

2.21 $\dfrac{\rho_0}{3\alpha r^2 \epsilon_0}\left[\exp(\alpha r^3) - \exp(\alpha a^3)\right]$

2.23 $\dfrac{\lambda}{8\pi\epsilon_0}\left[\dfrac{L^2}{x\left(x^2 - \dfrac{L^2}{4}\right)}\right]$; $-\dfrac{\lambda}{\pi\epsilon_0}\left[\dfrac{1}{y} - \dfrac{1}{\left(y^2 + \dfrac{L^2}{4}\right)^{1/2}}\right]$

2.28 $\dfrac{p_1 p_2}{4\pi\epsilon_0 r^3}\sin\theta$; $\dfrac{p_1 p_2}{2\pi\epsilon_0 r^3}\sin\theta$

2.29 $-3aQa_x + aQa_y$

2.30 $\frac{4}{3}\pi a^3 \sigma_0$

2.32   (a) $\epsilon_0(2xz + 6y^2 z^2 - xy^2)$ (b) $\epsilon_0(2 + 6xyz)$

2.34 16π

2.35 $6y - 6yz - 2y^3$; 0; $-\dfrac{2b^3}{r^3}\cos\theta$

Chapter 3

3.1 $18\,\text{V}$; $28{\cdot}5\,\text{V}$

3.2 $33{\cdot}3\,\text{S}$

3.3   (a) $Q_1 = 4\pi\epsilon_0\left(\dfrac{a_1 a'_2}{a'_2 - a_1}\right)(V_1 - V_2)$; $Q_2 = 4\pi\epsilon_0\left(\dfrac{a_1 a'_2}{a'_2 - a_1}\right)(V_2 - V_1)$
$\qquad\qquad + 4\pi\epsilon_0 a_2 V_2$

$\qquad$ (b) $Q_1 = -4\pi\epsilon_0\left(\dfrac{a'_2 a_1}{a'_2 - a_1}\right)V_2$; $Q_2 = 4\pi\epsilon_0\left(a_2 + \dfrac{a'_2 a_1}{a'_2 - a_1}\right)$

$\qquad$ (c) $Q_1 = -Q_2 = 4\pi\epsilon_0\left(\dfrac{a'_2 a_1}{a'_2 - a_1}\right)V_1$

3.4   (a) $1200\,\text{V}$ (b) $2{\cdot}8 \times 10^4\,\text{V m}^{-1}$ (c) $2{\cdot}6 \times 10^4\,\text{V m}^{-1}$
$\qquad$ (d) $16{\cdot}7\,\text{pF}$

3.5 $-0{\cdot}25\,\mu\text{C}$, $16{\cdot}83\,\text{kV}$; $14{\cdot}8\,\text{pF}$

3.10 $1{\cdot}44 \times 10^6\,\text{V m}^{-1}$, $0{\cdot}72 \times 10^6\,\text{V m}^{-1}$; $3{\cdot}68\,\text{pF}$

3.15 $0{\cdot}0198\,\mu\text{F}$

3.18 $100\,\text{V}$

3.19   (a) $4870\,\text{pF}$ (b) $4990\,\text{pF}$ (a) indeterminate   (b) $2680\,\text{V}$

Chapter 4

4.1 17·7 pF, 565 V   (a) 141 V   (b) 70·8 pF   (c) 0·25 μC m^{-2}
or 0·75 μC m^{-2}

4.2 (a) 311 V   (b) 0·75 μC m^{-2}, 3 μC m^{-2} (c) 322 pF   8.25 kV

4.5 $\dfrac{(\beta/\alpha)}{\log\dfrac{b}{a}}[a(2\pi-\theta)+\epsilon_r a\theta]$, where $\alpha = 2\pi a - a\theta(1-\epsilon_r)$
$\beta = 2\pi\epsilon_0 - \theta\epsilon_0(1-\epsilon_r)$

4.6 7 kV; 53·5 kV

4.7 (a) $6\mathbf{a}_x + 10\mathbf{a}_y - 3\mathbf{a}_z$, 69·23°   (b) 7·67, 45·5°   (c) 10·55, 30·75°

4.10 1·0282

4.11 $5·75 \times 10^{-29}$, $13·2 \times 10^{-7}$

4.12 1·205

4.13 $E' = 1·91E$

Chapter 5

5.8 11·8 kV

Chapter 6

6.2 2650 V

6.8 3°, 29·7°

Chapter 7

7.1 8·9 A mm^{-2}

7.6 $\dfrac{\mu_0 I^2}{4\pi D}(4\sqrt{3}-3)$ N m^{-1} on outer conductors, zero on central
conductor

7.14 $a = 2d$, for this separation the flux density at the midpoint is
$$B_z = \frac{8}{5\sqrt{5}}\frac{\mu_0 IN}{a}$$

7.16 $\dfrac{\mu_0 NI}{l}\cos\alpha\sin^2\alpha\log\dfrac{h}{h-l}$, where h = height of cones, α = semi-apex angle

7.18 $\dfrac{1}{2}e\omega a^2$, $\dfrac{\mu_0 e\omega}{4\pi a}$

7.22 $\dfrac{1}{5}Qa^2\omega$: $\dfrac{\mu_0\omega Q}{4\pi a}$

7.26 $\dfrac{QR\rho}{2\epsilon_0 m}[1-u^2\epsilon_0\mu_0]$, where ρ = charge density

Chapter 8

8.1 (a) $\mathbf{j}_M = 0$, $\mathbf{k}_M = M_0 \sin\theta \mathbf{a}_\phi$; $\rho_m = 0$, $\sigma_m = M_0 \cos\theta$

(b) $0, \dfrac{\mu_0 M_0 a^3}{3z^2}$ (c) $\dfrac{2}{3} \dfrac{\mu_0 M_0 a^3}{3z^3}$

8.2 (a) 1 (b) 0 (c) $\frac{1}{2}$

8.3 $0, \dfrac{\mu_0}{2\pi} \dfrac{(b^3 - a^3)}{r^3} \cos\theta$

8.6 $1{\cdot}15 \times 10^{-3}\,\text{Wb}$; $1{\cdot}213\,\text{A}$

8.7 $0{\cdot}262\,\text{A}$; $1{\cdot}25\,\text{A}$

8.8 $1{\cdot}625\,\text{A}$

8.9 $9{\cdot}95\,\text{A}$

8.14 $4\,\text{cm}$, $1{\cdot}11\,\text{cm}^2$

Chapter 9

9.2 $22{\cdot}4\,\text{V}$

9.5 $1{\cdot}255 \sin\dfrac{10\pi}{3}\,\text{TV}$

9.8 (a) $0{\cdot}167\mu\,\text{A m}^{-2}$ (b) $4{\cdot}62 \times 10^{-15}\,\text{A m}^{-2}$

(c) $2{\cdot}65 \times 10^{-9}\,\text{A m}^{-2}$

Chapter 10

10.2 $107\,\text{Kg}$; $17{\cdot}3\,\text{Kg}$

10.4 $4{\cdot}96 \times 10^6\,\text{A}$

10.5 816

10.7 $0{\cdot}0675\,\text{ms}$

Chapter 11

11.1 $-\dfrac{\lambda h}{\pi r^2}$ where $r = $ distance from λ to a point on the plane

11.10 Images: $-Q$ at height $-h$, $-\dfrac{Qa}{h}$ at $-\dfrac{a^2}{h}$ and $\dfrac{Qa}{h}$ at $\dfrac{a^2}{h}$

$$\text{Force} = \frac{Q^2}{4\pi\epsilon_0}\left[\frac{1}{4h^2} + \frac{4a^3 h^3}{(h^4 - a^4)^2}\right]$$

$$\text{Induced charge} = -Q\left[1 - \frac{h^2 - a^2}{h(h^2 + a^2)^{1/2}}\right]$$

INDEX